Spacecraft Attitude Control

A Linear Matrix Inequality Approach

Chuang Liu Xiaokui Yue
Keke Shi Zhaowei Sun

Science Press
Beijing

ELSEVIER

内 容 简 介

本书凝聚了作者在航天器姿态控制领域近十年的原创性研究成果，系统研究了多源复杂扰动下姿态稳定控制方法。全书共 11 章。第 1 章对线性矩阵不等式方法与航天器姿态动力学进行了介绍，为后续控制系统设计奠定理论基础；第 2~6 章介绍了刚体航天器姿态稳定控制方法，主要包括：状态反馈非脆弱控制、动态输出反馈非脆弱控制、基于中间状态观测器的容错时滞控制与容错非脆弱控制，以及基于干扰观测器的输入受限控制；第 7~9 章介绍了柔性航天器姿态稳定控制方法，主要包括：具有极点配置约束的改进混合 H_2/H_∞ 控制、非脆弱 H_∞ 控制，以及基于主动振动抑制的抗干扰控制；第 10 章介绍了航天器混沌姿态同步跟踪控制方法，并在第 11 章给出了欠驱动混沌姿态角速度稳定控制方法供读者参考。

本书可供航空航天、机械电子及控制相关专业的高等院校本科生和研究生学习参考，也是相关领域科研工作者和工程技术人员查阅或教学的有效工具。

The print edition is only for sale in Chinese mainland. Customers from outside of Chinese mainland please order the print book from: Elsevier.
ISBN of the Co-Publisher's edition: 978-0-323-99005-9

图书在版编目（CIP）数据

航天器姿态控制：一种线性矩阵不等式方法 = Spacecraft Attitude Control: A Linear Matrix Inequality Approach：英文 / 刘闯等著. — 北京：科学出版社，2022.6
ISBN 978-7-03-071978-2

Ⅰ. ①航… Ⅱ. ①刘… Ⅲ. ①航天器－姿态飞行控制－英文 Ⅳ. ①V448.22

中国版本图书馆 CIP 数据核字（2022）第 048369 号

责任编辑：徐杨峰 / 责任校对：谭宏宇
责任印制：黄晓鸣 / 封面设计：殷 靓

科学出版社 出版
北京东黄城根北街 16 号
邮政编码：100717
http：//www. sciencep. com

南京展望文化发展有限公司排版
广东虎彩云印刷有限公司印刷
科学出版社发行 各地新华书店经销

*

2022 年 6 月第 一 版 开本：特16（710 × 960）
2022 年 6 月第一次印刷 印张：24
字数：430 000

定价：200.00 元

（如有印装质量问题，我社负责调换）

Preface

The basic topic of this book is to solve problems arising from spacecraft attitude control systems using convex optimization. Uncertainties inherently occur in rigid spacecraft and flexible spacecraft during the attitude maneuvering process. Specifically for rigid spacecraft, it can lead to chaotic attitude motion if certain circumstances are encountered. In this book, we consider the problem of stability analysis and controller synthesis in the spacecraft attitude control system. We show that a wide variety of problems arising in spacecraft attitude control systems can be reduced to a handful of quasiconvex or standard convex optimization problems that involve linear matrix inequalities (LMIs). For a few special cases, analytic solutions to these problems may exist, but the main point of this book is that they can be solved numerically in all cases. These standard problems derived from spacecraft attitude control can be solved in polynomial time, and so are tractable, at least in a theoretical sense. Therefore we consider the convex optimization problems of spacecraft attitude control systems based on an LMI approach.

This book is intended for readers who are interested in the field of spacecraft attitude control or robust control systems as it provides an extensive review of research in this area and also systematically discusses the feasible solution against the background of various uncertainties that help to improve robust control performance. This book can be used as both a reference guide for practicing professionals and a textbook for undergraduate and graduate students in the field of spacecraft attitude control or control engineering. Some of the content of this book has been used to teach graduate students at Northwestern Polytechnical University, Xi'an, China, for the robust control for spacecraft attitude course. Furthermore, some MATLAB codes or the YALMIP toolbox are easily available on the Internet for beginners to acquaint themselves with the solution of LMI conditions.

The present generation of spacecraft should be capable of high-precision pointing and better robustness to external disturbances and various uncertainties. Consequently, multi-source complex disturbances, for example, external disturbances, model parameter uncertainties, controller gain perturbations, input delay, measurement errors, input constraints or saturation, and actuator fault signals, should be considered. Otherwise, they will have a considerable influence on the stability of the attitude control system leading to degradation or instability in the robust performance of the systems. Thus to understand the behavior of attitude control systems properly, it is important to include the aforementioned uncertainties in the mathematical representation that can lead to accurate stability analysis and consequently proper controller designs. The mathematical representation of the attitude control system with uncertainties turns out to be the state-space equations model. A primary motivation behind such research and investigation is the presence of various uncertainties in a wide variety of rigid and flexible spacecraft attitude control systems. This text is intended to develop the LMI theory of attitude stabilization control from first principles to practical algorithms, because very few of the existing texts on LMIs treat spacecraft attitude control in depth. This book emphasizes stability analysis and attitude controller synthesis of three-axis stabilized rigid and flexible spacecraft systems, so the reader can understand how the derived LMI theory is applied to actual orbiting spacecraft. We also illuminate some simplified analytical expressions that can serve as first-cut analysis for further research and are especially important in the initial design phase of a spacecraft attitude control system. In writing this book the authors hope to make a significant contribution toward expediting the process most newcomers must go through in assimilating and applying LMI approach to spacecraft attitude control or other control systems design.

Based on the Lyapunov functional approach in a linear matrix inequality framework, the stability analysis and controller design for spacecraft attitude control system is discussed, where Chapter 1 gives an introduction of basic knowledge, Chapters 2–6 discuss common rigid spacecraft, Chapters 7–9 discuss flexible spacecraft, and Chapters 10 and 11 discuss chaotic attitude control of rigid spacecraft. In particular, Chapter 1 introduces basic knowledge about linear matrix inequalities and spacecraft dynamics, which paves the way for the subsequent controller design. Chapter 2 investigates state feedback nonfragile controller design for spacecraft considering controller perturbations. Once the attitude information cannot be measured, the dynamic output feedback nonfragile controller can be a good choice to deal with attitude stabilization problems, and this results in the writing of Chapter 3, where a hybrid nonfragile controller in the case of coexisting additive and multiplicative perturbations is developed. Chapter 4 provides detailed derivations of the intermediate observer used to estimate attitude information and fault signals simultaneously, based on which a fault-tolerant delayed controller is developed, and a uniformly, ultimately bounded result is achieved. Chapter 5 covers fault-tolerant control and nonfragile control by incorporating controller perturbations into a fault-tolerant control process to retain a nonfragile performance. Chapter 6 provides a uniformly ultimately bounded condition to obtain the observer and controller gains simultaneously while input magnitude and rate constraints are satisfied. Chapter 7 describes the mathematical model of flexible spacecraft, and an improved mixed H_2/H_∞ control strategy is investigated to reduce conservatives in the analysis and synthesis of such problems with the selection of appropriate Lyapunov functional variables and poles assignment constraints. Chapter 8 extends the modal dimension of flexible spacecraft introduced in Chapter 7 to infinity and two vibration suppression schemes are introduced, and then a nonfragile control H_∞ controller is proposed. Based on an intermediate observer, input magnitude, and rate constraints and active vibration suppression scheme, Chapter 9 shows the sufficient conditions for an antidisturbance attitude controller, including some recent theoretical advancements on the control system design. Chapter 10 gives detailed analysis on chaotic attitude motion, for example, system characteristics and chaotic attractors, and an adaptive variable structure tracking controller is designed for chaotic attitude synchronization. Strictly speaking, the controller design in this chapter is not based on LMIs, but the chaotic attitude dynamical model established here can be used for a controller design based on LMIs, and Chapter 11 provides a good example. Subsequently, Chapter 11 discusses the angular velocity stabilization problem of underactuated spacecraft chaotic attitude motion, that is, single-axis and two-axis actuator failures, then a sliding mode control law is designed with corresponding matrices solved via LMI conditions.

This book is the assemblage of many years of research outcomes by the authors working on spacecraft attitude control designs. As this book approaches the finish line, the authors are indebted to numerous colleagues and students for contributions to various aspects of this work. In particular, we wish to express our gratitude to Honghua Dai and especially to Xibin Cao for encouraging us to write this book. Many students have provided excellent insights and recommendations to enhance the pedagogical value, as well as promoting the completion of this book. Although there are far too many students to name individually here, our sincere appreciation goes out to them. We do wish to acknowledge the significant contributions of the following individuals: Feng Wang and George Vukovich, who coauthored several chapters included in this book.

We also wish to thank Jianqiao Zhang and Ming Liu for their many discussions and insights throughout the development of this book. In addition, we are grateful to Mr. Yangfeng Xu at Science Press for his invitation to write a book with our choice of topic. Finally, our deepest and most sincere appreciation must be expressed to our families for their patience and understanding throughout the years while we prepared this text.

Contents

CHAPTER 1

Introduction of basic knowledge

1.1 Linear matrix inequalities

1.1.1 What are linear matrix inequalities?

Just as the name implies, linear matrix inequalities (LMIs) are matrix inequalities that are linear in the matrix variables. The concept of LMI has been widely recognized and accepted today. The story of LMIs begins in about 1890, when Lyapunov published his seminal work introducing what we now call Lyapunov theory. He showed that the differential equation

$$\frac{d}{dt}\boldsymbol{x}(t) = \boldsymbol{A}\boldsymbol{x}(t) \tag{1.1}$$

is stable if and only if there exists a positive definite matrix $\boldsymbol{P}$ such that

$$\boldsymbol{A}^T\boldsymbol{P} + \boldsymbol{P}\boldsymbol{A} < 0 \tag{1.2}$$

The requirement of $\boldsymbol{A}^T\boldsymbol{P} + \boldsymbol{P}\boldsymbol{A} < 0$ is what we now call a Lyapunov inequality on $\boldsymbol{P}$, which is the earliest special form of the LMI problem. It is no exaggeration to say that the Lyapunov inequality (1.2) was the first LMI used to analyze the stability of dynamic systems, which opened new doors for research of control engineering techniques.

The next major milestone occurred in the 1940s, where Lur'e, Postnikov, and other researchers of the Soviet Union applied Lyapunov's methods to some specific practical problems in control engineering, especially the stability verification of control systems with nonlinear inputs [1]. Although they did not explicitly form matrix inequalities, their stability criteria have the form of LMIs. Of course, this limited their application to relatively simple systems. The important role of LMIs in control theory was already recognized in the early 1960s, namely, by Yakubovich [2].

The positive-real (PR) lemma and extensions were intensively studied in the latter half of the 1960s, and were found to be related to the ideas of passivity, the small-gain criteria introduced by Zames and Sandberg, and quadratic optimal control. By 1970, it was known that the LMI appearing in the PR lemma could be solved not only by graphical means, but also by solving

Spacecraft Attitude Control. DOI: https://doi.org/10.1016/B978-0-323-99005-9.00001-8

a certain algebraic Riccati equation (ARE). In 1971 [3], on a research paper of quadratic optimal control, Willems obtained a famous LMI:

$$\begin{bmatrix} A^TP+PA+Q & PB+C^T \\ B^TP+C & R \end{bmatrix} \geq 0 \tag{1.3}$$

and pointed out that it can be solved by studying the symmetric solutions of the ARE

$$A^TP+PA-\left(PB+C^T\right)R^{-1}\left(B^TP+C\right)+Q=0 \tag{1.4}$$

This connection was observed earlier when the ARE was called the Lur'e resolving equation [4]. So, by 1971, researchers knew several methods for solving special types of LMIs: direct (for small systems), graphical methods, and by solving Lyapunov or Riccati equations. These methods are all "closed-form" or "analytic" solutions that can be used to solve special forms of LMIs. (Most control researchers and engineers consider the Riccati equation to have an "analytic" solution, since the standard algorithms that solve it are very predictable in terms of the effort required, which depends almost entirely on the problem size and not the particular problem data. Of course, it cannot be solved exactly in a finite number of arithmetic steps for systems of fifth and higher order.)

Since the mid-1970s, the next major advancement in trying to solve LMI has occurred:

The LMIs that arise in system and control theory can be formulated as convex optimization problems that are amenable to computer solution.

Although this is a simple statement, it leads to some important conclusions, the most important of which is that one can reliably solve many LMIs that have not been found with analytic solutions. Many researchers have devoted themselves to discussing this statement in depth, wherein Pyatnitskii and Skorodinskii [5] were perhaps the first researchers to make this point clearly and completely. They reduced the original problem of Lur'e to a convex optimization problem involving LMIs and solved it through the ellipsoid algorithm. In addition, they were the first to formulate the search for a Lyapunov function as a convex optimization problem, and then employed proper algorithms to solve the optimization problem.

The next stage of the LMIs' story is in the late 1980s and of great practical importance:

Powerful and efficient interior-point methods were developed to solve the LMIs that arise in system and control theory.

In 1984 [6], Karmarkar introduced a new linear programming algorithm that solves linear programs in polynomial-time, like the ellipsoid method but in contrast to it, is also very efficient in practice. Karmarkar's work spurred an enormous amount of work in the area of interior-point methods for LMIs. The first were Nesterov and Nemirovsky [7], who developed interior-point methods that apply directly to convex problems involving LMIs. Although there still remains much to be done in this area, several interior-point algorithms for LMI problems have been implemented and tested on specific families of LMIs that arise in control theory, including Alizadeh [8,9], Vandenberghe, and Boyd [10], etc. In addition, it is necessary to mention that, in the middle of the 1990s, the MATLAB-based LMI toolbox was produced by Gahinet et al. [11]. Since then, this toolbox has found so many applications and has performed a great role in developing LMI-based control theories and applications. Without efficient software packages, for example, LMI Toolbox, LMIs certainly would not have become so popular in the control community.

Since 2000, more and more applications of LMIs, both theoretical and practical, have appeared in a very diverse and dramatically fast way. One of the biggest features is that more books and papers on this topic have appeared while entering into the new millennium. The new century's first book on LMIs may be Ghaoui and Niculescu [12], which is in fact a volume on advances in LMI methods in control, which collects 17 separated topics from 33 contributors. The new book from Ostertag treated mono- and multivariable control and estimation to introduce some basic concepts and results of LMIs [13]. There are also some other special research topics related to LMI approaches, such as analysis and synthesis of multidimensional systems, analysis and control of fuzzy systems, dynamic output feedback control of uncertain nonlinear systems, etc.

In general, LMIs in practice, as a subdiscipline in systems and control, have now become very rich in content. Both LMIs and their corresponding toolboxes today have become key techniques for control engineering. LMI has a 100-years long story from its birth till now, and it will continue by virtue of its critical research value.

Having reviewed the development process of LMIs, one can now give a brief summary of the representation of LMI approaches for the analysis and design of control systems. For most researchers in the control community, perhaps the most famed type of LMIs is the following continuous-time Lyapunov matrix inequality

$$\boldsymbol{F}(\boldsymbol{P}) = \boldsymbol{A}^T\boldsymbol{P} + \boldsymbol{P}\boldsymbol{A} + \boldsymbol{Q} < 0 \tag{1.5}$$

where $\boldsymbol{A} \in \mathbb{R}^{m \times m}$, $\mathbf{Q} \in \mathbb{S}^m = \{\boldsymbol{M} | \boldsymbol{M} = \boldsymbol{M}^T \in \mathbb{R}^{m \times m}\}$ are given matrices, and $\boldsymbol{P} \in \mathbb{S}^m$ is the unknown matrix. Here "$<$" denotes symmetric negative definiteness. Then, with this well-known simple example, we should now give the general form of LMIs.

Definition 1.1: Let $\mathbf{Q} \in \mathbb{S}^n, \boldsymbol{D}, \boldsymbol{E}_i \in \mathbb{R}^{m \times n}, \boldsymbol{F}_i \in \mathbb{R}^{n \times n}, i = 1, \ldots, l$ then it is easy to verify that

$$\boldsymbol{L}(\boldsymbol{X}) = \mathbf{D}^T\mathbf{X} + \mathbf{X}^T\boldsymbol{D} + \sum_{i=1}^{l} (\boldsymbol{E}_i^T\boldsymbol{X}\boldsymbol{F}_i + \boldsymbol{F}_i^T\mathbf{X}^T\boldsymbol{E}_i) + \mathbf{Q} \tag{1.6}$$

is linear in the matrix $\boldsymbol{X} \in \mathbb{R}^{m \times n}$. Note that the matrix function $\boldsymbol{L}(\boldsymbol{X})$ is symmetric, which defines the inequality

$$\boldsymbol{L}(\mathbf{X}) < 0 \tag{1.7}$$

which holds in the sense that the matrix $\boldsymbol{L}(\boldsymbol{X})$ is symmetric negative definite. Thus (1.7) is a linear matrix inequality in the parameter matrix $\boldsymbol{X}$, and is called the general form of LMIs.

Specifically, when $m = n$, and the unknown parameter matrix $\boldsymbol{X}$ is symmetric, and the given LMI in (1.7) could be expanded as:

$$\boldsymbol{L}(\boldsymbol{X}) = \mathbf{D}^T\mathbf{X} + \boldsymbol{X}\boldsymbol{D} + \sum_{i=1}^{l} (\boldsymbol{E}_i^T\boldsymbol{X}\boldsymbol{F}_i + \boldsymbol{F}_i^T\boldsymbol{X}\boldsymbol{E}_i) + \mathbf{Q} < 0 \tag{1.8}$$

Further, when $\mathbf{D} = \boldsymbol{A}, \boldsymbol{D}, \boldsymbol{F}_i = \pm \boldsymbol{E}_i, i = 1, \ldots, l$, the above LMI becomes a special type of LMI in a parameter matrix $\boldsymbol{X} \in \mathbb{S}^m$, which takes the following form:

$$\boldsymbol{L}(\boldsymbol{X}) = \boldsymbol{A}^T\mathbf{X} + \boldsymbol{X}\boldsymbol{A} \pm 2\sum_{i=1}^{l} \boldsymbol{E}_i^T\boldsymbol{X}\boldsymbol{E}_i + \mathbf{Q} < 0 \tag{1.9}$$

The continuous-time Lyapunov matrix inequality (1.2) is a special form of (1.9) (i.e., $\boldsymbol{E}_i$ are zero-matrices). Moreover, we give the following discrete-time Lyapunov matrix inequality which is also a special form of (1.9):

$$\boldsymbol{A}^T\boldsymbol{P}\boldsymbol{A} - \boldsymbol{P} + \mathbf{Q} < 0 \tag{1.10}$$

Now, the general form of LIMs can be obtained. However, during the design or analysis process of controllers for spacecraft attitude systems, directly determining the general solution is not as easy as it seems. Therefore a standard form of LIMs should be provided for researchers to systemically summarize and handle various problems.

Definition 1.2: The standard form for an LMI is defined as follows:

$$\boldsymbol{A}(\boldsymbol{x}) = \boldsymbol{A}_0 + x_1\boldsymbol{A}_1 + \cdots + x_n\boldsymbol{A}_n < 0 \tag{1.11}$$

where $\boldsymbol{A}_i \in \mathbb{S}^m, i = 0, 1, \ldots, n$ are known appropriate-size matrices, and $x_i, i = 0, 1, \ldots, n$ are unknown scalars which are called decision variables.

Such an intuitive form of LMIs shown in Eq. (1.11) can simplify the difficulty of the general form of Eq. (1.7) to a great extent. For illustrative purposes, here are some simple examples.

Example 1.1: Let $x_1, x_2 \in \mathbb{R}$, and $\boldsymbol{A}(\boldsymbol{x}) = \boldsymbol{A}_0 + \boldsymbol{A}_1 x_1 + \boldsymbol{A}_2 x_2$, where

$$\boldsymbol{A}_0 = \begin{bmatrix} 1 & 0 \\ 0 & -1 \end{bmatrix}, \quad \boldsymbol{A}_1 = \begin{bmatrix} -1 & -1 \\ -1 & 4 \end{bmatrix}, \quad \boldsymbol{A}_2 = \begin{bmatrix} -1 & 1 \\ 1 & -2 \end{bmatrix} \tag{1.12}$$

Thus $\boldsymbol{A}(\boldsymbol{x})$ can be easily obtained as

$$\boldsymbol{A}(\boldsymbol{x}) = \begin{bmatrix} 1 - x_1 - x_2 & -x_1 + x_2 \\ -x_1 + x_2 & -1 + 4x_1 - 2x_2 \end{bmatrix} \tag{1.13}$$

then the inequality $\boldsymbol{A}(\boldsymbol{x}) < 0$ is equivalent to

$$\begin{aligned} & 1 - x_1 - x_2 < 0 \\ & \det\begin{bmatrix} 1 - x_1 - x_2 & -x_1 + x_2 \\ -x_1 + x_2 & -1 + 4x_1 - 2x_2 \end{bmatrix} = -5x_1^2 + 5x_1 + x_2^2 - x_2 - 1 > 0 \end{aligned} \tag{1.14}$$

Thus all $x_1, x_2 \in \mathbb{R}$ satisfying $\boldsymbol{A}(\boldsymbol{x}) < 0$ are determined by Eq. (1.14), which can be compactly written as

$$\begin{cases} -1 + x_1 + x_2 > 0 \\ \left(x_2 - \sqrt{5}x_1 + \dfrac{\sqrt{5} - 1}{2} \right)\left(x_2 + \sqrt{5}x_1 - \dfrac{\sqrt{5} + 1}{2} \right) > 0 \end{cases} \tag{1.15}$$

This equation set of inequalities will finally induce the feasible region of the (x_1, x_2) plane.

Example 1.2: The continuous-time Lyapunov matrix inequality (1.5) can be converted into a standard form. From Eq. (1.11), any real symmetric positive definite matrix can be resolved into the sum of a series of simple matrices, where each term is composed of the elements in the corresponding symmetric positions of the original matrix and the zeroes in the remaining positions. Thus it is clear that, letting $\boldsymbol{E}_i, i = 1, 2, \ldots, m(m+1)/2$ be a set of bases in $\mathbb{S}^m$, there exist x_i such that

$$\boldsymbol{P} = x_1\boldsymbol{E}_1 + x_2\boldsymbol{E}_2 + \cdots + x_d\boldsymbol{E}_d \tag{1.16}$$

Then,

$$\begin{aligned} \boldsymbol{F}(\boldsymbol{P}) = \boldsymbol{A}^T\boldsymbol{P} + \boldsymbol{P}\boldsymbol{A} + \mathbf{Q} = x_1\left(\boldsymbol{A}^T\boldsymbol{E}_1 + \boldsymbol{E}_1\boldsymbol{A}\right) \\ + x_2\left(\boldsymbol{A}^T\boldsymbol{E}_2 + \boldsymbol{E}_2\boldsymbol{A}\right) + \cdots + x_d\left(\boldsymbol{A}^T\boldsymbol{E}_d + \boldsymbol{E}_d\boldsymbol{A}\right) + \mathbf{Q} \end{aligned} \tag{1.17}$$

Hence, Eq. (1.5) is rewritten into the following LMIs

$$\boldsymbol{F}(\boldsymbol{x}) = \boldsymbol{F}_0 + x_1\boldsymbol{F}_1 + \cdots + x_d\boldsymbol{F}_d < 0 \tag{1.18}$$

with $\boldsymbol{F}_0 = \mathbf{Q}, \boldsymbol{F}_i = \boldsymbol{A}^T\boldsymbol{E}_i + \boldsymbol{E}_i\boldsymbol{A}, i = 1, 2, \ldots, d$.

The examples above have presented a classical solving method for a simple LMI of the standard form and a way of converting continuous-time Lyapunov matrix inequalities into standard LMIs. Similarly, the discrete-time Lyapunov matrix inequality (1.10) can also be converted into a standard LMI form as follows

$$\boldsymbol{L}\left(x_{ij}, i = 1, 2, \ldots, m, j = 1, 2, \ldots, n\right) = \mathbf{Q} + \sum_{i=1}^{m}\sum_{j=1}^{n} x_{ij}\mathbf{Q}_{ij} \tag{1.19}$$

Remark 1.1: When the strictly less symbol "$<$" in the LMI (1.11) is replaced by the not greater symbol "$\leq$," the LMI (1.7) becomes a non-strict one. Specifically, one emphasizes that many conclusions for strict LMIs also hold for nonstrict ones.

Example 1.3: Since the continuous-time Lyapunov matrix inequality has been transformed into an LMI expression in Example 1.2, consider the combined constraints of the form

$$\begin{cases} \boldsymbol{F}(\boldsymbol{x}) < 0 \\ \boldsymbol{x} = \boldsymbol{B}\boldsymbol{u} + \boldsymbol{\alpha}^T \end{cases} \tag{1.20}$$

where the affine function $\boldsymbol{F}: \mathbb{R}^n \to \mathbb{S}^m$, matrix $\boldsymbol{B} \in \mathbb{R}^{n \times r}$, and vector $\boldsymbol{\alpha} \in \mathbb{R}^n$ are given with $r \leq n$. Then, it can be obtained that Eq. (1.20) is in fact a constrained LMI problem. Denote

$$\boldsymbol{B} = \begin{bmatrix} b_{11} & b_{12} & \cdots & b_{1r} \\ b_{21} & b_{22} & \cdots & b_{2r} \\ \vdots & \vdots & \ddots & \vdots \\ b_{n1} & b_{n2} & \cdots & b_{nr} \end{bmatrix}, \quad \boldsymbol{\alpha} = \begin{bmatrix} \alpha_1 \\ \alpha_2 \\ \vdots \\ \alpha_n \end{bmatrix}, \quad \boldsymbol{u} = \begin{bmatrix} u_1 \\ u_2 \\ \vdots \\ u_r \end{bmatrix} \tag{1.21}$$

Then, x_i can be expanded as

$$x_i = \alpha_i + \sum_{k=1}^{r} b_{ik} u_k, \; i = 1, 2, \ldots, n. \tag{1.22}$$

Thus

$$\begin{aligned} \boldsymbol{F}(\boldsymbol{x}) &= \boldsymbol{F}_0 + \sum_{l=1}^{n} x_l \boldsymbol{F}_l = \boldsymbol{F}_0 + \sum_{l=1}^{n} \left(\alpha_l + \sum_{k=1}^{r} b_{lk} u_k \right) \boldsymbol{F}_l \\ &= \boldsymbol{F}_0 + \sum_{l=1}^{n} \alpha_l \boldsymbol{F}_l + \sum_{k=1}^{r} u_k \left(\sum_{l=1}^{n} b_{lk} \boldsymbol{F}_l \right) \\ &= \tilde{\boldsymbol{F}}_0 + u_1 \tilde{\boldsymbol{F}}_1 + \cdots + u_r \tilde{\boldsymbol{F}}_r \\ &= \tilde{\boldsymbol{F}}(\boldsymbol{u}) \end{aligned} \tag{1.23}$$

where $\tilde{\boldsymbol{F}}_0 = \boldsymbol{F}_0 + \sum_{l=1}^{n} \alpha_l \boldsymbol{F}_l, \tilde{\boldsymbol{F}}_k = \sum_{l=1}^{n} b_{lk} \boldsymbol{F}_l, k = 1, 2, \ldots, r$. This implies that Eq. (1.20) is satisfied for $\boldsymbol{x} \in \mathbb{R}^n$ if and only if the LMI $\tilde{\boldsymbol{F}}(\boldsymbol{u}) < 0$ is satisfied for $\boldsymbol{u} \in \mathbb{R}^r$.

To complete this subsection, we conclude by mentioning some common propositions related to LMI.

Proposition 1.1: Let $\boldsymbol{A}(\boldsymbol{x}) = \left[\boldsymbol{A}_{ij}(\boldsymbol{x})\right]_{q \times p}$, then, $\boldsymbol{A}(\boldsymbol{x}) < 0$ is an LMI in vector $\boldsymbol{x}$ only if $\boldsymbol{A}_{ij}(\boldsymbol{x}), i = 1, \ldots, q, \; j = 1, \ldots, p$ are all linear in $\boldsymbol{x}$.

Proposition 1.2: Let

$$\boldsymbol{A}_i(\boldsymbol{x}) < 0, i = 1, 2, \ldots, l \tag{1.24}$$

be a set of linear matrix functions in $\boldsymbol{x}$. Then,

1. Inequality (1.24) means a set of LMIs in $\boldsymbol{x}$ if and only if $diag(\boldsymbol{A}_1(\boldsymbol{x}), \ldots, \boldsymbol{A}_l(\boldsymbol{x})) < 0$ is an LMI in $\boldsymbol{x}$;

2. Inequality (1.24) means a set of LMIs in $\boldsymbol{x}$ implies that

$$\sum_{i=1}^{l} \alpha_i \boldsymbol{A}_i(\boldsymbol{x}) < 0 \tag{1.25}$$

is also an LMI in $\boldsymbol{x}$, where $\alpha_i \geq 0$ are a set of real scalars that are not simultaneously zero.

Proposition 1.3: Let

$$\boldsymbol{A}_i(\boldsymbol{y}_i), \quad i = 1, 2, ..., l \tag{1.26}$$

be a set of linear matrix functions in $\boldsymbol{y}_i$, and define $\boldsymbol{z} = \left[\boldsymbol{y}_1^T, \boldsymbol{y}_2^T, \ldots, \boldsymbol{y}_l^T\right]^T$. Then,

1. Inequality (1.26) are a set of LMIs in $\boldsymbol{y}_i$, if and only if $diag(\boldsymbol{A}_1(\boldsymbol{y}_1), \boldsymbol{A}_2(\boldsymbol{y}_2), \ldots, \boldsymbol{A}_l(\boldsymbol{y}_l)) < 0$ is an LMI in $\boldsymbol{z}$;
2. Inequality (1.26) are a set of LMIs in $\boldsymbol{y}_i$, implies that

$$\sum_{i=1}^{l} \alpha_i \boldsymbol{A}_i(\boldsymbol{y}_i) < 0 \tag{1.27}$$

is an LMI in $\boldsymbol{z}$, where $\alpha_i \geq 0$ are a set of real scalars that are not simultaneously zero.

1.1.2 Useful lemmas for linear matrix inequalities

As preliminarily given in the previous section, in this part, some technical lemmas will be directly presented, which are significant and useful for the following chapters of this book. For detailed proofs of subsequent lemmas, please refer to [14].

1. Generalized square inequalities

   To begin with, we present the following well-known square inequality: For arbitrary scalars x, y, and $\delta > 0$, there is

$$\left(\sqrt{\delta}x - \frac{1}{\sqrt{\delta}}y\right)^2 = \delta x^2 + \frac{1}{\delta}y^2 - 2xy \geq 0 \tag{1.28}$$

   from which one can immediately obtain

$$2xy \leq \delta x^2 + \frac{1}{\delta}y^2 \tag{1.29}$$

Thus in this subsection, several analogous inequalities will be given as fundamental roles in the LMIs' techniques.

Lemma 1.1: (Restriction-Free Inequality) Let $\boldsymbol{X}, \boldsymbol{Y} \in \mathbb{R}^{m \times n}$, $\boldsymbol{F} \in \mathbb{S}^m, \boldsymbol{F} > 0$, and $\delta > 0$ be a scalar, then

$$\boldsymbol{X}^T \boldsymbol{F} \boldsymbol{Y} + \boldsymbol{Y}^T \boldsymbol{F} \boldsymbol{X} \leq \delta \boldsymbol{X}^T \boldsymbol{F} \boldsymbol{X} + \delta^{-1} \boldsymbol{Y}^T \boldsymbol{F} \boldsymbol{Y} \tag{1.30}$$

Particularly, when $\boldsymbol{X} = \boldsymbol{x}$ and $\boldsymbol{Y} = \boldsymbol{y}$ are vectors, Eq. (1.30) can be rewritten as

$$2\boldsymbol{x}^T \boldsymbol{F} \boldsymbol{y} \leq \delta \boldsymbol{x}^T \boldsymbol{F} \boldsymbol{x} + \delta^{-1} \boldsymbol{y}^T \boldsymbol{F} \boldsymbol{y} \tag{1.31}$$

However, the lemma above may be restricted sometimes. Therefore here we give the definition of the following restriction set associated with several incoming results in this section:

$$\mathbb{F} = \left\{ \boldsymbol{F} | \boldsymbol{F} \in \mathbb{R}^{n \times n}, \boldsymbol{F}^T \boldsymbol{F} \leq \boldsymbol{I} \right\} \tag{1.32}$$

Lemma 1.2: Let $\boldsymbol{X} \in \mathbb{R}^{m \times n}, \boldsymbol{Y} \in \mathbb{R}^{n \times m}$. Then for arbitrary scalar $\delta > 0$, there holds

$$\boldsymbol{X} \boldsymbol{F} \boldsymbol{Y} + \boldsymbol{Y}^T \boldsymbol{F}^T \boldsymbol{X}^T \leq \delta \boldsymbol{X} \boldsymbol{X}^T + \delta^{-1} \boldsymbol{Y}^T \boldsymbol{Y}, \quad \forall \boldsymbol{F} \in \mathbb{F} \tag{1.33}$$

Lemma 1.3: Let $\boldsymbol{X} \in \mathbb{R}^{m \times n}, \boldsymbol{Y} \in \mathbb{R}^{n \times m}$. Then for arbitrary scalar $\delta > 0$, for any nonzero vectors $\boldsymbol{x}$ and $\boldsymbol{y}$, there holds,

$$2\boldsymbol{x}^T \boldsymbol{X} \boldsymbol{F} \boldsymbol{Y} \boldsymbol{y} \leq \delta \boldsymbol{x}^T \boldsymbol{X} \boldsymbol{X}^T \boldsymbol{x} + \delta^{-1} \boldsymbol{y}^T \boldsymbol{Y}^T \boldsymbol{Y} \boldsymbol{y}, \quad \forall \boldsymbol{F} \in \mathbb{F} \tag{1.34}$$

Lemma 1.4: For arbitrary nonzero vectors $\boldsymbol{x}, \boldsymbol{y} \in \mathbb{R}^n$, there is

$$\max_{\boldsymbol{F} \in \mathbb{F}} \left(\boldsymbol{x}^T \boldsymbol{F} \boldsymbol{y} \right)^2 = \left(\boldsymbol{x}^T \boldsymbol{x} \right) \left(\boldsymbol{y}^T \boldsymbol{y} \right) \tag{1.35}$$

Expressions from Eqs. (1.33) to (1.35) show us some crucial properties of restricted LMIs. The following variable elimination lemma often play an important role in certain analysis problems involving LMIs.

Lemma 1.5: (Variable Elimination Lemma) Let $X \in \mathbb{R}^{m \times n}$, $Y \in \mathbb{R}^{n \times m}$, $Q \in \mathbb{R}^{m \times m}$. Then

$$\mathbf{Q} + \boldsymbol{XFY} + \mathbf{Y}^T\boldsymbol{F}^T\mathbf{X}^T < 0, \quad \forall \boldsymbol{F} \in \mathbb{F} \tag{1.36}$$

if and only if there exists a scalar $\delta > 0$, such that

$$\mathbf{Q} + \delta\mathbf{X}\mathbf{X}^T + \delta^{-1}\mathbf{Y}^T\mathbf{Y} < 0 \tag{1.37}$$

2. Schur Complements
First, we give the definition of Schur complements.

Definition 1.3: Consider the partitioned matrix

$$\boldsymbol{A} = \begin{bmatrix} \boldsymbol{A}_{11} & \boldsymbol{A}_{12} \\ \boldsymbol{A}_{21} & \boldsymbol{A}_{22} \end{bmatrix} \tag{1.38}$$

(1) When $\boldsymbol{A}_{11}$ is nonsingular, $\boldsymbol{A}_{22} - \boldsymbol{A}_{21}\boldsymbol{A}_{11}^{-1}\boldsymbol{A}_{12}$ is called the Schur complement of $\boldsymbol{A}_{11}$ in $\boldsymbol{A}$, and denoted by $\boldsymbol{S}_{\mathrm{sh}}(\boldsymbol{A}_{11})$;

(2) When $\boldsymbol{A}_{22}$ is nonsingular, $\boldsymbol{A}_{11} - \boldsymbol{A}_{12}\boldsymbol{A}_{22}^{-1}\boldsymbol{A}_{21}$ is called the Schur complement of $\boldsymbol{A}_{22}$ in $\boldsymbol{A}$, and denoted by $\boldsymbol{S}_{\mathrm{sh}}(\boldsymbol{A}_{22})$.

Lemma 1.6: Let $\cong$ represent the equivalence relation between two matrices. Then, for the partitioned matrix Eq. (1.38), the following conclusions hold.

1. When $\boldsymbol{A}_{11}$ is nonsingular,

$$\boldsymbol{A} \cong \begin{bmatrix} \boldsymbol{A}_{11} & 0 \\ 0 & \boldsymbol{A}_{22} - \boldsymbol{A}_{21}\boldsymbol{A}_{11}^{-1}\boldsymbol{A}_{12} \end{bmatrix} = \begin{bmatrix} \boldsymbol{A}_{11} & 0 \\ 0 & \boldsymbol{S}_{ch}(\boldsymbol{A}_{11}) \end{bmatrix} \tag{1.39}$$

and hence $\boldsymbol{A}$ is nonsingular if and only if $\boldsymbol{S}_{\mathrm{sh}}(\boldsymbol{A}_{11})$ is nonsingular, and

$$\det \boldsymbol{A} = \det \boldsymbol{A}_{11} \det \boldsymbol{S}_{\mathrm{ch}}(\boldsymbol{A}_{11}) \tag{1.40}$$

2. When $\boldsymbol{A}_{22}$ is nonsingular,

$$\boldsymbol{A} \cong \begin{bmatrix} \boldsymbol{A}_{11} - \boldsymbol{A}_{12}\boldsymbol{A}_{22}^{-1}\boldsymbol{A}_{21} & 0 \\ 0 & \boldsymbol{A}_{22} \end{bmatrix} = \begin{bmatrix} \boldsymbol{S}_{\mathrm{ch}}(\boldsymbol{A}_{22}) & 0 \\ 0 & \boldsymbol{A}_{22} \end{bmatrix} \tag{1.41}$$

and hence $\boldsymbol{A}$ is nonsingular if and only if $\boldsymbol{S}_{\mathrm{sh}}(\boldsymbol{A}_{22})$ is nonsingular, and

$$\det \boldsymbol{A} = \det \boldsymbol{A}_{22} \det \boldsymbol{S}_{\mathrm{ch}}(\boldsymbol{A}_{22}) \tag{1.42}$$

Based on the well-defined Schur complements, and in light of Lemma 2.6, we will next provide the renowned matrix inversion lemma.

Lemma 1.7: For the partitioned matrix (1.39), the following conclusions hold.

1. When A_{11} is nonsingular,

$$A^{-1}=\begin{bmatrix} A_{11}^{-1}+A_{11}^{-1}A_{12}S_{ch}^{-1}(A_{11})A_{21}A_{11}^{-1} & -A_{11}^{-1}A_{12}S_{ch}^{-1}(A_{11}) \\ -S_{ch}^{-1}(A_{11})A_{21}A_{11}^{-1} & S_{ch}^{-1}(A_{11}) \end{bmatrix} \tag{1.43}$$

2. When A_{22} is nonsingular,

$$A^{-1}=\begin{bmatrix} S_{ch}^{-1}(A_{22}) & -S_{ch}^{-1}(A_{22})A_{12}A_{22}^{-1} \\ -A_{22}^{-1}A_{12}S_{ch}^{-1}(A_{22}) & A_{22}^{-1}+A_{22}^{-1}A_{21}S_{ch}^{-1}(A_{22})A_{12}A_{22}^{-1} \end{bmatrix} \tag{1.44}$$

Corollary 1.1: Let both A_{11} and A_{22} be non-singular matrices of appropriate dimensions. Then,

$$\begin{aligned}(A_{11}-A_{12}A_{22}^{-1}A_{21})^{-1} &= A_{11}^{-1}+A_{11}^{-1}A_{12}\left(A_{22}-A_{21}A_{11}^{-1}A_{12}\right)^{-1}A_{21}A_{11}^{-1} \\ (A_{22}-A_{21}A_{11}^{-1}A_{12})^{-1} &= A_{22}^{-1}+A_{22}^{-1}A_{21}\left(A_{11}-A_{12}A_{22}^{-1}A_{21}\right)^{-1}A_{12}A_{22}^{-1}\end{aligned} \tag{1.45}$$

Until now, some basic lemmas and definitions have been provided for LMIs analysis. In the next step, maybe the most important lemma for actual control problems relating LMIs, that is, the Schur complement lemma, will be given to readers. In fact, it plays a fundamental role aimed at converting nonlinear matrix inequalities into LMIs.

Lemma 1.8: **(Schur complement lemma)** Define a symmetric partitioned matrix as

$$A=\begin{bmatrix} A_{11} & A_{12} \\ A_{12}^T & A_{22} \end{bmatrix} \tag{1.46}$$

Then,

$$A<0 \Leftrightarrow A_{11}<0, S_{ch}(A_{11})<0 \Leftrightarrow A_{22}<0, S_{ch}(A_{22})<0 \tag{1.47}$$

or

$$A>0 \Leftrightarrow A_{11}>0, S_{ch}(A_{11})>0 \Leftrightarrow A_{22}>0, S_{ch}(A_{22})>0 \tag{1.48}$$

Corollary 1.2: Let

$$\boldsymbol{A} = \begin{bmatrix} \boldsymbol{A}_{11} & \boldsymbol{A}_{12} & \cdots & \boldsymbol{A}_{1r} \\ \boldsymbol{A}_{12}^{\mathrm{T}} & \boldsymbol{A}_{22} & \cdots & \boldsymbol{A}_{2r} \\ \vdots & \vdots & \ddots & \vdots \\ \boldsymbol{A}_{1r}^{\mathrm{T}} & \boldsymbol{A}_{2r}^{\mathrm{T}} & \cdots & \boldsymbol{A}_{rr} \end{bmatrix} \tag{1.49}$$

Then, $\boldsymbol{A} < 0$ implies $\boldsymbol{A}_{ii} < 0, i = 1, 2, ..., r$.

3. Variable elimination in LMIs

In many deduction processes involving LMIs, the techniques of elimination of variables are commonly used. This subsection introduces some user-friendly lemmas that serve as the basic tools for related processing techniques.

Lemma 1.9: Let

$$\boldsymbol{Z} = \begin{bmatrix} \boldsymbol{Z}_{11} & \boldsymbol{Z}_{12} \\ \boldsymbol{Z}_{12}^{\mathrm{T}} & \boldsymbol{Z}_{22} \end{bmatrix}, \quad \boldsymbol{Z}_{11} \in \mathbb{R}^{n \times n} \tag{1.50}$$

be symmetric. Then, there exists a symmetric matrix $\boldsymbol{X}$ such that

$$\begin{bmatrix} \boldsymbol{Z}_{11} - \boldsymbol{X} & \boldsymbol{Z}_{12} & \boldsymbol{X} \\ \boldsymbol{Z}_{12}^{T} & \boldsymbol{Z}_{22} & 0 \\ \boldsymbol{X} & 0 & -\boldsymbol{X} \end{bmatrix} < 0 \tag{1.51}$$

if and only if

$$\boldsymbol{Z} = \begin{bmatrix} \boldsymbol{Z}_{11} & \boldsymbol{Z}_{12} \\ \boldsymbol{Z}_{12}^{T} & \boldsymbol{Z}_{22} \end{bmatrix} < 0 \tag{1.52}$$

Lemma 1.10: Let $\boldsymbol{Z}_{ij}, i = 1, 2, 3, j = i, ..., 3$ be given matrices of appropriate dimensions. Then, there exists a matrix $\boldsymbol{X}$ such that

$$\begin{bmatrix} \boldsymbol{Z}_{11} & \boldsymbol{Z}_{12} & \boldsymbol{Z}_{13} \\ \boldsymbol{Z}_{12}^{T} & \boldsymbol{Z}_{22} & \boldsymbol{Z}_{23} + \boldsymbol{X}^{T} \\ \boldsymbol{Z}_{13}^{T} & \boldsymbol{Z}_{23}^{T} + \boldsymbol{X} & \boldsymbol{Z}_{33} \end{bmatrix} < 0 \tag{1.53}$$

if and only if

$$\begin{bmatrix} \boldsymbol{Z}_{11} & \boldsymbol{Z}_{12} \\ \boldsymbol{Z}_{12}^{T} & \boldsymbol{Z}_{22} \end{bmatrix} < 0, \begin{bmatrix} \boldsymbol{Z}_{11} & \boldsymbol{Z}_{13} \\ \boldsymbol{Z}_{13}^{T} & \boldsymbol{Z}_{33} \end{bmatrix} < 0 \tag{1.54}$$

In this case, the aforementioned matrix $\boldsymbol{X}$ is given by

$$\boldsymbol{X} = \boldsymbol{Z}_{13}^T \boldsymbol{Z}_{11}^{-1} \boldsymbol{Z}_{12} - \boldsymbol{Z}_{23}^T \tag{1.55}$$

The two lemmas above in this subsection focus on treating the elimination of variables in a partitioned matrix. Next, we are going to present the condition for eliminating a variable in LMIs through orthogonal complements. First, one should provide the definition of orthogonal complements.

Definition 1.4: Let $\boldsymbol{A} \in \mathbb{R}^{m \times n}$. Then, $\boldsymbol{M}_a$ is called a left orthogonal complement of $\boldsymbol{A}$ if it satisfies

$$\boldsymbol{M}_a \boldsymbol{A} = 0, \quad \text{rank}(\boldsymbol{M}_a) = m - \text{rank}(\boldsymbol{A}) \tag{1.56}$$

and $\boldsymbol{N}_a$ is called a right orthogonal complement of $\boldsymbol{A}$ if it satisfies

$$\boldsymbol{A}\boldsymbol{N}_a = 0, \quad \text{rank}(\boldsymbol{N}_a) = n - \text{rank}(\boldsymbol{A}) \tag{1.57}$$

Based on the upper definition, the following projection lemma makes use of the concept of orthogonal complement.

Lemma 1.11: Let $\boldsymbol{P}, \boldsymbol{Q}$ and $\boldsymbol{H} = \boldsymbol{H}^T$ be given matrices of appropriate dimensions, $\boldsymbol{N}_p$ and $\boldsymbol{N}_q$ be the right orthogonal complements of $\boldsymbol{P}, \boldsymbol{Q}$, respectively. Then, there exists a matrix $\boldsymbol{X}$ such that

$$\boldsymbol{H} + \boldsymbol{P}^T \boldsymbol{X}^T \boldsymbol{Q} + \boldsymbol{Q}^T \boldsymbol{X} \boldsymbol{P} < 0 \tag{1.58}$$

if and only if

$$\boldsymbol{N}_p^T \boldsymbol{H} \boldsymbol{N}_p < 0, \quad \boldsymbol{N}_q^T \boldsymbol{H} \boldsymbol{N}_q < 0 \tag{1.59}$$

Further, the so-called reciprocal projection lemma will be presented as an advancement of Lemma 1.11, which is proposed in [15].

Lemma 1.12: For a given symmetric matrix $\boldsymbol{\Psi} \in \mathbb{S}^n$, there exists a matrix $\boldsymbol{S} \in \mathbb{R}^{n \times n}$ satisfying

$$\boldsymbol{\Psi} + \boldsymbol{S}^T + \boldsymbol{S} < 0 \tag{1.60}$$

if and only if, for an arbitrarily fixed symmetric matrix $\boldsymbol{P} \in \mathbb{S}^n$, there exists a matrix $\boldsymbol{W} \in \mathbb{R}^{n \times n}$ satisfying

$$\begin{bmatrix} \boldsymbol{\Psi} + \boldsymbol{P} - (\boldsymbol{W}^T + \boldsymbol{W}) & \boldsymbol{S}^T + \boldsymbol{W}^T \\ \boldsymbol{S} + \boldsymbol{W} & -\boldsymbol{P} \end{bmatrix} < 0 \tag{1.61}$$

In short, all of these aforementioned lemmas are going to be used for stability analysis of spacecraft dynamics in subsequent chapters of this book, and some of them may be employed quite frequently. In fact, they are also widely used in the literature for control systems by many researchers [16,17].

1.1.3 Advantages of linear matrix inequalities

After finishing the introduction of LMI's expressions, as those examples have been given in order from Section 1.1.1, plenty of control problems can be transformed to that involving LMIs, and solving those will bring us more fascinating results. Here shows some advantages of application of LMIs.

- *Global optimal solution and numerical reliability*

From Section 1.1.1, we obtained that many optimization problems involving LMIs have the following features:

1. They are convex, and hence have solvable global optimal solutions;
2. They can be solved numerically efficiently and reliably;
3. They are always solvable even when their sizes are very large.

So, for an LMI problem, a solution can always be obtained with acceptable precision due to existence of efficient algorithms as long as the problem has a solution.

- *Multiple objective design*

Another outstanding advantage of the LMI method for control system analysis and design is that it can greatly reduce the difficulty of controller design for multiobjective systems. The general multiobjective control problems are full of challenges and have not been sufficiently solved so far. In implementing multiobjective control tasks, LMIs can be applied to comprehensive problems that fully consider certain design specifications ranging from H_2 and H_∞ performance to regional pole placement, asymptotic tracking or regulation, and settling time or saturation constraints. With LMI approaches, as long as the design objectives can be converted into certain LMI constraints, the solution to the multiobjective design problem is straightforward.

Besides the design specifications mentioned above, researchers may have other design objectives in interest such as passivity, asymptotic disturbance rejection, time-domain constraints, and constraints on the closed-loop pole location. In addition, these objectives can be specified on different channels of the closed-loop system. When all objectives are formulated in terms of a common Lyapunov function, controller design amounts to solving a system of linear matrix inequalities.

- *Mature software tools*

  For solving problems related to LMIs, there have been several mature and widely accepted software tools, that is, YALMIP, CVX, and the MATLAB LMI toolbox. These packages have all been implemented and tested on specific families of LMIs that arise in control theory and have been found to be very efficient. With the help of these standard and open-source software packages, an application of the LMI approach to a control system analysis and design today has only been a problem of converting an analysis or design problem into an LMI problem. Then, the problem can be handed over to some standard software package, and the next step for us to do is to analyze the results provided by the package. This is why we are saying that LMI today has been a widely used technique.

1.1.4 Some standard linear matrix inequalities problems

As has been mentioned in Section 1.1.3 that the top-two advantages of LMI approaches are their global optimality and numerical reliability. In this subsection, firstly, we give some related definitions in connection with optimization issues; and, secondly, we establish theoretical support for these two important features of LMIs. To do this, in this part, we will recall some basic knowledge about optimization theory and try to theoretically summarize categories of standard LMIs problems. Finally, it leads to the important conclusion that LMIs are special types of convex optimization problems.

1. Definitions and properties

To draw the final conclusion as what has been written above, certain basic mathematical concepts involved in optimization theory should first be presented. Thus let's start from the definition of convex sets.

Definition 1.5: A set $\mathbb{F}$ is called convex if for any $x_1, x_2 \in \mathbb{F}$ and $0 \leq \theta \leq 1$ there holds

$$\theta x_1 + (1-\theta)x_2 \in \mathbb{F} \tag{1.62}$$

Definition 1.6: Let $\mathbb{F} \subset \mathbb{R}^n$, then the intersection of all convex sets in $\mathbb{R}^n$ containing $\mathbb{F}$ is called the convex hull of $\mathbb{F}$ and is denoted by conv($\mathbb{F}$).

Definition 1.7: A convex combination of points $z_1, z_2, ..., z_m$ from $\mathbb{R}^n$ is a linear combination

$$\alpha_1 z_1 + \alpha_2 z_2 + \cdots + \alpha_m z_m \tag{1.63}$$

with

$$\alpha_1 + \alpha_2 + \cdots + \alpha_m = 1 \tag{1.64}$$

and

$$\alpha_i \geq 0, \quad i = 1, 2, \ldots, m \tag{1.65}$$

Definition 1.8: Let $\mathbb{F} \subset \mathbb{R}^n$, and for some $x_0 \subset \mathbb{R}^n$ and $\varepsilon > 0$, denote

$$\mathbb{B}_\epsilon(x_0) = \{x | \|x - x_0\| < \epsilon\} \tag{1.66}$$

Then,

- x_0 is called an interior point of $\mathbb{F}$ if there exists a scalar $\varepsilon > 0$ such that $\mathbb{B}_\varepsilon(x_0) \supset \mathbb{F}$;
- $\mathbb{F}$ is called open if every inner point of $\mathbb{F}$ is an interior point;
- x_0 is called a boundary point of $\mathbb{F}$ if for arbitrary small scalar $\varepsilon > 0$ there holds

$$\mathbb{B}_\epsilon(x_0) \cap \mathbb{F} \neq \begin{cases} \varnothing \\ \mathbb{B}_\epsilon(x_0) \end{cases} \tag{1.67}$$

  and the set formed by all the boundary points of $\mathbb{F}$ is called the boundary set of $\mathbb{F}$, denoted by $\mathbb{F}_B$;
- $\mathbb{F}$ is called closed if it contains all its boundary points;
- $\mathbb{F}$ is called bounded if there exists a scalar $\gamma > 0$ such that $\mathbb{B}_\gamma(x_0) \supset \mathbb{F}$;
- $\mathbb{F}$ is called compact if it is both bounded and closed;
- When $\mathbb{F}$ is convex, $x_0 \in \mathbb{F}$ is called an extreme point if there are no two distinct points x_1 and x_2 in $\mathbb{F}$ such that $x_0 = \theta x_1 + (1 - \theta)x_2$ for $0 < \theta < 1$, and the set formed by all the extreme points of $\mathbb{F}$ is denoted by $\mathbb{F}_E$.

Definition 1.9: Let f: $\Omega \subset \mathbb{R}^n \to \mathbb{R}$. Then,

- f is called a convex function if Ω is a convex set and

$$f(\theta x + (1 - \theta)y) \leq \theta f(x) + (1 - \theta)f(y) \tag{1.68}$$

for all $x, y \in \Omega$ and $0 \leq \theta \leq 1$;

- f is called strictly convex if (1.68) holds strictly for all $x, y \in \Omega$, and $0 < \theta < 1$;
- f is called concave if $-f$ is convex.

In addition to the aforementioned fundamental definitions in this part, a significant lemma will be directly provided to reveal the most important feature of the convex optimization. The following lemma is going to solve the optimality of this problem:

$$\begin{cases} \min & f_0(x) \\ \text{s.t.} & x \in \Omega \end{cases} \tag{1.69}$$

where $\Omega \subset \mathbb{R}^n$ is a convex set, while f_0 is a convex function defined on Ω.

Lemma 1.13: Any locally optimal point of the convex optimization problem (1.69) is globally optimal. Furthermore, the problem has a unique globally optimal point when the objective function $f_0(x)$ is strictly convex.

2. LMI problems

After introducing several basic definitions and properties about convex optimization, since many actual optimal issues are restricted objectively, we are going to discuss a problem in the following form with an LMI constraint:

$$\begin{cases} \min f_0(\boldsymbol{x}) \\ \text{s.t. } \boldsymbol{x} \in \{\boldsymbol{x} | \boldsymbol{A}(\boldsymbol{x}) < 0\} \cap \mathbb{F} \end{cases} \tag{1.70}$$

with $\boldsymbol{A}(\boldsymbol{x}) < 0$ being an LMI, $f_0(\boldsymbol{x})$ denotes a convex function, and $\mathbb{F}$ is a convex set. First, the convexity of an LMI constraint should be discussed with the following expression:

$$\boldsymbol{A}(\boldsymbol{x}) = \boldsymbol{A}_0 + x_1\boldsymbol{A}_1 + \cdots + x_n\boldsymbol{A}_n < 0 \tag{1.71}$$

The forthcoming theorem is about to confirm that an LMI constraint actually defines a convex set. But before that, we need to give a lemma that establishes the connection between an LMI and the max eigenvalue of a matrix.

Lemma 1.14: Let $\boldsymbol{M}$ be a symmetric matrix. Then,

$$\lambda_{\max}(\boldsymbol{M}) \le t \Leftrightarrow \boldsymbol{M} - t\boldsymbol{I} \le 0 \tag{1.72}$$

Theorem 1.1: Let $\boldsymbol{A}_i \in \mathbb{S}^m, i = 0, 1, \ldots, n$, and $\boldsymbol{A}(\boldsymbol{x}) = \boldsymbol{A}_0 + x_1\boldsymbol{A}_1 + \cdots + x_n\boldsymbol{A}_n$. Then,

1. $\Phi_0 = \{\boldsymbol{x} | \boldsymbol{A}(\boldsymbol{x}) < 0\}$ is an open convex set.
2. $\Phi_c = \{\boldsymbol{x} | \boldsymbol{A}(\boldsymbol{x}) \le 0\}$ is a closed convex set.

Proof: Choose

$$x = \begin{bmatrix} x_1 & x_2 & \cdots & x_n \end{bmatrix}^T \in \Phi_0, y = \begin{bmatrix} y_1 & y_2 & \cdots & y_n \end{bmatrix}^T \in \Phi_0 \tag{1.73}$$

Then,

$$A(x) = A_0 + x_1 A_1 + \cdots + x_n A_n < 0 \tag{1.74}$$

and

$$A(y) = A_0 + y_1 A_1 + \cdots + y_n A_n < 0 \tag{1.75}$$

Thus for any $0 \le \theta \le 1$, there holds

$$\begin{aligned} &A(\theta x + (1-\theta)y) \\ &= A_0 + (\theta x_1 + (1-\theta)y_1)A_1 + \cdots + (\theta x_n + (1-\theta)y_n)A_n \\ &= \theta A_0 + \theta x_1 A_1 + \cdots + \theta x_n A_n + (1-\theta)A_0 + (1-\theta)y_1 A_1 + \cdots + (1-\theta)y_n A_n \\ &= \theta A(x) + (1-\theta)A(y) \\ &< 0 \end{aligned} \tag{1.76}$$

This proves the convexity of Φ_0. Similarly, one can obtain the convexity of Φ_c. Next, it discusses the openness of Φ_0. Arbitrarily choose $x^* \in \Phi_0$, then $A(x^*) < 0$. Let

$$\Delta x = \begin{bmatrix} \Delta x_1 & \Delta x_2 & \cdots & \Delta x_n \end{bmatrix}^T \tag{1.77}$$

Then,

$$\begin{aligned} A(x^* + \Delta x) &= A(x^*) + \Delta x_1 A_1 + \cdots + \Delta x_n A_n \\ &= A(x^*) + \Delta A(\Delta x) \end{aligned} \tag{1.78}$$

Since $\Delta A(\Delta x)$ is linear with respect to Δx, it is not difficult to have

$$\lim_{||\Delta x|| \to 0} ||\Delta A(\Delta x)|| = 0 \tag{1.79}$$

Further, considering the Lemma 1.13, and noticing $\lambda_{\max}(A(x^*)) < 0$ and the fact that the eigenvalues of a matrix are continuous with respect to the matrix elements, one can conclude that there exists a scalar $\varepsilon > 0$, satisfying

$$\lambda_{\max}(A(x^* + \Delta x)) < 0, \ \|\Delta x\| < \epsilon \tag{1.80}$$

which is equivalent to

$$A(x) < 0, \quad \left\| x - x^* \right\| < \epsilon \tag{1.81}$$

This implies that

$$\{x| \|x - x^*\| < \epsilon\} \subset \Phi_0 \tag{1.82}$$

Thus Φ_0 is an open set. Similarly, it can be verified that $\overline{\Phi}_c = \{x|A(x) > 0\}$ is an open set.

In the light of Theorem 1.1 and Lemma 1.13, it follows from the earlier result that the optimization problem (1.70) with an LMI constraint is in fact a convex one; thus it leads to the feature that any local minimum is a global one, or it has a unique globally optimal point when the function $f_0(x)$ is strictly convex.

The next theorem shows us that an LMI holds over a compact and convex set if and only if it holds on the extreme points of the compact and convex set.

Theorem 1.2: Let $\mathbb{F}$ be a compact and convex set, $A_i \in \mathbb{S}^m, i = 0, 1, \ldots, n$ and $A(x) = A_0 + x_1 A_1 + \cdots + x_n A_n$. Then, there holds

$$A(x) < 0, \quad x \in \mathbb{F} \Leftrightarrow A(x) < 0, \quad x \in \mathbb{F}_E \tag{1.83}$$

Proof: Since $\mathbb{F}_E \subset \mathbb{F}$, one can arbitrarily choose $x \in \mathbb{F}, x \notin \mathbb{F}_E$. Besides, the fact that $\mathbb{F}$ is compact and convex leads to $\mathbb{F} = \mathrm{conv}(\mathbb{F}_E)$. Therefore there exist $e_i \in \mathbb{F}_E, i = 1, 2, \ldots, k$ such that

$$x = \alpha_1 e_1 + \alpha_2 e_2 + \cdots + \alpha_k e_k \tag{1.84}$$

where,

$$\alpha_i \geq 0, i = 1, 2, \ldots, k, \quad \sum_{i=1}^{k} \alpha_i = 1 \tag{1.85}$$

Since $A(x)$ is linear in x, and assume that $A(e_i) < 0, i = 1, 2, \ldots, k$, there is

$$A(x) = \sum_{i=1}^{k} \alpha_i A(e_i) < 0 \tag{1.86}$$

Thus the condition is sufficient.

With the help of the above two theorems, now we will present three standard convex optimization problems with LMI constraints.

For convenience, we briefly introduce the following standard problems involving LMIs. In general, many actual problems can be transformed into one of these standard forms. In the forthcoming content, we set the rules in advance that all of $\boldsymbol{A}(\boldsymbol{x})$, $\boldsymbol{B}(\boldsymbol{x})$, $\boldsymbol{C}(\boldsymbol{x})$ and $\boldsymbol{L}(\boldsymbol{x})$, $\boldsymbol{R}(\boldsymbol{x})$ are assumed to be matrix functions of appropriate dimensions, which are all linear in vector $\boldsymbol{x}$. In addition, corresponding functions of MATLAB Toolbox will also be provided for readers.

- *Feasibility problem*

  Find a solution $\boldsymbol{x} \in \mathbb{R}^n$ satisfying the following LMI:

$$\boldsymbol{A}(\boldsymbol{x}) < \boldsymbol{B}(\boldsymbol{x})$$

  The corresponding function for solving this feasibility problem in the MATLAB LMI Toolbox is "feasp," which solves the following auxiliary convex optimization with LMI constraints:

$$\begin{cases} \min t \\ \text{s.t.}\, \boldsymbol{A}(\boldsymbol{x}) < \boldsymbol{B}(\boldsymbol{x}) + t\boldsymbol{I} \end{cases} \tag{1.87}$$

  where the scalar t are the decision variables.

- *Convex minimization problem*

  Given a convex function $f(\boldsymbol{x})$, find a solution $\boldsymbol{x} \in \mathbb{R}^n$ to the following minimization problem with LMI constraints:

$$\begin{cases} \min f(\boldsymbol{x}) \\ \text{s.t.}\, \boldsymbol{A}(\boldsymbol{x}) < \boldsymbol{B}(\boldsymbol{x}) \end{cases} \tag{1.88}$$

  The corresponding function for solving this problem in the MATLAB LMI Toolbox is "mincx," which solves the LMI problem:

$$\begin{cases} \min \boldsymbol{c}^T \boldsymbol{x} \\ \text{s.t.}\, \boldsymbol{L}(\boldsymbol{x}) < \boldsymbol{R}(\boldsymbol{x}) \end{cases} \tag{1.89}$$

  where $\boldsymbol{x}$ is the vector of decision variable.

- *Generalized Eigenvalue problem*

  Find a solution $\boldsymbol{x} \in \mathbb{R}^n$ to the following minimization problem

$$\begin{cases} \min \lambda \\ \text{s.t.}\, \boldsymbol{A}(\boldsymbol{x}) < \lambda \boldsymbol{B}(\boldsymbol{x}) \\ 0 < \boldsymbol{B}(\boldsymbol{x}) \\ \boldsymbol{C}(\boldsymbol{x}) < 0 \end{cases} \tag{1.90}$$

where $\boldsymbol{x}$ is the vector of decision variable. The corresponding function for solving this problem in the MATLAB LMI Toolbox is "gevp."

In practical applications, such facts show us that an optimization problem with LMI constraints and a convex objective is a special convex optimization problem. Thus many problems involving LMIs can be classified into these three types. Besides, the MATLAB LMI toolbox provides a very efficient and convenient package for solving these three types of problems involving LMIs. So, no wonder LMIs today have attracted so much attention in the control community.

1.2 Spacecraft attitude kinematics and dynamics

In this section, we are going to present the fundamental background knowledge of the spacecraft attitude kinematics and dynamics. Spacecraft kinematics and dynamics belong to a rich subject involving a variety of topics from modeling and control theory.

The attitude of a spacecraft, that is, its orientation in space, is an important concept in spacecraft dynamics and control. Attitude motion is approximately decoupled from orbital motion, so that the two subjects are typically treated separately. More precisely, the orbital motion does have a significant effect on the attitude motion, but the attitude motion has a less significant effect on the orbital motion. Operationally, the most important aspect of attitude dynamics is attitude determination. The reason for formulating and studying the dynamics problem is so that these operational tasks can be performed accurately and efficiently. Firstly, the deterministic algorithm to compute the six orbital elements (i.e., range $\boldsymbol{\rho}$, range rate $\dot{\boldsymbol{\rho}}$, azimuth $\boldsymbol{A}_z$, azimuth rate $\dot{\boldsymbol{A}}_z$, elevation $\boldsymbol{E}_l$, and elevation rate $\dot{\boldsymbol{E}}_l$) from these six measurements is well-known and can be found in most astrodynamics texts [18]. However, since measurement noise is inevitable, it is not practical to use only six measurements, and statistical methods are normally used, incorporating a large number of observations. Secondly, researchers usually determine the attitude, which can be described by three parameters such as Euler angles, by measuring the directions from the spacecraft to some known points of interest. This book mainly orients towards spacecraft attitude control problems with LMIs; thus basic knowledge of kinematics & dynamics will only be briefly introduced in this section. For elaboration, plenty of books record reliable and accurate knowledge of spacecraft attitude dynamics from 1960 to now.

In addition, for the sake of the reader, the difference between dynamics and kinematics can be made more clearly through the following expression:

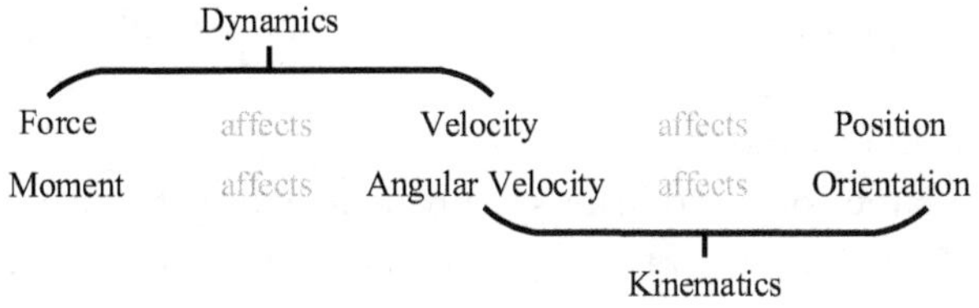

As noted above, the study of spacecraft motion can be decomposed into the study of kinematics and dynamics. For the translational motion of a particle of mass m, this decomposition amounts to expressing Newton's second law,

$$m\ddot{\boldsymbol{r}} = \boldsymbol{f} \tag{1.91}$$

a 2nd-order vector differential equation, as the two 1st-order vector differential equations

$$\begin{aligned} \dot{\boldsymbol{r}} &= \boldsymbol{p}/m \\ \dot{\boldsymbol{p}} &= \boldsymbol{f} \end{aligned} \tag{1.92}$$

Here $\boldsymbol{r}$ is the position vector of the particle relative to an inertial origin O, $\boldsymbol{p} = m\dot{\boldsymbol{r}}$ is the linear momentum of the particle, and $\boldsymbol{f}$ is the sum of all the forces acting on the particle. The first equation in Eq. (1.92) is the kinematics differential equation, describing how position changes for a given velocity; the second equation in Eq. (1.92) is the dynamics differential equation, describing how velocity changes for a given force.

This section begins with the development of attitude representations, including some of the variables that can be used to describe attitude motion. Then we develop the differential equations that describe attitude kinematics & dynamics, which are equivalent to Eq. (1.92), for a given angular velocity.

1.2.1 Attitude representations

This subsection begins with a discussion of basic concepts of spacecraft attitude kinematics. This discussion introduces various representations of the attitude or orientation of spacecraft. Based on that, we will further discuss the kinematics of column vectors in rotating reference frames.

We begin by discussing reference frames, vectors, and their representations in reference frames. The problem of representing vectors in different reference frames leads to the development of rotations, rotation matrices, and various ways of representing rotation matrices, including Euler angles, Euler parameters, and quaternions.

1. Reference frame and rotations

   A reference frame, or coordinate system, is generally taken to be ternary unit vectors that are mutually perpendicular. An equivalent definition is that a reference frame is a triad of orthonormal vectors. The reason that reference frames are so important in spacecraft attitude dynamics is that orientation in a reference frame is completely equivalent to that of a rigid body, since the rigid body model is an excellent approximation for studying attitude dynamics [19].

The orthonormal property of a reference frame's base vectors is defined by the dot products of the vectors with each other. Specifically, for a set of orthonormal base vectors, the dot products satisfy

$$\hat{\boldsymbol{i}}_i \cdot \hat{\boldsymbol{i}}_j = \begin{cases} 1 & \text{if} \quad i = j \\ 0 & \text{if} \quad i \neq j \end{cases} \tag{1.93}$$

where $\{\hat{\boldsymbol{i}}_1, \hat{\boldsymbol{i}}_2, \hat{\boldsymbol{i}}_3\}$ denotes an inertial frame. The right-handed or dextral property of a reference frame's base vectors is defined by the cross products of the vectors with each other. Specifically, for a right-handed set of orthonormal base vectors, the cross products satisfy

$$\begin{array}{lll} \hat{\boldsymbol{i}}_1 \times \hat{\boldsymbol{i}}_1 = \boldsymbol{0} & \hat{\boldsymbol{i}}_1 \times \hat{\boldsymbol{i}}_2 = \hat{\boldsymbol{i}}_3 & \hat{\boldsymbol{i}}_1 \times \hat{\boldsymbol{i}}_3 = -\hat{\boldsymbol{i}}_2 \\ \hat{\boldsymbol{i}}_2 \times \hat{\boldsymbol{i}}_1 = -\hat{\boldsymbol{i}}_3 & \hat{\boldsymbol{i}}_2 \times \hat{\boldsymbol{i}}_2 = \boldsymbol{0} & \hat{\boldsymbol{i}}_2 \times \hat{\boldsymbol{i}}_3 = \hat{\boldsymbol{i}}_1 \\ \hat{\boldsymbol{i}}_3 \times \hat{\boldsymbol{i}}_1 = \hat{\boldsymbol{i}}_2 & \hat{\boldsymbol{i}}_3 \times \hat{\boldsymbol{i}}_2 = -\hat{\boldsymbol{i}}_1 & \hat{\boldsymbol{i}}_3 \times \hat{\boldsymbol{i}}_3 = \boldsymbol{0} \end{array} \tag{1.94}$$

For the sake of expression and in terms of Eq. (1.94), here give the definition of the cross product for three-component vectors $\boldsymbol{x} = (x_1, x_2, x_3)^T$ and $\boldsymbol{y} = (y_1, y_2, y_3)^T$ as

$$\boldsymbol{x} \times \boldsymbol{y} = \begin{bmatrix} x_2 y_3 - x_3 y_2 \\ x_3 y_1 - x_1 y_3 \\ x_1 y_2 - x_2 y_1 \end{bmatrix} = -\boldsymbol{y} \times \boldsymbol{x} \tag{1.95}$$

The cross product can also be obtained using matrix multiplication:

$$\boldsymbol{x} \times \boldsymbol{y} = [\boldsymbol{x} \times]\boldsymbol{y} \tag{1.96}$$

where $[\boldsymbol{x} \times]$ is the cross product matrix and can be defined as

$$[\boldsymbol{x} \times] \equiv \begin{bmatrix} 0 & -x_3 & x_2 \\ x_3 & 0 & -x_1 \\ -x_2 & x_1 & 0 \end{bmatrix} \tag{1.97}$$

Note that $[\boldsymbol{x} \times]$ satisfies the skew-symmetry property $[\boldsymbol{x} \times]$ $\boldsymbol{T} = -[\boldsymbol{x} \times]$. Further, the cross product and the cross-product matrix obey the following relations [20]:

$$\begin{aligned}
&\boldsymbol{x}\cdot(\boldsymbol{y}\times\mathbf{z})=(\boldsymbol{x}\times\boldsymbol{y})\cdot\boldsymbol{z}\\
&[\boldsymbol{x}\times][\boldsymbol{y}\times]=-(\boldsymbol{x}\cdot\boldsymbol{y})\boldsymbol{I}+\boldsymbol{y}\boldsymbol{x}^T\\
&[\boldsymbol{x}\times][\boldsymbol{y}\times]-[\boldsymbol{y}\times][\boldsymbol{x}\times]=\boldsymbol{y}\boldsymbol{x}^T-\boldsymbol{x}\boldsymbol{y}^T=[(\boldsymbol{x}\times\boldsymbol{y})\times]\\
&\mathrm{adj}[\boldsymbol{x}\times]=\boldsymbol{x}\boldsymbol{x}^T\\
&|\boldsymbol{x}\times\boldsymbol{y}|^2=|\boldsymbol{x}|^2|\boldsymbol{y}|^2-(\boldsymbol{x}\cdot\boldsymbol{y})^2\\
&|\boldsymbol{x}\times\boldsymbol{y}|=|\boldsymbol{x}||\boldsymbol{y}|\sin\theta
\end{aligned} \tag{1.98}$$

Next, a handy way to write a vector in terms of its components and the base vectors is to write it as the product of two vectors, one the component vector, and the other a vector containing the base unit vectors, such as

$$\boldsymbol{v}=\begin{bmatrix} v_1 & v_2 & v_3 \end{bmatrix}\begin{Bmatrix} \widehat{\boldsymbol{i}}_1 \\ \widehat{\boldsymbol{i}}_2 \\ \widehat{\boldsymbol{i}}_3 \end{Bmatrix}=\boldsymbol{v}_i^T\{\widehat{\boldsymbol{i}}\} \tag{1.99}$$

Using this notation, one can write $\boldsymbol{v}$ in terms of different frames as

$$\boldsymbol{v}=\boldsymbol{v}_i^T\{\widehat{\boldsymbol{i}}\}=\boldsymbol{v}_o^T\{\widehat{\boldsymbol{o}}\}=\boldsymbol{v}_b^T\{\widehat{\boldsymbol{b}}\} \tag{1.100}$$

where $\{\hat{\boldsymbol{i}}\}$, $\{\hat{\boldsymbol{o}}\}$ and $\{\hat{\boldsymbol{b}}\}$ denote basis vectors in their corresponding reference frames F_i, F_b, F_o, respectively. Equations above involve determining the components of the same vector in different frames that have been reoriented or rotated. Each equal-sign connects two expressions with rotations, which are the subject of the next topic.

Consider Eq. (1.100), for example, we seek a way to express $\{\hat{\boldsymbol{i}}\}$ in terms of $\{\hat{\boldsymbol{b}}\}$, say $\{\hat{\boldsymbol{i}}\}=\boldsymbol{R}\{\hat{\boldsymbol{b}}\}$, where $\boldsymbol{R}\in\mathbb{R}^{3\times3}$ is called a transformation matrix. Then, there holds

$$\boldsymbol{v}=\boldsymbol{v}_i^T\{\widehat{\boldsymbol{i}}\}=\boldsymbol{v}_i^T\boldsymbol{R}\{\widehat{\boldsymbol{b}}\}=\boldsymbol{v}_b^T\{\widehat{\boldsymbol{b}}\} \tag{1.101}$$

Thus to express the rotation, we just need to determine $\boldsymbol{R}$ and solve the linear system of equations defined by Eq. (1.101). Further, since the elements of any three-dimensional vector can be represented as the product of the length of the vector and the cosine of the Angle between the vector and corresponding basis vector. Thus Eq. (1.99) can be expanded into

$$\boldsymbol{v}=v_1\hat{\boldsymbol{i}}_1+v_2\hat{\boldsymbol{i}}_2+v_3\hat{\boldsymbol{i}}_3=v\left(\cos\alpha_1\hat{\boldsymbol{i}}_1+\cos\alpha_2\hat{\boldsymbol{i}}_2+\cos\alpha_3\hat{\boldsymbol{i}}_3\right) \tag{1.102}$$

Base on this, if we rewrite the components of $\boldsymbol{R}$ as $\boldsymbol{R}_{ij}$, where i denotes the row and j denotes the column, then (1.101) may be expanded to

$$\begin{aligned}\widehat{\boldsymbol{i}}_1 &= \boldsymbol{R}_{11}\widehat{\boldsymbol{b}}_1 + \boldsymbol{R}_{12}\widehat{\boldsymbol{b}}_2 + \boldsymbol{R}_{13}\widehat{\boldsymbol{b}}_3 \\ \widehat{\boldsymbol{i}}_2 &= \boldsymbol{R}_{21}\widehat{\boldsymbol{b}}_1 + \boldsymbol{R}_{22}\widehat{\boldsymbol{b}}_2 + \boldsymbol{R}_{23}\widehat{\boldsymbol{b}}_3 \\ \widehat{\boldsymbol{i}}_3 &= \boldsymbol{R}_{31}\widehat{\boldsymbol{b}}_1 + \boldsymbol{R}_{32}\widehat{\boldsymbol{b}}_2 + \boldsymbol{R}_{33}\widehat{\boldsymbol{b}}_3\end{aligned} \tag{1.103}$$

Comparing these expressions with Eq. (1.102), thus $\boldsymbol{R}$ is a matrix of direction cosines, and is frequently referred to as the DCM (direction cosine matrix).

2. Euler angles

Computing the nine direction cosines of DCM is one way to construct a rotation matrix, but it may be too ineffective. One of the easiest ways to visualize is the Euler angle approach. Euler reasoned that any rotation from one frame to another can be visualized as a sequence of three simple rotations about base vectors. Let us consider the rotation from F_i to F_b through a sequence of three angles θ_1, θ_2, and θ_3.

Begin with a simple rotation about the $\hat{\boldsymbol{i}}_3$ axis, we denote the resulting reference frame as $F_{i'}$. Using the rules developed above for constructing $\boldsymbol{R}^{i'i}$, it is easy to show that the correct rotation matrix is

$$\boldsymbol{R}^{i'i} = \boldsymbol{R}_3(\theta_1) = \begin{bmatrix} \cos\theta_1 & \sin\theta_1 & 0 \\ -\sin\theta_1 & \cos\theta_1 & 0 \\ 0 & 0 & 1 \end{bmatrix} \tag{1.104}$$

so that

$$\boldsymbol{v}_{i'} = \boldsymbol{R}_3(\theta_1)\boldsymbol{v}_i \tag{1.105}$$

Similarly, there are three possibilities for the first simple rotation in an Euler angle sequence. By the same token, we directly present the following representations of the other two rotations, that is

$$\boldsymbol{R}^{i''i'} = \boldsymbol{R}_2(\theta_2) = \begin{bmatrix} \cos\theta_2 & 0 & -\sin\theta_2 \\ 0 & 1 & 0 \\ \sin\theta_2 & 0 & \cos\theta_2 \end{bmatrix} \tag{1.106}$$

so that

$$\boldsymbol{v}_{i''} = \boldsymbol{R}_2(\theta_2)\boldsymbol{v}_{i'} = \boldsymbol{R}_2(\theta_2)\boldsymbol{R}_3(\theta_1)\boldsymbol{v}_i \tag{1.107}$$

and

$$\boldsymbol{R}^{bi''} = \boldsymbol{R}_1(\theta_3) = \begin{bmatrix} 1 & 0 & 0 \\ 0 & \cos\theta_3 & \sin\theta_3 \\ 0 & -\sin\theta_3 & \cos\theta_3 \end{bmatrix} \tag{1.108}$$

Finally, the overall rotational motion should be expressed as

$$\boldsymbol{v}_b = \boldsymbol{R}_1(\theta_3)\boldsymbol{v}_{i''} = \boldsymbol{R}_1(\theta_3)\boldsymbol{R}_2(\theta_2)\boldsymbol{R}_3(\theta_1)\boldsymbol{v}_i \tag{1.109}$$

It is not difficult to find that, for a given rotational motion of a reference frame, if one can keep track of the three Euler angles, then one can track the changing orientation of the frame. Note on Euler angle sequences, recall that there were three axes to choose from for the first rotation, two to choose from for the second rotation, and two to choose from for the third rotation. Thus there are twelve possible sequences of Euler angles. These are commonly referred to by the axes that are used. For example, the sequence used above is called a "3−2−1" sequence, because we first rotate about the "3" axis, then about the "2" axis, and finally about the "1" axis. It is also possible for the third rotation to be of the same type as the first, then we could use a "3−2−3" sequence. Definitely, because of its intuitive and simple expression, the Euler Angle is widely used in a variety of posture description tasks.

3. Euler's theorem and quaternions

The Euler angle sequence approach to describing the relative orientation of two frames is reasonably easy to develop and visualize, but it is not the most useful approach for spacecraft dynamics because of the phenomenon of "Gimbal Lock." Another of Euler's contributions is the theorem that tells us that only one rotation is necessary to reorient one frame to another. This theorem is known as Euler's Theorem and is formally stated as:

Theorem 1.3: The most general motion of a rigid body with a fixed point is a rotation about a fixed axis.

Thus instead of using three simple rotations (and three angles) to keep track of rotational motion, we only need to use a single rotation (and a single angle) about the "fixed axis" mentioned in the theorem. This axis, denoted $\hat{\boldsymbol{a}}$, is called the Euler axis, or the eigenaxis, and the angle, denoted Φ, is called the Euler angle, or the Euler principal angle.

The same rotation from $F_{i'}$ to F_b, about axis $\hat{\boldsymbol{a}}$ through angle Φ, it is possible to express the rotation matrix $\boldsymbol{R}^{bi}$, in terms of $\hat{\boldsymbol{a}}$ and Φ, just as we expressed $\boldsymbol{R}^{bi}$ in terms of the Euler angles in the previous section. Note that since the rotation is about $\hat{\boldsymbol{a}}$, the Euler axis vector has the same components in $F_{i'}$ and F_b, that is,

$$\boldsymbol{R}^{bi}\boldsymbol{a} = \boldsymbol{a} \tag{1.110}$$

and

$$R^{bi} = \cos\Phi \mathbf{1} + (1 - \cos\Phi)aa^T - \sin\Phi[a\times] \tag{1.111}$$

where a is the column matrix of the components of $\hat{a}$ in either $F_{i'}$ or F_b. (1.110) provides the justification for the term eigen-axis for the Euler axis, since this equation defines a as the eigenvector of R^{bi} associated with the eigenvalue 1.

Corollary 1.3: Every rotation matrix has one eigenvalue that is, unity.

Thus given an Euler axis and Euler angle, one can easily compute the rotation matrix. Besides, the component matrix a and the angle Φ, for a given rotation matrix R, also need to be computed, that is

$$\begin{aligned} \Phi &= \cos^{-1}\left[\frac{1}{2}(\text{trace}R - 1)\right] \\ [a\times] &= \frac{1}{2\sin\Phi}\left(R^T - R\right) \end{aligned} \tag{1.112}$$

Except Euler Angles and Euler's Theorem, there are several other efficient approaches to represent the attitude of spacecraft. We finally introduce one of the most widely applied attitude representations: the quaternions, which is composed of four new variables in terms of a and Φ, that is,

$$\begin{aligned} q &= a\sin\frac{\Phi}{2} \\ q_4 &= \cos\frac{\Phi}{2} \end{aligned} \tag{1.113}$$

where $q \in \mathbb{R}^3$ forms the Euler axis component of the quaternion, also called the vector component, and the q_4 is called the scalar component. Scholars usually use the notation $\bar{q}$ to denote the vector containing all four variables, that is,

$$\bar{q} = \begin{bmatrix} q^T & q_4 \end{bmatrix}^T \tag{1.114}$$

For a given a and Φ, we give the rotation matrix in terms of $\bar{q}$ representing the corresponding rotational motion between two reference frames without derivation

$$R = \left(q_4^2 - q^T q\right)\mathbf{1} + 2qq^T - 2q_4[q\times] \tag{1.115}$$

and we also give the respective expression of $\overline{q}$ in terms of the components of $\boldsymbol{R}$:

$$
\begin{aligned}
q_4 &= \pm\frac{1}{2}\sqrt{1+\operatorname{trace}\boldsymbol{R}} \\
q &= \frac{1}{4q_4}\begin{bmatrix} \boldsymbol{R}_{23}-\boldsymbol{R}_{32} \\ \boldsymbol{R}_{31}-\boldsymbol{R}_{13} \\ \boldsymbol{R}_{12}-\boldsymbol{R}_{21} \end{bmatrix}
\end{aligned}
\tag{1.116}
$$

We now summarize all \three basic ways to express the rotation matrix in Eqs. (1.109), (1.111), and (1.115). The Euler Angle description of attitude suggests the possibility of representing a rotation by the vector quantity $\Phi\hat{\boldsymbol{a}}$. If there are two sequential rotations, then the net rotation could be expressed by the sum of vectors (i.e., $\Phi_1\hat{\boldsymbol{a}}_1+\Phi_2\hat{\boldsymbol{a}}_2$). We all know that the vector addition is commutative; however, it is easy to see that the actual rotation resulting from the two individual rotations does depend on the order of the rotations. Thus the commutation of rotation order is different from that of vector addition.

1.2.2 Attitude kinematics

The previous subsection provided several different ways to describe the attitude, or orientation, of one reference frame with respect to another, in terms of attitude variables. In this part, we will develop the kinematics differential equations for the spacecraft attitude.

Firstly, we should discuss the representation of angular velocity. An illustrative example developed in that section was for a "3−2−1" Euler angle sequence. Thus it should start from the rotation of one frame $F_{i'}$ with respect to another one F_i (the "3" rotation). Then, the angular velocity is

$$\boldsymbol{\omega}^{i'i}=\dot{\theta}_1\widehat{\boldsymbol{i}}'_3=\dot{\theta}_1\widehat{\boldsymbol{i}}_3 \tag{1.117}$$

Note the ordering of the superscripts in this expression. Also, note that this vector quantity has the same components in either frame; that is,

$$\boldsymbol{\omega}^{i'i}_{i'}=\boldsymbol{\omega}^{i'i}_{i}=\begin{bmatrix}0\\0\\\dot{\theta}_1\end{bmatrix} \tag{1.118}$$

Similarly, one can obtain the "2" and "1" rotations (from $F_{i'}$ to $F_{i''}$ then to F_b)

$$\begin{aligned}\boldsymbol{\omega}^{i''i'} &= \dot{\theta}_2\widehat{\boldsymbol{i}}_2'' = \dot{\theta}_2\widehat{\boldsymbol{i}}_2', \quad \boldsymbol{\omega}_{i''}^{i''i'} = \boldsymbol{\omega}_{i'}^{i''i'} = \begin{bmatrix} 0 \\ \dot{\theta}_2 \\ 0 \end{bmatrix}, \\ \boldsymbol{\omega}^{bi''} &= \dot{\theta}_3\widehat{\boldsymbol{b}}_1 = \dot{\theta}_3\widehat{\boldsymbol{i}}_1'', \quad \boldsymbol{\omega}_b^{bi''} = \boldsymbol{\omega}_{i''}^{bi''} = \begin{bmatrix} \dot{\theta}_3 \\ 0 \\ 0 \end{bmatrix}\end{aligned} \tag{1.119}$$

Thus the angular velocities for simple rotations are also simple angular velocities. Mathematically, angular velocity vectors add in the following way:

$$\boldsymbol{\omega}^{bi} = \boldsymbol{\omega}^{bi''} + \boldsymbol{\omega}^{i''i'} + \boldsymbol{\omega}^{i'i} \tag{1.120}$$

To perform computations involving angular velocities, we must choose a reference frame and express all these vectors in that reference frame and add them together. Notice that in Eqs. (1.118) and (1.119), the components of these vectors are given in different reference frames. To add them, we must transform them all to the same frame, which, in most attitude dynamics applications, usually choose the body frame F_b. Look back to Eq. (1.120), the second and the third terms on the right-hand side are given in $F_{i''}$ and $F_{i'}$. Therefore by virtue of rotational matrices introduced in Section 1.2.1, both $\boldsymbol{\omega}^{i''i'}$ and $\boldsymbol{\omega}^{i'i}$ should be transformed as $\boldsymbol{\omega}_b^{i''i'} = \boldsymbol{R}^{bi''}\boldsymbol{\omega}_{i''}^{i''i'} = \boldsymbol{R}_1(\theta_3)\boldsymbol{\omega}_{i''}^{i''i'}$ and $\boldsymbol{\omega}_b^{i'i} = \boldsymbol{R}^{bi'}\boldsymbol{\omega}_{i'}^{i'i} = \boldsymbol{R}_1(\theta_3)\boldsymbol{R}_2(\theta_2)\boldsymbol{\omega}_{i'}^{i'i}$. Now we have all the angular velocity vectors of Eq. (1.120) expressed in F_b and can add them together:

$$\begin{aligned}\boldsymbol{\omega}_b^{bi} &= \boldsymbol{\omega}_b^{bi''} + \boldsymbol{\omega}_b^{i''i'} + \boldsymbol{\omega}_b^{i'i} \\ &= \begin{bmatrix} \dot{\theta}_3 \\ 0 \\ 0 \end{bmatrix} + \begin{bmatrix} 0 \\ \cos\theta_3\dot{\theta}_2 \\ -\sin\theta_3\dot{\theta}_2 \end{bmatrix} + \begin{bmatrix} -\sin\theta_2\dot{\theta}_1 \\ \cos\theta_2\sin\theta_3\dot{\theta}_1 \\ \cos\theta_2\cos\theta_3\dot{\theta}_1 \end{bmatrix} \\ &= \begin{bmatrix} -\sin\theta_2 & 0 & 1 \\ \cos\theta_2\sin\theta_3 & \cos\theta_3 & 0 \\ \cos\theta_2\cos\theta_3 & -\sin\theta_3 & 0 \end{bmatrix}\begin{bmatrix} \dot{\theta}_1 \\ \dot{\theta}_2 \\ \dot{\theta}_3 \end{bmatrix} = \boldsymbol{S}(\boldsymbol{\theta})\dot{\boldsymbol{\theta}}\end{aligned} \tag{1.121}$$

If the $\dot{\boldsymbol{\theta}}$ is known, then it can be integrated to determine the $\boldsymbol{\theta}$, and the components of $\boldsymbol{\omega}$ can be determined. This integration is entirely analogous to knowing $\dot{x}, \dot{y}$ and $\dot{z}$, and integrating these to determine the position x, y, and z. The velocity is determined from the kinetics equations of motion as in Eq. (1.92). Similarly, for rotational motion, the kinetics equations of motion are used to determine the angular velocity, which is in turn used to determine the $\dot{\boldsymbol{\theta}}$s, but not vice versa.

Solving Eq. (1.121) for $\dot{\boldsymbol{\theta}}$ needs an inversion of $\boldsymbol{S}(\boldsymbol{\theta})$, which is straight forward to obtain:

$$\dot{\boldsymbol{\theta}} = \begin{bmatrix} 0 & \sin\theta_3/\cos\theta_2 & \cos\theta_3/\cos\theta_2 \\ 0 & \cos\theta_3 & -\sin\theta_3 \\ 1 & \sin\theta_3\sin\theta_2/\cos\theta_2 & \cos\theta_3\sin\theta_2/\cos\theta_2 \end{bmatrix} \begin{bmatrix} \omega_1 \\ \omega_2 \\ \omega_3 \end{bmatrix} = \boldsymbol{S}^{-1}\boldsymbol{\omega} \quad (1.122)$$

If one knows the $\boldsymbol{\omega}$ as a function of time, and have initial conditions for the three Euler angles, then one can integrate Eq. (1.122) to obtain $\boldsymbol{\theta}$ as a function of time. It should be mentioned that some of the elements of matrix $\boldsymbol{S}^{-1}$ become large when θ_2 approaches $\pi/2$, and indeed become infinite when $\theta_2 = \pi/2$. This phenomenon is usually called "Kinematic Singularity," which is one of the difficulties associated with applying Euler angles as attitude variables. It can also be seen that even though the angular velocity is small, the Euler angle rates can become quite large. Besides, for different Euler angle sequences, the kinematic singularity occurs at different points. Another difficulty is the huge computing resources for computing the sines and cosines necessary to integrate Eq. (1.122).

Hence, as we have introduced in Section 1.2.1, two approaches may avoid the drawback mentioned above. Now, let's provide the differential equations relating $(\boldsymbol{a}, \Phi)$ and $(\overline{\boldsymbol{q}}, \boldsymbol{\omega})$. For the Euler axis/angle set of attitude variables, there holds

$$\begin{aligned} \dot{\Phi} &= \boldsymbol{a}^T\boldsymbol{\omega} \\ \dot{\boldsymbol{a}} &= \frac{1}{2}\left([\boldsymbol{a}\times] - \cot\frac{\Phi}{2}[\boldsymbol{a}\times][\boldsymbol{a}\times]\right)\boldsymbol{\omega} \end{aligned} \quad (1.123)$$

where the kinematic singularity in Eq. (1.123) occurs evidently when $\Phi = 0$ *or* 2π, and both of these correspond to $\boldsymbol{R} = \boldsymbol{I}$ (i.e., two reference frames are identical). Therefore it is reasonably straightforward to deal with the singularity. For quaternions, the kinematic equations of motion are

$$\dot{\overline{\boldsymbol{q}}} = \frac{1}{2}\begin{bmatrix} [\boldsymbol{q}\times] + q_4\mathbf{1} \\ -\boldsymbol{q}^T \end{bmatrix}\boldsymbol{\omega} = \mathbf{Q}(\overline{\boldsymbol{q}})\boldsymbol{\omega} \quad (1.124)$$

where no kinematic singularities associated with $\dot{\overline{\boldsymbol{q}}}$, and no more trigonometric functions are needed to evaluate. For these reasons, the expression of quaternion as an attitude variable is favored in most satellite attitude dynamics applications.

So far, introductions of the three most typical kinematics equations for spacecraft attitude have been done, which denote research backgrounds for controller design in this book. However, most spacecraft attitude dynamics & control textbooks cover kinematics only as a part of the dynamics presentation, with no exception in this one.

1.2.3 Attitude dynamics

After discussing attitude kinematics of spacecraft, in this subsection, we shall focus on attitude dynamics. As mentioned before, for rotational motion, kinematics is the study of the change in orientation for a given angular velocity, and dynamics pay attention to how moments cause changes in angular velocity. The spacecraft attitude dynamics is derived from rigid body dynamic equations, where important environmental torques including gravity gradients, magnetics, aerodynamics, etc., affect spacecraft attitude dynamics. Among them, the gravity gradient torque occupies the most significant status. Thus this part only includes the development of the gravity gradient torque.

We assume a rigid spacecraft in orbit about a spherical primary, and every differential mass element of the body is subject to Newton's Universal Gravitational Law:

$$d\boldsymbol{f}_g = -\frac{GMdm}{r^2}\hat{\boldsymbol{e}}_r \tag{1.125}$$

where G is the universal gravitational constant, M is the mass of the spherical primary, dm denotes the mass of any unit mass element of the body in orbit, r is the radial distance from the mass center of the primary to the mass element, $\boldsymbol{f}_g$ denotes the gravity acting on body, and $\hat{\boldsymbol{e}}_r$ is a unit vector from the mass center of the primary to the mass element. Further, the position vector to a differential mass element can be written as the sum of the position vector from the primary to the mass center of the body and the vector from the mass center of the body to the differential mass element:

$$\boldsymbol{r} = {}^{c}_{O}\boldsymbol{r} + {}_{c}\boldsymbol{r} \tag{1.126}$$

Integrating Eq. (1.125) and substituting Eq. (1.126), the force acting on every part of the body can be expressed as

$$\boldsymbol{f}_g = -\int_B \frac{GM}{\left|{}^{c}_{O}\boldsymbol{r}^3 + {}_{c}\boldsymbol{r}^3\right|}\left({}_{c_o}\boldsymbol{r} + {}_{c}\boldsymbol{r}\right)dm \tag{1.127}$$

where the vector ${}_c\boldsymbol{r}$ is the only variable that depends on the differential mass element. However, in general, this integral cannot be computed in closed form. The usual approach is to assume that the radius of the orbit is much greater than the size of the body, that is, $\left|{}_O^c\boldsymbol{r}\right| \gg \left|{}_c\boldsymbol{r}\right|$. With the help of Taylor expansion, one has

$$\frac{{}_O^c\boldsymbol{r} + {}_c\boldsymbol{r}}{{}_O^c\boldsymbol{r}^3 + {}_c\boldsymbol{r}^3} = \frac{{}_O^c\boldsymbol{r} + {}_c\boldsymbol{r}}{{}_O^c\boldsymbol{r}^3}\left(1 - 3\frac{{}_O^c\boldsymbol{r}_c\boldsymbol{r}}{{}_O^c\boldsymbol{r}^2} + \text{H.O.T.}\right) \approx \frac{{}_O^c\boldsymbol{r} + {}_c\boldsymbol{r}}{{}_O^c\boldsymbol{r}^3} \tag{1.128}$$

Substituting Eq. (1.128) into Eq. (1.127) yields

$$\boldsymbol{f}_g = -\int_B \frac{GM({}_O^c\boldsymbol{r} + {}_c\boldsymbol{r})}{{}_O^c\boldsymbol{r}^3} dm = -\frac{GMm}{{}_O^c\boldsymbol{r}^3}\left({}_O^c\boldsymbol{r}\right) \tag{1.129}$$

Applying Newton's Second Law, we obtain:

$$\ddot{\boldsymbol{r}} + \frac{GM}{r^3}\boldsymbol{r} = \boldsymbol{0} \tag{1.130}$$

Then, developing the moment about the mass center due to gravitational forces, this can be expressed as an integral over the body:

$$\boldsymbol{g}_g^c = -\int_B {}_c\boldsymbol{r} \times d\boldsymbol{f}_g \tag{1.131}$$

Usually, the facts that the moment and moment of inertia are about the mass center and that the vectors are expressed in F_b are more comprehensible. Thus, applying the same assumption $\left|{}_O^c\boldsymbol{r}\right| \gg \left|{}_c\boldsymbol{r}\right|$, one can simplify the notation to express the approximate moment in a body-fixed reference frame, that is,

$$\boldsymbol{g}_g = 3\frac{GM}{r^3}[\boldsymbol{o}_3 \times]\boldsymbol{I}\boldsymbol{o}_3 \tag{1.132}$$

where $\boldsymbol{o}_3$ is the third column of the rotation matrix $\boldsymbol{R}^{bo}$. This torque affects the motion of all orbiting bodies, including all kinds of spacecraft. For a rigid spacecraft in a central gravitational field, the equations of motion may be approximated as

$$\begin{aligned} &\ddot{\boldsymbol{r}} + \frac{GM}{r^3}\boldsymbol{r} = \boldsymbol{0} \\ &\dot{\boldsymbol{r}} = 3\frac{GM}{r^3}\hat{\boldsymbol{o}}_3 \times \boldsymbol{I}\cdot\hat{\boldsymbol{o}}_3 \end{aligned} \tag{1.133}$$

where $\boldsymbol{r}$ is the position vector of the mass center of the body with respect to the center of the gravitational primary, $\boldsymbol{I} = diag(\boldsymbol{I}_1, \boldsymbol{I}_2, \boldsymbol{I}_3)$ is the moment of inertia matrix, and $\boldsymbol{h}$ is the angular momentum of the body about its mass center. The vector $\hat{\boldsymbol{o}}_3$ is the nadir vector; that is, $\hat{\boldsymbol{o}}_3 = -\boldsymbol{r}/r$.

Typically, researchers write these equations in terms of the principal body-frame components of these vectors. Thus the rotational equations for a rigid body subject only to gravitational forces and moments can be written as

$$\boldsymbol{I}\dot{\boldsymbol{\omega}} = -[\boldsymbol{\omega}\times]\boldsymbol{I}\boldsymbol{\omega} + 3\frac{GM}{r^3}[\boldsymbol{o}_3\times]\boldsymbol{I}\boldsymbol{o}_3 \tag{1.134}$$

To investigate the dynamics of small motions of spacecraft, we utilize a 1–2–3 rotational sequence from F_o to F_b, that is,

$$\boldsymbol{R}^{bo} = \boldsymbol{R}_3(\theta_3)\boldsymbol{R}_2(\theta_2)\boldsymbol{R}_1(\theta_1) = \begin{bmatrix} c_2c_3 & s_1s_2c_3 + c_1s_3 & s_1s_3 - c_1s_2c_3 \\ -s_2s_3 & c_1c_3 - s_1s_2s_3 & s_1c_3 + c_1s_2s_3 \\ s_2 & -s_1c_2 & c_1c_2 \end{bmatrix} \tag{1.135}$$

If we assume that all angles are small, that is, $\sin\theta_i \approx \theta_i, \cos\theta_i \approx 1$, then this rotational matrix becomes

$$\boldsymbol{R}^{bo} \approx \boldsymbol{I} - [\boldsymbol{\theta}\times]^{bo} \tag{1.136}$$

Thus $\boldsymbol{o}_3 = [-\theta_2, \theta_1, 1]^T$. This approximation may be used immediately to compute the gravity gradient torque (for small angles with orbital frequency $\boldsymbol{\omega}_c$) as

$$\boldsymbol{g}_g = 3\omega_c^2[\boldsymbol{o}_3\times]\boldsymbol{I}\boldsymbol{o}_3 = 3\omega_c^2\begin{bmatrix} (I_3 - I_2)\theta_1 \\ (I_3 - I_1)\theta_2 \\ 0 \end{bmatrix} \tag{1.137}$$

Assuming small angles and small angular rates, the angular velocity of F_b with respect to F_o is simply $\boldsymbol{\omega}^{bo} = \dot{\boldsymbol{\theta}}$, then, applying $\boldsymbol{\omega} = \boldsymbol{S}(\boldsymbol{\theta})\dot{\boldsymbol{\theta}}$ and $\boldsymbol{\omega}^{bi} = \boldsymbol{\omega}^{bo} + \boldsymbol{\omega}^{oi}$, one can obtain the expression in F_b, that is,

$$\boldsymbol{\omega}^{bi} = \begin{bmatrix} \dot{\theta}_1 \\ \dot{\theta}_2 \\ \dot{\theta}_3 \end{bmatrix} + \begin{bmatrix} 1 & \theta_3 & -\theta_2 \\ -\theta_3 & 1 & \theta_1 \\ \theta_2 & -\theta_1 & 1 \end{bmatrix}\begin{bmatrix} 0 \\ -\omega_c \\ 0 \end{bmatrix} = \begin{bmatrix} \dot{\theta}_1 - \omega_c\theta_3 \\ \dot{\theta}_2 - \omega_c \\ \dot{\theta}_3 + \omega_c\theta_1 \end{bmatrix} \tag{1.138}$$

Further,

$$\dot{\boldsymbol{\omega}} = \begin{bmatrix} \ddot{\theta}_1 - \omega_c\dot{\theta}_3 \\ \ddot{\theta}_2 \\ \ddot{\theta}_3 + \omega_c\dot{\theta}_1 \end{bmatrix} \tag{1.139}$$

Now, substituting Eqs. (1.137), (1.138), and (1.139) into (1.134) yields the attitude dynamics of orbital bodies subject only to the gravity torque, that is,

$$\begin{aligned} &I_1\ddot{\theta}_1 + (I_2 - I_3 - I_1)\omega_c\dot{\theta}_3 - 4(I_3 - I_2)\omega_c^2\theta_1 = 0 \\ &I_2\ddot{\theta}_2 + 3\omega_c^2(I_1 - I_3)\theta_2 = 0 \\ &I_3\ddot{\theta}_3 + (I_3 + I_1 - I_2)\omega_c\dot{\theta}_1 + (I_2 - I_1)\omega_c^2\theta_3 = 0 \end{aligned} \tag{1.140}$$

Finally, consider the situation that external disturbance and control torques affect the fixed-body spacecraft, the overall attitude dynamics of spacecraft can be obtained as

$$\begin{aligned} &I_1\ddot{\theta}_1 + 4(I_2 - I_3)\omega_c^2\theta_1 + (I_2 - I_3 - I_1)\omega_c\dot{\theta}_3 = T_{cx} + T_{dx} \\ &I_2\ddot{\theta}_2 + 3\omega_c^2(I_1 - I_3)\theta_2 = T_{cy} + T_{dy} \\ &I_3\ddot{\theta}_3 + (I_2 - I_1)\omega_c^2\theta_3 + (I_3 + I_1 - I_2)\omega_c\dot{\theta}_1 = T_{cz} + T_{dz} \end{aligned} \tag{1.141}$$

where T_{cx}, T_{cy} and T_{cz} denote the three components of control input torque, and T_{dx}, T_{dy} and T_{dz} denote the three components of external disturbance torque.

References

[1] Lur'e AI. Some non-linear problems in the theory of automatic control: nekotorye nelineinye Zadachi Teorii avtomaticheskogo regulirovaniya (Gos. Isdat. Tekh. Teor. Lit., 1951, USSR) A Translation from the Russian. HM Stationery Office; 1957.

[2] Yakubovich VA. The solution of some matrix inequalities encountered in automatic control theory//Doklady Akademii Nauk. Russian Academy of Sciences 1962;143 (6):1304–7.

[3] Willems J. Least squares stationary optimal control and the algebraic Riccati equation. IEEE Transactions on Automatic Control 1971;16(6):621–34.

[4] Yakubovich VA. Dichotomy and absolute stability of nonlinear systems with periodically nonstationary linear part. Systems & Control Letters 1988;11(3):221–8.

[5] Pyatnitskiy YS, Skorodinskiy VI. Numerical methods of Lyapunov function construction and their application to the absolute stability problem. Systems & Control Letters 1982;2(2):130–5.

[6] Karmarkar N. A new polynomial-time algorithm for linear programming. In: Proceedings of the sixteenth annual ACM symposium on Theory of computing 1984;302–311.

[7] Nesterov JE. Self-concordant functions and polynomial-time methods in convex programming. Report, Central Economic and Mathematic Institute, USSR Acad. Sci 1989;.

[8] Alizadeh-Dehkharghani F. Combinatorial optimization with interior point methods and semidefinite matrices 1991; University of Minnesota.

[9] Alizadeh F. Optimization over the positive-definite cone: interior point methods and combinatorial applications. Advances in Optimization and Parallel Computing 1992;.

[10] Vandenberghe L, Boyd S. A primal—dual potential reduction method for problems involving matrix inequalities. Mathematical Programming 1995;69(1):205–36.

[11] Gahinet P, Nemirovski A, Laub AJ, et al. The LMI control toolbox. For use with Matlab. User's Guide. Natick, MA: The MathWorks; 1995.
[12] Ghaoui LE, Niculescu SI. Advances in linear matrix inequality methods in control. Society for Industrial and Applied Mathematics 2000;.
[13] Ostertag E. Mono-and multivariable control and estimation: linear, quadratic and LMI methods. Springer Science & Business Media. 2011.
[14] Scherer C, Weiland S. Linear matrix inequalities in control. Lecture Notes, Dutch Institute for Systems and Control, Delft, The Netherlands 2000;3(2).
[15] Apkarian P, Tuan HD, Bernussou J. Analysis, eigenstructure assignment and H/sub 2/multichannel synthesis with enhanced LMI characterizations. In: Proceedings of the 39th IEEE conference on decision and control (Cat. No. 00CH37187). IEEE; 2000. vol. 2, p. 1489–1494.
[16] VanAntwerp JG, Braatz RD. A tutorial on linear and bilinear matrix inequalities. Journal of Process Control 2000;10(4):363–85.
[17] Suplin V, Fridman E, Shaked U. H/sub/spl infin//control of linear uncertain time-delay systems-a projection approach. IEEE Transactions on Automatic Control 2006;51(4):680–5.
[18] Modi VJ. Attitude dynamics of satellites with flexible appendages-a brief review. Journal of Spacecraft and Rockets 1974;11(11):743–51.
[19] Chevallier DP, Lerbet J. Multi-body kinematics and dynamics with lie groups 2017; Elsevier.
[20] Potter MC, Lessing JL, Aboufadel EF. The theory of matrices. Advanced Engineering Mathematics 2019;200–70 Springer, Cham.

CHAPTER 2

State feedback nonfragile control

2.1 Introduction

Many uncertain factors such as gyro drift, external disturbances, and controller perturbations may damage and affect the performance of on-orbit service spacecraft. Due to the increasing requirements for high performance and functionality, spacecraft attitude control systems have become more sophisticated. In the threatening space environment, the complex system inevitably confronts various disturbances. Thus accurate control for the spacecraft attitude system is significant to accomplish space missions such as observation, navigation, and communication, etc. Parameter uncertainty has been taken into consideration in many systems, for example, microgrid systems [1], double pendulum [2], and finite sampling in Markov state models [3]. Methods centered on robust control have attracted considerable attention over the past few years, which has been widely applied in near space vehicles with fuzzy models [4], atomic force microscopy [5], and active magnetic bearing systems [6], etc. A tremendous amount of scientific work based on control theory has been completed, see Refs. [7,8].

The robust H_∞ controller can guarantee that the closed-loop system satisfies certain H_∞ performance when bounded parameter uncertainty or unmodeled dynamics exist [9,10]. Nevertheless, the realization effect of a robust controller depends on the precise realization of the controller. The actuator or controller parameters may show some variations due to uncertain factors. The traditional feedback control methods, for example, H_∞, H_2, μ, l_1, are sensitive to small variations [11]. It has been shown that even a very small perturbation on controller parameters may lead to performance degradation or even destabilize the closed-loop system. Sensitivity analysis of the H_∞ quadratic stability problem for continuous-time systems was performed to show that proper methods can lead to tight perturbation bounds [12]. Thus it is necessary to design a nonfragile controller that is, robust against its own parametric variations. This problem has been widely investigated by many researchers, who have applied control theory to nonfragile control problems with different requirements. For example, a nonfragile controller was designed by solving a pair of indefinite algebraic Riccati equations for a

Spacecraft Attitude Control. DOI: https://doi.org/10.1016/B978-0-323-99005-9.00002-X

known linear time-invariant system in Ref. [13]. A nonfragile procedure was introduced in Ref. [14] to explore the synchronization problem of neural networks with time-varying delay. However, the spacecraft attitude control system is sophisticated and parametric uncertainties have not been taken into consideration in the existing studies. Moreover, they are not actually related to spacecraft attitude control problems in the studies mentioned above.

In this chapter, our goal is to develop an effective method to solve the spacecraft attitude control problem subject to H_∞ performance constraint, quadratic stability, external disturbances, controller perturbation, and control input saturation. Considering two types of controller perturbations including additive perturbation and multiplicative perturbation, corresponding theorems are derived according to Lyapunov theory, respectively. The main contribution of this chapter is to propose the robust state feedback nonfragile attitude control method for a class of spacecraft attitude control problems with parameter uncertainty, gyro drift, external disturbances, controller perturbation, H_∞ performance constraint, quadratic stability, and input saturation. Based on Lyapunov theory, sufficient conditions for the existence of robust state feedback nonfragile controller are given by use of linear matrix inequalities (LMIs).

The structure of the rest of the chapter is organized as follows. The following section introduces the spacecraft attitude dynamics equation and converts it into a state-space form, and the corresponding robust control problem is formulated. Then, the design method for the robust state feedback nonfragile controller based on LMIs is proposed with respect to additive perturbation and multiplicative perturbation. This is followed by the simulations of robust state feedback nonfragile controller and mixed H_2/H_∞ controller under conditions of controller perturbation and model parameter uncertainty, which illustrate the effectiveness of the proposed approach. Finally, some conclusions are drawn.

2.2 Problem formulation

2.2.1 Attitude dynamics modeling

The spacecraft is assumed to be a rigid body with actuators that provide torque about three mutually perpendicular axes. The three axes define a body-fixed frame. The attitude dynamics equation is [15]

$$\boldsymbol{I}_b\dot{\boldsymbol{\omega}} + \boldsymbol{\omega}\times(\boldsymbol{I}_b\boldsymbol{\omega}) = \boldsymbol{T}_c+\boldsymbol{T}_g+\boldsymbol{T}_d \tag{2.1}$$

where $\boldsymbol{I}_b$ and $\boldsymbol{\omega}$ denote the inertia matrix and the angular velocity and $\boldsymbol{T}_c$, $\boldsymbol{T}_g$, and $\boldsymbol{T}_d$ denote the control input torque, gravity gradient torque, and the external disturbance torque, respectively.

The kinematics of the spacecraft can be described by [16]

$$\begin{bmatrix} \omega_x \\ \omega_y \\ \omega_z \end{bmatrix} = \begin{bmatrix} 1 & 0 & -\sin\theta \\ 0 & \cos\phi & \sin\phi\cos\theta \\ 0 & -\sin\phi & \cos\phi\cos\theta \end{bmatrix} \begin{bmatrix} \dot{\phi} \\ \dot{\theta} \\ \dot{\psi} \end{bmatrix} - \omega_0 \begin{bmatrix} \cos\theta\sin\psi \\ \sin\phi\sin\theta\sin\psi + \cos\phi\cos\psi \\ \cos\phi\sin\theta\sin\psi - \sin\phi\cos\psi \end{bmatrix} \tag{2.2}$$

where ω_x, ω_y, and ω_z denote the three components of angular velocity; ϕ, θ, and ψ denote the three components of attitude angle, that is, roll, pitch, and yaw attitude angle, and ω_0 is the orbital frequency.

For spacecraft orbiting in a circle, the gravity gradient torque could be represented in terms of attitude angles [17], that is,

$$\boldsymbol{T}_g = 3\omega_0^2 \begin{bmatrix} (I_z - I_y)sin\phi cos\phi cos^2\theta \\ (I_z - I_x)cos\phi\sin\theta cos\theta \\ (I_x - I_y)\sin\phi\sin\theta cos\theta \end{bmatrix} \tag{2.3}$$

For small Euler angles, Eqs. (2.2) and (2.3) are approximated as Eqs. (2.4) and (2.5) below.

$$\begin{bmatrix} \omega_x \\ \omega_y \\ \omega_z \end{bmatrix} = \begin{bmatrix} \dot{\phi} - \omega_0\psi \\ \dot{\theta} - \omega_0 \\ \dot{\psi} + \omega_0\phi \end{bmatrix} \tag{2.4}$$

$$\boldsymbol{T}_g = \begin{bmatrix} 3\omega_0^2(I_z - I_y)\phi \\ 3\omega_0^2(I_z - I_x)\theta \\ 0 \end{bmatrix} \tag{2.5}$$

Provided the body-fixed frame is chosen to be a principal-axis frame, then, substituting Eqs. (2.4) and (2.5) into Eq. (2.1) gives

$$\begin{cases} I_x\ \ddot{\phi} + 4(I_y - I_z)\omega_0^2\phi + (I_y - I_z - I_x)\omega_0\dot{\psi} = T_{cx} + T_{dx} \\ I_y\ \ddot{\theta} + 3\omega_0^2(I_x - I_z)\theta = T_{cy} + T_{dy} \\ I_z\ \ddot{\psi} + (I_y - I_x)\omega_0^2\psi + (I_x + I_z - I_y)\omega_0\dot{\phi} = T_{cz} + T_{dz} \end{cases} \tag{2.6}$$

where I_x, I_y, and I_z denote the three components of inertia matrix; T_{cx}, T_{cy}, and T_{cz} denote the three components of control input torque, and T_{dx}, T_{dy}, and T_{dz} denote the three components of external disturbance torque.

Let

$$\boldsymbol{x} = [\phi\ \theta\ \psi\ \dot{\phi}\ \dot{\theta}\ \dot{\psi}]^T, \boldsymbol{y} = [\phi\ \theta\ \psi\ \omega_x\ \omega_y\ \omega_z]^T,$$

$$\boldsymbol{u} = [T_{cx}\ T_{cy}\ T_{cz}]^T,\ \boldsymbol{w} = [T_{dx}\ T_{dy}\ T_{dz}]^T,$$

Then, Eq. (2.6) can be written as

$$\begin{cases} \dot{\boldsymbol{x}}(t) = \boldsymbol{A}\boldsymbol{x}(t) + \boldsymbol{B}_1\boldsymbol{u}(t) + \boldsymbol{B}_2\boldsymbol{w}(t) \\ \boldsymbol{y}(t) = \boldsymbol{C}_1\boldsymbol{x}(t) + \boldsymbol{D}_2 \\ \boldsymbol{z}(t) = \boldsymbol{C}_2\boldsymbol{x}(t) \end{cases} \tag{2.7}$$

where,

$$\boldsymbol{A} = \begin{bmatrix} 0 & 0 & 0 & 1 & 0 & 0 \\ 0 & 0 & 0 & 0 & 1 & 0 \\ 0 & 0 & 0 & 0 & 0 & 1 \\ \boldsymbol{A}_{41} & 0 & 0 & 0 & 0 & -\omega_0 I_x^{-1}(I_y - I_x - I_z) \\ 0 & \boldsymbol{A}_{52} & 0 & 0 & 0 & 0 \\ 0 & 0 & \boldsymbol{A}_{63} & \omega_0 I_z^{-1}(I_y - I_x - I_z) & 0 & 0 \end{bmatrix}$$

$$\boldsymbol{A}_{41} = -4\omega_0^2 I_x^{-1}(I_y - I_z), \boldsymbol{A}_{52} = -3\omega_0^2 I_y^{-1}(I_x - I_z), \boldsymbol{A}_{63} = -\omega_0^2 I_z^{-1}(I_y - I_x)$$

$$\boldsymbol{B}_1 = \boldsymbol{B}_2 = \begin{bmatrix} \boldsymbol{0}_{3\times 3} & diag(I_x^{-1}, I_y^{-1}, I_z^{-1}) \end{bmatrix}^T,$$

$$\boldsymbol{D}_2 = \begin{bmatrix} \boldsymbol{0}_{1\times 4} & -\omega_0 & 0 \end{bmatrix}^T, \ \boldsymbol{C}_1 = \begin{bmatrix} \boldsymbol{I}_{3\times 3} & \boldsymbol{0}_{3\times 3} \\ \boldsymbol{B} & \boldsymbol{I}_{3\times 3} \end{bmatrix},$$

$$\boldsymbol{B} = \begin{bmatrix} 0 & 0 & -\omega_0 \\ 0 & 0 & 0 \\ \omega_0 & 0 & 0 \end{bmatrix}, \ \boldsymbol{C}_2 = [\boldsymbol{I}_{6\times 6}]$$

The following important aspects should be considered simultaneously.

1. Model parameter uncertainty

   Due to the detection errors and various uncertain factors in space, the attitude angle and angular velocity of on-orbit servicing spacecraft cannot be determined online accurately, and the system has certain model parameter uncertainty. Accounting for parameter uncertainty, $\boldsymbol{A}$ will be transformed into $\boldsymbol{A} + \Delta\boldsymbol{A}$, where $\Delta\boldsymbol{A}$ denotes model parameter uncertainty and satisfies the matching condition, that is,

$$\Delta\boldsymbol{A}(t) = \boldsymbol{M}_1\boldsymbol{F}_1(t)\boldsymbol{N}_1, \ \left\|\boldsymbol{F}_1(t)\right\| \leq 1, \ \forall t$$

   where $\boldsymbol{M}_1$ and $\boldsymbol{N}_1$ are known constant real matrices of appropriate dimension and $\boldsymbol{F}_1$(t) is an unknown matrix function with Lebesgue measurable elements.

2. Gyro drift

   Gyro drift means the slow precession of gyro spin axis caused by disturbance torque, which can be divided into two types, one is caused by regular systematic disturbance torque, and the other is caused by irregular random factors. An important indicator in the measurement of the gyroscope is gyro drift rate [18] with bounded L_2 norm, and it can be confined into a certain range with the amplitude defined as constant drift $\boldsymbol{g}(t)$.

3. Input saturation

   Two types of constraints are usually considered in the controller design, one is the time constraint related to the maneuvering process, and the other is an energy constraint related to the actuator. Taking the power capacity of the actuator into consideration, the actual control input is easily saturated, so control input optimization is very important in practice. Therefore the control input should be designed as small as possible to reduce energy consumption. A positive scalar λ_0 is introduced satisfying

$$\|\boldsymbol{u}\|_2^2 \leq \lambda_0 \tag{2.8}$$

   where λ_0 can be regarded as an optimization indicator, which denotes the theoretical upper boundary of control input torque.

   When the spacecraft makes attitude maneuver, if the controller gain matrix $\boldsymbol{K}$ cannot guarantee that theoretical control input is less than the saturation value of actual control input, the saturation treatment of theoretical control input should be made to generate the actual control input $sat(\boldsymbol{u})$ with its form

$$sat(\boldsymbol{u}) = \begin{bmatrix} sat(u_1) & sat(u_2) & sat(u_3) \end{bmatrix}^T$$

   satisfying

$$sat(u_i) = \begin{cases} u_{mi} & u_i > u_{mi} \\ u_i & -u_{mi} \leq u_i \leq u_{mi} \\ -u_{mi} & u_i < -u_{mi} \end{cases} \tag{2.9}$$

   where u_{mi} $(i = 1, 2, 3)$ is the upper boundary of control input that the actuator can provide.

4. Controller gain perturbation

The actuator is often affected by mounting error and a number of unknown factors, which lead to the problem of actuator perturbation or nominal controller perturbation, which are named as controller gain perturbation. Assuming the actuator perturbation is $\boldsymbol{\Xi}$, the nominal controller gain matrix is $\boldsymbol{K}$, and adding $\boldsymbol{\Xi}$ into the control system yields the controller gain matrix $\boldsymbol{K}(\boldsymbol{I} + \boldsymbol{\Xi})$ as

$$\boldsymbol{K}(\boldsymbol{I} + \boldsymbol{\Xi}) = \boldsymbol{K} + \boldsymbol{K}\boldsymbol{\Xi} = \boldsymbol{K} + \Delta\boldsymbol{K} \tag{2.10}$$

$\Delta\boldsymbol{K}$ means the controller gain perturbation generated by the actuator, satisfying

$$\Delta\boldsymbol{K}\Delta\boldsymbol{K}^T \leq \eta_0^2\boldsymbol{I} \tag{2.11}$$

$\Delta\boldsymbol{K}$ means controller gain perturbation, two cases are considered as follows:

$\Delta\boldsymbol{K}$ means additive perturbation,

$$\Delta\boldsymbol{K}=\boldsymbol{M}_2\boldsymbol{F}_2(t)\boldsymbol{N}_2,\ \left\|\boldsymbol{F}_2(t)\right\|\leq 1 \tag{2.12}$$

$\Delta\boldsymbol{K}$ means multiplicative perturbation

$$\Delta\boldsymbol{K}=\boldsymbol{M}_3\boldsymbol{F}_3(t)\boldsymbol{N}_3\boldsymbol{K},\ \left\|\boldsymbol{F}_3(t)\right\|\leq 1 \tag{2.13}$$

The definitions of $\boldsymbol{M}_2$, $\mathbf{N}_2$, $\boldsymbol{M}_3$, $\mathbf{N}_3$, $\boldsymbol{F}_2(t)$, and $\boldsymbol{F}_3(t)$ are similar to that of $\Delta\boldsymbol{A}$.

Design controller

$$\boldsymbol{u}(t)=(\boldsymbol{K}+\Delta\boldsymbol{K})\boldsymbol{x}(t) \tag{2.14}$$

which is also called nonfragile controller, because it is able to tolerate some uncertainties in the controller.

The resulting spacecraft closed-loop attitude control system can be written as:

$$\begin{cases}\dot{\boldsymbol{x}}(t)=(\boldsymbol{A}+\Delta\boldsymbol{A}+\boldsymbol{B}_1\boldsymbol{K}+\boldsymbol{B}_1\Delta\boldsymbol{K})\boldsymbol{x}(t)+\boldsymbol{B}_2\boldsymbol{w}(t)\\ \boldsymbol{y}(t)=\mathbf{C}_1\boldsymbol{x}(t)+\boldsymbol{D}_1\boldsymbol{g}(t)+\boldsymbol{D}_2\\ \boldsymbol{z}(t)=\mathbf{C}_2\boldsymbol{x}(t)\end{cases} \tag{2.15}$$

where $\boldsymbol{D}_1=\left[\boldsymbol{0}_{3\times 3}\quad diag(I_x^{-1},I_y^{-1},I_z^{-1})\right]^{\mathrm{T}}$.

2.2.2 Control objective

The objective of this chapter is to seek a nonfragile controller such that the resulting system (2.15) satisfies the following conditions:

1. The closed-loop system (2.15) ($\boldsymbol{w}(t)$=0) is quadratically stable.
2. For a given scalar $\gamma>0$, the performance $\left\|\mathbf{G}_{zw}(s)\right\|_{\infty}<\gamma$ is guaranteed, where $\mathbf{G}_{zw}(s)$ denotes the closed-loop transfer function from $\boldsymbol{w}$ to $\boldsymbol{z}$.
3. The control input is limited, that is, it cannot exceed its saturation value under optimal conditions.

Definition 1: If there is a symmetric positive definite matrix $\boldsymbol{P}>0$ and a positive constant α, for arbitrary uncertainty, the time derivative of Lyapunov function $V(\boldsymbol{x},t)$ satisfies

$$\dot{V}(\boldsymbol{x},t)\leq-\alpha\left\|\boldsymbol{x}(t)\right\|_2^2 \tag{2.16}$$

Then, system (2.15) ($\boldsymbol{w}(t)=0$) is quadratically stable.

2.3 State feedback nonfragile control law

2.3.1 Some lemmas

Before the main results are introduced, we first introduce the following essential preliminary lemmas.

Lemma 2.1: [17] Let $\boldsymbol{H}, \boldsymbol{E} \in \mathbb{R}^{m \times n}, \boldsymbol{F}(t) \in \mathbb{S}^m$, and $\boldsymbol{F}(t)$ satisfies $\|\boldsymbol{F}(t)\| \leq 1$, then for a scalar $\xi > 0$,

$$\boldsymbol{H}\boldsymbol{F}(t)\boldsymbol{E} + \boldsymbol{E}^{\mathrm{T}}\boldsymbol{F}(t)^{\mathrm{T}}\boldsymbol{H}^{\mathrm{T}} \leq \xi^{-1}\boldsymbol{H}\boldsymbol{H}^{\mathrm{T}} + \xi\boldsymbol{E}^{\mathrm{T}}\boldsymbol{E}$$

Lemma 2.2: (Schur complement lemma) Let the partitioned matrix

$$\boldsymbol{A} = \begin{bmatrix} \boldsymbol{A}_{11} & \boldsymbol{A}_{12} \\ \boldsymbol{A}_{12}^T & \boldsymbol{A}_{22} \end{bmatrix}$$

be symmetric. Then

$$\boldsymbol{A} < 0 \Leftrightarrow \boldsymbol{A}_{11} < 0, \boldsymbol{A}_{22} - \boldsymbol{A}_{12}^{\mathrm{T}}\boldsymbol{A}_{11}^{-1}\boldsymbol{A}_{12} < 0 \Leftrightarrow \boldsymbol{A}_{22} < 0, \boldsymbol{A}_{11} - \boldsymbol{A}_{12}\boldsymbol{A}_{22}^{-1}\boldsymbol{A}_{12}^T < 0$$

or

$$\boldsymbol{A} > 0 \Leftrightarrow \boldsymbol{A}_{11} > 0, \boldsymbol{A}_{22} - \boldsymbol{A}_{12}^T\boldsymbol{A}_{11}^{-1}\boldsymbol{A}_{12} > 0 \Leftrightarrow \boldsymbol{A}_{22} > 0, \boldsymbol{A}_{11} - \boldsymbol{A}_{12}\boldsymbol{A}_{22}^{-1}\boldsymbol{A}_{12}^T > 0$$

Lemma 2.3: [17] Assume that $\boldsymbol{M}$, $\boldsymbol{N}$ are real matrices of appropriate dimension, then, for arbitrary ε, one has

$$\begin{bmatrix} 0 & \boldsymbol{N}\boldsymbol{M}^T \\ \boldsymbol{M}\boldsymbol{N}^T & 0 \end{bmatrix} \leq \begin{bmatrix} \varepsilon\boldsymbol{N}\boldsymbol{N}^T & 0 \\ 0 & \varepsilon^{-1}\boldsymbol{M}\boldsymbol{M}^T \end{bmatrix}$$

Lemma 2.4: [19] Consider the closed-loop system described by Eq. (2.7), the H_2 performance and H_∞ performance are satisfied simultaneously if and only if there exist matrices $\boldsymbol{Q}, \boldsymbol{N}_2, \boldsymbol{W}$, symmetric matrix $\boldsymbol{Z}$, and symmetric positive definite matrix $\boldsymbol{X}$ such that

$$\begin{bmatrix} -\boldsymbol{Q} - \boldsymbol{Q}^T & (\boldsymbol{Q}^T\boldsymbol{A} + \boldsymbol{N}_2) + \boldsymbol{X} & \boldsymbol{Q}^T\boldsymbol{B}_2 & \boldsymbol{Q}^T \\ (\boldsymbol{Q}^T\boldsymbol{A} + \boldsymbol{N}_2)^T + \boldsymbol{X} & -\boldsymbol{X} & 0 & 0 \\ \boldsymbol{B}_2^T\boldsymbol{Q} & 0 & -\boldsymbol{I} & 0 \\ \boldsymbol{Q} & 0 & 0 & -\boldsymbol{X} \end{bmatrix} < 0$$

$$\begin{bmatrix} \boldsymbol{X} & \boldsymbol{C}_2^T \\ \boldsymbol{C}_2 & \boldsymbol{Z} \end{bmatrix} > 0$$

$$Trace(\boldsymbol{Z}) < \gamma_2^2$$

$$\begin{bmatrix} \boldsymbol{A}\boldsymbol{X} + \boldsymbol{B}_1\boldsymbol{W} + (\boldsymbol{A}\boldsymbol{X} + \boldsymbol{B}_1\boldsymbol{W})^T & \boldsymbol{B}_2 & (\boldsymbol{C}_1\boldsymbol{X} + \boldsymbol{D}_1\boldsymbol{W})^T \\ \boldsymbol{B}_2^T & -\gamma_\infty\boldsymbol{I} & \boldsymbol{D}_2^T \\ \boldsymbol{C}_1\boldsymbol{X} + \boldsymbol{D}_1\boldsymbol{W} & \boldsymbol{D}_2 & -\gamma_\infty\boldsymbol{I} \end{bmatrix} < 0 \tag{2.17}$$

By minimizing $c_\infty\gamma_\infty + c_2\gamma_2^2$, one can obtain the state feedback controller gain matrix $\boldsymbol{K} = \boldsymbol{W}\boldsymbol{X}^{-1}$.

2.3.2 Sufficient conditions under additive perturbation

Theorem 2.1 gives the existence conditions of the controller gain matrix $\boldsymbol{K}$ in terms of additive perturbation.

Theorem 2.1: When controller gain perturbation is in the form of Eq. (2.12), for given $\xi_1 > 0$, $\xi_2 > 0$ and $\gamma > 0$, uncertain system (2.15) is quadratically stable under the robust nonfragile controller (2.14), $\boldsymbol{z}(t)$ satisfies H_∞ performance constraint, and control input constraint is satisfied if there exists a symmetric positive definite matrix $\boldsymbol{X}$ and matrix $\boldsymbol{W}$, such that the following LMIs hold:

$$\begin{bmatrix} \boldsymbol{AX} + \boldsymbol{B}_1\boldsymbol{W} + \boldsymbol{XA}^T + \boldsymbol{W}^T\boldsymbol{B}_1^T & \boldsymbol{M}_1 & \boldsymbol{XN}_1^T & \boldsymbol{B}_1\boldsymbol{M}_2 & \boldsymbol{XN}_2^T & \boldsymbol{XC}_2^T & \boldsymbol{B}_2 \\ \boldsymbol{M}_1^T & -\xi_1^{-1}\boldsymbol{I} & 0 & 0 & 0 & 0 & 0 \\ \boldsymbol{N}_1\boldsymbol{X} & 0 & -\xi_1\boldsymbol{I} & 0 & 0 & 0 & 0 \\ (\boldsymbol{B}_1\boldsymbol{M}_2)^T & 0 & 0 & -\xi_2^{-1}\boldsymbol{I} & 0 & 0 & 0 \\ \boldsymbol{N}_2\boldsymbol{X} & 0 & 0 & 0 & -\xi_2\boldsymbol{I} & 0 & 0 \\ \boldsymbol{C}_2\boldsymbol{X} & 0 & 0 & 0 & 0 & -\boldsymbol{I} & 0 \\ \boldsymbol{B}_2^T & 0 & 0 & 0 & 0 & 0 & -\gamma^2\boldsymbol{I} \end{bmatrix} < 0 \tag{2.18}$$

$$\begin{bmatrix} -\gamma_0\boldsymbol{I} & \boldsymbol{x}(0)^T \\ \boldsymbol{x}(0) & -\boldsymbol{X} \end{bmatrix} < 0 \tag{2.19}$$

$$\begin{bmatrix} -\boldsymbol{X} & \boldsymbol{W}^T & \boldsymbol{X}^T \\ \boldsymbol{W} & -\lambda_0\gamma_0^{-1}\boldsymbol{I} + \varepsilon^{-1}\eta_0^2\boldsymbol{I} & 0 \\ \boldsymbol{X} & 0 & -\varepsilon^{-1}\boldsymbol{I} \end{bmatrix} < 0 \tag{2.20}$$

By minimizing λ_0, the optimal feasible solution of LMIs (2.18)–(2.20) can be obtained, and the state feedback controller gain matrix $\boldsymbol{K} = \boldsymbol{W}\boldsymbol{X}^{-1}$.

Proof: At first, one shows that the system (2.15) is quadratically stable.

Define the Lyapunov function

$$V(t) = \boldsymbol{x}(t)^T\boldsymbol{P}\boldsymbol{x}(t)$$

satisfying

$$V(t) < \gamma_0$$

where $\boldsymbol{P} = \boldsymbol{X}^{-1}$.

Then

$$\begin{aligned}
\dot{V}(t) &= \dot{\boldsymbol{x}}(t)^T \boldsymbol{P}\boldsymbol{x}(t) + \boldsymbol{x}(t)^T \boldsymbol{P}\dot{\boldsymbol{x}}(t) \\
&= \boldsymbol{x}(t)^T (\boldsymbol{A}+\Delta\boldsymbol{A}+\boldsymbol{B}_1\boldsymbol{K}+\boldsymbol{B}_1\Delta\boldsymbol{K})^T \boldsymbol{P}\boldsymbol{x}(t) + \boldsymbol{x}(t)^T \boldsymbol{P}(\boldsymbol{A}+\Delta\boldsymbol{A}+\boldsymbol{B}_1\boldsymbol{K}+\boldsymbol{B}_1\Delta\boldsymbol{K})\boldsymbol{x}(t) \\
&= \boldsymbol{x}(t)^T (\boldsymbol{A}+\boldsymbol{M}_1\boldsymbol{F}_1(t)\boldsymbol{N}_1+\boldsymbol{B}_1\boldsymbol{K}+\boldsymbol{B}_1\boldsymbol{M}_2\boldsymbol{F}_2(t)\boldsymbol{N}_2)^T \boldsymbol{P}\boldsymbol{x}(t) + \\
&\quad \boldsymbol{x}(t)^T \boldsymbol{P}(\boldsymbol{A}+\boldsymbol{M}_1\boldsymbol{F}_1(t)\boldsymbol{N}_1+\boldsymbol{B}_1\boldsymbol{K}+\boldsymbol{B}_1\boldsymbol{M}_2\boldsymbol{F}_2(t)\boldsymbol{N}_2)\boldsymbol{x}(t) \\
&= \boldsymbol{x}(t)^T \left[(\boldsymbol{A}+\boldsymbol{B}_1\boldsymbol{K})^T + (\boldsymbol{M}_1\boldsymbol{F}_1(t)\boldsymbol{N}_1+\boldsymbol{B}_1\boldsymbol{M}_2\boldsymbol{F}_2(t)\boldsymbol{N}_2)^T\right] \boldsymbol{P}\boldsymbol{x}(t) + \\
&\quad \boldsymbol{x}(t)^T \boldsymbol{P}[(\boldsymbol{A}+\boldsymbol{B}_1\boldsymbol{K}) + (\boldsymbol{M}_1\boldsymbol{F}_1(t)\boldsymbol{N}_1+\boldsymbol{B}_1\boldsymbol{M}_2\boldsymbol{F}_2(t)\boldsymbol{N}_2)]\boldsymbol{x}(t) \\
&= \boldsymbol{x}(t)^T \begin{bmatrix} (\boldsymbol{A}+\boldsymbol{B}_1\boldsymbol{K})^T \boldsymbol{P} + (\boldsymbol{M}_1\boldsymbol{F}_1(t)\boldsymbol{N}_1+\boldsymbol{B}_1\boldsymbol{M}_2\boldsymbol{F}_2(t)\boldsymbol{N}_2)^T \boldsymbol{P} \\ +\boldsymbol{P}(\boldsymbol{A}+\boldsymbol{B}_1\boldsymbol{K}) + \boldsymbol{P}(\boldsymbol{M}_1\boldsymbol{F}_1(t)\boldsymbol{N}_1+\boldsymbol{B}_1\boldsymbol{M}_2\boldsymbol{F}_2(t)\boldsymbol{N}_2) \end{bmatrix} \boldsymbol{x}(t) \\
&= \boldsymbol{x}(t)^T \begin{bmatrix} (\boldsymbol{A}+\boldsymbol{B}_1\boldsymbol{K})^T \boldsymbol{P} + \boldsymbol{P}(\boldsymbol{A}+\boldsymbol{B}_1\boldsymbol{K}) + \boldsymbol{P}\boldsymbol{M}_1\boldsymbol{F}_1(t)\boldsymbol{N}_1 + \boldsymbol{N}_1^T\boldsymbol{F}_1(t)^T\boldsymbol{M}_1^T\boldsymbol{P} \\ +\boldsymbol{P}\boldsymbol{B}_1\boldsymbol{M}_2\boldsymbol{F}_2(t)\boldsymbol{N}_2 + \boldsymbol{N}_2^T\boldsymbol{F}_2(t)^T\boldsymbol{M}_2^T\boldsymbol{B}_1^T\boldsymbol{P} \end{bmatrix} \boldsymbol{x}(t) \\
&\le \boldsymbol{x}(t)^T \begin{bmatrix} (\boldsymbol{A}+\boldsymbol{B}_1\boldsymbol{K})^T \boldsymbol{P} + \boldsymbol{P}(\boldsymbol{A}+\boldsymbol{B}_1\boldsymbol{K}) + \xi_1\boldsymbol{P}\boldsymbol{M}_1\boldsymbol{M}_1^T\boldsymbol{P} + \xi_1^{-1}\boldsymbol{N}_1^T\boldsymbol{N}_1 \\ +\xi_2\boldsymbol{P}\boldsymbol{B}_1\boldsymbol{M}_2\boldsymbol{M}_2^T\boldsymbol{B}_1^T\boldsymbol{P} + \xi_2^{-1}\boldsymbol{N}_2^T\boldsymbol{N}_2 \end{bmatrix} \boldsymbol{x}(t)
\end{aligned}$$

Considering inequality (2.18), multiply both sides by $diag\{\boldsymbol{P}, \boldsymbol{I}, \boldsymbol{I}, \boldsymbol{I}, \boldsymbol{I}, \boldsymbol{I}, \boldsymbol{I}\}$ simultaneously, and inequality (2.21) can be obtained.

$$\begin{bmatrix} \boldsymbol{P}\boldsymbol{A}+\boldsymbol{P}\boldsymbol{B}_1\boldsymbol{K}+\boldsymbol{A}^T\boldsymbol{P}+\boldsymbol{K}^T\boldsymbol{B}_1^T\boldsymbol{P} & \boldsymbol{P}\boldsymbol{M}_1 & \boldsymbol{N}_1^T & \boldsymbol{P}\boldsymbol{B}_1\boldsymbol{M}_2 & \boldsymbol{N}_2^T & \boldsymbol{C}_2^T & \boldsymbol{P}\boldsymbol{B}_2 \\ \boldsymbol{M}_1^T\boldsymbol{P} & -\xi_1^{-1}\boldsymbol{I} & 0 & 0 & 0 & 0 & 0 \\ \boldsymbol{N}_1 & 0 & -\xi_1\boldsymbol{I} & 0 & 0 & 0 & 0 \\ (\boldsymbol{B}_1\boldsymbol{M}_2)^T\boldsymbol{P} & 0 & 0 & -\xi_2^{-1}\boldsymbol{I} & 0 & 0 & 0 \\ \boldsymbol{N}_2 & 0 & 0 & 0 & -\xi_2\boldsymbol{I} & 0 & 0 \\ \boldsymbol{C}_2 & 0 & 0 & 0 & 0 & -\boldsymbol{I} & 0 \\ \boldsymbol{B}_2^T\boldsymbol{P} & 0 & 0 & 0 & 0 & 0 & -\gamma^2\boldsymbol{I} \end{bmatrix} < 0 \tag{2.21}$$

Let

$$\boldsymbol{M}_0 = (\boldsymbol{A}+\boldsymbol{B}_1\boldsymbol{K})^T\boldsymbol{P} + \boldsymbol{P}(\boldsymbol{A}+\boldsymbol{B}_1\boldsymbol{K}) + \xi_1\boldsymbol{P}\boldsymbol{M}_1\boldsymbol{M}_1^T\boldsymbol{P}$$

$$+\xi_1^{-1}\boldsymbol{N}_1^T\boldsymbol{N}_1 + \xi_2\boldsymbol{P}\boldsymbol{B}_1\boldsymbol{M}_2\boldsymbol{M}_2^T\boldsymbol{B}_1^T\boldsymbol{P} + \xi_2^{-1}\boldsymbol{N}_2^T\boldsymbol{N}_2$$

According to inequality (2.21), one knows $\boldsymbol{M}_0 < 0$, furthermore,

$$\dot{V}(t) \le \boldsymbol{x}(t)^T\boldsymbol{M}_0\boldsymbol{x}(t) \le \lambda_{\max}(\boldsymbol{M}_0)\boldsymbol{x}(t)^T\boldsymbol{x}(t)$$

Let $\alpha = -\lambda_{\max}(\boldsymbol{M}_0) > 0$, then,

$$\dot{V}(t) \leq -\alpha \left\| \boldsymbol{x}(t) \right\|_2^2$$

that is, system (2.15) is quadratically stable under the conditions of a robust nonfragile controller.

Then, one shows the output $\boldsymbol{z}(t)$ that satisfies the H_∞ performance constraint.

To establish the L_2 [0, ∞) norm bound $\gamma^2 \left\| \boldsymbol{w}(t) \right\|_2^2$, consider the following functional:

$$J = \int_0^\infty [\boldsymbol{z}(t)^T \boldsymbol{z}(t) - \gamma^2 \boldsymbol{w}(t)^T \boldsymbol{w}(t)] dt$$

As the closed-loop system has quadratic stability, for arbitrary nonzero $\boldsymbol{w}(t) \in L_2[0, \infty)$, let $\boldsymbol{x}(0) = 0$, then,

$$\begin{aligned}
J &= \int_0^\infty [\boldsymbol{z}(t)^T \boldsymbol{z}(t) - \gamma^2 \boldsymbol{w}(t)^T \boldsymbol{w}(t) + \dot{V}(\boldsymbol{x}(t))] dt - V(\infty) + V(0) \\
&\leq \int_0^\infty \{\boldsymbol{x}(t)^T \boldsymbol{C}_2^T \boldsymbol{C}_2 \boldsymbol{x}(t) - \gamma^2 \boldsymbol{w}(t)^T \boldsymbol{w}(t) + \boldsymbol{x}(t)^T [(\boldsymbol{A}+\boldsymbol{B}_1\boldsymbol{K})^T \boldsymbol{P} + \boldsymbol{P}(\boldsymbol{A}+\boldsymbol{B}_1\boldsymbol{K}) + \xi_1 \boldsymbol{P}\boldsymbol{M}_1\boldsymbol{M}_1^T\boldsymbol{P} + \\
&\quad \xi_1^{-1}\boldsymbol{N}_1^T\boldsymbol{N}_1 + \xi_2 \boldsymbol{P}\boldsymbol{B}_1\boldsymbol{M}_2\boldsymbol{M}_2^T\boldsymbol{B}_1^T\boldsymbol{P} + \xi_2^{-1}\boldsymbol{N}_2^T\boldsymbol{N}_2]\boldsymbol{x}(t) + \boldsymbol{w}(t)^T\boldsymbol{B}_2^T\boldsymbol{P}\boldsymbol{x}(t) + \boldsymbol{x}(t)^T\boldsymbol{P}\boldsymbol{B}_2\boldsymbol{w}(t)\} dt \\
&\leq \int_0^\infty \begin{bmatrix} \boldsymbol{x}(t)^T & \boldsymbol{w}(t)^T \end{bmatrix} \begin{bmatrix} \begin{matrix} (\boldsymbol{A}+\boldsymbol{B}_1\boldsymbol{K})^T\boldsymbol{P} + \boldsymbol{P}(\boldsymbol{A}+\boldsymbol{B}_1\boldsymbol{K}) + \xi_1\boldsymbol{P}\boldsymbol{M}_1\boldsymbol{M}_1^T\boldsymbol{P} \\ + \xi_1^{-1}\boldsymbol{N}_1^T\boldsymbol{N}_1 + \xi_2\boldsymbol{P}\boldsymbol{B}_1\boldsymbol{M}_2\boldsymbol{M}_2^T\boldsymbol{B}_1^T\boldsymbol{P} + \xi_2^{-1}\boldsymbol{N}_2^T\boldsymbol{N}_2 + \boldsymbol{C}_2^T\boldsymbol{C}_2 \end{matrix} & \boldsymbol{P}\boldsymbol{B}_2 \\ \boldsymbol{B}_2^T\boldsymbol{P} & -\gamma^2\boldsymbol{I} \end{bmatrix} \begin{bmatrix} \boldsymbol{x}(t) \\ \boldsymbol{w}(t) \end{bmatrix} dt
\end{aligned}$$

According to Lemma 2.2 and inequality (2.21), $J < 0$ holds, that is, $\boldsymbol{z}$ (t) satisfies H_∞ performance constraint.

Finally, one shows the control input $\boldsymbol{u}(t)$ is constrained, that is, it meets constraint condition (2.8).

Then inequality (2.20) is equal to

$$\begin{bmatrix} -\boldsymbol{X} + \varepsilon \boldsymbol{X}^T \boldsymbol{X} & \boldsymbol{W}^T \\ \boldsymbol{W} & -\lambda_0 \gamma_0^{-1} \boldsymbol{I} + \varepsilon^{-1} \eta_0^2 \boldsymbol{I} \end{bmatrix} < 0 \tag{2.22}$$

Combine inequalities (2.11) and (2.22), one has

$$\begin{bmatrix} -\boldsymbol{X} + \varepsilon \boldsymbol{X}^T \boldsymbol{X} & \boldsymbol{W}^T \\ \boldsymbol{W} & -\lambda_0 \gamma_0^{-1} \boldsymbol{I} + \varepsilon^{-1} \Delta\boldsymbol{K} \Delta\boldsymbol{K}^T \end{bmatrix} < 0 \tag{2.23}$$

Substituting $\boldsymbol{W} = \boldsymbol{K}\boldsymbol{X}$ into inequality (2.23) yields

$$\begin{bmatrix} -\boldsymbol{X} + \varepsilon \boldsymbol{X}^T \boldsymbol{X} & (\boldsymbol{K}\boldsymbol{X})^T \\ \boldsymbol{K}\boldsymbol{X} & -\lambda_0 \gamma_0^{-1} \boldsymbol{I} + \varepsilon^{-1} \Delta\boldsymbol{K} \Delta\boldsymbol{K}^T \end{bmatrix} < 0 \tag{2.24}$$

Multiplying by $diag\{\boldsymbol{P}, \boldsymbol{I}\}$ at both sides of inequality (2.24) simultaneously, inequality (2.25) can be obtained.

$$\begin{bmatrix} -\boldsymbol{P}+\varepsilon\boldsymbol{I} & \boldsymbol{K}^T \\ \boldsymbol{K} & -\lambda_0\gamma_0^{-1}\boldsymbol{I}+\varepsilon^{-1}\Delta\boldsymbol{K}\Delta\boldsymbol{K}^T \end{bmatrix} < 0 \tag{2.25}$$

According to Lemma 2.3, one has

$$\begin{bmatrix} -\boldsymbol{P} & (\boldsymbol{K}+\Delta\boldsymbol{K})^T \\ \boldsymbol{K}+\Delta\boldsymbol{K} & -\lambda_0\gamma_0^{-1}\boldsymbol{I} \end{bmatrix} < 0$$

According to Lemma 2.2, one has

$$\gamma_0(\boldsymbol{K}+\Delta\boldsymbol{K})^T(\boldsymbol{K}+\Delta\boldsymbol{K}) < \lambda_0\boldsymbol{P}$$

Then,

$$\gamma_0\boldsymbol{x}(t)^T(\boldsymbol{K}+\Delta\boldsymbol{K})^T(\boldsymbol{K}+\Delta\boldsymbol{K})\boldsymbol{x}(t) < \lambda_0\boldsymbol{x}(t)^T\boldsymbol{P}\boldsymbol{x}(t)$$

that is,

$$\gamma_0\left\|\boldsymbol{u}(t)\right\|_2^2 < \lambda_0\boldsymbol{x}(t)^T\boldsymbol{P}\boldsymbol{x}(t) \tag{2.26}$$

According to inequality (2.16),

$$\dot{V}(t) \leq 0 \tag{2.27}$$

Integrating inequality (2.27), it can be obtained

$$\int_0^t \dot{V}(\tau) \leq 0$$

that is,

$$V(t) \leq V(0)$$

Based on the definition of Lyapunov function,

$$\boldsymbol{x}(t)^T\boldsymbol{P}\boldsymbol{x}(t) \leq \boldsymbol{x}(0)^T\boldsymbol{P}\boldsymbol{x}(0)$$

According to inequality (2.19), one has

$$\boldsymbol{x}(0)^T\boldsymbol{P}\boldsymbol{x}(0) < \gamma_0\boldsymbol{I} \tag{2.28}$$

Then,

$$x(t)^T P x(t) < \gamma_0 I \tag{2.29}$$

Combine inequalities (2.26) and (2.29) to obtain

$$\left\| u(t) \right\|_2^2 < \lambda_0 \tag{2.30}$$

2.3.3 Sufficient conditions under multiplicative perturbation

Theorem 2.2 gives the existence conditions of the controller gain matrix $\boldsymbol{K}$ in terms of multiplicative perturbation.

Theorem 2.2: When controller gain perturbation is the form of Eq. (2.13), for given $\xi_1 > 0$, $\xi_2 > 0$ and $\gamma > 0$, uncertainty system (2.15) is quadratically stable under conditions of robust nonfragile controller (2.14), $z(t)$ satisfies H_∞ performance constraint, and control input constraint is satisfied if there exists a symmetric positive definite matrix $\boldsymbol{X}$ and matrix $\boldsymbol{W}$, such that the following LMIs hold:

$$\begin{bmatrix} AX + B_1 W + XA^T + W^T B_1^T & M_1 & XN_1^T & B_1 M_3 & W^T N_3^T & XC_2^T & B_2 \\ M_1^T & -\xi_1^{-1} I & 0 & 0 & 0 & 0 & 0 \\ N_1 X & 0 & -\xi_1 I & 0 & 0 & 0 & 0 \\ (B_1 M_3)^T & 0 & 0 & -\xi_2^{-1} I & 0 & 0 & 0 \\ N_3 W & 0 & 0 & 0 & -\xi_2 I & 0 & 0 \\ C_2 X & 0 & 0 & 0 & 0 & -I & 0 \\ B_2^T & 0 & 0 & 0 & 0 & 0 & -\gamma^2 I \end{bmatrix} < 0 \tag{2.31}$$

$$\begin{bmatrix} -\gamma_0 I & x(0)^T \\ x(0) & -X \end{bmatrix} < 0 \tag{2.32}$$

$$\begin{bmatrix} -X & W^T & X^T \\ W & -\lambda_0 \gamma_0^{-1} I + \varepsilon^{-1} \eta_0^2 I & 0 \\ X & 0 & -\varepsilon^{-1} I \end{bmatrix} < 0 \tag{2.33}$$

By minimizing λ_0, the optimal feasible solution of LMIs (2.31)–(2.33) can be obtained, and the state feedback controller gain matrix $\boldsymbol{K} = \boldsymbol{W}\boldsymbol{X}^{-1}$.

Proof: At first, one shows that the system (2.15) is quadratically stable. The proof process is similar to that of additive perturbation.

Define Lyapunov function

$$V(t) = \boldsymbol{x}(t)^T\boldsymbol{P}\boldsymbol{x}(t)$$

where $\boldsymbol{P} = \boldsymbol{X}^{-1}$. Then,

$$\begin{aligned}
\dot V(t) &= \dot{\boldsymbol{x}}(t)^T\boldsymbol{P}\boldsymbol{x}(t) + \boldsymbol{x}(t)^T\boldsymbol{P}\dot{\boldsymbol{x}}(t)\\
&= \boldsymbol{x}(t)^T(\boldsymbol{A}+\Delta\boldsymbol{A}+\boldsymbol{B}_1\boldsymbol{K}+\boldsymbol{B}_1\Delta\boldsymbol{K})^T\boldsymbol{P}\boldsymbol{x}(t) + \boldsymbol{x}(t)^T\boldsymbol{P}(\boldsymbol{A}+\Delta\boldsymbol{A}+\boldsymbol{B}_1\boldsymbol{K}+\boldsymbol{B}_1\Delta\boldsymbol{K})\boldsymbol{x}(t)\\
&= \boldsymbol{x}(t)^T(\boldsymbol{A}+\boldsymbol{M}_1\boldsymbol{F}_1(t)\boldsymbol{N}_1+\boldsymbol{B}_1\boldsymbol{K}+\boldsymbol{B}_1\boldsymbol{M}_3\boldsymbol{F}_3(t)\boldsymbol{N}_3\boldsymbol{K})^T\boldsymbol{P}\boldsymbol{x}(t) +\\
&\quad \boldsymbol{x}(t)^T\boldsymbol{P}(\boldsymbol{A}+\boldsymbol{M}_1\boldsymbol{F}_1(t)\boldsymbol{N}_1+\boldsymbol{B}_1\boldsymbol{K}+\boldsymbol{B}_1\boldsymbol{M}_3\boldsymbol{F}_3(t)\boldsymbol{N}_3\boldsymbol{K})\boldsymbol{x}(t)\\
&= \boldsymbol{x}(t)^T\left[(\boldsymbol{A}+\boldsymbol{B}_1\boldsymbol{K})^T+(\boldsymbol{M}_1\boldsymbol{F}_1(t)\boldsymbol{N}_1+\boldsymbol{B}_1\boldsymbol{M}_3\boldsymbol{F}_3(t)\boldsymbol{N}_3\boldsymbol{K})^T\right]\boldsymbol{P}\boldsymbol{x}(t) +\\
&\quad \boldsymbol{x}(t)^T\boldsymbol{P}[(\boldsymbol{A}+\boldsymbol{B}_1\boldsymbol{K})+(\boldsymbol{M}_1\boldsymbol{F}_1(t)\boldsymbol{N}_1+\boldsymbol{B}_1\boldsymbol{M}_3\boldsymbol{F}_3(t)\boldsymbol{N}_3\boldsymbol{K})]\boldsymbol{x}(t)\\
&= \boldsymbol{x}(t)^T\begin{bmatrix}(\boldsymbol{A}+\boldsymbol{B}_1\boldsymbol{K})^T\boldsymbol{P}+(\boldsymbol{M}_1\boldsymbol{F}_1(t)\boldsymbol{N}_1+\boldsymbol{B}_1\boldsymbol{M}_3\boldsymbol{F}_3(t)\boldsymbol{N}_3\boldsymbol{K})^T\boldsymbol{P}\\ +\boldsymbol{P}(\boldsymbol{A}+\boldsymbol{B}_1\boldsymbol{K})+\boldsymbol{P}(\boldsymbol{M}_1\boldsymbol{F}_1(t)\boldsymbol{N}_1+\boldsymbol{B}_1\boldsymbol{M}_3\boldsymbol{F}_3(t)\boldsymbol{N}_3\boldsymbol{K})\end{bmatrix}\boldsymbol{x}(t)\\
&= \boldsymbol{x}(t)^T\begin{bmatrix}(\boldsymbol{A}+\boldsymbol{B}_1\boldsymbol{K})^T\boldsymbol{P}+\boldsymbol{P}(\boldsymbol{A}+\boldsymbol{B}_1\boldsymbol{K})+\boldsymbol{P}\boldsymbol{M}_1\boldsymbol{F}_1(t)\boldsymbol{N}_1+\boldsymbol{N}_1^T\boldsymbol{F}_1(t)^T\boldsymbol{M}_1^T\boldsymbol{P}\\ +\boldsymbol{P}\boldsymbol{B}_1\boldsymbol{M}_3\boldsymbol{F}_3(t)\boldsymbol{N}_3\boldsymbol{K}+\boldsymbol{K}^T\boldsymbol{N}_3^T\boldsymbol{F}_3(t)^T\boldsymbol{M}_3^T\boldsymbol{B}_1^T\boldsymbol{P}\end{bmatrix}\boldsymbol{x}(t)\\
&\le \boldsymbol{x}(t)^T\begin{bmatrix}(\boldsymbol{A}+\boldsymbol{B}_1\boldsymbol{K})^T\boldsymbol{P}+\boldsymbol{P}(\boldsymbol{A}+\boldsymbol{B}_1\boldsymbol{K})+\xi_1\boldsymbol{P}\boldsymbol{M}_1\boldsymbol{M}_1^T\boldsymbol{P}+\xi_1^{-1}\boldsymbol{N}_1^T\boldsymbol{N}_1\\ +\xi_2\boldsymbol{P}\boldsymbol{B}_1\boldsymbol{M}_3\boldsymbol{M}_3^T\boldsymbol{B}_1^T\boldsymbol{P}+\xi_2^{-1}\boldsymbol{K}^T\boldsymbol{N}_3^T\boldsymbol{N}_3\boldsymbol{K}\end{bmatrix}\boldsymbol{x}(t)
\end{aligned}$$

Considering inequality (2.31), multiply both sides by $diag\{\boldsymbol{P},\boldsymbol{I},\boldsymbol{I},\boldsymbol{I},\boldsymbol{I},\boldsymbol{I},\boldsymbol{I}\}$ simultaneously, and inequality (2.34) can be obtained,

$$\begin{bmatrix}
\boldsymbol{PA}+\boldsymbol{PB}_1\boldsymbol{K}+\boldsymbol{A}^{\mathrm T}\boldsymbol{P}+\boldsymbol{K}^{\mathrm T}\boldsymbol{B}_1^{\mathrm T}\boldsymbol{P} & \boldsymbol{PM}_1 & \boldsymbol{N}_1^{\mathrm T} & \boldsymbol{PB}_1\boldsymbol{M}_3 & \boldsymbol{K}^{\mathrm T}\boldsymbol{N}_3^{\mathrm T} & \boldsymbol{C}_2^{\mathrm T} & \boldsymbol{PB}_2\\
\boldsymbol{M}_1^{\mathrm T}\boldsymbol{P} & -\xi_1^{-1}\boldsymbol{I} & 0 & 0 & 0 & 0 & 0\\
\boldsymbol{N}_1 & 0 & -\xi_1\boldsymbol{I} & 0 & 0 & 0 & 0\\
(\boldsymbol{B}_1\boldsymbol{M}_3)^{\mathrm T}\boldsymbol{P} & 0 & 0 & -\xi_2^{-1}\boldsymbol{I} & 0 & 0 & 0\\
\boldsymbol{N}_3\boldsymbol{K} & 0 & 0 & 0 & -\xi_2\boldsymbol{I} & 0 & 0\\
\boldsymbol{C}_2 & 0 & 0 & 0 & 0 & -\boldsymbol{I} & 0\\
\boldsymbol{B}_2^{\mathrm T}\boldsymbol{P} & 0 & 0 & 0 & 0 & 0 & -\gamma^2\boldsymbol{I}
\end{bmatrix} < 0 \tag{2.34}$$

Let

$$\begin{aligned}\tilde{\boldsymbol{M}}_0 &= (\boldsymbol{A}+\boldsymbol{B}_1\boldsymbol{K})^T\boldsymbol{P}+\boldsymbol{P}(\boldsymbol{A}+\boldsymbol{B}_1\boldsymbol{K})+\xi_1\boldsymbol{P}\boldsymbol{M}_1\boldsymbol{M}_1^T\boldsymbol{P}+\xi_1^{-1}\boldsymbol{N}_1^T\boldsymbol{N}_1\\
&\quad +\xi_2\boldsymbol{P}\boldsymbol{B}_1\boldsymbol{M}_3\boldsymbol{M}_3^T\boldsymbol{B}_1^T\boldsymbol{P}+\xi_2^{-1}\boldsymbol{K}^T\boldsymbol{N}_3^T\boldsymbol{N}_3\boldsymbol{K}\end{aligned}$$

According to inequality (2.34), one knows $\tilde{\boldsymbol{M}}_0 < 0$.
Furthermore,

$$\dot V(t) \le \boldsymbol{x}(t)^T\tilde{\boldsymbol{M}}_0\boldsymbol{x}(t) \le \lambda_{\max}(\tilde{\boldsymbol{M}}_0)\boldsymbol{x}(t)^T\boldsymbol{x}(t)$$

Let

$$\alpha = -\lambda_{\max}(\tilde{\boldsymbol{M}}_0) > 0,$$

then,

$$\dot{V}(t) \leq -\alpha \left\| \boldsymbol{x}(t) \right\|_2^2$$

that is, system (2.15) is quadratically stable under the conditions of a robust nonfragile controller.

Then, one shows output $\boldsymbol{z}(t)$ satisfies H_∞ performance constraint.

To establish the L_2 [0, ∞) norm bound $\gamma^2 \left\| \boldsymbol{w}(t) \right\|_2^2$, consider the following functional:

$$J = \int_0^\infty [\boldsymbol{z}(t)^T \boldsymbol{z}(t) - \gamma^2 \boldsymbol{w}(t)^T \boldsymbol{w}(t)] dt$$

As the closed-loop system has quadratic stability, for arbitrary nonzero $\boldsymbol{w}(t) \in L_2[0, \infty)$, let $\boldsymbol{x}$ (0) = 0, then,

$$\begin{aligned} J &= \int_0^\infty [\boldsymbol{z}(t)^T \boldsymbol{z}(t) - \gamma^2 \boldsymbol{w}(t)^T \boldsymbol{w}(t) + \dot{V}(\boldsymbol{x}(t))] dt - V(\infty) + V(0) \\ &\leq \int_0^\infty \{\boldsymbol{x}(t)^T \boldsymbol{C}_2^T \boldsymbol{C}_2 \boldsymbol{x}(t) - \gamma^2 \boldsymbol{w}(t)^T \boldsymbol{w}(t) + \boldsymbol{x}(t)^T [(\boldsymbol{A}+\boldsymbol{B}_1\boldsymbol{K})^T \boldsymbol{P} + \boldsymbol{P}(\boldsymbol{A}+\boldsymbol{B}_1\boldsymbol{K}) + \xi_1 \boldsymbol{P}\boldsymbol{M}_1\boldsymbol{M}_1^T\boldsymbol{P} + \\ &\quad \xi_1^{-1}\boldsymbol{N}_1^T\boldsymbol{N}_1 + \xi_2 \boldsymbol{P}\boldsymbol{B}_1\boldsymbol{M}_3\boldsymbol{M}_3^T\boldsymbol{B}_1^T\boldsymbol{P} + \xi_2^{-1}\boldsymbol{K}^T\boldsymbol{N}_3^T\boldsymbol{N}_3\boldsymbol{K}]\boldsymbol{x}(t) + \boldsymbol{w}(t)^T\boldsymbol{B}_2^T\boldsymbol{P}\boldsymbol{x}(t) + \boldsymbol{x}(t)^T\boldsymbol{P}\boldsymbol{B}_2\boldsymbol{w}(t)\} dt \\ &\leq \int_0^\infty \begin{bmatrix} \boldsymbol{x}(t)^T & \boldsymbol{w}(t)^T \end{bmatrix} \begin{bmatrix} \begin{array}{c} (\boldsymbol{A}+\boldsymbol{B}_1\boldsymbol{K})^T\boldsymbol{P} + \boldsymbol{P}(\boldsymbol{A}+\boldsymbol{B}_1\boldsymbol{K}) + \xi_1\boldsymbol{P}\boldsymbol{M}_1\boldsymbol{M}_1^T\boldsymbol{P} \\ +\xi_1^{-1}\boldsymbol{N}_1^T\boldsymbol{N}_1 + \xi_2\boldsymbol{P}\boldsymbol{B}_1\boldsymbol{M}_3\boldsymbol{M}_3^T\boldsymbol{B}_1^T\boldsymbol{P} + \xi_2^{-1}\boldsymbol{K}^T\boldsymbol{N}_3^T\boldsymbol{N}_3\boldsymbol{K} + \boldsymbol{C}_2^T\boldsymbol{C}_2 \end{array} & \boldsymbol{P}\boldsymbol{B}_2 \\ \boldsymbol{B}_2^T\boldsymbol{P} & -\gamma^2\boldsymbol{I} \end{bmatrix} \begin{bmatrix} \boldsymbol{x}(t) \\ \boldsymbol{w}(t) \end{bmatrix} dt \end{aligned}$$

According to Lemma 2.2 and inequality (2.34), $J < 0$ holds, that is, $\boldsymbol{z}$ (t) satisfies H_∞ performance constraint.

Finally, one shows the control input $\boldsymbol{u}(t)$ is constrained, that is, it meets constraint condition (2.8), the proving process is the same as that under conditions of additive perturbation, omitted here.

2.4 Simulation test

In this section, the effectiveness of the proposed robust state feedback nonfragile controller is illustrated with simulations on a spacecraft attitude system. In comparison, the simulation results of mixed H_2/H_∞ controller based on extended LMI are also considered.

The moment of inertia is assumed as $I_x = 20$ kg m^2, $I_y = 18$ kg m^2, $I_z = 15$ kg m^2, and the orbit height is 300 km. Attitude angles and angular velocities are measured with sensors and gyros. Choose the initial state as $\boldsymbol{x}$ (0) = [0.08 rad, 0.06 rad, 0.06 rad, 0.01 rad s^{-1}, 0.01 rad s^{-1}, 0.01 rad s^{-1}], the upper boundary of control input saturation is $u_{m1} = u_{m2} = u_{m3} = 0.25$ Nm, the constant value of gyro drift [20] can be assumed as $\boldsymbol{d}(t) = 6 \times [10^{-5}\ 10^{-5}$

$10^{-5}]^T$, and the expected state is zero. The disturbance torques are white noise modeled as

$$w(t) = \begin{bmatrix} 5\cos(\omega_0 t) \\ 5\cos(\omega_0 t+\pi/4) \\ 5\cos(\omega_0 t+\pi/3) \end{bmatrix} \times 10^{-4} Nm$$

Meanwhile, choose

$$\begin{aligned}
&\boldsymbol{M}_1 = \begin{bmatrix} 0.8 & 1.1 & 1.3 & 1.5 & 1.6 & 1.8 \end{bmatrix}^T, \\
&\boldsymbol{N}_1 = \begin{bmatrix} -0.1 & -0.2 & -0.3 & -0.4 & -0.2 & 1 \end{bmatrix}, \\
&\boldsymbol{F}_1(t) = \sin(100\omega_0 t), \boldsymbol{M}_2 = \begin{bmatrix} 1 & 1 & 1 \end{bmatrix}^T, \boldsymbol{N}_2 = \begin{bmatrix} 0.1 & 0.1 & 0.1 & 0.1 & 0.1 & 0.1 \end{bmatrix}, \\
&\boldsymbol{F}_2(t) = \sin(100\omega_0 t+\pi/4), \boldsymbol{M}_3 = \begin{bmatrix} 0.1 & 0.1 & 0.1 \end{bmatrix}^T, \boldsymbol{N}_3 = 10^{-3} \times \boldsymbol{1}_{6\times 3}, \\
&\boldsymbol{F}_3(t) = \begin{bmatrix} 0.3 & 0.4 & 0.5 & 0.6 & 0.5 & 0.4 \end{bmatrix} \cos(0.012t), \xi_1 = 0.025, \xi_2 = 0.022, \\
&\gamma = 0.1, \varepsilon = 0.001, \eta_0 = 0.1.
\end{aligned}$$

2.4.1 Simulation results under additive perturbation

First, we consider the simulation under the condition of additive perturbation in controller gain. By solving the convex optimization problem represented by Theorem 2.1, the gain matrix for the state feedback controller $\boldsymbol{K}$ is obtained as follows:

$$\boldsymbol{K} = \begin{bmatrix} -21.3156 & -13.8040 & -67.5413 & -49.3180 & -11.3360 & 24.2525 \\ 6.8685 & -40.5112 & -105.5648 & -19.3245 & -46.8950 & 13.7111 \\ 14.3558 & -8.5390 & -127.3973 & 1.4976 & -14.4843 & -99.5766 \end{bmatrix}$$

Related matrices $\boldsymbol{W}$ and $\boldsymbol{X}$ are

$$\boldsymbol{W} = \begin{bmatrix} -2.7160 & -3.8649 & -1.4515 & -16.7351 & 10.0528 & -0.4256 \\ -2.5107 & -4.3986 & -1.3046 & 6.6194 & -16.1121 & 3.2466 \\ -1.6668 & -2.5767 & -1.1477 & -2.4447 & 1.2365 & -8.4331 \end{bmatrix}$$

$$\boldsymbol{X} = \begin{bmatrix} 0.6001 & -0.0830 & 0.1059 & -0.3761 & 0.1159 & -0.0476 \\ 0.0830 & 0.6324 & -0.0535 & 0.1389 & -0.3767 & 0.0850 \\ 0.1059 & -0.0535 & 0.0663 & -0.1105 & -0.0291 & -0.0509 \\ -0.3761 & 0.1389 & -0.1105 & 0.7711 & -0.3381 & 0.1606 \\ 0.1159 & -0.3767 & -0.0291 & -0.3381 & 0.8738 & -0.0583 \\ -0.0476 & 0.0850 & -0.0509 & 0.1606 & -0.0583 & 0.1466 \end{bmatrix}$$

Then, we obtain a nonfragile controller $\boldsymbol{K} + \Delta\boldsymbol{K}$, apply the state feedback controller in the spacecraft attitude control system (2.15), and apply saturation treatment of theoretical control input torque, where the results of attitude angle, angular velocity, and control input torque can be seen from Figs. 2.1–2.3. It can be observed that the closed-loop system can reach stable state within 15 seconds, and the control input is bounded within 0.25 Nm.

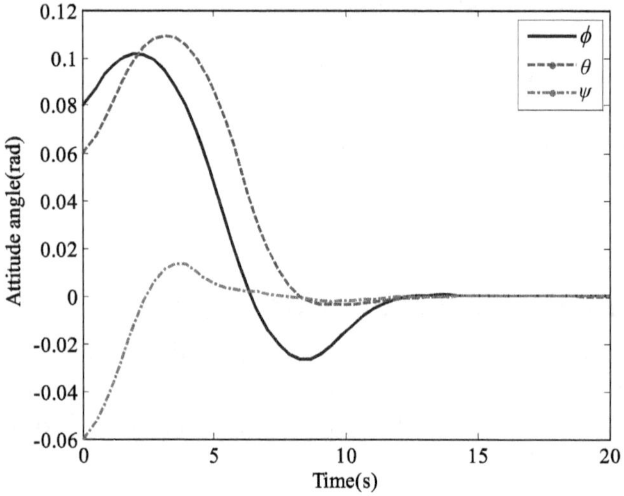

Figure 2.1 Attitude angle with additive perturbation.

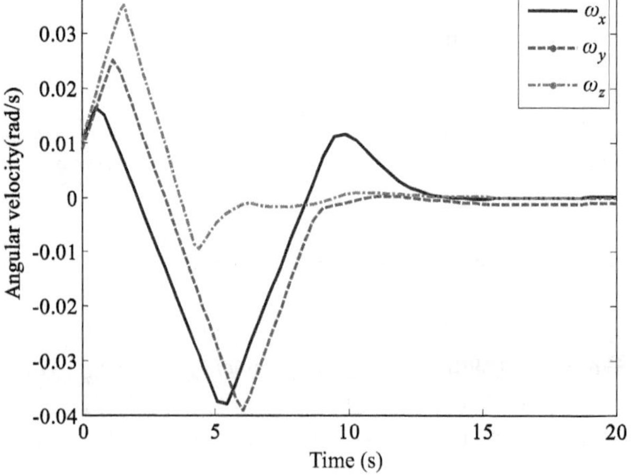

Figure 2.2 Angular velocity with additive perturbation.

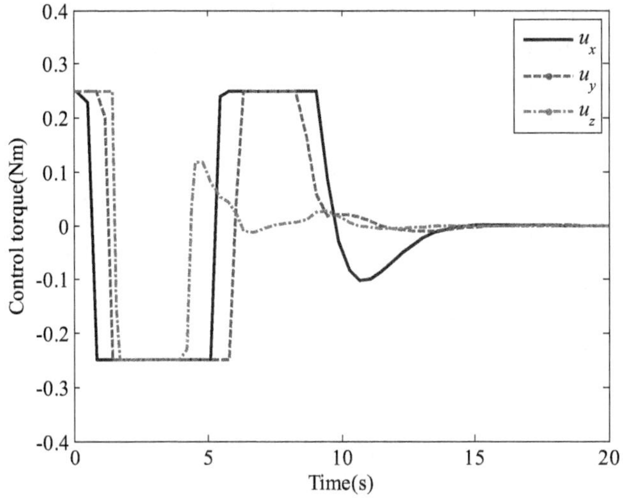

Figure 2.3 Control input torque with additive perturbation.

2.4.2 Simulation results under multiplicative perturbation

Then, we consider the simulation under the condition of multiplicative perturbation in controller gain. By solving the convex optimization problem represented by Theorem 2.2, the gain matrix for the state feedback controller $\boldsymbol{K}$ is obtained as follows:

$$\boldsymbol{K}=\begin{bmatrix} -21.3487 & -10.7701 & -58.5217 & -45.0491 & -7.8565 & 17.8708 \\ 3.1249 & -38.2941 & -92.4687 & -25.0868 & -50.7376 & 5.0388 \\ 7.3569 & -4.8572 & -73.8587 & 3.5166 & -10.2772 & -89.5867 \end{bmatrix}$$

Related matrices $\boldsymbol{W}$ and $\boldsymbol{X}$ are

$$\boldsymbol{W}=\begin{bmatrix} -3.9977 & -2.8769 & -1.7065 & -17.0590 & 11.3090 & -1.2131 \\ -2.5875 & -3.1999 & -1.2132 & 4.8411 & -15.6724 & 2.1141 \\ -1.8165 & -2.0865 & -1.2689 & -1.9304 & 2.3614 & -9.0012 \end{bmatrix}$$

$$\boldsymbol{X}=\begin{bmatrix} 0.6517 & -0.0673 & 0.0962 & -0.3673 & 0.1449 & -0.0329 \\ -0.0673 & 0.6861 & -0.0579 & 0.1076 & -0.3988 & 0.0783 \\ 0.0962 & -0.0579 & 0.1033 & -0.1431 & -0.0499 & -0.0599 \\ -0.3673 & 0.1076 & -0.1431 & 0.8413 & -0.3369 & 0.1752 \\ 0.1449 & -0.3988 & -0.0499 & -0.3369 & 0.8700 & -0.0647 \\ -0.0329 & 0.0783 & -0.0599 & 0.1752 & -0.0647 & 0.1572 \end{bmatrix}$$

Then, a nonfragile controller $\boldsymbol{K}+\Delta\boldsymbol{K}$ is obtained, apply the state feedback controller in the spacecraft attitude control system (2.15), and similarly apply saturation treatment of the theoretical control input torque, and the results of attitude angle, angular velocity, and control input torque can be seen from (Figs. 2.4–2.6). It can be observed that the closed-loop

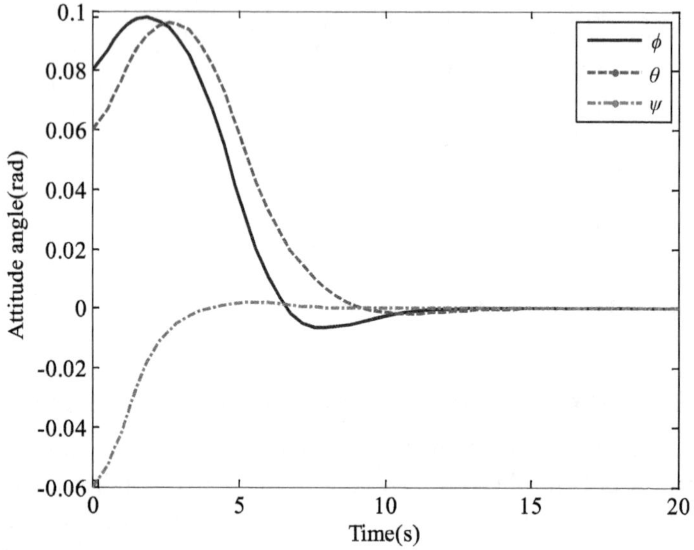

Figure 2.4 Attitude angle with multiplicative perturbation.

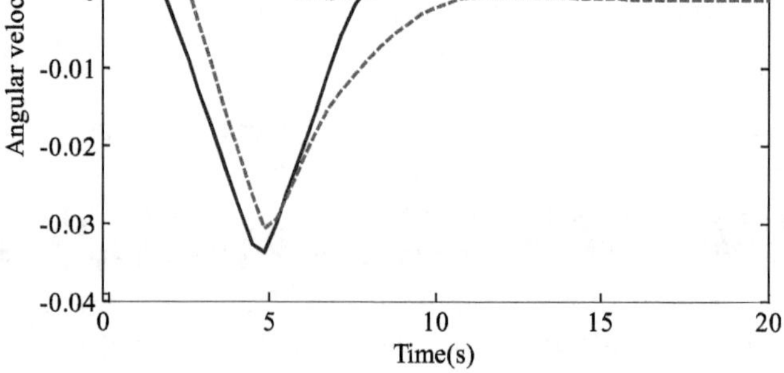

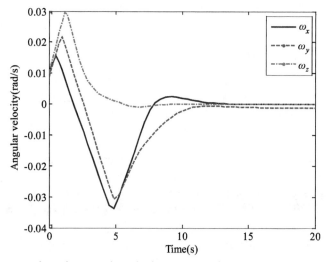

Figure 2.5 Angular velocity with multiplicative perturbation.

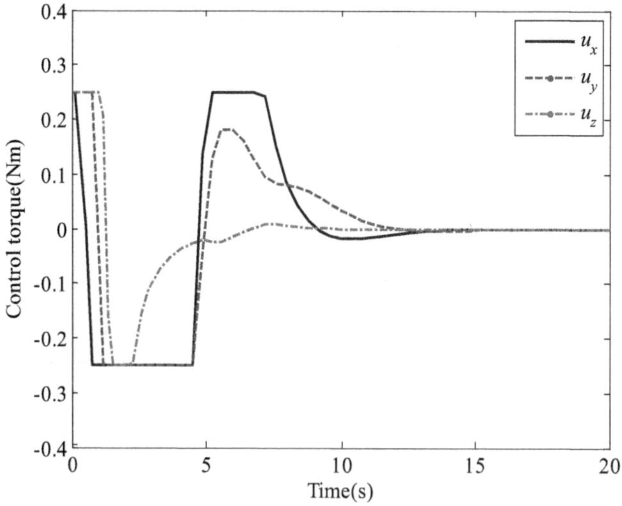

Figure 2.6 Control input torque with multiplicative perturbation.

system can reach steady state within 15 seconds, and the control input is bounded within 0.25 Nm.

2.4.3 Simulation results using a mixed H_2/H_∞ controller

In comparison, we consider the simulation of the mixed H_2/H_∞ controller based on extended LMI represented in Lemma 2.4. By solving the convex optimization problem represented by Lemma 2.4, the state feedback controller $\boldsymbol{K}$ is obtained as follows:

$$\boldsymbol{K} = \begin{bmatrix} -5.3098 & 0 & 0.0047 & -11.5544 & 0 & -0.0003 \\ 0 & -5.1507 & 0 & 0 & -10.6709 & 0 \\ -0.0039 & 0 & -4.8695 & 0.0022 & 0 & -9.2559 \end{bmatrix}$$

Related matrices $\boldsymbol{W}$ and $\boldsymbol{X}$ are

$$\boldsymbol{W} = 1.0 \times 10^5 \begin{bmatrix} -6.8057 & 0 & 0.0041 & -9.1904 & 0 & 0.0006 \\ 0 & -6.8839 & 0 & 0 & -8.8988 & 0 \\ -0.0043 & 0 & -6.9907 & 0.0010 & 0 & -8.3626 \end{bmatrix}$$

$$\boldsymbol{X} = 1.0 \times 10^5 \begin{bmatrix} 2.6282 & 0 & 0 & -0.6187 & 0 & -0.0008 \\ 0 & 2.6156 & 0 & 0 & -0.6174 & 0 \\ 0 & 0 & 2.6106 & 0.0008 & 0 & -0.6181 \\ -0.6187 & 0 & 0.0008 & 1.0797 & 0 & 0 \\ 0 & -0.6174 & 0 & 0 & 1.1320 & 0 \\ -0.0008 & 0 & -0.6181 & 0 & 0 & 1.2287 \end{bmatrix}$$

This demonstrates that the mixed H_2/H_∞ controller has certain nonfragility about the controller gain perturbation. To validate this conclusion, we simulated under the condition of additive perturbation and multiplicative perturbation of controller gain without model parameter uncertainty $\Delta \boldsymbol{A}$, respectively. The simulation results are shown from Figs. 2.7 to 2.12. It can be observed that the mixed H_2/H_∞

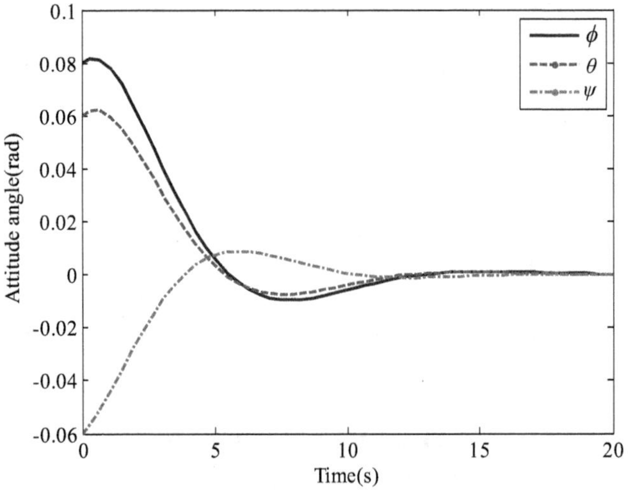

Figure 2.7 Attitude angle with additive perturbation.

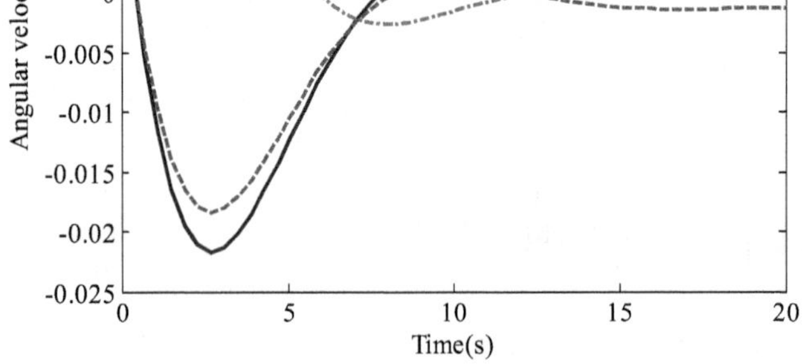

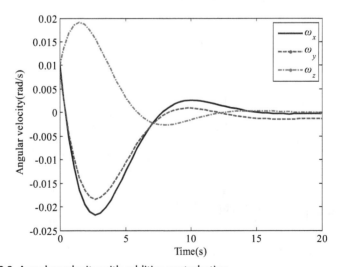

Figure 2.8 Angular velocity with additive perturbation.

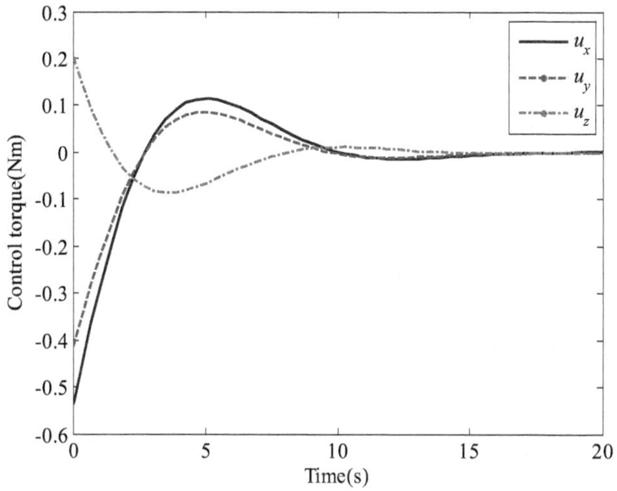

Figure 2.9 Control input torque with additive perturbation.

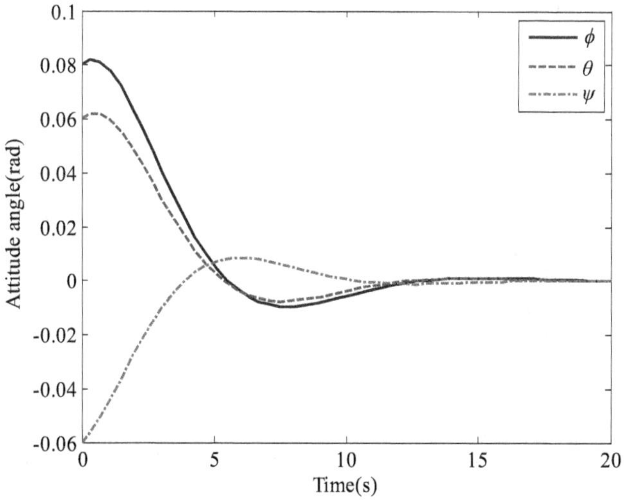

Figure 2.10 Attitude angle with multiplicative perturbation.

controller can guarantee the stability of the closed-loop system with certain nonfragility about the additive perturbation and multiplicative perturbation of controller gain, but it has a higher requirement about the actuator.

To test the control performance of the mixed H_2/H_∞ controller when model parameter uncertainty $\Delta \boldsymbol{A}$ is considered, one can obtain the

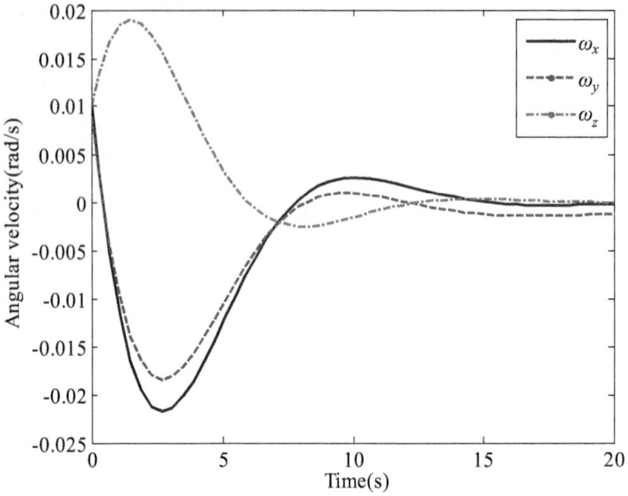

Figure 2.11 Angular velocity with multiplicative perturbation.

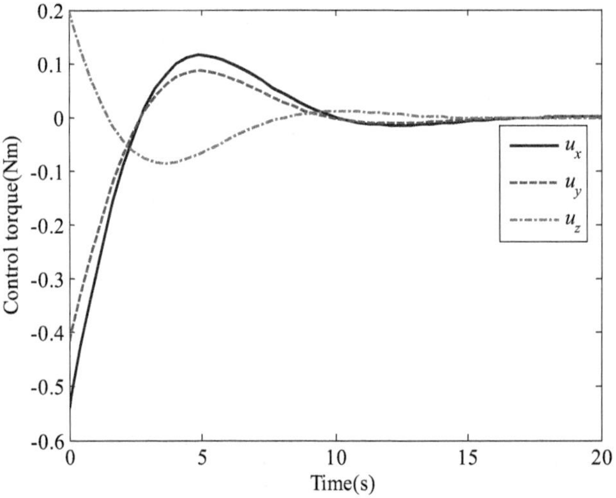

Figure 2.12 Control input torque with multiplicative perturbation.

following results as seen from Figs. 2.13 to 2.15. It can be seen that the mixed H_2/H_∞ controller can guarantee the stability of the control system between 20 and 35 seconds. However, after 35 seconds, the performance of spacecraft attitude control system will have a sharp decline, and the system will become unstable, while the control input torque will rapidly increase, leading to eventual spacecraft failure.

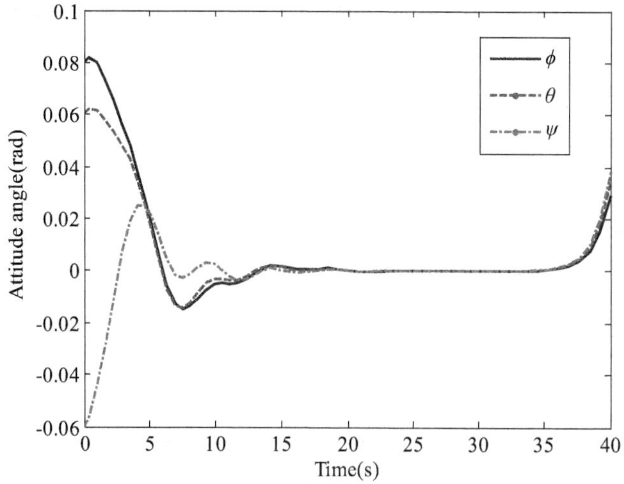

Figure 2.13 Attitude angle with parameter uncertainty.

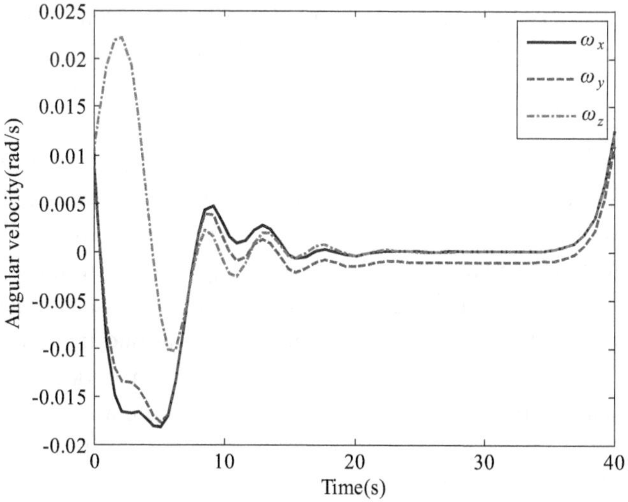

Figure 2.14 Angular velocity with parameter uncertainty.

2.5 Conclusions

This chapter proposes a robust state feedback nonfragile attitude control method for uncertain spacecraft with input saturation. The controller design is subject to the constraints of H_∞ performance, quadratic stability, and control input saturation. By using LMI techniques resulting from

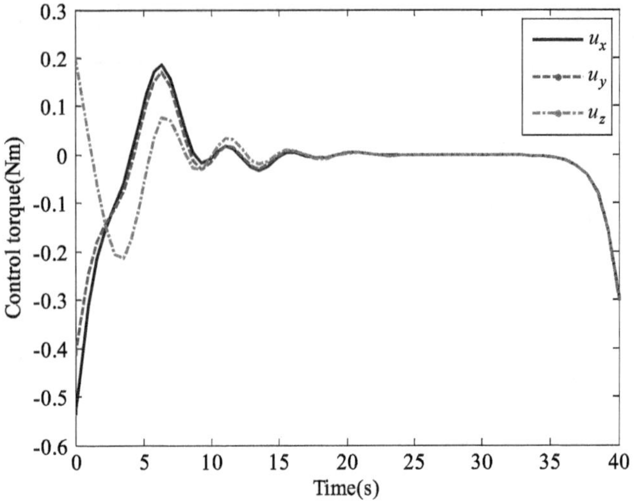

Figure 2.15 Control input torque with parameter uncertainty.

Lyapunov theory, the spacecraft attitude control problem is transformed into a convex optimization problem with LMI constraints. The simulation results show that the mixed H_2/H_∞ controller based on the extended LMI method guarantee system stability when only controller gain perturbation is considered. Nevertheless, when parametric uncertainty is considered, it will lead to eventual failure of spacecraft. However, the robust state feedback nonfragile attitude controller proposed in this chapter exhibits superior performance, for it is not only robust about the model parameter uncertainty, but also nonfragile about the controller gain perturbation no matter whether it is additive or multiplicative. In comparison, the practicability and effectiveness of the proposed method on uncertain spacecraft attitude control system is prominent.

References

[1] Parvizimosaed M, Farmani F, Rahimi-Kian A, et al. A multi-objective optimization for energy management in a renewable micro-grid system: a data mining approach. Journal of Renewable and Sustainable Energy 2014;6(2):023127.

[2] Ananievski IM. Control of double pendulum with uncertain parameters. In: Applications of mathematics in engineering and economics: 33rd international conference, vol. 946, issue 1. AIP Publishing; 2007. p. 3–10.

[3] Bowman GR. Improved coarse-graining of Markov state models via explicit consideration of statistical uncertainty. The Journal of Chemical Physics 2012;137(13):134111.

[4] Gao Z, Jiang B, Qi R, et al. Robust reliable control for a near space vehicle with parametric uncertainties and actuator faults. International Journal of Systems Science 2011;42(12):2113–24.
[5] Chuang N, Petersen IR, Pota HR. Robust H_∞ control in fast atomic force microscopy. Asian Journal of Control 2013;15(3):872–87.
[6] Lee DH, Park JB, Joo YH, et al. Robust H_∞ control for uncertain nonlinear active magnetic bearing systems via Takagi-Sugeno fuzzy models. International Journal of Control, Automation and Systems 2010;8(3):636–46.
[7] Lin P, Jia Y, Li L. Distributed robust H_∞ consensus control in directed networks of agents with time-delay. Systems & Control Letters 2008;57(8):643–53.
[8] Ren J, Zhang Q. Robust H_∞ control for uncertain descriptor systems by proportional–derivative state feedback. International Journal of Control 2010;83(1):89–96.
[9] Mahmoud MS, Zribi M. H_∞-controllers for time-delay systems using linear matrix inequalities. Journal of Optimization Theory and Applications 1999;100(1):89–122.
[10] Isidori A, Kang W. H_∞ control via measurement feedback for general nonlinear systems. IEEE Transactions on Automatic Control 1995;40(3):466–72.
[11] Keel LH, Bhattacharyya SP. Robust, fragile, or optimal? IEEE Transactions on Automatic Control 1997;42(8):1098–105.
[12] Yonchev A, Gilev B, Penev D. Sensitivity analysis of the continuous-time LMI-based H_∞ quadratic stability problem. American Institute of Physics Conference Series 2010;1293(1):3–12.
[13] Yang GH, Wang JL. Nonfragile H_∞ output feedback controller design for linear systems. Journal of Dynamic Systems, Measurement, and Control 2003;125(1):117–23.
[14] Fang M, Park JH. Non-fragile synchronization of neural networks with time-varying delay and randomly occurring controller gain fluctuation. Applied Mathematics and Computation 2013;219(15):8009–17.
[15] Hughes PC. Spacecraft attitude dynamics. New York: Dover; 2004. p. 282–3.
[16] Xiao B, Yin S, Kaynak O. Attitude stabilization control of flexible satellites with high accuracy: an estimator-based approach. IEEE/ASME Transactions on Mechatronics 2017;22(1):349–58.
[17] Yang X, Gao H, Shi P, et al. Robust H_∞ control for a class of uncertain mechanical systems. International Journal of Control 2010;83(7):1303–24.
[18] Pan H, Yang B, Wang L. Measurement of random gyro drift establishment of mathematical model. Aerospace Shanghai 2003;20(3):20–3.
[19] Shi K, Liu C, Wang F, et al. Mixed H_2/H_∞ state feedback attitude control of microsatellite based on extended LMI method. Journal of Harbin Institute of Technology (New Series) 2016;23(5):15–22.
[20] Gao C, Zhao Q, Duan G. Robust actuator fault diagnosis scheme for satellite attitude control systems. Journal of the Franklin Institute 2013;350(9):2560–80.

CHAPTER 3

Dynamic output feedback nonfragile control

3.1 Introduction

To meet the requirements of future space missions, the high-precision and high-stability attitude control problem of spacecraft has become increasingly important. However, many uncertain factors such as environmental and nonenvironmental disturbances, parameter uncertainty, and other nonlinear perturbations widely act upon spacecraft [1,2]. Moreover, the structural perturbations of the actuator, which can also be regarded as controller perturbations on the role of influence, usually occur in maneuvers, which seriously affect attitude control performance [3,4]. These problems pose a huge challenge for attitude control system designers, and thus good control approaches are required to improve robust performance and control accuracy to solve these problems.

State feedback controllers require all the state information, which is difficult to meet for high-order systems, and some components of the state vector may be virtual and cannot be measured. If the output is always able to be obtained, then the output feedback controller embodies its own advantages. Over the last decades, many researchers have performed extensive studies on the spacecraft attitude control system. In the 1980s, the attitude stabilization problem of flexible spacecraft was discussed in [5–7]. Afterward, to improve robust and nonfragile performance in the presence of external disturbances and model parameter uncertainty, many control methods were studied for the spacecraft attitude control system. Jovan et al. [8] proposed two globally stable control algorithms for robust stabilization of spacecraft in the presence of control input saturation, parameter uncertainty, and external disturbances. Hu et al. [9] proposed the robust adaptive variable structure output feedback control algorithm for stabilization of a three-axis stabilized flexible spacecraft in the presence of parameter uncertainty, external disturbances, and control input nonlinearity. Tayebi [10] proposed a dynamic output feedback control approach based on quaternions for spacecraft attitude tracking problems, but he did not consider the problem of input limit and nonlinear perturbation of the controller, which

Spacecraft Attitude Control. DOI: https://doi.org/10.1016/B978-0-323-99005-9.00003-1

brought difficulties in engineering practice. The attitude tracking control problem of rigid spacecraft without knowing angular velocity has been studied in [11–13], in which the observer was adopted to estimate the angular velocity used in the control law. The adaptive output feedback control method for the rotational maneuver and vibration suppression of flexible spacecraft only considering the output variable (pitch angle) was studied in [14]. The robust H_∞ output feedback control approach based on linear matrix inequality (LMI) was studied to achieve attitude stabilization of a flexible spacecraft in [15]. Taking actuator misalignment into account, the mixed finite-time control and control allocation method for attitude stabilization of a rigid spacecraft were investigated in [16]. In [17], a robust adaptive terminal sliding mode control method was used in the relative position and attitude tracking control problem during autonomous rendezvous and docking of two spacecraft. In addition, the robust nonfragile control method has been widely applied in many other systems. The robust output tracking control problem of nonlinear multiinput multioutput systems has been discussed in [18] and [19], with sliding mode technique and fuzzy approach, respectively. The problem of a nonfragile H_∞ static output feedback controller design for a class of continuous-time semi-Markovian jump linear systems has been studied in [20]. The nonfragile finite-time filtering problem has been studied for a class of nonlinear Markovian jumping systems with time delays and uncertainties in [21]. The nonfragile controller design for an offshore steel jacket platform with nonlinear perturbation was addressed in [22]. The problem of nonfragile H_∞ dynamic output feedback control for uncertain Takagi-Sugeno fuzzy systems with time-varying state delay was studied in [23]. However, the above references mainly investigated the additive perturbation problems which often appear in the system's initial operational phase for minute adjustment of parameters. In addition, multiplicative perturbation will appear in the decay phase of the controller performance.

The objective of this chapter is to propose the dynamic output feedback nonfragile control (DOFNFC) method for a class of spacecraft attitude stabilization problems with parameter uncertainty, gyro drift, external disturbances, H_∞ performance constraint, quadratic stability and input saturation. This chapter only considers constant gyro drift, and the original spacecraft attitude control system is easy to transform into a new form that is easier for the controller design. By constructing a new state variable, one can generate a new closed-loop system. Based on the Lyapunov theory, with respect to additive and multiplicative perturbation of the controller, sufficient conditions for the existence of two types of

DOFNFC are given by use of LMIs. For comparison, the standard output feedback H_∞ controller (SOFHC) is also designed.

The remaining structure of this chapter is organized as follows. The following section is to introduce spacecraft attitude dynamics equations and convert them into a state-space form, and the corresponding DOFNFC problem is formulated. In Section 3.3, the DOFNFC design method based on LMIs is proposed with respect to additive perturbation and multiplicative perturbation, and a hybrid nonfragile controller is developed to deal with coexisting additive and multiplicative perturbations. In Section 3.4, numerical simulations of DOFNFC are made. For comparison, the simulations of SOFHC are also performed to illustrate the effectiveness of the proposed nonfragile approach. Finally, some conclusions are drawn.

3.2 Problem formulation

3.2.1 Attitude system description

To simplify the control problem, three-axis rotation is considered, and the rotational dynamics of the rigid spacecraft considering external disturbances is described as follow

$$\boldsymbol{I}_b\dot{\boldsymbol{\omega}} + \boldsymbol{\omega} \times (\boldsymbol{I}_b\boldsymbol{\omega}) = \boldsymbol{T}_c + \boldsymbol{T}_g + \boldsymbol{T}_d \tag{3.1}$$

where $\boldsymbol{I}_b$ and $\boldsymbol{\omega}$ denote the inertia matrix and the angular velocity, $\boldsymbol{T}_c$, $\boldsymbol{T}_g$ and $\boldsymbol{T}_d$ denote the control input, gravitational torque, and external disturbance torque, respectively.

Provided the body-fixed frame is chosen to be a principal-axis frame, under the condition of small angles, Eq. (3.1) can be converted into the form of components

$$\begin{cases} I_x\ddot{\phi} + 4(I_y - I_z)\omega_0^2\phi + (I_y - I_z - I_x)\omega_0\dot{\psi} = T_{cx} + T_{dx} \\ I_y\ddot{\theta} + 3\omega_0^2(I_x - I_z)\theta = T_{cy} + T_{dy} \\ I_z\ddot{\psi} + (I_y - I_x)\omega_0^2\psi + (I_x + I_z - I_y)\omega_0\dot{\phi} = T_{cz} + T_{dz} \end{cases} \tag{3.2}$$

where I_x, I_y, and I_z denote the three components of inertia matrix; ω_x, ω_y, and ω_z denote the three components of angular velocity; T_{cx}, T_{cy}, and T_{cz} denote the three components of control input; T_{dx}, T_{dy}, and T_{dz} denote the three components of external disturbance torque; ω_0 represents the orbital angular velocity of spacecraft; and Euler angles ϕ, θ and ψ are roll, pitch and yaw attitude angle, respectively.

Let

$$\boldsymbol{x}=\begin{bmatrix}\phi & \theta & \psi & \dot{\phi} & \dot{\theta} & \dot{\psi}\end{bmatrix}^T \boldsymbol{y}=\begin{bmatrix}\phi & \theta & \psi & \omega_x & \omega_y & \omega_z\end{bmatrix}^T$$

$$\boldsymbol{u}=\begin{bmatrix}T_{cx} & T_{cy} & T_{cz}\end{bmatrix}^T,\ \boldsymbol{w}=\begin{bmatrix}T_{dx} & T_{dy} & T_{dz}\end{bmatrix}^T,$$

Then, Eq. (3.2) can be written as

$$\begin{cases}\dot{\boldsymbol{x}}(t)=\boldsymbol{A}\boldsymbol{x}(t)+\boldsymbol{B}_1\boldsymbol{u}(t)+\boldsymbol{B}_2\boldsymbol{w}(t)\\ \boldsymbol{y}(t)=\boldsymbol{C}_1\boldsymbol{x}(t)+\boldsymbol{D}_2\\ \boldsymbol{z}(t)=\boldsymbol{C}_2\boldsymbol{x}(t)\end{cases} \tag{3.3}$$

where,

$$\boldsymbol{A}=\begin{bmatrix}0 & 0 & 0 & 1 & 0 & 0\\ 0 & 0 & 0 & 0 & 1 & 0\\ 0 & 0 & 0 & 0 & 0 & 1\\ A_{41} & 0 & 0 & 0 & 0 & -\omega_0 I_x^{-1}(I_y-I_x-I_z)\\ 0 & A_{52} & 0 & 0 & 0 & 0\\ 0 & 0 & A_{63} & \omega_0 I_z^{-1}(I_y-I_x-I_z) & 0 & 0\end{bmatrix}$$

$$A_{41}=-4\omega_0^2 I_x^{-1}(I_y-I_z),\ A_{52}=-3\omega_0^2 I_y^{-1}(I_x-I_z),\ A_{63}=-\omega_0^2 I_z^{-1}(I_y-I_x)$$

$$\boldsymbol{B}_1=\boldsymbol{B}_2=[\boldsymbol{0}_{3\times 3}\ \ \mathrm{diag}(I_x^{-1},I_y^{-1},I_z^{-1})]^T,$$

$$\boldsymbol{D}_2=\begin{bmatrix}\boldsymbol{0}_{1\times 4} & -\omega_0 & 0\end{bmatrix}^T,\ \boldsymbol{C}_1=\begin{bmatrix}\boldsymbol{I}_{3\times 3} & \boldsymbol{0}_{3\times 3}\\ \boldsymbol{B} & \boldsymbol{I}_{3\times 3}\end{bmatrix},$$

$$\boldsymbol{B}=\begin{bmatrix}0 & 0 & -\omega_0\\ 0 & 0 & 0\\ \omega_0 & 0 & 0\end{bmatrix},\ \boldsymbol{C}_2=[\boldsymbol{I}_{6\times 6}]$$

Due to the detection errors and various uncertain factors in space, the attitude angle and angular velocity of on-orbit servicing spacecraft cannot be accurately determined online, and the system shows certain model parameter uncertainty. Considering the parameter uncertainty, $\boldsymbol{A}$ will be transformed into $\boldsymbol{A}+\Delta\boldsymbol{A}$, where $\Delta\boldsymbol{A}$ denotes model parameter uncertainty and satisfies matching condition, that is,

$$\Delta\boldsymbol{A}(t)=\boldsymbol{M}_1\boldsymbol{F}_1(t)\boldsymbol{N}_1,\boldsymbol{F}_1(t)^T\boldsymbol{F}_1(t)\leq\boldsymbol{I},\ \ \forall t$$

where, $\boldsymbol{M}_1$ and $\boldsymbol{N}_1$ are known constant real matrices of appropriate dimension, $\boldsymbol{F}_1$(t) is an unknown matrix function with Lebesgue measurable elements.

Gyro drift means the slow precession of gyro spin axis caused by disturbance torque, which can be divided into two types, one is caused by regular systematic disturbance torque, and the other is caused by irregular random factors. An important indicator in accuracy measurement of the gyroscope is gyro drift rate with bounded L_2 norm, and it can be confined into a certain range with the amplitude defined as constant drift $\boldsymbol{g}(t)$.

Taking parameter uncertainty and gyro drift into account, the resulting spacecraft attitude control system can be written as follows:

$$\begin{cases} \dot{\boldsymbol{x}}(t) = (\boldsymbol{A} + \Delta\boldsymbol{A})\boldsymbol{x}(t) + \boldsymbol{B}_1\boldsymbol{u}(t) + \boldsymbol{B}_2\boldsymbol{w}(t) \\ \boldsymbol{y}(t) = \boldsymbol{C}_1\boldsymbol{x}(t) + \boldsymbol{D}_1\boldsymbol{g}(t) + \boldsymbol{D}_2 \\ \boldsymbol{z}(t) = \boldsymbol{C}_2\boldsymbol{x}(t) \end{cases} \tag{3.4}$$

$\boldsymbol{D}_1$, $\boldsymbol{D}_2$ are known constant coefficient matrices, where $\boldsymbol{D}_1 = [\boldsymbol{0}_{3\times 3}\ \text{diag}(\boldsymbol{I}_x^{-1}, \boldsymbol{I}_y^{-1}, \boldsymbol{I}_z^{-1})]^T$. For the construction convenience of controller design, new output $\widehat{\boldsymbol{y}}(t)$ is constructed with the form:

$$\widehat{\boldsymbol{y}}(t) = \boldsymbol{y}(t) - \boldsymbol{D}_1\boldsymbol{g}(t) - \boldsymbol{D}_2$$

When considering control input constraint directly, the new spacecraft rotational dynamics equations can be shown as:

$$\begin{cases} \dot{\boldsymbol{x}}(t) = (\boldsymbol{A} + \Delta\boldsymbol{A})\boldsymbol{x}(t) + \boldsymbol{B}_1\text{sat}(\boldsymbol{u}(t)) + \boldsymbol{B}_2\boldsymbol{w}(t) \\ \widehat{\boldsymbol{y}}(t) = \boldsymbol{C}_1\boldsymbol{x}(t) \\ \boldsymbol{z}(t) = \boldsymbol{C}_2\boldsymbol{x}(t) \end{cases} \tag{3.5}$$

where, $sat(\boldsymbol{u})$ means the actual control input after saturation processing with the form:

$$sat(\boldsymbol{u}) = \begin{bmatrix} sat(u_1) & sat(u_2) & sat(u_3) \end{bmatrix}^T$$

satisfying

$$sat(u_i) = \begin{cases} u_{mi} & u_i > u_{mi} \\ u_i & -u_{mi} \leq u_i \leq u_{mi} \\ -u_{mi} & u_i < -u_{mi} \end{cases} \tag{3.6}$$

This can guarantee continuous control input, and $u_{mi}(i = 1, 2, 3)$ depends on the actuator ability, from which one knows $\text{sat}(u_i) = (1/c_i)u_i$, $sat(\boldsymbol{u}) = \boldsymbol{C}\boldsymbol{u}$, where, $\boldsymbol{C} = diag[1/c_1\ 1/c_2\ 1/c_3]$, $c_i \geq 1,\ i = 1, 2, 3$.

3.2.2 Nonfragile control problem

Assuming the actuator uncertainty is $\boldsymbol{\Xi}$, and the standard controller gain is $\boldsymbol{K}$, adding $\boldsymbol{\Xi}$ into the system, the nominal controller gain becomes $\boldsymbol{K}(\boldsymbol{I}+\boldsymbol{\Xi})$, that is,

$$\boldsymbol{K}(\boldsymbol{I}+\boldsymbol{\Xi})=\boldsymbol{K}+\boldsymbol{K}\boldsymbol{\Xi}=\boldsymbol{K}+\Delta\boldsymbol{K} \tag{3.7}$$

where, $\Delta\boldsymbol{K}$ is the gain perturbation induced by the actuator.

With respect to Eq. (3.5), design a dynamic output feedback controller as

$$\begin{cases}\dot{\widehat{\boldsymbol{x}}}(t)=\boldsymbol{A}_c\widehat{\boldsymbol{x}}(t)+\boldsymbol{B}_c\widehat{\boldsymbol{y}}(t)\\ \boldsymbol{u}(t)=(\boldsymbol{K}+\Delta\boldsymbol{K})\widehat{\boldsymbol{x}}(t)\end{cases} \tag{3.8}$$

where $\Delta\boldsymbol{K}$ can be divided into two types, that is, additive perturbation represented by (3.9) and multiplicative perturbation represented by (3.10), the former often appears in the system's initial operational phase, and the latter often appears in the decay phase of controller performance.

$$\Delta\boldsymbol{K}=\boldsymbol{M}_2\boldsymbol{F}_2(t)\boldsymbol{N}_2,\quad \boldsymbol{F}_2(t)^T\boldsymbol{F}_2(t)\le\boldsymbol{I} \tag{3.9}$$

$$\Delta\boldsymbol{K}=\boldsymbol{M}_3\boldsymbol{F}_3(t)\boldsymbol{N}_3\boldsymbol{K},\quad \boldsymbol{F}_3(t)^T\boldsymbol{F}_3(t)\le\boldsymbol{I} \tag{3.10}$$

The definition of $\boldsymbol{M}_2$, $\boldsymbol{N}_2$, $\boldsymbol{M}_3$, $\boldsymbol{N}_3$, $\boldsymbol{F}_2(t)$, and $\boldsymbol{F}_3(t)$ is similar to that of $\Delta\boldsymbol{A}$.

Then, substituting Eq. (3.8) into Eq. (3.5) yields

$$\begin{cases}\zeta(t)=(\widehat{\boldsymbol{A}}+\Delta\widehat{\boldsymbol{A}}(t))\zeta(t)+\widehat{\boldsymbol{B}}\boldsymbol{w}(t)\\ \boldsymbol{z}(t)=\widehat{\boldsymbol{C}}\zeta(t)\end{cases} \tag{3.11}$$

The derived process is as follows.

First, substitute the second equation in Eq. (3.8) into the state equation in Eq. (3.5), then

$$\begin{aligned}\dot{\boldsymbol{x}}(t)&=(\boldsymbol{A}+\Delta\boldsymbol{A})\boldsymbol{x}(t)+\boldsymbol{B}_1\boldsymbol{u}(t)+\boldsymbol{B}_2\boldsymbol{w}(t)\\&=(\boldsymbol{A}+\Delta\boldsymbol{A})\boldsymbol{x}(t)+\boldsymbol{B}_1(\boldsymbol{K}+\Delta\boldsymbol{K})\widehat{\boldsymbol{x}}(t)+\boldsymbol{B}_2\boldsymbol{w}(t)\end{aligned} \tag{3.12}$$

Substitute the second equation in Eq. (3.5) into the first equation in Eq. (3.8), we get

$$\dot{\widehat{\boldsymbol{x}}}(t)=\boldsymbol{A}_c\widehat{\boldsymbol{x}}(t)+\boldsymbol{B}_c\widehat{\boldsymbol{y}}(t)=\boldsymbol{A}_c\widehat{\boldsymbol{x}}(t)+\boldsymbol{B}_c\boldsymbol{C}_1\boldsymbol{x}(t) \tag{3.13}$$

Let $\boldsymbol{\zeta}(t) = \begin{bmatrix} \boldsymbol{x}(t)^T & \widehat{\boldsymbol{x}}(t)^T \end{bmatrix}^T$, then

$$\dot{\boldsymbol{\zeta}}(t) = \begin{bmatrix} \dot{\boldsymbol{x}}(t) \\ \dot{\widehat{\boldsymbol{x}}}(t) \end{bmatrix} = \left\{ \begin{bmatrix} \boldsymbol{A} & \boldsymbol{B}_1\boldsymbol{K} \\ \boldsymbol{B}_c\boldsymbol{C}_1 & \boldsymbol{A}_c \end{bmatrix} + \begin{bmatrix} \Delta\boldsymbol{A}(t) & \boldsymbol{B}_1\Delta\boldsymbol{K} \\ 0 & 0 \end{bmatrix} \right\} \begin{bmatrix} \boldsymbol{x}(t) \\ \widehat{\boldsymbol{x}}(t) \end{bmatrix} + \begin{bmatrix} \boldsymbol{B}_2 \\ 0 \end{bmatrix} \boldsymbol{w}(t) \tag{3.14}$$

$$\boldsymbol{z}(t) = \boldsymbol{C}_2\boldsymbol{x}(t) = \begin{bmatrix} \boldsymbol{C}_2 & 0 \end{bmatrix} \boldsymbol{\zeta}(t) \tag{3.15}$$

Therefore, the coefficient matrices in Eq. (3.11) are

$$\widehat{\boldsymbol{A}} = \begin{bmatrix} \boldsymbol{A} & \boldsymbol{B}_1\boldsymbol{K} \\ \boldsymbol{B}_c\boldsymbol{C}_1 & \boldsymbol{A}_c \end{bmatrix},\ \Delta\widehat{\boldsymbol{A}}(t) = \begin{bmatrix} \Delta\boldsymbol{A}(t) & \boldsymbol{B}_1\Delta\boldsymbol{K} \\ 0 & 0 \end{bmatrix},\ \widehat{\boldsymbol{B}} = \begin{bmatrix} \boldsymbol{B}_2 \\ 0 \end{bmatrix}, \widehat{\boldsymbol{C}} = \begin{bmatrix} \boldsymbol{C}_2 & 0 \end{bmatrix}$$

Then, one can transform $\Delta\widehat{\boldsymbol{A}}(t)$ into the form of Eq. (3.16), that is,

$$\Delta\widehat{\boldsymbol{A}}(t) = \widehat{\boldsymbol{M}}\boldsymbol{F}(t)\widehat{\boldsymbol{N}} \tag{3.16}$$

When $\Delta\boldsymbol{K}$ represents additive perturbation,

$$\Delta\widehat{\boldsymbol{A}}(t) = \begin{bmatrix} \Delta\boldsymbol{A}(t) & \boldsymbol{B}_1\Delta\boldsymbol{K} \\ 0 & 0 \end{bmatrix} = \begin{bmatrix} \boldsymbol{M}_1\boldsymbol{F}_1(t)\boldsymbol{N}_1 & \boldsymbol{B}_1\boldsymbol{M}_2\boldsymbol{F}_2(t)\boldsymbol{N}_2 \\ 0 & 0 \end{bmatrix}$$

$$= \begin{bmatrix} \boldsymbol{M}_1 & \boldsymbol{B}_1\boldsymbol{M}_2 \\ 0 & 0 \end{bmatrix} \begin{bmatrix} \boldsymbol{F}_1(t) & 0 \\ 0 & \boldsymbol{F}_2(t) \end{bmatrix} \begin{bmatrix} \boldsymbol{N}_1 & 0 \\ 0 & \boldsymbol{N}_2 \end{bmatrix}$$

When ΔK represents multiplicative perturbation,

$$\Delta\widehat{\boldsymbol{A}}(t) = \begin{bmatrix} \Delta\boldsymbol{A}(t) & \boldsymbol{B}_1\Delta\boldsymbol{K} \\ 0 & 0 \end{bmatrix} = \begin{bmatrix} \boldsymbol{M}_1\boldsymbol{F}_1(t)\boldsymbol{N}_1 & \boldsymbol{B}_1\boldsymbol{M}_3\boldsymbol{F}_3(t)\boldsymbol{N}_3\boldsymbol{K} \\ 0 & 0 \end{bmatrix}$$

$$= \begin{bmatrix} \boldsymbol{M}_1 & \boldsymbol{B}_1\boldsymbol{M}_3 \\ 0 & 0 \end{bmatrix} \begin{bmatrix} \boldsymbol{F}_1(t) & 0 \\ 0 & \boldsymbol{F}_3(t) \end{bmatrix} \begin{bmatrix} \boldsymbol{N}_1 & 0 \\ 0 & \boldsymbol{N}_3\boldsymbol{K} \end{bmatrix}$$

With respect to the second equation in Eq. (3.8), one has

$$\begin{aligned} \boldsymbol{u}(t) = (\boldsymbol{K}+\Delta\boldsymbol{K})\widehat{\boldsymbol{x}}(t) &= \begin{bmatrix} \boldsymbol{0}_{3\times 6} & \boldsymbol{K}+\Delta\boldsymbol{K} \end{bmatrix} \boldsymbol{\zeta}(t) \\ &= \left\{ \begin{bmatrix} \boldsymbol{0}_{3\times 6} & \boldsymbol{K} \end{bmatrix} + \begin{bmatrix} \boldsymbol{0}_{3\times 6} & \Delta\boldsymbol{K} \end{bmatrix} \right\} \boldsymbol{\zeta}(t) \end{aligned} \tag{3.17}$$

Let

$$\widehat{\boldsymbol{K}} = \begin{bmatrix} \boldsymbol{0}_{3\times 6} & \boldsymbol{K} \end{bmatrix}, \Delta\widehat{\boldsymbol{K}} = \begin{bmatrix} \boldsymbol{0}_{3\times 6} & \Delta\boldsymbol{K} \end{bmatrix}$$

Then Eq. (3.17) is transformed into

$$u(t) = (\widehat{K}+\Delta\widehat{K})\zeta(t) \tag{3.18}$$

that is, DOFNFC has been transformed into the form of state feedback controller with new state variable, and the design method can become easier.

Here,

$$\Delta\widehat{K}\Delta\widehat{K}^T = \begin{bmatrix} 0_{3\times 6} & \Delta K \end{bmatrix}\begin{bmatrix} 0_{6\times 3} \\ \Delta K^T \end{bmatrix} = \Delta K \Delta K^T \le \eta_0^2 I \tag{3.19}$$

The actuator in the spacecraft exhibits many nonlinear problems, of which actuator saturation is the most common. Due to the limited physical characteristics of the actuator or limiting magnitude artificially for security reasons, the characteristics of the controller will be affected, leading to the control input constraints. So there exists control input $u_{\max}$, satisfying

$$\|u\|_2 \le u_{max} \tag{3.20}$$

where, $u_{\max}$ means the upper bound of the control input provided by the actuator.

3.2.3 Control objective

The objective of this chapter is to design the DOFNFC represented by Eq. (3.8), so that the spacecraft attitude control system represented by Eq. (3.5) has quadratic stability, where the controlled output $z(t)$ satisfies H_∞ performance constraint and the control input meets (3.20).

Definition 3.1: If there is a symmetric positive definite matrix $P>0$ and a positive constant α, for arbitrary uncertainty, the time derivative of Lyapunov function $V(x,t)$ satisfies

$$\dot{V}(x,t) \le -\alpha\|x(t)\|_2^2 \tag{3.21}$$

Then, the system represented by Eq. (3.11) ($w(t)=0$) is quadratically stable.

3.3 Dynamic output feedback nonfragile control law design

3.3.1 Some lemmas

Lemma 3.1: (Schur complement lemma) For partitioned symmetric matrix

$$A = \begin{bmatrix} A_{11} & A_{12} \\ A_{12}^T & A_{22} \end{bmatrix}$$

the following inequalities are equivalent:

$$A < 0 \Leftrightarrow A_{11} < 0, A_{22} - A_{12}^T A_{11}^{-1} A_{12} < 0 \Leftrightarrow A_{22} < 0, A_{11} - A_{12} A_{22}^{-1} A_{12}^T < 0$$

Lemma 3.2: [24] Let $H, E \in \mathbb{R}^{m \times n}, F(t) \in \mathbb{S}^m$, $F(t)$ satisfies $F(t)^T F(t) \leq I$, for scalar $\xi > 0$, the following holds:

$$HF(t)E + E^T F(t)^T H^T \leq \xi H H^T + \xi^{-1} E^T E \tag{3.22}$$

Lemma 3.3: Considering the system (3.11) with controller (3.18), and the control input satisfies inequality (3.20), if there exists symmetric positive definite matrix P_1, matrix V, such that

$$\begin{bmatrix} -u_{\max}^2 I & -(\widehat{K}+\Delta\widehat{K}) \\ -(\widehat{K}+\Delta\widehat{K})^T & -\rho^{-1} P_1 \end{bmatrix} \leq 0 \tag{3.23}$$

$$\begin{bmatrix} -\rho I & -\zeta^T(0) V \\ -V^T \zeta(0) & P_1 - (V+V^T) \end{bmatrix} < 0 \tag{3.24}$$

Proof: Choose Lyapunov function as

$$U(\zeta(t)) = \zeta(t)^T P_1 \zeta(t)$$

satisfying inequality

$$U(\zeta(t)) \leq \rho$$

where ρ is a given positive constant, define $\Lambda(\boldsymbol{P}_1, \rho) = \{\zeta(t)| \zeta(t)^T \boldsymbol{P}_1 \zeta(t) \leq \rho\}$, for $\|\boldsymbol{u}\|_2 \leq u_{\max}$, define

$$\Lambda(\widehat{\boldsymbol{K}}+\Delta\widehat{\boldsymbol{K}}) = \left\{\zeta(t) \middle| \zeta(t)^T (\widehat{\boldsymbol{K}}+\Delta\widehat{\boldsymbol{K}})^T (\widehat{\boldsymbol{K}}+\Delta\widehat{\boldsymbol{K}}) \zeta(t) \leq u_{\max}^2 \right\}$$

so control input constraint can be ensured with (3.25).

$$\Lambda(\boldsymbol{P}_1, \rho) \subset \Lambda(\widehat{\boldsymbol{K}}+\Delta\widehat{\boldsymbol{K}}) \tag{3.25}$$

According to [25], (3.25) holds if and only if

$$\left(\widehat{\boldsymbol{K}}+\Delta\widehat{\boldsymbol{K}}\right)\left(\frac{\boldsymbol{P}_1}{\rho}\right)^{-1}\left(\widehat{\boldsymbol{K}}+\Delta\widehat{\boldsymbol{K}}\right)^T \leq u_{\max}^2 \boldsymbol{I}$$

With Lemma 3.1, one can get inequality (3.23), inequality (3.24) will be proved next.

When $\dot{U}(\zeta(t)) < 0$, $\zeta(t)^T \boldsymbol{P}_1 \zeta(t) < \zeta(0)^T \boldsymbol{P}_1 \zeta(0)$,

Let $\zeta(0)^T \boldsymbol{P}_1 \zeta(0) \leq \rho$, with Lemma 3.1, we get

$$\begin{bmatrix} -\rho \boldsymbol{I} & -\zeta^T(0) \\ -\zeta(0) & -\boldsymbol{P}_1^{-1} \end{bmatrix} < 0$$

Multiplied by diag$\{\boldsymbol{I}\ \boldsymbol{V}^T\}$ and its transposed on both sides, one has

$$\begin{bmatrix} -\rho \boldsymbol{I} & -\zeta^T(0)\boldsymbol{V} \\ -\boldsymbol{V}^T\zeta(0) & -\boldsymbol{V}^T\boldsymbol{P}_1^{-1}\boldsymbol{V} \end{bmatrix} < 0 \tag{3.26}$$

With $(\boldsymbol{P}_1 - \boldsymbol{V})^T \boldsymbol{P}_1^{-1} (\boldsymbol{P}_1 - \boldsymbol{V}) \geq 0$, we get

$$-\boldsymbol{V}^T\boldsymbol{P}_1^{-1}\boldsymbol{V} \leq \boldsymbol{P}_1 - (\boldsymbol{V}^T + \boldsymbol{V})$$

If inequality (3.26) holds, only the following LMI is satisfied:

$$\begin{bmatrix} -\rho \boldsymbol{I} & -\zeta^T(0)\boldsymbol{V} \\ -\boldsymbol{V}^T\zeta(0) & \boldsymbol{P}_1 - (\boldsymbol{V}+\boldsymbol{V}^T) \end{bmatrix} < 0$$

Lemma 3.4: For given $\xi > 0, \gamma > 0$, the system (3.11) has quadratic stability, $(\boldsymbol{w}(t) = 0)$, the control input meets inequality (3.20), and controlled output $\boldsymbol{z}(t)$ satisfies H_∞ performance constraint

$$\|\boldsymbol{z}(t)\|_2^2 < \gamma^2 \|\boldsymbol{w}(t)\|_2^2 \tag{3.27}$$

if there exists symmetric matrix $\boldsymbol{P} > 0$, matrices $\boldsymbol{X}$ and $\boldsymbol{W}$, such that

$$\begin{bmatrix} \widehat{A}^T P + P\widehat{A} & P\widehat{M} & \widehat{N}^T & \widehat{C}^T & P\widehat{B} \\ \widehat{M}^T P & -\xi^{-1} I & 0 & 0 & 0 \\ \widehat{N} & 0 & -\xi I & 0 & 0 \\ \widehat{C} & 0 & 0 & -I & 0 \\ \widehat{B}^T P & 0 & 0 & 0 & -\gamma^2 I \end{bmatrix} < 0 \tag{3.28}$$

$$\begin{bmatrix} (\gamma_1 \eta_0^2 - u_{\max}^2) I & -W & 0 \\ -W^T & -\rho^{-1}\eta_1 P & X^T \\ 0 & X & -\gamma_1 I \end{bmatrix} < 0 \tag{3.29}$$

$$\begin{bmatrix} -\rho I & -\eta_2 \zeta^T(0) \\ -\eta_2 \zeta(0) & \eta_1 P - \eta_2 (X + X^T) \end{bmatrix} < 0 \tag{3.30}$$

Proof: First, we prove that the system (3.11) has quadratic stability.

Choose Lyapunov function as

$$V(\zeta(t)) = \zeta(t)^T P \zeta(t)$$

When $w(t)=0$,

$$\begin{aligned} \dot{V}(\zeta(t)) &= \dot{\zeta}(t)^T P \zeta(t) + \zeta(t)^T P \dot{\zeta}(t) \\ &= \zeta(t)^T (\widehat{A}^T + \Delta\widehat{A}(t)^T) P \zeta(t) + \zeta(t)^T P (\widehat{A} + \Delta\widehat{A}(t)) \zeta(t) \\ &= \zeta(t)^T (\widehat{A}^T + \Delta\widehat{A}(t)^T) P \zeta(t) + \zeta(t)^T P (\widehat{A} + \Delta\widehat{A}(t)) \zeta(t) \\ &= \zeta(t)^T \left[(\widehat{A}^T + \Delta\widehat{A}(t)^T) P + P (\widehat{A} + \Delta\widehat{A}(t)) \right] \zeta(t) \\ &= \zeta(t)^T \left[\widehat{A}^T P + P\widehat{A} + P\widehat{M} F(t) \widehat{N} + \widehat{N}^T F(t)^T \widehat{M}^T P \right] \zeta(t) \\ &\le \zeta(t)^T (\widehat{A}^T P + P\widehat{A} + \xi P \widehat{M}\widehat{M}^T P + \xi^{-1} \widehat{N}^T \widehat{N}) \zeta(t) \end{aligned}$$

Let $Q_0 = \widehat{A}^T P + P\widehat{A} + \xi P\widehat{M}\widehat{M}^T P + \xi^{-1}\widehat{N}^T \widehat{N}$ With Lemma 3.1 and inequality (3.28), one knows $Q_0 < 0$.

For $\dot{V}(\zeta(t)) \le \zeta(t)^T Q_0 \zeta(t) \le \lambda_{\max}(Q_0) \zeta(t)^T \zeta(t)$, where, $\lambda_{\max}(Q_0)$ means the maximum eigenvalue of Q_0.

Let $\alpha = -\lambda_{\max}(Q_0) > 0$, then $\dot{V}(\zeta(t)) \le -\alpha \| \zeta(t) \|_2^2$ which means that the system (3.11) has quadratic stability.

Second, we prove that $z(t)$ satisfies H_∞ performance constraint.

To establish the norm bound $\gamma^2\|w(t)\|_2^2$ of $L_2[0,\infty)$, considering functional

$$J=\int_0^{\infty}[z(t)^T z(t)-\gamma^2 w(t)^T w(t)]dt$$

For arbitrary nonzero $w(t)\in L_2[0,\infty)$, let $\zeta(0)=0$, then

$$J=\int_0^{\infty}[z(t)^T z(t)-\gamma^2 w(t)^T w(t)+\dot{V}(\zeta(t))]dt-V(\infty)+V(0)$$

$$\le\int_0^{\infty}[\zeta(t)^T\hat{C}^T\hat{C}\zeta(t)-\gamma^2 w(t)^T w(t)+\zeta(t)^T(\hat{A}^T P+P\hat{A}$$

$$+\xi P\hat{M}\hat{M}^T P+\xi^{-1}\hat{N}^T\hat{N})\zeta(t)+w(t)^T\hat{B}^T P\zeta(t)+\zeta(t)^T P\hat{B}w(t)]dt$$

$$\le\int_0^{\infty}[\zeta(t)^T w(t)^T]\begin{bmatrix}\hat{A}^T P+P\hat{A}+\xi P\hat{M}\hat{M}^T P+\xi^{-1}\hat{N}^T\hat{N}+\hat{C}^T\hat{C} & P\hat{B}\\ \hat{B}^T P & -\gamma^2 I\end{bmatrix}\begin{bmatrix}\zeta(t)\\ w(t)\end{bmatrix}dt$$

With Lemma 3.1 and inequality (3.28), one has $J<0$, that is, $z(t)$ satisfies H_∞ performance constraint $\|z(t)\|_2^2<\gamma^2\|w(t)\|_2^2$.

Finally, we prove control input meets inequality (3.20).

According to Lemma 3.1, inequality (3.29) is equivalent to inequality (3.31),

$$\begin{bmatrix}(\gamma_1\eta_0^2-u_{\max}^2)I & -W\\ -W^T & \gamma_1^{-1}X^T X-\rho^{-1}\eta_1 P\end{bmatrix}<0 \tag{3.31}$$

With Lemma 3.3 and inequality (3.19), one knows inequalities (3.32) and (3.33) hold,

$$\begin{bmatrix}0 & -\Delta\hat{K}X\\ -(\Delta\hat{K}X)^T & 0\end{bmatrix}\le\begin{bmatrix}\gamma_1\Delta\hat{K}\Delta\hat{K}^T & 0\\ 0 & \gamma_1^{-1}X^T X\end{bmatrix} \tag{3.32}$$

$$\begin{bmatrix}0 & -\Delta\hat{K}X\\ -(\Delta\hat{K}X)^T & 0\end{bmatrix}\le\begin{bmatrix}\gamma_1\eta_0^2 I & 0\\ 0 & \gamma_1^{-1}X^T X\end{bmatrix} \tag{3.33}$$

With inequalities (3.31) and (3.33), inequality (3.34) holds,

$$\begin{bmatrix}-u_{\max}^2 I & -W-\Delta\hat{K}X\\ -W^T-(\Delta\hat{K}X)^T & -\rho^{-1}\eta_1 P\end{bmatrix}<0 \tag{3.34}$$

Substitute $\boldsymbol{W} = \widehat{\boldsymbol{K}}\boldsymbol{X}$ into inequality (3.34), one has

$$\begin{bmatrix} -u_{\max}^2 \boldsymbol{I} & -(\widehat{\boldsymbol{K}} + \Delta\widehat{\boldsymbol{K}})\boldsymbol{X} \\ -\boldsymbol{X}^T(\widehat{\boldsymbol{K}} + \Delta\widehat{\boldsymbol{K}})^T & -\rho^{-1}\eta_1 \boldsymbol{P} \end{bmatrix} < 0 \tag{3.35}$$

Multiplied by diag$\{\boldsymbol{I}\ \boldsymbol{X}^{-T}\}$ and its transpose on both sides of inequality (3.35), one has

$$\begin{bmatrix} -u_{\max}^2 \boldsymbol{I} & -(\widehat{\boldsymbol{K}} + \Delta\widehat{\boldsymbol{K}}) \\ -(\widehat{\boldsymbol{K}} + \Delta\widehat{\boldsymbol{K}})^T & -\rho^{-1}\eta_1 \boldsymbol{X}^{-T}\boldsymbol{P}\boldsymbol{X}^{-1} \end{bmatrix} < 0$$

Let $\boldsymbol{P}_1 = \eta_1 \boldsymbol{X}^{-T}\boldsymbol{P}\boldsymbol{X}^{-1}$, one has

$$\begin{bmatrix} -u_{\max}^2 \boldsymbol{I} & -(\widehat{\boldsymbol{K}} + \Delta\widehat{\boldsymbol{K}}) \\ -(\widehat{\boldsymbol{K}} + \Delta\widehat{\boldsymbol{K}})^T & -\rho^{-1}\boldsymbol{P}_1 \end{bmatrix} < 0 \tag{3.36}$$

Similarly, multiplied by diag$\{\boldsymbol{I}\ \boldsymbol{X}^{-T}\}$ and then transposed on both sides of inequality (3.30), one has

$$\begin{bmatrix} -\rho \boldsymbol{I} & -\eta_2 \boldsymbol{\zeta}^T(0)\boldsymbol{X}^{-1} \\ -\eta_2 \boldsymbol{X}^{-T}\boldsymbol{\zeta}(0) & \eta_1 \boldsymbol{X}^{-T}\boldsymbol{P}\boldsymbol{X}^{-1} - \eta_2 \boldsymbol{X}^{-T}(\boldsymbol{X}+\boldsymbol{X}^T)\boldsymbol{X}^{-1} \end{bmatrix} < 0$$

that is,

$$\begin{bmatrix} -\rho \boldsymbol{I} & -\eta_2 \boldsymbol{\zeta}^T(0)\boldsymbol{X}^{-1} \\ -\eta_2 \boldsymbol{X}^{-T}\boldsymbol{\zeta}(0) & \boldsymbol{P}_1 - \eta_2 \boldsymbol{X}^{-T} - \eta_2 \boldsymbol{X}^{-1} \end{bmatrix} < 0 \tag{3.37}$$

Let $\boldsymbol{V} = \eta_2 \boldsymbol{X}^{-1}$, inequality (3.37) is transformed into

$$\begin{bmatrix} -\rho \boldsymbol{I} & -\boldsymbol{\zeta}^T(0)\boldsymbol{V} \\ -\boldsymbol{V}^T\boldsymbol{\zeta}(0) & \boldsymbol{P}_1 - (\boldsymbol{V}+\boldsymbol{V}^T) \end{bmatrix} < 0 \tag{3.38}$$

With inequalities (3.36) and (3.38) and Lemma 3.3, one knows $\|\boldsymbol{u}\|_2 \le u_{\max}$.

LMIs (3.29) and (3.30) can guarantee the control input constraint. For $\boldsymbol{\zeta}(0)$ is a twelve-dimensional column vector, the number of related parameters in solving LMIs is too many which leads to difficulty, and the chosen parameters may lead to feasibility failure for the LMIs in Lemma 3.4. While ensuring control input constraint denoted by inequality (3.20), one can use Eq. (3.6) to solve the problem of input saturation. Thus the controller should satisfy the following three condition simultaneously:

1. Closed-loop system (3.11) has quadratic stability($\boldsymbol{w}(t)=0$);
2. Control input constraint satisfies inequality (3.20);
3. Controlled output $\boldsymbol{z}(t)$ satisfies H_∞ performance constraint.

3.3.2 Controller design under additive perturbation

Theorem 3.1: When controller gain perturbation has form (3.9), for given $\xi > 0, \gamma > 0$, the system (3.5) has quadratic stability under the condition of controller (3.8), controlled output $\boldsymbol{z}(t)$ satisfies H_∞ performance constraint, if there exist symmetric matrices $\boldsymbol{P}_{11} > 0$, $\boldsymbol{Q}_{11} > 0$, matrices $\boldsymbol{Q}_{21}, \boldsymbol{A}, \boldsymbol{B}$ and $\boldsymbol{K}$, such that:

$$\begin{bmatrix} \boldsymbol{G}_{11} & \boldsymbol{G}_{12} & \boldsymbol{M}_1 & \boldsymbol{B}_1\boldsymbol{M}_2 & \boldsymbol{Q}_{11}\boldsymbol{N}_1^T & \boldsymbol{Q}_{21}^T\boldsymbol{N}_2^T & \boldsymbol{Q}_{11}\boldsymbol{C}_2^T & \boldsymbol{B}_2 \\ \boldsymbol{G}_{21} & \boldsymbol{G}_{22} & \boldsymbol{P}_{11}\boldsymbol{M}_1 & \boldsymbol{P}_{11}\boldsymbol{B}_1\boldsymbol{M}_2 & \boldsymbol{N}_1^T & 0 & \boldsymbol{C}_2^T & \boldsymbol{P}_{11}\boldsymbol{B}_2 \\ \boldsymbol{M}_1^T & \boldsymbol{M}_1^T\boldsymbol{P}_{11} & -\xi^{-1}\boldsymbol{I} & 0 & 0 & 0 & 0 & 0 \\ \boldsymbol{M}_2^T\boldsymbol{B}_1^T & \boldsymbol{M}_2^T\boldsymbol{B}_1^T\boldsymbol{P}_{11} & 0 & -\xi^{-1}\boldsymbol{I} & 0 & 0 & 0 & 0 \\ \boldsymbol{N}_1\boldsymbol{Q}_{11} & \boldsymbol{N}_1 & 0 & 0 & -\xi\boldsymbol{I} & 0 & 0 & 0 \\ \boldsymbol{N}_2\boldsymbol{Q}_{21} & 0 & 0 & 0 & 0 & -\xi\boldsymbol{I} & 0 & 0 \\ \boldsymbol{C}_2\boldsymbol{Q}_{11} & \boldsymbol{C}_2 & 0 & 0 & 0 & 0 & -\boldsymbol{I} & 0 \\ \boldsymbol{B}_2^T & \boldsymbol{B}_2^T\boldsymbol{P}_{11} & 0 & 0 & 0 & 0 & 0 & -\gamma^2\boldsymbol{I} \end{bmatrix} < 0 \tag{3.39}$$

where,

$$\begin{aligned} \boldsymbol{G}_{11} &= \boldsymbol{A}\boldsymbol{Q}_{11} + \boldsymbol{Q}_{11}\boldsymbol{A}^T + \boldsymbol{B}_1\boldsymbol{K} + \boldsymbol{K}^T\boldsymbol{B}_1^T \\ \boldsymbol{G}_{12} &= \boldsymbol{A} + \boldsymbol{A}^T \\ \boldsymbol{G}_{21} &= \boldsymbol{A}^T + \boldsymbol{A} \\ \boldsymbol{G}_{22} &= \boldsymbol{P}_{11}\boldsymbol{A} + \boldsymbol{A}^T\boldsymbol{P}_{11} + \boldsymbol{B}\boldsymbol{C}_1 + \boldsymbol{C}_1^T\boldsymbol{B}^T \end{aligned} \tag{3.40}$$

and

$$\begin{aligned} \bar{\boldsymbol{K}} &= \boldsymbol{K}\boldsymbol{Q}_{21} \\ \bar{\boldsymbol{B}} &= \boldsymbol{P}_{12}\boldsymbol{B}_c \\ \bar{\boldsymbol{A}} &= \boldsymbol{P}_{11}\boldsymbol{A}\boldsymbol{Q}_{11} + \bar{\boldsymbol{B}}\boldsymbol{C}_1\boldsymbol{Q}_{11} + \boldsymbol{P}_{11}\boldsymbol{B}_1\bar{\boldsymbol{K}} + \boldsymbol{P}_{12}\boldsymbol{A}_c\boldsymbol{Q}_{21} \end{aligned}$$

When LMIs (3.29), (3.30) and (3.39) are feasible, the theoretical DOFNFC is

$$\begin{aligned} \boldsymbol{A}_c &= \boldsymbol{P}_{12}^{-1}(\bar{\boldsymbol{A}} - \boldsymbol{P}_{11}\boldsymbol{A}\boldsymbol{Q}_{11} - \bar{\boldsymbol{B}}\boldsymbol{C}_1\boldsymbol{Q}_{11} - \boldsymbol{P}_{11}\boldsymbol{B}_1\bar{\boldsymbol{K}})\boldsymbol{Q}_{21}^{-1} \\ \boldsymbol{B}_c &= \boldsymbol{P}_{12}^{-1}\bar{\boldsymbol{B}} \\ \boldsymbol{K} &= \bar{\boldsymbol{K}}\boldsymbol{Q}_{21}^{-1} \end{aligned}$$

Proof: First, the components of $\boldsymbol{P}$ and $\boldsymbol{P}^{-1}$ are introduced,

$$\boldsymbol{P}=\begin{bmatrix}\boldsymbol{P}_{11} & \boldsymbol{P}_{12}\\ \boldsymbol{P}_{12}^{T} & \boldsymbol{P}_{22}\end{bmatrix},\boldsymbol{P}^{-1}=\begin{bmatrix}\boldsymbol{Q}_{11} & \boldsymbol{Q}_{12}\\ \boldsymbol{Q}_{21} & \boldsymbol{Q}_{22}\end{bmatrix}$$

where, $\boldsymbol{P}_{11},\boldsymbol{P}_{22},\boldsymbol{Q}_{11}$ and $\boldsymbol{Q}_{22}$ are symmetric positive definite matrices. For $\boldsymbol{P}\boldsymbol{P}^{-1}=\boldsymbol{I}$, one has

$$\begin{bmatrix}\boldsymbol{P}_{11} & \boldsymbol{P}_{12}\\ \boldsymbol{P}_{12}^{T} & \boldsymbol{P}_{22}\end{bmatrix}\begin{bmatrix}\boldsymbol{Q}_{11}\\ \boldsymbol{Q}_{21}\end{bmatrix}=\begin{bmatrix}\boldsymbol{I}\\ 0\end{bmatrix}$$

that is,

$$\begin{matrix}\boldsymbol{P}_{11}\boldsymbol{Q}_{11}+\boldsymbol{P}_{12}\boldsymbol{Q}_{21}=\boldsymbol{I}, & \boldsymbol{Q}_{11}\boldsymbol{P}_{11}+\boldsymbol{Q}_{21}^{T}\boldsymbol{P}_{12}^{T}=\boldsymbol{I}\\ \boldsymbol{P}_{12}^{T}\boldsymbol{Q}_{11}+\boldsymbol{P}_{22}\boldsymbol{Q}_{21}=0, & \boldsymbol{Q}_{11}\boldsymbol{P}_{12}+\boldsymbol{Q}_{21}^{T}\boldsymbol{P}_{22}=0\end{matrix}$$

Here, controller gain perturbation has form (3.9), so

$$\begin{aligned}&\widehat{\boldsymbol{A}}=\begin{bmatrix}\boldsymbol{A} & \boldsymbol{B}_1\boldsymbol{K}\\ \boldsymbol{B}_c\boldsymbol{C}_1 & \boldsymbol{A}_c\end{bmatrix},\ \widehat{\boldsymbol{B}}=\begin{bmatrix}\boldsymbol{B}_2\\ 0\end{bmatrix},\widehat{\boldsymbol{C}}=\begin{bmatrix}\boldsymbol{C}_2 & 0\end{bmatrix},\\ &\widehat{\boldsymbol{M}}=\begin{bmatrix}\boldsymbol{M}_1 & \boldsymbol{B}_1\boldsymbol{M}_2\\ 0 & 0\end{bmatrix},\widehat{\boldsymbol{N}}=\begin{bmatrix}\boldsymbol{N}_1 & 0\\ 0 & \boldsymbol{N}_2\end{bmatrix}\end{aligned}$$

Combine inequality (3.28), and show it in components, one has

$$\begin{bmatrix}\boldsymbol{A}^{T}\boldsymbol{P}_{11}+\boldsymbol{C}_1^{T}\boldsymbol{B}_c^{T}\boldsymbol{P}_{12}^{T}+\boldsymbol{P}_{11}\boldsymbol{A}+\boldsymbol{P}_{12}\boldsymbol{B}_c\boldsymbol{C}_1 & \boldsymbol{A}^{T}\boldsymbol{P}_{12}+\boldsymbol{C}_1^{T}\boldsymbol{B}_c^{T}\boldsymbol{P}_{22}+\boldsymbol{P}_{11}\boldsymbol{B}_1\boldsymbol{K}+\boldsymbol{P}_{12}\boldsymbol{A}_c & \boldsymbol{P}_{11}\boldsymbol{M}_1 & \boldsymbol{P}_{11}\boldsymbol{B}_1\boldsymbol{M}_2 & \boldsymbol{N}_1^{T} & 0 & \boldsymbol{C}_2^{T} & \boldsymbol{P}_{11}\boldsymbol{B}_2\\ \boldsymbol{K}^{T}\boldsymbol{B}_1^{T}\boldsymbol{P}_{11}+\boldsymbol{A}_c^{T}\boldsymbol{P}_{12}^{T}+\boldsymbol{P}_{12}^{T}\boldsymbol{A}+\boldsymbol{P}_{22}\boldsymbol{B}_c\boldsymbol{C}_1 & \boldsymbol{K}^{T}\boldsymbol{B}_1^{T}\boldsymbol{P}_{12}+\boldsymbol{A}_c^{T}\boldsymbol{P}_{22}+\boldsymbol{P}_{12}^{T}\boldsymbol{B}_1\boldsymbol{K}+\boldsymbol{P}_{22}\boldsymbol{A}_c & \boldsymbol{P}_{12}^{T}\boldsymbol{M}_1 & \boldsymbol{P}_{12}^{T}\boldsymbol{B}_1\boldsymbol{M}_2 & 0 & \boldsymbol{N}_2^{T} & 0 & \boldsymbol{P}_{12}^{T}\boldsymbol{B}_2\\ \boldsymbol{M}_1^{T}\boldsymbol{P}_{11} & \boldsymbol{M}_1^{T}\boldsymbol{P}_{12} & -\xi^{-1}\boldsymbol{I} & 0 & 0 & 0 & 0 & 0\\ \boldsymbol{M}_2^{T}\boldsymbol{B}_1^{T}\boldsymbol{P}_{11} & \boldsymbol{M}_2^{T}\boldsymbol{B}_1^{T}\boldsymbol{P}_{12} & 0 & -\xi^{-1}\boldsymbol{I} & 0 & 0 & 0 & 0\\ \boldsymbol{N}_1 & 0 & 0 & 0 & -\xi\boldsymbol{I} & 0 & 0 & 0\\ 0 & \boldsymbol{N}_2 & 0 & 0 & 0 & -\xi\boldsymbol{I} & 0 & 0\\ \boldsymbol{C}_2 & 0 & 0 & 0 & 0 & 0 & -\boldsymbol{I} & 0\\ \boldsymbol{B}_2^{T}\boldsymbol{P}_{11} & \boldsymbol{B}_2^{T}\boldsymbol{P}_{12} & 0 & 0 & 0 & 0 & 0 & -\gamma^2\boldsymbol{I}\end{bmatrix}<0 \tag{3.41}$$

Let

$$\boldsymbol{S}=\begin{bmatrix}\hat{\boldsymbol{Q}} & 0\\ 0 & \boldsymbol{I}_{6\times 6}\end{bmatrix},\hat{\boldsymbol{Q}}=\begin{bmatrix}\boldsymbol{Q}_{11} & \boldsymbol{I}\\ \boldsymbol{Q}_{21} & 0\end{bmatrix} \tag{3.42}$$

Multiply by $\boldsymbol{S}$ and $\boldsymbol{S}^{T}$ on both sides of inequality (3.42), that is,

$$\boldsymbol{S}^{T}\begin{bmatrix}\boldsymbol{A}^{T}\boldsymbol{P}_{11}+\boldsymbol{C}_1^{T}\boldsymbol{B}_c^{T}\boldsymbol{P}_{12}^{T}+\boldsymbol{P}_{11}\boldsymbol{A}+\boldsymbol{P}_{12}\boldsymbol{B}_c\boldsymbol{C}_1 & \boldsymbol{A}^{T}\boldsymbol{P}_{12}+\boldsymbol{C}_1^{T}\boldsymbol{B}_c^{T}\boldsymbol{P}_{22}+\boldsymbol{P}_{11}\boldsymbol{B}_1\boldsymbol{K}+\boldsymbol{P}_{12}\boldsymbol{A}_c & \boldsymbol{P}_{11}\boldsymbol{M}_1 & \boldsymbol{P}_{11}\boldsymbol{B}_1\boldsymbol{M}_2 & \boldsymbol{N}_1^{T} & 0 & \boldsymbol{C}_2^{T} & \boldsymbol{P}_{11}\boldsymbol{B}_2\\ \boldsymbol{K}^{T}\boldsymbol{B}_1^{T}\boldsymbol{P}_{11}+\boldsymbol{A}_c^{T}\boldsymbol{P}_{12}^{T}+\boldsymbol{P}_{12}^{T}\boldsymbol{A}+\boldsymbol{P}_{22}\boldsymbol{B}_c\boldsymbol{C}_1 & \boldsymbol{K}^{T}\boldsymbol{B}_1^{T}\boldsymbol{P}_{12}+\boldsymbol{A}_c^{T}\boldsymbol{P}_{22}+\boldsymbol{P}_{12}^{T}\boldsymbol{B}_1\boldsymbol{K}+\boldsymbol{P}_{22}\boldsymbol{A}_c & \boldsymbol{P}_{12}^{T}\boldsymbol{M}_1 & \boldsymbol{P}_{12}^{T}\boldsymbol{B}_1\boldsymbol{M}_2 & 0 & \boldsymbol{N}_2^{T} & 0 & \boldsymbol{P}_{12}^{T}\boldsymbol{B}_2\\ \boldsymbol{M}_1^{T}\boldsymbol{P}_{11} & \boldsymbol{M}_1^{T}\boldsymbol{P}_{12} & -\xi^{-1}\boldsymbol{I} & 0 & 0 & 0 & 0 & 0\\ \boldsymbol{M}_2^{T}\boldsymbol{B}_1^{T}\boldsymbol{P}_{11} & \boldsymbol{M}_2^{T}\boldsymbol{B}_1^{T}\boldsymbol{P}_{12} & 0 & -\xi^{-1}\boldsymbol{I} & 0 & 0 & 0 & 0\\ \boldsymbol{N}_1 & 0 & 0 & 0 & -\xi\boldsymbol{I} & 0 & 0 & 0\\ 0 & \boldsymbol{N}_2 & 0 & 0 & 0 & -\xi\boldsymbol{I} & 0 & 0\\ \boldsymbol{C}_2 & 0 & 0 & 0 & 0 & 0 & -\boldsymbol{I} & 0\\ \boldsymbol{B}_2^{T}\boldsymbol{P}_{11} & \boldsymbol{B}_2^{T}\boldsymbol{P}_{12} & 0 & 0 & 0 & 0 & 0 & -\gamma^2\boldsymbol{I}\end{bmatrix}\boldsymbol{S}<0 \tag{3.43}$$

One can get inequality (3.44).

$$\begin{bmatrix} G_{11} & G_{12} & Q_{11}P_{11}M_1+Q_{21}^TP_{12}^TM_1 & Q_{11}P_{11}B_1M_2+Q_{21}^TP_{12}^TB_1^TM_2 & Q_{11}N_1^T & Q_{21}^TN_2^T & Q_{11}C_2^T & Q_{11}P_{11}B_2+Q_{21}^TP_{12}^TB_2 \\ G_{21} & G_{22} & P_{11}M_1 & P_{11}B_1M_2 & N_1^T & 0 & C_2^T & P_{11}B_2 \\ M_1^TP_{11}Q_{11}+M_1^TP_{12}Q_{21} & M_1^TP_{11} & -\xi^{-1}I & 0 & 0 & 0 & 0 & 0 \\ M_2^TB_1^TP_{11}Q_{11}+M_2^TB_1^TP_{12}Q_{21} & M_2^TB_1^TP_{11} & 0 & -\xi^{-1}I & 0 & 0 & 0 & 0 \\ N_1Q_{11} & N_1 & 0 & 0 & -\xi I & 0 & 0 & 0 \\ N_2Q_{21} & 0 & 0 & 0 & 0 & -\xi I & 0 & 0 \\ C_2Q_{11} & C_2 & 0 & 0 & 0 & 0 & -I & 0 \\ B_2^TP_{11}Q_{11}+B_2^TP_{12}Q_{21} & B_2^TP_{11} & 0 & 0 & 0 & 0 & 0 & -\gamma^2I \end{bmatrix} < 0 \tag{3.44}$$

where,

$$\begin{aligned} G_{11} &= Q_{11}A^TP_{11}Q_{11}+Q_{11}C_1^TB_c^TP_{12}^TQ_{11}+Q_{11}P_{11}AQ_{11} \\ &\quad +Q_{11}P_{12}B_cCQ_{11}+Q_{21}^TK^TB_1^TP_{11}Q_{11}+Q_{21}^TA_c^TP_{12}^TQ_{11} \\ &\quad +Q_{21}^TP_{12}^TAQ_{11}+Q_{21}^TP_{22}B_cC_1Q_{11}+Q_{11}A^TP_{12}Q_{21} \\ &\quad +Q_{11}C_1^TB_c^TP_{22}Q_{21}+Q_{11}P_{11}B_1KQ_{21}+Q_{11}P_{12}A_cQ_{21} \\ &\quad +Q_{21}^TK^TB_1^TP_{12}Q_{21}+Q_{21}^TA_c^TP_{22}Q_{21} \\ &\quad +Q_{21}^TP_{12}^TB_1KQ_{21}+Q_{21}^TP_{22}A_cQ_{21} \\ G_{12} &= Q_{11}A^TP_{11}+Q_{11}C_1^TB_c^TP_{12}^T+Q_{11}P_{11}A+Q_{11}P_{12}B_cC_1 \\ &\quad +Q_{21}^TK^TB_1^TP_{11}+Q_{21}^TA_c^TP_{12}^T+Q_{21}^TP_{12}^TA+Q_{21}^TP_{22}B_cC_1 \\ G_{21} &= A^TP_{11}Q_{11}+C_1^TB_c^TP_{12}^TQ_{11}+P_{11}AQ_{11}+P_{12}B_cC_1Q_{11} \\ &\quad +A^TP_{12}Q_{21}+C_1^TB_c^TP_{22}Q_{21}+P_{11}B_1KQ_{21}+P_{12}A_cQ_{21} \\ G_{22} &= A^TP_{11}+C_1^TB_c^TP_{12}^T+P_{11}A+P_{12}B_cC_1 \end{aligned} \tag{3.45}$$

Through the merger of similar items, inequalities (3.44) and (3.45) can be transformed into inequalities (3.46) and (3.47), that is,

$$\begin{bmatrix} G_{11} & G_{12} & (Q_{11}P_{11}+Q_{21}^TP_{12}^T)M_1 & (Q_{11}P_{11}+Q_{21}^TP_{12}^T)B_1M_2 & Q_{11}N_1^T & Q_{21}^TN_2^T & Q_{11}C_2^T & (Q_{11}P_{11}+Q_{21}^TP_{12}^T)B_2 \\ G_{21} & G_{22} & P_{11}M_1 & P_{11}B_1M_2 & N_1^T & 0 & C_2^T & P_{11}B_2 \\ M_1^T(P_{11}Q_{11}+P_{12}Q_{21}) & M_1^TP_{11} & -\xi^{-1}I & 0 & 0 & 0 & 0 & 0 \\ M_2^TB_1^T(P_{11}Q_{11}+P_{12}Q_{21}) & M_2^TB_1^TP_{11} & 0 & -\xi^{-1}I & 0 & 0 & 0 & 0 \\ N_1Q_{11} & N_1 & 0 & 0 & -\xi I & 0 & 0 & 0 \\ N_2Q_{21} & 0 & 0 & 0 & 0 & -\xi I & 0 & 0 \\ C_2Q_{11} & C_2 & 0 & 0 & 0 & 0 & -I & 0 \\ B_2^T(P_{11}Q_{11}+P_{12}Q_{21}) & B_2^TP_{11} & 0 & 0 & 0 & 0 & 0 & -\gamma^2I \end{bmatrix} < 0 \tag{3.46}$$

$$\begin{aligned} G_{11} &= (Q_{11}P_{11}+Q_{21}^TP_{12}^T)AQ_{11}+Q_{11}A^T(P_{11}Q_{11}+P_{12}Q_{21}) \\ &\quad +(Q_{11}P_{11}+Q_{21}^TP_{12}^T)B_1KQ_{21}+Q_{21}^TK^TB_1^T(P_{11}Q_{11}+P_{12}Q_{21}) \\ &\quad +Q_{11}C_1^TB_c^T(P_{12}^TQ_{11}+P_{22}Q_{21})+Q_{21}^TA_c^T(P_{12}^TQ_{11}+P_{22}Q_{21}) \\ &\quad +(Q_{21}^TP_{22}+Q_{11}P_{12})(B_cC_1Q_{11}+A_cQ_{21}) \\ G_{12} &= (Q_{11}P_{11}+Q_{21}^TP_{12}^T)A+Q_{11}A^TP_{11}+Q_{11}C_1^TB_c^TP_{12}^T \\ &\quad +(Q_{11}P_{12}+Q_{21}^TP_{22})B_cC_1+Q_{21}^TK^TB_1^TP_{11}+Q_{21}^TA_c^TP_{12}^T \\ G_{21} &= A^T(P_{11}Q_{11}+P_{12}Q_{21})+C_1^TB_c^T(P_{12}^TQ_{11}+P_{22}Q_{21}) \\ &\quad +P_{11}AQ_{11}+P_{12}B_cC_1Q_{11}+P_{11}B_1KQ_{21}+P_{12}A_cQ_{21} \\ G_{22} &= P_{11}A+A^TP_{11}+C_1^TB_c^TP_{12}^T+P_{12}B_cC_1 \end{aligned} \tag{3.47}$$

For

$$\begin{array}{ll} \boldsymbol{P}_{11}\boldsymbol{Q}_{11} + \boldsymbol{P}_{12}\boldsymbol{Q}_{21} = \boldsymbol{I}, & \boldsymbol{Q}_{11}\boldsymbol{P}_{11} + \boldsymbol{Q}_{21}^T\boldsymbol{P}_{12}^T = \boldsymbol{I} \\ \boldsymbol{P}_{12}^T\boldsymbol{Q}_{11} + \boldsymbol{P}_{22}\boldsymbol{Q}_{21} = 0, & \boldsymbol{Q}_{11}\boldsymbol{P}_{12} + \boldsymbol{Q}_{21}^T\boldsymbol{P}_{22} = 0 \end{array}$$

Let

$$\bar{\boldsymbol{K}} = \boldsymbol{K}\boldsymbol{Q}_{21}\ \bar{\boldsymbol{B}} = \boldsymbol{P}_{12}\boldsymbol{B}_c,$$

$$\bar{\boldsymbol{A}} = \boldsymbol{P}_{11}\boldsymbol{A}\boldsymbol{Q}_{11} + \bar{\boldsymbol{B}}\boldsymbol{C}_1\boldsymbol{Q}_{11} + \boldsymbol{P}_{11}\boldsymbol{B}_1\bar{\boldsymbol{K}} + \boldsymbol{P}_{12}\boldsymbol{A}_c\boldsymbol{Q}_{21}$$

Then, inequalities (3.46) and (3.47) can be transformed into inequalities (3.39) and (3.40).

When $\Delta\boldsymbol{K} = 0$, Theorem 3.1 becomes SOFHC problem with LMI form:

$$\begin{bmatrix} \boldsymbol{A}\boldsymbol{Q}_{11} + \boldsymbol{Q}_{11}\boldsymbol{A}^T + \boldsymbol{B}_1\bar{\boldsymbol{K}} + \bar{\boldsymbol{K}}^T\boldsymbol{B}_1^T & \boldsymbol{A} + \bar{\boldsymbol{A}}^T & \boldsymbol{M}_1 & \boldsymbol{Q}_{11}\boldsymbol{N}_1^T & \boldsymbol{Q}_{11}\boldsymbol{C}_2^T & \boldsymbol{B}_2 \\ \boldsymbol{A}^T + \bar{\boldsymbol{A}} & \boldsymbol{P}_{11}\boldsymbol{A} + \boldsymbol{A}^T\boldsymbol{P}_{11} + \bar{\boldsymbol{B}}\boldsymbol{C}_1 + \boldsymbol{C}_1^T\bar{\boldsymbol{B}}^T & \boldsymbol{P}_{11}\boldsymbol{M}_1 & \boldsymbol{N}_1^T & \boldsymbol{C}_2^T & \boldsymbol{P}_{11}\boldsymbol{B}_2 \\ \boldsymbol{M}_1^T & \boldsymbol{M}_1^T\boldsymbol{P}_{11} & -\xi^{-1}\boldsymbol{I} & 0 & 0 & 0 \\ \boldsymbol{N}_1\boldsymbol{Q}_{11} & \boldsymbol{N}_1 & 0 & -\xi\boldsymbol{I} & 0 & 0 \\ \boldsymbol{C}_2\boldsymbol{Q}_{11} & \boldsymbol{C}_2 & 0 & 0 & -\boldsymbol{I} & 0 \\ \boldsymbol{B}_2^T & \boldsymbol{B}_2^T\boldsymbol{P}_{11} & 0 & 0 & 0 & -\gamma^2\boldsymbol{I} \end{bmatrix} < 0 \tag{3.48}$$

When LMIs (3.29), (3.30) and (3.48) are feasible, one can obtain the SOFHC:

$$\begin{aligned} \boldsymbol{A}_c &= \boldsymbol{P}_{12}^{-1}\left(\bar{\boldsymbol{A}} - \boldsymbol{P}_{11}\boldsymbol{A}\boldsymbol{Q}_{11} - \bar{\boldsymbol{B}}\boldsymbol{C}_1\boldsymbol{Q}_{11} - \boldsymbol{P}_{11}\boldsymbol{B}_1\bar{\boldsymbol{K}}\right)\boldsymbol{Q}_{21}^{-1} \\ \boldsymbol{B}_c &= \boldsymbol{P}_{12}^{-1}\bar{\boldsymbol{B}} \\ \boldsymbol{K} &= \bar{\boldsymbol{K}}\boldsymbol{Q}_{21}^{-1} \end{aligned}$$

where, $\boldsymbol{P}_{12}$ and $\boldsymbol{Q}_{21}$ are arbitrary invertible matrices satisfying $\boldsymbol{P}_{11}\boldsymbol{Q}_{11} + \boldsymbol{P}_{12}\boldsymbol{Q}_{21} = \boldsymbol{I}$.

3.3.3 Controller design under multiplicative perturbation

Theorem 3.2: When controller gain perturbation has form (3.10), for given $\xi > 0$, $\gamma > 0$, the system (3.5) has quadratic stability under the condition of controller (3.8), controlled output $z(t)$ satisfies H_∞ performance constraint, if there exist symmetric matrices $\boldsymbol{P}_{11} > 0$, $\boldsymbol{Q}_{11} > 0$, matrices $\boldsymbol{Q}_{21}, \bar{\boldsymbol{A}}, \bar{\boldsymbol{B}}$ and $\bar{\boldsymbol{K}}$, such that:

$$\begin{bmatrix} \mathbf{G}_{11} & \mathbf{G}_{12} & \boldsymbol{M}_1 & \boldsymbol{B}_1\boldsymbol{M}_3 & \mathbf{Q}_{11}\boldsymbol{N}_1^T & \bar{\boldsymbol{K}}^T\boldsymbol{N}_3^T & \mathbf{Q}_{11}\boldsymbol{C}_2^T & \boldsymbol{B}_2 \\ \mathbf{G}_{21} & \mathbf{G}_{22} & \boldsymbol{P}_{11}\boldsymbol{M}_1 & \boldsymbol{P}_{11}\boldsymbol{B}_1\boldsymbol{M}_3 & \boldsymbol{N}_1^T & 0 & \boldsymbol{C}_2^T & \boldsymbol{P}_{11}\boldsymbol{B}_2 \\ \boldsymbol{M}_1^T & \boldsymbol{M}_1^T\boldsymbol{P}_{11} & -\xi^{-1}\boldsymbol{I} & 0 & 0 & 0 & 0 & 0 \\ \boldsymbol{M}_3^T\boldsymbol{B}_1^T & \boldsymbol{M}_3^T\boldsymbol{B}_1^T\boldsymbol{P}_{11} & 0 & -\xi^{-1}\boldsymbol{I} & 0 & 0 & 0 & 0 \\ \boldsymbol{N}_1\mathbf{Q}_{11} & \boldsymbol{N}_1 & 0 & 0 & -\xi\boldsymbol{I} & 0 & 0 & 0 \\ \boldsymbol{N}_3\bar{\boldsymbol{K}}^T & 0 & 0 & 0 & 0 & -\xi\boldsymbol{I} & 0 & 0 \\ \boldsymbol{C}_2\mathbf{Q}_{11} & \boldsymbol{C}_2 & 0 & 0 & 0 & 0 & -\boldsymbol{I} & 0 \\ \boldsymbol{B}_2^T & \boldsymbol{B}_2^T\boldsymbol{P}_{11} & 0 & 0 & 0 & 0 & 0 & -\gamma^2\boldsymbol{I} \end{bmatrix} < 0 \tag{3.49}$$

where,

$$\begin{aligned} \mathbf{G}_{11} &= \boldsymbol{A}\mathbf{Q}_{11} + \mathbf{Q}_{11}\boldsymbol{A}^T + \boldsymbol{B}_1\bar{\boldsymbol{K}} + \bar{\boldsymbol{K}}^T\boldsymbol{B}_1^T \\ \mathbf{G}_{12} &= \boldsymbol{A} + \bar{\boldsymbol{A}}^T \\ \mathbf{G}_{21} &= \boldsymbol{A}^T + \bar{\boldsymbol{A}} \\ \mathbf{G}_{22} &= \boldsymbol{P}_{11}\boldsymbol{A} + \boldsymbol{A}^T\boldsymbol{P}_{11} + \bar{\boldsymbol{B}}\boldsymbol{C}_1 + \boldsymbol{C}_1^T\bar{\boldsymbol{B}}^T \end{aligned} \tag{3.50}$$

and

$$\begin{aligned} \bar{\boldsymbol{K}} &= \boldsymbol{K}\mathbf{Q}_{21} \\ \bar{\boldsymbol{B}} &= \boldsymbol{P}_{12}\boldsymbol{B}_c \\ \bar{\boldsymbol{A}} &= \boldsymbol{P}_{11}\boldsymbol{A}\mathbf{Q}_{11} + \bar{\boldsymbol{B}}\boldsymbol{C}_1\mathbf{Q}_{11} + \boldsymbol{P}_{11}\boldsymbol{B}_1\bar{\boldsymbol{K}} + \boldsymbol{P}_{12}\boldsymbol{A}_c\mathbf{Q}_{21} \end{aligned}$$

When LMIs (3.29), (3.30) and (3.49) are feasible, the theoretical DOFNFC is:

$$\begin{aligned} \boldsymbol{A}_c &= \boldsymbol{P}_{12}^{-1}\left(\bar{\boldsymbol{A}} - \boldsymbol{P}_{11}\boldsymbol{A}\mathbf{Q}_{11} - \bar{\boldsymbol{B}}\boldsymbol{C}_1\mathbf{Q}_{11} - \boldsymbol{P}_{11}\boldsymbol{B}_1\bar{\boldsymbol{K}}\right)\mathbf{Q}_{21}^{-1} \\ \boldsymbol{B}_c &= \boldsymbol{P}_{12}^{-1}\bar{\boldsymbol{B}} \\ \boldsymbol{K} &= \bar{\boldsymbol{K}}\mathbf{Q}_{21}^{-1} \end{aligned}$$

where, $\boldsymbol{P}_{12}$ and $\mathbf{Q}_{21}$ are arbitrary invertible matrices satisfying $\boldsymbol{P}_{11}\mathbf{Q}_{11} + \boldsymbol{P}_{12}\mathbf{Q}_{21} = \boldsymbol{I}$.

Proof: Here, ΔK means multiplicative gain perturbation,

$$\widehat{\boldsymbol{M}} = \begin{bmatrix} \boldsymbol{M}_1 & \boldsymbol{B}_1\boldsymbol{M}_3 \\ 0 & 0 \end{bmatrix}, \quad \widehat{\boldsymbol{N}} = \begin{bmatrix} \boldsymbol{N}_1 & 0 \\ 0 & \boldsymbol{N}_3\boldsymbol{K} \end{bmatrix}$$

When $\Delta\boldsymbol{K}$ means additive gain perturbation, one has

$$\widehat{\boldsymbol{M}} = \begin{bmatrix} \boldsymbol{M}_1 & \boldsymbol{B}_1\boldsymbol{M}_2 \\ 0 & 0 \end{bmatrix}, \quad \widehat{\boldsymbol{N}} = \begin{bmatrix} \boldsymbol{N}_1 & 0 \\ 0 & \boldsymbol{N}_2 \end{bmatrix}$$

Theorem 3.1 and Theorem 3.2 are derived from Lemma 3.4, the difference is only the form of $\widehat{\boldsymbol{M}}$ and $\widehat{\boldsymbol{N}}$, and the other parameters are the same, so the proving process is similar to that of Theorem 3.1, only $\boldsymbol{M}_2 \to \boldsymbol{M}_3, \boldsymbol{N}_2 \to \boldsymbol{N}_3\boldsymbol{K}$ should be done.

3.3.4 Controller design under coexisting additive and multiplicative perturbations

Due to the unexpected variations in the environment, limitations of measurement technique, and aging of measuring equipment, there may appear measurement errors $\boldsymbol{v}(t) \in L_2[0, \infty)$ between the system outputs and measurement outputs, especially for electronic measurements. In this case, Eq. (3.11) becomes

$$\begin{cases} \dot{\zeta}(t) = (\widehat{\boldsymbol{A}} + \Delta\widehat{\boldsymbol{A}}(t))\zeta(t) + \widehat{\boldsymbol{B}}\boldsymbol{w}(t) \\ \boldsymbol{z}(t) = \widehat{\boldsymbol{C}}\zeta(t) + \boldsymbol{v}(t) \end{cases} \tag{3.51}$$

To improve tolerance to uncertainties in the controller, the nonfragility of the controller has been taken into account as $\Delta\boldsymbol{K}$ represents possible controller gain perturbation. Introducing a scalar variable $\bar{\rho}$ such that increasing $\bar{\rho}$ will lead to system performance degradation, which may represent the maximum allowed perturbation parameter. In this case, the controller is designed as

$$\begin{cases} \dot{\widehat{\boldsymbol{x}}}(t) = \boldsymbol{A}_c\widehat{\boldsymbol{x}}(t) + \boldsymbol{B}_c\widehat{\boldsymbol{y}}(t) \\ \boldsymbol{u}(t) = -(\boldsymbol{K} + \bar{\rho}\Delta\boldsymbol{K})\widehat{\boldsymbol{x}}(t) \end{cases} \tag{3.52}$$

Then, one has

$$\widehat{\boldsymbol{A}} = \begin{bmatrix} \boldsymbol{A} & -\boldsymbol{B}_1\boldsymbol{K} \\ \boldsymbol{B}_c\boldsymbol{C}_1 & \boldsymbol{A}_c \end{bmatrix},\ \Delta\widehat{\boldsymbol{A}}(t) = \begin{bmatrix} \Delta\boldsymbol{A}_\rho(t) & -\bar{\rho}\boldsymbol{B}_1\Delta\boldsymbol{K} \\ 0 & 0 \end{bmatrix},\ \widehat{\boldsymbol{B}} = \begin{bmatrix} \boldsymbol{B}_2 \\ 0 \end{bmatrix}, \widehat{\boldsymbol{C}} = \begin{bmatrix} \boldsymbol{C}_2 & 0 \end{bmatrix}$$

When the additive perturbation $\Delta\boldsymbol{K}_a$ and multiplicative perturbation $\Delta\boldsymbol{K}_m$ happen together, one has

$$\begin{aligned}
\Delta\widehat{\boldsymbol{A}}(t) &= \begin{bmatrix} \Delta\boldsymbol{A}_P(t) & -\bar{\rho}\boldsymbol{B}_1(\Delta\boldsymbol{K}_a + \Delta\boldsymbol{K}_m) \\ 0 & 0 \end{bmatrix} \\
&= \begin{bmatrix} 0.5\Delta\boldsymbol{A}_P(t) & -\bar{\rho}\boldsymbol{B}_1\Delta\boldsymbol{K}_a \\ 0 & 0 \end{bmatrix} + \begin{bmatrix} 0.5\Delta\boldsymbol{A}_P(t) & -\bar{\rho}\boldsymbol{B}_1\Delta\boldsymbol{K}_m \\ 0 & 0 \end{bmatrix} \\
&= \begin{bmatrix} 0.5\boldsymbol{M}_1\boldsymbol{F}_1(t)\boldsymbol{N}_1 & -\bar{\rho}\boldsymbol{B}_1\boldsymbol{M}_2\boldsymbol{F}_2(t)\boldsymbol{N}_2 \\ 0 & 0 \end{bmatrix} + \begin{bmatrix} 0.5\boldsymbol{M}_1\boldsymbol{F}_1(t)\boldsymbol{N}_1 & -\bar{\rho}\boldsymbol{B}_1\boldsymbol{M}_3\boldsymbol{F}_3(t)\boldsymbol{N}_3\boldsymbol{K} \\ 0 & 0 \end{bmatrix} \\
&= \begin{bmatrix} 0.5\boldsymbol{M}_1 & -\bar{\rho}\boldsymbol{B}_1\boldsymbol{M}_2 \\ 0 & 0 \end{bmatrix}\begin{bmatrix} \boldsymbol{F}_1(t) & 0 \\ 0 & \boldsymbol{F}_2(t) \end{bmatrix}\begin{bmatrix} \boldsymbol{N}_1 & 0 \\ 0 & \boldsymbol{N}_2 \end{bmatrix} \\
&\quad + \begin{bmatrix} 0.5\boldsymbol{M}_1 & -\bar{\rho}\boldsymbol{B}_1\boldsymbol{M}_3 \\ 0 & 0 \end{bmatrix}\begin{bmatrix} \boldsymbol{F}_1(t) & 0 \\ 0 & \boldsymbol{F}_3(t) \end{bmatrix}\begin{bmatrix} \boldsymbol{N}_1 & 0 \\ 0 & \boldsymbol{N}_3\boldsymbol{K} \end{bmatrix}
\end{aligned} \tag{3.53}$$

Then, $\Delta\widehat{A}(t)$ has the following form

$$\Delta\widehat{A}(t) = \widehat{M}_1\widehat{F}_1(t)\widehat{N}_1 + \widehat{M}_2\widehat{F}_2(t)\widehat{N}_2 \tag{3.54}$$

According to the second equation of controller (3.52), one has

$$u(t) = -(K+\bar{\rho}\,\Delta K)\widehat{x}(t) = -\left\{\begin{bmatrix} 0_{3\times6} & K \end{bmatrix} + \begin{bmatrix} 0_{3\times6} & \bar{\rho}\,\Delta K \end{bmatrix}\right\}\zeta(t) \tag{3.55}$$

Defining $\widehat{K} = \begin{bmatrix} 0_{3\times6} & K \end{bmatrix}$ and $\Delta\widehat{K} = \begin{bmatrix} 0_{3\times6} & \Delta K \end{bmatrix}$, Eq. (3.55) becomes

$$u(t) = -(\widehat{K}+\bar{\rho}\,\Delta\widehat{K})\zeta(t) \tag{3.56}$$

Before developing the main results, the following Theorem 3.3 is introduced first.

Theorem 3.3: For given positive constants ξ and γ, the closed-loop system (3.51) has quadratic stability ($\hat{w}(t) = 0$), and the output vector $z(t)$ satisfies H_∞ performance constraint, that is, $\|z(t)\|_2^2 < \gamma^2\|\hat{w}(t)\|_2^2$, if there exist symmetric positive matrix P, matrices X and W such that

$$\begin{bmatrix} \widehat{A}^T P + P\widehat{A} & P\widehat{M}_1 & \widehat{N}_1^T & P\widehat{M}_2 & \widehat{N}_2^T & \widehat{C}^T & P\widehat{B} & \widehat{C}^T \\ * & -\xi_1^{-1}I & 0 & 0 & 0 & 0 & 0 & 0 \\ * & * & -\xi_1 I & 0 & 0 & 0 & 0 & 0 \\ * & * & * & -\xi_2^{-1}I & 0 & 0 & 0 & 0 \\ * & * & * & * & -\xi_1 I & 0 & 0 & 0 \\ * & * & * & * & * & -I & 0 & 0 \\ * & * & * & * & * & * & -\gamma^2 I & 0 \\ * & * & * & * & * & * & * & (1-\gamma^2)I \end{bmatrix} < 0 \tag{3.57}$$

Proof: Choose a Lyapunov function candidate as

$$V(t) = \zeta(t)^T P\zeta(t)$$

Taking the first derivative of $V(t)$ with respect to time yields

$$\begin{aligned} \dot{V}(t) &= \dot{\zeta}(t)^T P\zeta(t) + \zeta(t)^T P\dot{\zeta}(t) \\ &= \zeta(t)^T\left[(\widehat{A}+\Delta\widehat{A}(t))^T P + P(\widehat{A}+\Delta\widehat{A}(t))\right]\zeta(t) \\ &= \zeta(t)^T\left[\widehat{A}^T P + P\widehat{A} + P\widehat{M}_1\widehat{F}_1(t)\widehat{N}_1 + \widehat{N}_1^T\widehat{F}(t)_1^T\widehat{M}_1^T P + P\widehat{M}_2\widehat{F}_2(t)\widehat{N}_2 + \widehat{N}_2^T\widehat{F}(t)_2^T\widehat{M}_2^T P\right]\zeta(t) \\ &\le \zeta(t)^T(\widehat{A}^T P + P\widehat{A} + \xi_1 P\widehat{M}_1\widehat{M}_1^T P + \xi_1^{-1}\widehat{N}_1^T\widehat{N}_1 + \xi_2 P\widehat{M}_2\widehat{M}_2^T P + \xi_2^{-1}\widehat{N}_2^T\widehat{N}_2)\zeta(t) \end{aligned}$$

Defining

$$\mathbf{Q}_0 = \widehat{\boldsymbol{A}}^T \boldsymbol{P} + \boldsymbol{P}\widehat{\boldsymbol{A}} + \xi_1 \boldsymbol{P}\widehat{\boldsymbol{M}}_1 \widehat{\boldsymbol{M}}_1^T \boldsymbol{P} + \xi_1^{-1} \widehat{\mathbf{N}}_1^T \widehat{\mathbf{N}}_1 + \xi_2 \boldsymbol{P}\widehat{\boldsymbol{M}}_2 \widehat{\boldsymbol{M}}_2^T \boldsymbol{P} + \xi_2^{-1} \widehat{\mathbf{N}}_2^T \widehat{\mathbf{N}}_2$$

According to Schur complement lemma and inequality (3.57), one has $\mathbf{Q}_0 < 0$. As $\dot{V}(t) \le \lambda_{\max}(\mathbf{Q}_0)\zeta(t)^T\zeta(t)$, where $\lambda_{\max}(\mathbf{Q}_0)$ denotes the maximum eigenvalue of $\mathbf{Q}_0$. Let $\alpha = -\lambda_{\max}(\mathbf{Q}_0) > 0$, one has

$$\dot{V}(t) \le -\alpha \left\| \zeta(t) \right\|_2^2 \tag{3.58}$$

Thus the closed-loop system (3.51) is quadratically stable under controller (3.52).

Step 2: One shows that the output $\boldsymbol{z}(t)$ satisfies H_∞ performance constraint.

To establish the L_2 $[0, \infty)$ norm bound $\gamma^2 \left\| \hat{\boldsymbol{w}}(t) \right\|_2^2$, consider the following functional:

$$J = \int_0^\infty [\boldsymbol{z}(t)^T \boldsymbol{z}(t) - \gamma^2 \hat{\boldsymbol{w}}(t)^T \hat{\boldsymbol{w}}(t)] dt$$

As the closed-loop system has quadratic stability, for arbitrary nonzero $\hat{\boldsymbol{w}}(t) \in L_2[0, \infty)$, under zero initial conditions, one has

$$\begin{aligned}
J &= \int_0^\infty \left[\boldsymbol{z}(t)^T \boldsymbol{z}(t) - \gamma^2 \hat{\boldsymbol{w}}(t)^T \hat{\boldsymbol{w}}(t) + \dot{V}(t)\right] dt - V(\infty) + V(0) \\
&\le \int_0^\infty \left[\zeta(t)^T \widehat{\boldsymbol{C}}^T \widehat{\boldsymbol{C}} \zeta(t) + 2\zeta(t)^T \widehat{\boldsymbol{C}}^T \boldsymbol{v}(t) + \boldsymbol{v}(t)^T \boldsymbol{v}(t) - \gamma^2 (\boldsymbol{w}(t)^T \boldsymbol{w}(t) + \boldsymbol{v}(t)^T \boldsymbol{v}(t)) + \zeta(t)^T \mathbf{Q}_0 \zeta(t) + 2\zeta(t)^T \boldsymbol{P}\widehat{\boldsymbol{B}} \boldsymbol{w}(t)\right] dt \\
&\le \int_0^\infty [\zeta(t)^T \boldsymbol{w}(t)^T \boldsymbol{v}(t)^T] \begin{bmatrix} \mathbf{Q}_0 + \widehat{\boldsymbol{C}}^T \widehat{\boldsymbol{C}} & \boldsymbol{P}\widehat{\boldsymbol{B}} & \widehat{\boldsymbol{C}}^T \\ * & -\gamma^2 \boldsymbol{I} & 0 \\ * & * & (1-\gamma^2)\boldsymbol{I} \end{bmatrix} \begin{bmatrix} \zeta(t) \\ \boldsymbol{w}(t) \\ \boldsymbol{v}(t) \end{bmatrix} dt
\end{aligned}$$

According to Schur complement lemma and inequality (3.57), $J < 0$ holds, that is, $\boldsymbol{z}(t)$ satisfies H_∞ performance constraint.

This completes the proof.

Based on Theorem 3.3, Theorem 3.4 now gives the existence condition of the hybrid nonfragile controller under coexisting additive and multiplicative perturbations.

Theorem 3.4: When the additive and multiplicative perturbations coexist, for given constants $\xi > 0$, $\bar{\rho} > 0$, and $\gamma > 0$, the attitude control system represented by Eq. (3.51) has quadratic stability under controller (3.52), $\boldsymbol{z}(t)$ satisfies H_∞ performance constraint, if there exist symmetric positive definite matrices $\boldsymbol{P}_{11}$, $\mathbf{Q}_{11}$ and matrices $\mathbf{Q}_{21}$, $\overline{\boldsymbol{A}}$, $\overline{\boldsymbol{B}}$ and $\overline{\boldsymbol{K}}$ such that

$$\begin{bmatrix} \boldsymbol{G}_{11} & \boldsymbol{G}_{12} & 0.5\boldsymbol{M}_1 & -\overline{\rho}\boldsymbol{B}_1\boldsymbol{M}_2 & \boldsymbol{Q}_{11}\boldsymbol{N}_1^T & \boldsymbol{Q}_{21}^T\boldsymbol{N}_2^T & 0.5\boldsymbol{M}_1 & -\overline{\rho}\boldsymbol{B}_1\boldsymbol{M}_3 & \boldsymbol{Q}_{11}\boldsymbol{N}_1^T & \overline{\boldsymbol{K}}^T\boldsymbol{N}_3^T & \boldsymbol{Q}_{11}\boldsymbol{C}_2{}^T & \boldsymbol{B}_2 & \boldsymbol{Q}_{11}\boldsymbol{C}_2{}^T \\ * & \boldsymbol{G}_{22} & 0.5\boldsymbol{P}_{11}\boldsymbol{M}_1 & -\overline{\rho}\boldsymbol{P}_{11}\boldsymbol{B}_1\boldsymbol{M}_2 & \boldsymbol{N}_1^T & 0 & 0.5\boldsymbol{P}_{11}\boldsymbol{M}_1 & -\overline{\rho}\boldsymbol{P}_{11}\boldsymbol{B}_1\boldsymbol{M}_3 & \boldsymbol{N}_1^T & 0 & \boldsymbol{C}_2{}^T & \boldsymbol{P}_{11}\boldsymbol{B}_2 & \boldsymbol{C}_2{}^T \\ * & * & -\xi_1{}^{-1}\boldsymbol{I} & 0 & 0 & 0 & 0 & 0 & 0 & 0 & 0 & 0 & 0 \\ * & * & * & -\xi_1{}^{-1}\boldsymbol{I} & 0 & 0 & 0 & 0 & 0 & 0 & 0 & 0 & 0 \\ * & * & * & * & -\xi_1\boldsymbol{I} & 0 & 0 & 0 & 0 & 0 & 0 & 0 & 0 \\ * & * & * & * & * & -\xi_1\boldsymbol{I} & 0 & 0 & 0 & 0 & 0 & 0 & 0 \\ * & * & * & * & * & * & -\xi_2{}^{-1}\boldsymbol{I} & 0 & 0 & 0 & 0 & 0 & 0 \\ * & * & * & * & * & * & * & -\xi_2{}^{-1}\boldsymbol{I} & 0 & 0 & 0 & 0 & 0 \\ * & * & * & * & * & * & * & * & -\xi_2\boldsymbol{I} & 0 & 0 & 0 & 0 \\ * & * & * & * & * & * & * & * & * & -\xi_2\boldsymbol{I} & 0 & 0 & 0 \\ * & * & * & * & * & * & * & * & * & * & -\boldsymbol{I} & 0 & 0 \\ * & * & * & * & * & * & * & * & * & * & * & -\gamma^2\boldsymbol{I} & 0 \\ * & * & * & * & * & * & * & * & * & * & * & * & (1-\gamma^2)\boldsymbol{I} \end{bmatrix} < 0 \tag{3.59}$$

where,

$$\begin{aligned} \boldsymbol{G}_{11} &= \boldsymbol{A}\boldsymbol{Q}_{11} + \boldsymbol{Q}_{11}\boldsymbol{A}^T + \boldsymbol{B}_1\overline{\boldsymbol{K}} + \overline{\boldsymbol{K}}^T\boldsymbol{B}_1{}^T \\ \boldsymbol{G}_{12} &= \boldsymbol{A} + \overline{\boldsymbol{A}}^T \\ \boldsymbol{G}_{22} &= \boldsymbol{P}_{11}\boldsymbol{A} + \boldsymbol{A}^T\boldsymbol{P}_{11} + \overline{\boldsymbol{B}}\boldsymbol{C}_1 + \boldsymbol{C}_1^T\overline{\boldsymbol{B}}^T \end{aligned}$$

and

$$\begin{aligned} \overline{\boldsymbol{K}} &= -\boldsymbol{K}\boldsymbol{Q}_{21} \\ \overline{\boldsymbol{B}} &= \boldsymbol{P}_{12}\boldsymbol{B}_c \\ \overline{\boldsymbol{A}} &= \boldsymbol{P}_{11}\boldsymbol{A}\boldsymbol{Q}_{11} + \overline{\boldsymbol{B}}\boldsymbol{C}_1\boldsymbol{Q}_{11} + \boldsymbol{P}_{11}\boldsymbol{B}_1\overline{\boldsymbol{K}} + \boldsymbol{P}_{12}\boldsymbol{A}_c\boldsymbol{Q}_{21} \end{aligned}$$

When inequality (3.59) is feasible, the related matrices for output feedback controller are

$$\begin{aligned} \boldsymbol{A}_c &= \boldsymbol{P}_{12}^{-1}\left(\overline{\boldsymbol{A}} - \boldsymbol{P}_{11}\boldsymbol{A}\boldsymbol{Q}_{11} - \overline{\boldsymbol{B}}\boldsymbol{C}_1\boldsymbol{Q}_{11} - \boldsymbol{P}_{11}\boldsymbol{B}_1\overline{\boldsymbol{K}}\right)\boldsymbol{Q}_{21}^{-1} \\ \boldsymbol{B}_c &= \boldsymbol{P}_{12}^{-1}\overline{\boldsymbol{B}} \\ \boldsymbol{K} &= -\overline{\boldsymbol{K}}\boldsymbol{Q}_{21}^{-1} \end{aligned}$$

Proof: The component forms of $\boldsymbol{P}$ and $\boldsymbol{P}^{-1}$ are

$$\boldsymbol{P} = \begin{bmatrix} \boldsymbol{P}_{11} & \boldsymbol{P}_{12} \\ * & \boldsymbol{P}_{22} \end{bmatrix}, \boldsymbol{P}^{-1} = \begin{bmatrix} \boldsymbol{Q}_{11} & \boldsymbol{Q}_{12} \\ * & \boldsymbol{Q}_{22} \end{bmatrix}$$

where $\boldsymbol{P}_{11}, \boldsymbol{P}_{22}, \boldsymbol{Q}_{11}$ and $\boldsymbol{Q}_{22}$ are symmetric positive matrices.

For $\boldsymbol{P}\boldsymbol{P}^{-1} = \boldsymbol{I}$, one has

$$\begin{bmatrix} \boldsymbol{P}_{11} & \boldsymbol{P}_{12} \\ * & \boldsymbol{P}_{22} \end{bmatrix}\begin{bmatrix} \boldsymbol{Q}_{11} \\ \boldsymbol{Q}_{21} \end{bmatrix} = \begin{bmatrix} \boldsymbol{I} \\ 0 \end{bmatrix}$$

that is,

$$\begin{aligned} \boldsymbol{P}_{11}\boldsymbol{Q}_{11} + \boldsymbol{P}_{12}\boldsymbol{Q}_{21} &= \boldsymbol{I}, \quad \boldsymbol{Q}_{11}\boldsymbol{P}_{11} + \boldsymbol{Q}_{21}^T\boldsymbol{P}_{12}^T = \boldsymbol{I} \\ \boldsymbol{P}_{12}^T\boldsymbol{Q}_{11} + \boldsymbol{P}_{22}\boldsymbol{Q}_{21} &= 0, \quad \boldsymbol{Q}_{11}\boldsymbol{P}_{12} + \boldsymbol{Q}_{21}^T\boldsymbol{P}_{22} = 0 \end{aligned} \tag{3.60}$$

When the additive and multiplicative perturbations coexist, one has

$$\widehat{A} = \begin{bmatrix} A & -B_1K \\ B_cC_1 & A_c \end{bmatrix},\ \widehat{B} = \begin{bmatrix} B_2 \\ 0 \end{bmatrix},\ \widehat{C} = \begin{bmatrix} C_2 & 0 \end{bmatrix}$$

$$\widehat{M}_1 = \begin{bmatrix} 0.5M_1 & -\overline{\rho}B_1M_2 \\ 0 & 0 \end{bmatrix},\ \widehat{N}_1 = \begin{bmatrix} N_1 & 0 \\ 0 & N_2 \end{bmatrix}$$

$$\widehat{M}_2 = \begin{bmatrix} 0.5M_1 & -\overline{\rho}B_1M_3 \\ 0 & 0 \end{bmatrix},\ \widehat{N}_2 = \begin{bmatrix} N_1 & 0 \\ 0 & N_3K \end{bmatrix}$$

Substituting the above component forms into inequality (3.57) yields

$$\begin{bmatrix}
\overline{S}_{11} & \overline{S}_{12} & 0.5P_{11}M_1 & -\overline{\rho}P_{11}B_1M_2 & N_1^T & 0 & 0.5P_{11}M_1 & -\overline{\rho}P_{11}B_1M_3 & N_1^T & 0 & C_2^T & P_{11}B_2 & C_2^T \\
* & \overline{S}_{22} & 0.5P_{12}^TM_1 & -\overline{\rho}P_{12}^TB_1M_2 & 0 & N_2^T & 0.5P_{12}^TM_1 & -\overline{\rho}P_{12}^TB_1M_3 & 0 & -K^TN_3^T & 0 & P_{12}^TB_2 & 0 \\
* & * & -\xi_1^{-1}I & 0 & 0 & 0 & 0 & 0 & 0 & 0 & 0 & 0 & 0 \\
* & * & * & -\xi_1^{-1}I & 0 & 0 & 0 & 0 & 0 & 0 & 0 & 0 & 0 \\
* & * & * & * & -\xi_1I & 0 & 0 & 0 & 0 & 0 & 0 & 0 & 0 \\
* & * & * & * & * & -\xi_1I & 0 & 0 & 0 & 0 & 0 & 0 & 0 \\
* & * & * & * & * & * & -\xi_2^{-1}I & 0 & 0 & 0 & 0 & 0 & 0 \\
* & * & * & * & * & * & * & -\xi_2^{-1}I & 0 & 0 & 0 & 0 & 0 \\
* & * & * & * & * & * & * & * & -\xi_2I & 0 & 0 & 0 & 0 \\
* & * & * & * & * & * & * & * & * & -\xi_2I & 0 & 0 & 0 \\
* & * & * & * & * & * & * & * & * & * & -I & 0 & 0 \\
* & * & * & * & * & * & * & * & * & * & * & -\gamma^2I & 0 \\
* & * & * & * & * & * & * & * & * & * & * & * & (1-\gamma^2)I
\end{bmatrix} < 0 \tag{3.61}$$

where,

$$\overline{S}_{11} = A^TP_{11} + C_1^TB_c^TP_{12}^T + P_{11}A + P_{12}B_cC_1$$

$$\overline{S}_{12} = A^TP_{12} + C_1^TB_c^TP_{22} - P_{11}B_1K + P_{12}A_c$$

$$\overline{S}_{22} = -K^TB_1^TP_{12} + A_c^TP_{22} - P_{12}^TB_1K + P_{22}A_c$$

Defining $S = \begin{bmatrix} \tilde{Q} & 0_{2\times 11} \\ 0_{11\times 2} & I_{11\times 11} \end{bmatrix}$, where, $0_{2\times 11}$ denotes a block-zero matrix, the same to $0_{11\times 2}$, and $\tilde{Q} = \begin{bmatrix} Q_{11} & I \\ Q_{21} & 0 \end{bmatrix}$.

Pre- and post-multiplying inequality (3.61) by S^T and its transpose respectively, one has

$$\begin{bmatrix}
G_{11} & G_{12} & 0.5G_{13} & G_{14} & Q_{11}N_1^T & Q_{21}^TN_2^T & 0.5G_{13} & G_{18} & Q_{11}N_1^T & -Q_{21}^TK^TN_3^T & Q_{11}C_2^T & G_{112} & Q_{11}C_2^T \\
* & G_{22} & 0.5P_{11}M_1 & -\overline{\rho}P_{11}B_1M_2 & N_1^T & 0 & 0.5P_{11}M_1 & -\overline{\rho}P_{11}B_1M_3 & N_1^T & 0 & C_2^T & P_{11}B_2 & C_2^T \\
* & * & -\xi_1^{-1}I & 0 & 0 & 0 & 0 & 0 & 0 & 0 & 0 & 0 & 0 \\
* & * & * & -\xi_1^{-1}I & 0 & 0 & 0 & 0 & 0 & 0 & 0 & 0 & 0 \\
* & * & * & * & -\xi_1I & 0 & 0 & 0 & 0 & 0 & 0 & 0 & 0 \\
* & * & * & * & * & -\xi_1I & 0 & 0 & 0 & 0 & 0 & 0 & 0 \\
* & * & * & * & * & * & -\xi_2^{-1}I & 0 & 0 & 0 & 0 & 0 & 0 \\
* & * & * & * & * & * & * & -\xi_2^{-1}I & 0 & 0 & 0 & 0 & 0 \\
* & * & * & * & * & * & * & * & -\xi_2I & 0 & 0 & 0 & 0 \\
* & * & * & * & * & * & * & * & * & -\xi_2I & 0 & 0 & 0 \\
* & * & * & * & * & * & * & * & * & * & -I & 0 & 0 \\
* & * & * & * & * & * & * & * & * & * & * & -\gamma^2I & 0 \\
* & * & * & * & * & * & * & * & * & * & * & * & (1-\gamma^2)I
\end{bmatrix} < 0 \tag{3.62}$$

where,

$$\begin{aligned}
\mathbf{G}_{11} = {} & \mathbf{Q}_{11}\boldsymbol{A}^T\boldsymbol{P}_{11}\mathbf{Q}_{11} + \mathbf{Q}_{11}\boldsymbol{C}_1^T\boldsymbol{B}_c^T\boldsymbol{P}_{12}^T\mathbf{Q}_{11} + \mathbf{Q}_{11}\boldsymbol{P}_{11}\boldsymbol{A}\mathbf{Q}_{11} \\
& +\mathbf{Q}_{11}\boldsymbol{P}_{12}\boldsymbol{B}_c\boldsymbol{C}_1\mathbf{Q}_{11} - \mathbf{Q}_{21}^T\boldsymbol{K}^T\boldsymbol{B}_1^T\boldsymbol{P}_{11}\mathbf{Q}_{11} + \mathbf{Q}_{21}^T\boldsymbol{A}_c^T\boldsymbol{P}_{12}^T\mathbf{Q}_{11} \\
& +\mathbf{Q}_{21}^T\boldsymbol{P}_{12}^T\boldsymbol{A}\mathbf{Q}_{11} + \mathbf{Q}_{21}^T\boldsymbol{P}_{22}\boldsymbol{B}_c\boldsymbol{C}_1\mathbf{Q}_{11} + \mathbf{Q}_{11}\boldsymbol{A}^T\boldsymbol{P}_{12}\mathbf{Q}_{21} \\
& +\mathbf{Q}_{11}\boldsymbol{C}^T\boldsymbol{B}_c^T\boldsymbol{P}_{22}\mathbf{Q}_{21} - \mathbf{Q}_{11}\boldsymbol{P}_{11}\boldsymbol{B}_1\boldsymbol{K}\mathbf{Q}_{21} + \mathbf{Q}_{11}\boldsymbol{P}_{12}\boldsymbol{A}_c\mathbf{Q}_{21} \\
& - \mathbf{Q}_{21}^T\boldsymbol{K}^T\boldsymbol{B}_1^T\boldsymbol{P}_{12}\mathbf{Q}_{21} + \mathbf{Q}_{21}^T\boldsymbol{A}_c^T\boldsymbol{P}_{22}\mathbf{Q}_{21} \\
& - \mathbf{Q}_{21}^T\boldsymbol{P}_{12}^T\boldsymbol{B}_1\boldsymbol{K}\mathbf{Q}_{21} + \mathbf{Q}_{21}^T\boldsymbol{P}_{22}\boldsymbol{A}_c\mathbf{Q}_{21} \\
\mathbf{G}_{12} = {} & \mathbf{Q}_{11}\boldsymbol{A}^T\boldsymbol{P}_{11} + \mathbf{Q}_{11}\boldsymbol{C}_1^T\boldsymbol{B}_c^T\boldsymbol{P}_{12}^T + \mathbf{Q}_{11}\boldsymbol{P}_{11}\boldsymbol{A} + \mathbf{Q}_{11}\boldsymbol{P}_{12}\boldsymbol{B}_c\boldsymbol{C}_1 \\
& - \mathbf{Q}_{21}^T\boldsymbol{K}^T\boldsymbol{B}_1^T\boldsymbol{P}_{11} + \mathbf{Q}_{21}^T\boldsymbol{A}_c^T\boldsymbol{P}_{12}^T + \mathbf{Q}_{21}^T\boldsymbol{P}_{12}^T\boldsymbol{A} + \mathbf{Q}_{21}^T\boldsymbol{P}_{22}\boldsymbol{B}_c\boldsymbol{C}_1
\end{aligned}$$

$$\begin{aligned}
\mathbf{G}_{13} &= \mathbf{Q}_{11}\boldsymbol{P}_{11}\boldsymbol{M}_1 + \mathbf{Q}_{21}^T\boldsymbol{P}_{12}^T\boldsymbol{M}_1 \\
\mathbf{G}_{14} &= -\overline{\rho}\mathbf{Q}_{11}\boldsymbol{P}_{11}\boldsymbol{B}_1\boldsymbol{M}_2 - \overline{\rho}\mathbf{Q}_{21}^T\boldsymbol{P}_{12}^T\boldsymbol{B}_1\boldsymbol{M}_2 \\
\mathbf{G}_{18} &= -\overline{\rho}\mathbf{Q}_{11}\boldsymbol{P}_{11}\boldsymbol{B}_1\boldsymbol{M}_3 - \overline{\rho}\mathbf{Q}_{21}^T\boldsymbol{P}_{12}^T\boldsymbol{B}_1\boldsymbol{M}_3 \\
\mathbf{G}_{112} &= \mathbf{Q}_{11}\boldsymbol{P}_{11}\boldsymbol{B}_2 + \mathbf{Q}_{21}^T\boldsymbol{P}_{12}^T\boldsymbol{B}_2 \\
\mathbf{G}_{22} &= \boldsymbol{A}^T\boldsymbol{P}_{11} + \boldsymbol{C}_1^T\boldsymbol{B}_c^T\boldsymbol{P}_{12}^T + \boldsymbol{P}_{11}\boldsymbol{A} + \boldsymbol{P}_{12}\boldsymbol{B}_c\boldsymbol{C}_1
\end{aligned}$$

Defining $\overline{\boldsymbol{A}} = \boldsymbol{P}_{11}\boldsymbol{A}\mathbf{Q}_{11} + \overline{\boldsymbol{B}}\boldsymbol{C}_1\mathbf{Q}_{11} + \boldsymbol{P}_{11}\boldsymbol{B}_1\overline{\boldsymbol{K}} + \boldsymbol{P}_{12}\boldsymbol{A}_c\mathbf{Q}_{21}$, $\overline{\boldsymbol{K}} = -\boldsymbol{K}\mathbf{Q}_{21}$, $\overline{\boldsymbol{B}} = \boldsymbol{P}_{12}\boldsymbol{B}_c$, according to Eq. (3.60), inequality (3.62) is reduced to

$$\begin{bmatrix}
\mathbf{G}_{11} & \mathbf{G}_{12} & 0.5\boldsymbol{M}_1 & -\overline{\rho}\boldsymbol{B}_1\boldsymbol{M}_2 & \mathbf{Q}_{11}\boldsymbol{N}_1^T & \mathbf{Q}_{21}^T\boldsymbol{N}_2^T & 0.5\boldsymbol{M}_1 & -\overline{\rho}\boldsymbol{B}_1\boldsymbol{M}_3 & \mathbf{Q}_{11}\boldsymbol{N}_1^T & -\boldsymbol{K}^T\boldsymbol{N}_3^T & \mathbf{Q}_{11}\boldsymbol{C}_2^T & \boldsymbol{B}_2 & \mathbf{Q}_{11}\boldsymbol{C}_2^T \\
* & \mathbf{G}_{22} & 0.5\boldsymbol{P}_{11}\boldsymbol{M}_1 & -\overline{\rho}\boldsymbol{P}_{11}\boldsymbol{B}_1\boldsymbol{M}_2 & \boldsymbol{N}_1^T & 0 & 0.5\boldsymbol{P}_{11}\boldsymbol{M}_1 & -\overline{\rho}\boldsymbol{P}_{11}\boldsymbol{B}_1\boldsymbol{M}_3 & \boldsymbol{N}_1^T & 0 & \boldsymbol{C}_2^T & \boldsymbol{P}_{11}\boldsymbol{B}_2 & \boldsymbol{C}_2^T \\
* & * & -\xi_1{}^{-1}\boldsymbol{I} & 0 & 0 & 0 & 0 & 0 & 0 & 0 & 0 & 0 & 0 \\
* & * & * & -\xi_1{}^{-1}\boldsymbol{I} & 0 & 0 & 0 & 0 & 0 & 0 & 0 & 0 & 0 \\
* & * & * & * & -\xi_1\boldsymbol{I} & 0 & 0 & 0 & 0 & 0 & 0 & 0 & 0 \\
* & * & * & * & * & -\xi_1\boldsymbol{I} & 0 & 0 & 0 & 0 & 0 & 0 & 0 \\
* & * & * & * & * & * & -\xi_2{}^{-1}\boldsymbol{I} & 0 & 0 & 0 & 0 & 0 & 0 \\
* & * & * & * & * & * & * & -\xi_2{}^{-1}\boldsymbol{I} & 0 & 0 & 0 & 0 & 0 \\
* & * & * & * & * & * & * & * & -\xi_2\boldsymbol{I} & 0 & 0 & 0 & 0 \\
* & * & * & * & * & * & * & * & * & -\xi_2\boldsymbol{I} & 0 & 0 & 0 \\
* & * & * & * & * & * & * & * & * & * & -\boldsymbol{I} & 0 & 0 \\
* & * & * & * & * & * & * & * & * & * & * & -\gamma^2\boldsymbol{I} & 0 \\
* & * & * & * & * & * & * & * & * & * & * & * & (1-\gamma^2)\boldsymbol{I}
\end{bmatrix} < 0$$

where,

$$\mathbf{G}_{11} = \boldsymbol{A}\mathbf{Q}_{11} + \mathbf{Q}_{11}\boldsymbol{A}^T + \boldsymbol{B}_1\overline{\boldsymbol{K}} + \overline{\boldsymbol{K}}^T\boldsymbol{B}_1^T$$

$$\mathbf{G}_{12} = \boldsymbol{A} + \overline{\boldsymbol{A}}^T$$

$$\mathbf{G}_{22} = \boldsymbol{P}_{11}\boldsymbol{A} + \boldsymbol{A}^T\boldsymbol{P}_{11} + \overline{\boldsymbol{B}}\boldsymbol{C}_1 + \boldsymbol{C}_1^T\overline{\boldsymbol{B}}^T$$

This completes the proof.

3.4 Simulation test

In this section, the effectiveness of the proposed DOFNFC under additive and multiplicative perturbations is illustrated, respectively, with simulations on a spacecraft attitude system. For comparison, the simulation results of SOFHC are also considered.

The moment of inertia is assumed as $I_x = 20\ \text{kg m}^2$, $I_y = 18\ \text{kg m}^2$, $I_z = 15\ \text{kg m}^2$, and the orbit height is 300 km. Attitude angles and angular velocities are measured with sensors and gyros. Choose the initial state $\boldsymbol{x}(0) = [0.08\ \text{rad},\ 0.06\ \text{rad},\ -0.06\ \text{rad},\ 0.01\ \text{rad s}^{-1},\ 0.01\ \text{rad s}^{-1},\ 0.01\ \text{rad s}^{-1}]^T$, the upper saturation boundary of control input is $u_{m1} = u_{m2} = u_{m3} = = 0.5\ \text{Nm}$, the expected state is zero. The disturbance torques are white noise modeled as

$$\boldsymbol{w}(t) = \begin{bmatrix} 5\cos(\omega_0 t) \\ 5\cos(\omega_0 t + \pi/4) \\ 5\cos(\omega_0 t + \pi/3) \end{bmatrix} \times 10^{-4}\text{Nm}$$

where ω_0 means orbital angular velocity, and the initial value of controller variable $\widehat{\boldsymbol{x}}(t)$ is set as $\widehat{\boldsymbol{x}}(0) = 0.5\boldsymbol{x}(0)$.

$$\begin{aligned} \boldsymbol{M}_1 &= \begin{bmatrix} 1 & 1 & 1 & 1 & 1 & 1 \end{bmatrix}^T, \boldsymbol{F}_1(t) = \sin(100\omega_0 t), \\ \boldsymbol{N}_1 &= 0.01 \times \begin{bmatrix} 1 & 1 & 1 & 1 & 1 & 1 \end{bmatrix}, \xi = 0.025, \gamma = 0.1 \end{aligned}$$

Aiming at two types of gain perturbations, the simulations were made, respectively.

3.4.1 Simulation results under additive perturbation

Choose $\boldsymbol{M}_2 = \begin{bmatrix} 1 & 1 & 1 \end{bmatrix}^T, \boldsymbol{N}_2 = 0.18 \times \begin{bmatrix} 1 & 1 & 1 & 1 & 1 & 1 \end{bmatrix}$, $\boldsymbol{F}_2(t) = \sin(100\omega_0 t + \pi/4)$, with Theorem 3.1, using LMI toolbox yields the related matrices of controller as follows

$$\boldsymbol{A}_c = 1.0 \times 10^5 \times$$
$$\begin{bmatrix} -1.5366 & -1.5366 & -1.5366 & -1.5402 & -1.5350 & -1.5341 \\ -1.5951 & -1.5951 & -1.5951 & -1.5939 & -1.5984 & -1.5926 \\ -1.7061 & -1.7061 & -1.7061 & -1.5939 & -1.5984 & -1.5926 \\ 1.6555 & 1.6555 & 1.6555 & 1.6535 & 1.6566 & 1.6560 \\ 1.6213 & 1.6213 & 1.6213 & 1.6228 & 1.6189 & 1.6219 \\ 1.5564 & 1.5564 & 1.5564 & 1.5580 & 1.5577 & 1.5532 \end{bmatrix}$$

$$B_c = 1.0 \times 10^5 \times \begin{bmatrix} 5.8491 & 5.0452 & 4.9028 & 1.8500 & 0.7541 & 0.5453 \\ 5.1171 & 5.7889 & 4.8855 & 0.8820 & 1.7938 & 0.5444 \\ 5.0836 & 4.9958 & 5.7037 & 0.8802 & 0.7523 & 1.7293 \\ -5.1707 & -5.3554 & -5.2601 & -0.9388 & -1.2456 & -1.1133 \\ -5.4295 & -5.0898 & -5.2703 & -1.3356 & -0.8068 & -1.1139 \\ -5.4491 & -5.3842 & -4.9610 & -1.3367 & -1.2466 & -0.5903 \end{bmatrix}$$

$$K = 1.0 \times 10^3 \times \begin{bmatrix} -2.5563 & -2.5563 & -2.5563 & -2.5665 & -2.5515 & -2.5510 \\ -2.3878 & -2.3878 & -2.3878 & -2.3837 & -2.3967 & -2.3830 \\ -2.1154 & -2.1154 & -2.1154 & -2.1120 & -2.1118 & -2.1225 \end{bmatrix}$$

Then, the corresponding DOFNFC can be obtained with its form (3.8). Apply it in the spacecraft attitude control system with model parameter uncertainty and additive gain perturbation, and with saturation processing, the corresponding attitude angle, angular velocity and control input are shown in Figs. 3.1–3.3 from which one can see that when considering model parameter uncertainty and controller gain perturbation simultaneously, the designed DOFNFC can make attitude stabilization of

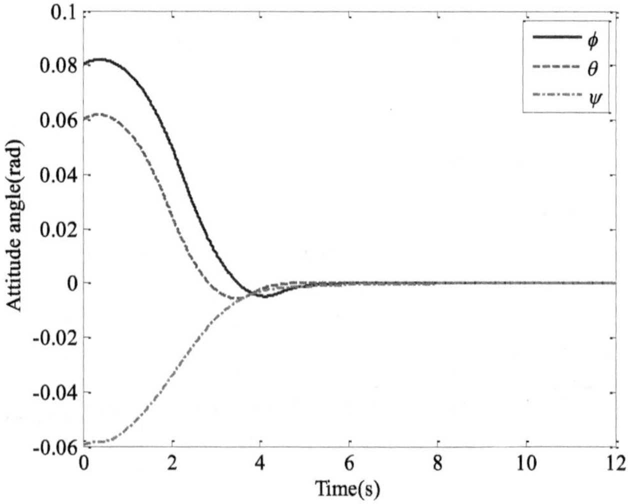

Figure 3.1 Attitude angle using dynamic output feedback nonfragile control.

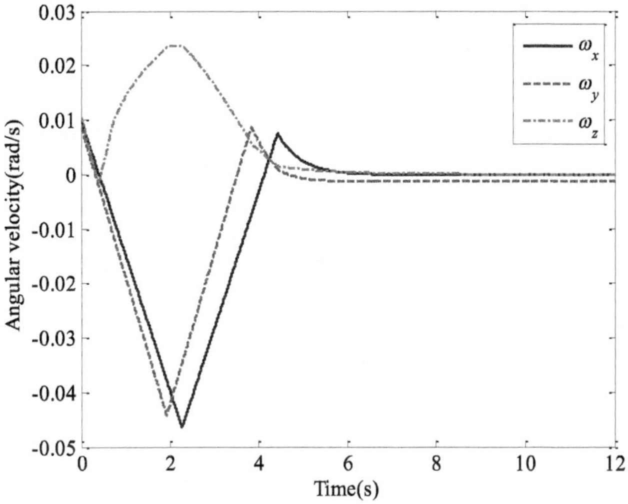

Figure 3.2 Angular velocity using dynamic output feedback nonfragile control.

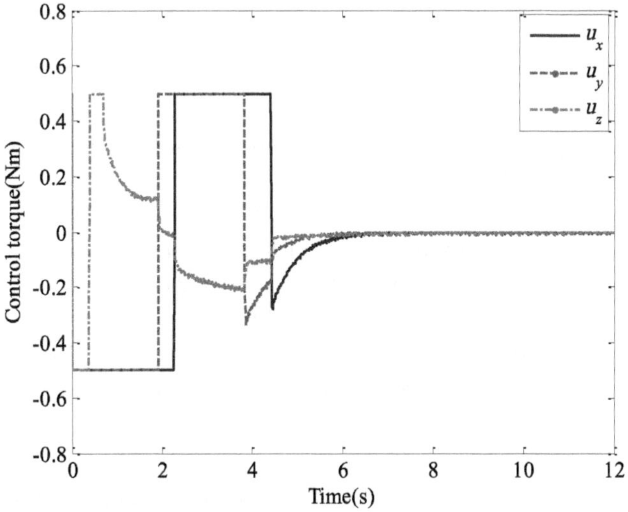

Figure 3.3 Control input using dynamic output feedback nonfragile control.

spacecraft within 6 seconds, and the control input is continuous and limited.

Without considering controller gain perturbation, with inequality (3.48), one can get the solution of SOFHC and choose appropriate invertible matrix $\mathbf{Q}_{21}$, and the related matrices can be obtained:

$$A_c = \begin{bmatrix} -1.8016 & -0.3055 & -0.2488 & -24.8551 & -8.7279 & -9.0537 \\ -0.4793 & -1.4124 & -0.2252 & -7.4247 & -28.1031 & -9.1334 \\ -0.4705 & -0.2712 & -1.1229 & -7.2893 & -8.7573 & -30.7468 \\ -0.0077 & 0.0104 & 0.0120 & -19.1910 & -1.1488 & -1.1598 \\ -0.0043 & 0.0864 & 0.0262 & -1.0175 & -19.9665 & -1.2101 \\ 0.0096 & 0.0297 & 0.1947 & -1.0403 & -1.1683 & -21.0404 \end{bmatrix}$$

$$B_c = \begin{bmatrix} 594.3602 & 344.3452 & 337.5966 & 400.0509 & 173.1159 & 177.9446 \\ 358.3366 & 560.9769 & 332.3763 & 152.8975 & 451.1895 & 178.6847 \\ 351.3463 & 334.6577 & 559.9031 & 149.9143 & 172.5050 & 493.5949 \\ 104.4814 & 30.8548 & 30.9396 & 110.1051 & 21.4220 & 21.6743 \\ 31.2972 & 98.4479 & 27.9384 & 19.2386 & 122.2646 & 22.1700 \\ 27.5281 & 25.4730 & 96.5595 & 19.1865 & 21.1318 & 139.7244 \end{bmatrix}$$

$$K = \begin{bmatrix} -0.0810 & -0.0016 & -0.0015 & -15.4758 & -0.0159 & -0.0126 \\ -0.0015 & -0.0729 & -0.0015 & -0.0176 & -13.9287 & -0.0131 \\ -0.0013 & -0.0012 & -0.0608 & -0.0168 & -0.0157 & -11.6080 \end{bmatrix}$$

To illustrate the influence of additive gain perturbation on SOFHC, simulations are made under the condition whether additive gain perturbation exists or not, respectively.

Case 3.1: Considering model parameter uncertainty without additive gain perturbation

In this case, apply the SOFHC in attitude control system with saturation processing, the corresponding attitude angle, angular velocity and control input are shown in Figs. 3.4–3.6. It can be seen that without considering controller gain perturbation, the designed SOFHC can make attitude stabilization within 8 seconds, with continuous and limited control input.

Case 3.2: Considering model parameter uncertainty and additive gain perturbation

In this case, apply the SOFHC in the attitude control system and with saturation processing, the corresponding attitude angle, angular velocity and control input are shown in Figs. 3.7–3.9. It can be seen that the designed SOFHC cannot achieve attitude stabilization. The attitude angle reaches the first peak of approximately 45 rad at about 32 seconds, and

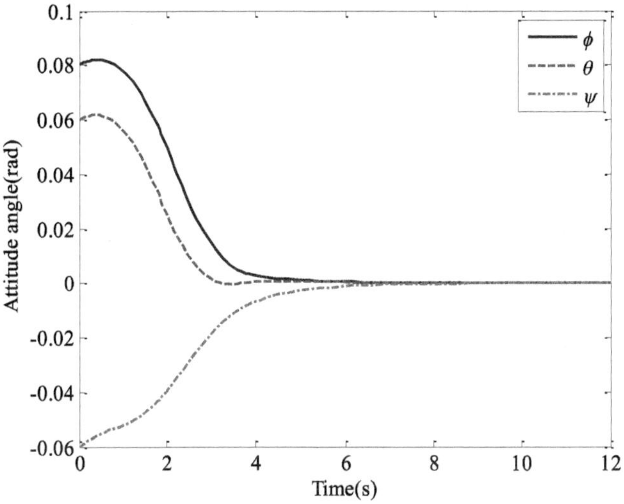

Figure 3.4 Attitude angle only considering model parameter uncertainty using standard output feedback H_∞ controller.

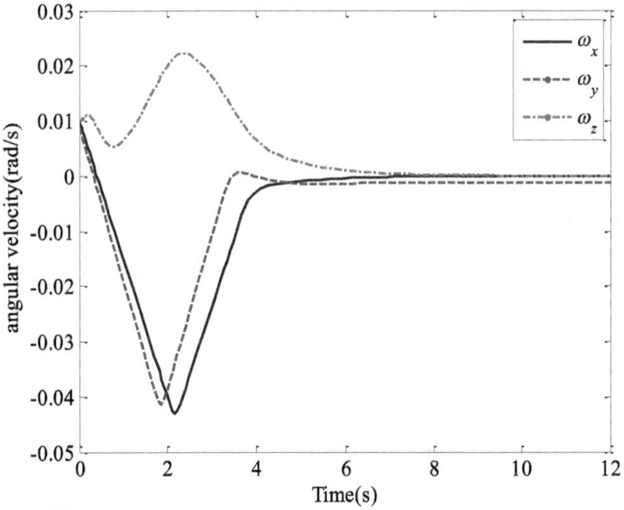

Figure 3.5 Angular velocity only considering model parameter uncertainty using standard output feedback H_∞ controller.

the second peak of approximately -2588 rad at about 87 seconds; the angular velocity reaches the first peak of approximately 3 rad s^{-1} at about 27 seconds, and the second peak of approximately -6 rad s^{-1} at about 42 seconds. At this time, the spacecraft has lost stability, and control

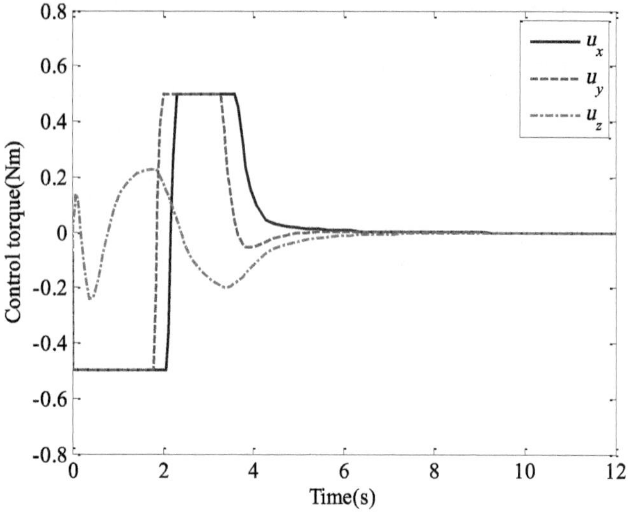

Figure 3.6 Control input only considering model parameter uncertainty using standard output feedback H_∞ controller.

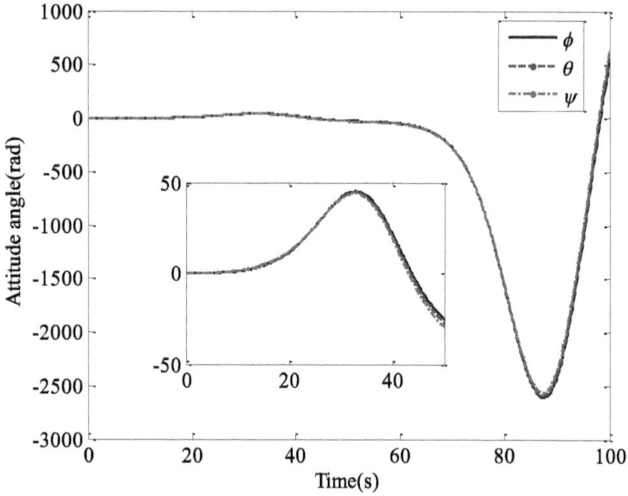

Figure 3.7 Attitude angle considering model parameter uncertainty and additive gain perturbation using standard output feedback H_∞ controller.

performance sharply declines. As time increases, the attitude angle and angular velocity peak will continue to increase, indicating that the spacecraft rotation will occur, and the rotation frequency will increase with longer duration.

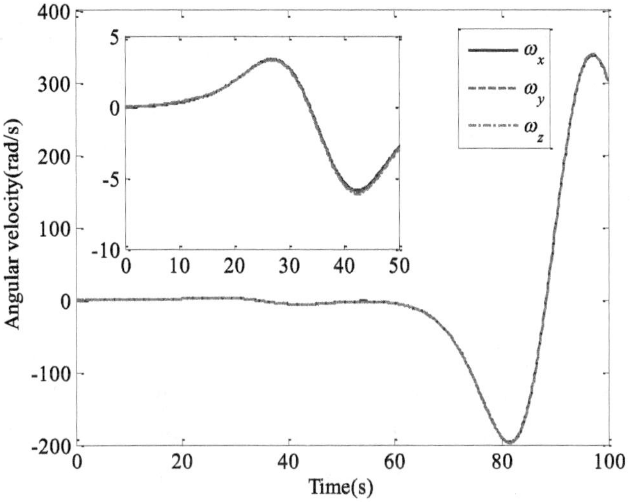

Figure 3.8. Angular velocity considering model parameter uncertainty and additive gain perturbation using standard output feedback H_∞ controller.

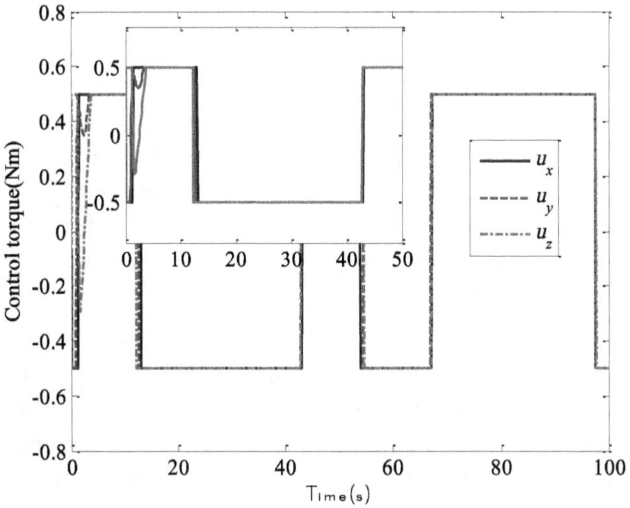

Figure 3.9 Control input considering model parameter uncertainty and additive gain perturbation using standard output feedback H_∞ controller.

3.4.2 Simulation results under multiplicative perturbation

To illustrate the influence of different multiplicative perturbation value of controller gain on DOFNFC and SOFHC, and verify the performance of controller aiming at multiplicative perturbation, choose different values

Table 3.1 Two sets of each parameter value.

Set	M_3	k	N_3	u_m/Nm
1	[118;118;118]	90	$0.00119*\boldsymbol{1}_{3\times3}$	0.5
2	[121;121;121]	120	$0.00121*\boldsymbol{1}_{3\times3}$	0.5

of $\boldsymbol{M}_3$, $\boldsymbol{F}_3$ and $\boldsymbol{N}_3$ for simulation, which can be divided into two types (Table 3.1). The form of $\boldsymbol{F}_3$ is:

$$\boldsymbol{F}_3(t) = \begin{bmatrix} \sin(k\omega_0 t) & \sin(k\omega_0 t + \pi/4) & \sin(k\omega_0 t + \pi/2) \end{bmatrix}$$

To test the influence of $\boldsymbol{M}_3$, $\boldsymbol{F}_3$ and $\boldsymbol{N}_3$ on the final state of spacecraft, two groups of numerical simulations are performed with respect to DOFNFC and SOFHC in each simulation. Specifically, each simulation condition is as follows:

C1: $\boldsymbol{M}_3 = [118;118;118]$, $\boldsymbol{N}_3 = 0.00119*\boldsymbol{1}_{3\times3}$, $k = 120$
C2: $\boldsymbol{M}_3 = [118;118;118]$, $\boldsymbol{N}_3 = 0.00119*\boldsymbol{1}_{3\times3}$, $k = 90$
C3: $\boldsymbol{M}_3 = [121;121;121]$, $\boldsymbol{N}_3 = 0.00119*\boldsymbol{1}_{3\times3}$, $k = 120$
C4: $\boldsymbol{M}_3 = [118;118;118]$, $\boldsymbol{N}_3 = 0.00121*\boldsymbol{1}_{3\times3}$, $k = 120$

Obviously, C1 can be regarded as a reference group, comparing C2 and C1 can witness the frequency variation influence of $\boldsymbol{F}_3$, comparing C3 and C1 can witness the frequency variation influence of $\boldsymbol{M}_3$, comparing C4 and C1 can witness the frequency variation influence of $\boldsymbol{N}_3$, and the self-comparative simulation results can witness the influence of multiplicative gain perturbation on DOFNFC and SOFHC to show the performance of nonfragile controller.

For a spacecraft attitude control system with nonlinear perturbations, when other parameters are specified, among the three parameters of $\boldsymbol{M}_3$, k and $\boldsymbol{N}_3$, only $\boldsymbol{M}_3$ and $\boldsymbol{N}_3$ can affect the related nonfragile controller parameters $\boldsymbol{A}_c$, $\boldsymbol{B}_c$ and $\boldsymbol{K}$, and the related matrices of SOFHC has been solved in the previous section. Thus the nonfragile controllers of C1 and C2 are the same, defined as controller 1, the nonfragile controller of C3 corresponds to another controller, defined as controller 2, the nonfragile controller of C4 corresponds to another controller, defined as controller 3, and SOFHC is defined as controller 4.

For C1 and C2, the related matrices are:

$$\boldsymbol{A}_c = \begin{bmatrix} 11.5225 & 15.3050 & 19.2044 & 35.3815 & -16.4417 & -154.8536 \\ 11.1895 & 13.2113 & 17.5017 & 68.5202 & -55.3221 & -141.7215 \\ 9.5026 & 11.9071 & 14.3843 & 60.3003 & -12.3285 & -166.4944 \\ 14.9337 & 17.8018 & 22.0279 & 40.7528 & -21.8686 & -169.7755 \\ 13.2494 & 16.4787 & 19.9471 & 67.3127 & -53.8964 & -153.4269 \\ 11.1864 & 13.6265 & 17.2943 & 57.5396 & -16.0020 & -165.9698 \end{bmatrix}$$

$$
\boldsymbol{B}_c = 1.0 \times 10^3 \times \begin{bmatrix} 1.8943 & 1.8368 & 2.3614 & 6.2745 & 6.2543 & 6.7721 \\ 1.5259 & 1.9994 & 2.2017 & 5.4572 & 6.1059 & 6.2215 \\ 1.3755 & 1.5436 & 2.2736 & 4.7421 & 4.9975 & 5.8017 \\ 1.3368 & 1.4736 & 2.0399 & 6.3735 & 6.5807 & 7.1539 \\ 1.1233 & 1.4273 & 1.8426 & 5.6468 & 6.1044 & 6.4755 \\ 0.9378 & 1.1124 & 1.6648 & 4.7612 & 5.0252 & 5.6386 \end{bmatrix}
$$

$$
\boldsymbol{K} = \begin{bmatrix} -0.0455 & 0.0299 & 0.0250 & -22.6514 & 14.4292 & 12.0299 \\ 0.0268 & -0.0467 & 0.0201 & 12.9534 & -23.1513 & 9.7197 \\ 0.0184 & 0.0166 & -0.0452 & 8.9548 & 8.0532 & -22.3070 \end{bmatrix}
$$

C1:

First, aiming at C1, simulations are made with respect to controller 1 and controller 4, the corresponding attitude angle, angular velocity and control input are shown in Figs. 3.10–3.12. It can be observed that both controller 1 and controller 4 can attain attitude stabilization of spacecraft within 8 seconds, but the error peak will appear with less than 5×10^{-4} rad at about 30 seconds, and the time interval of two adjacent error peak is about 45 seconds, and the angular velocity and control input are also similar, which indicates that under C1, the multiplicative gain perturbation of controller affects spacecraft attitude little, which can be neglected.

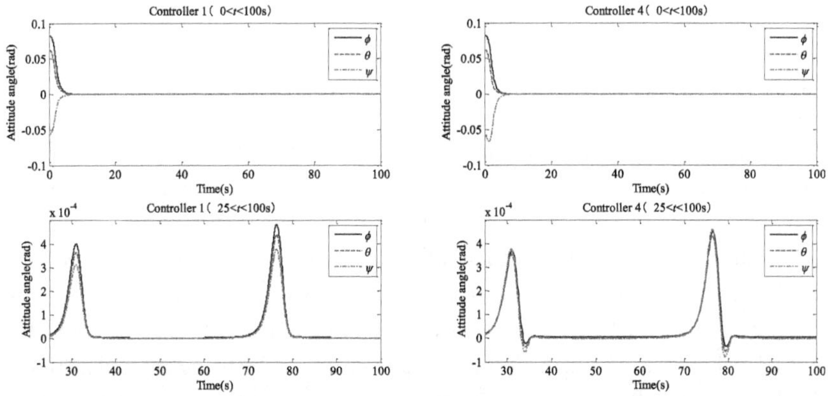

Figure 3.10 Corresponding attitude angle comparison of controller 1 and controller 4 under C1.

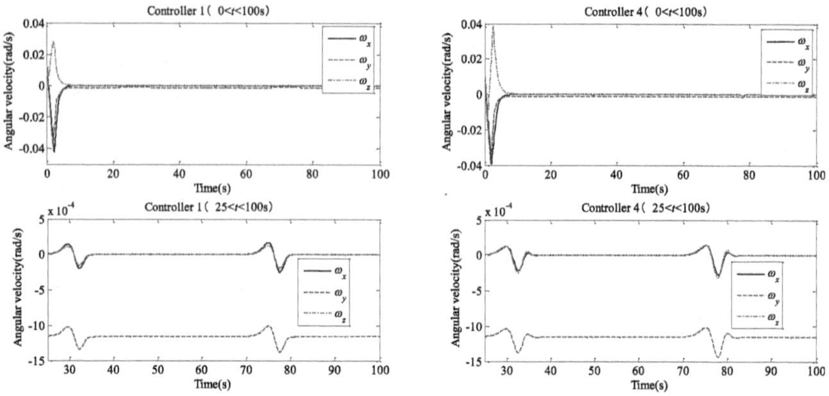

Figure 3.11 Corresponding angular velocity comparison of controller 1 and controller 4 under C1.

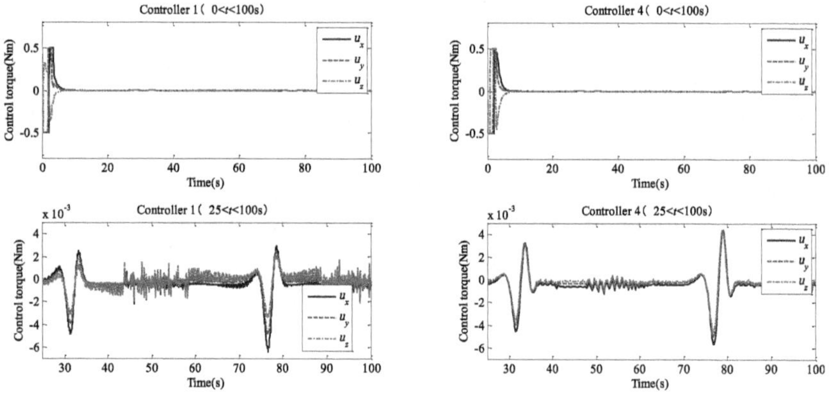

Figure 3.12 Corresponding control input comparison of controller 1 and controller 4 under C1.

C2:

Second, aiming at C2, simulations are made with respect to controller 1 and controller 4, the corresponding attitude angle, angular velocity and control input are shown in Figs. 3.13–3.15. It can be observed that both controller 1 and controller 4 can make attitude stabilization of spacecraft within 8 seconds, but the error peak will appear with less than 6×10^{-4} rad at about 40 seconds, and the time interval of two adjacent error peak is about 60 seconds, and the angular velocity and control input are also similar, which indicates that under C2, the multiplicative gain perturbation of controller affects spacecraft attitude little, which can be neglected.

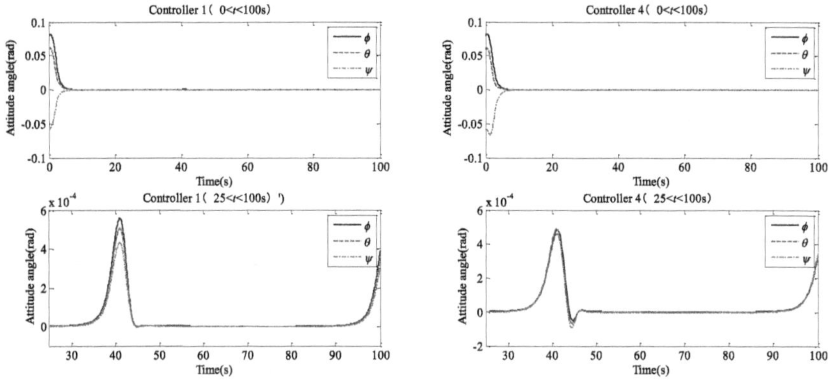

Figure 3.13 Corresponding attitude angle comparison of controller 1 and controller 4 under C2.

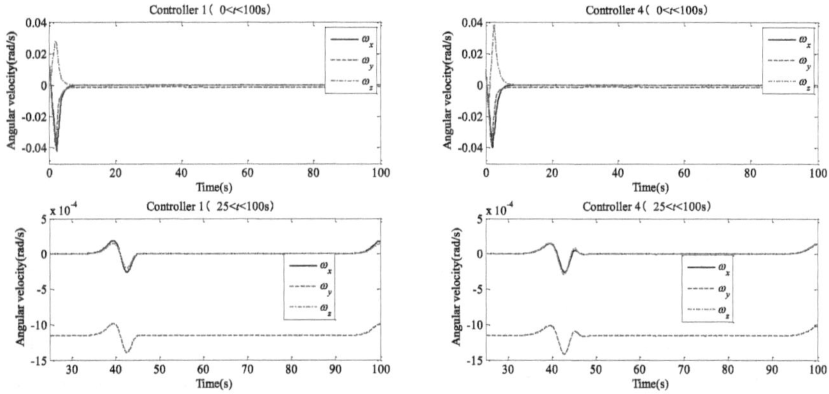

Figure 3.14 Corresponding angular velocity comparison of controller 1 and controller 4 under C2.

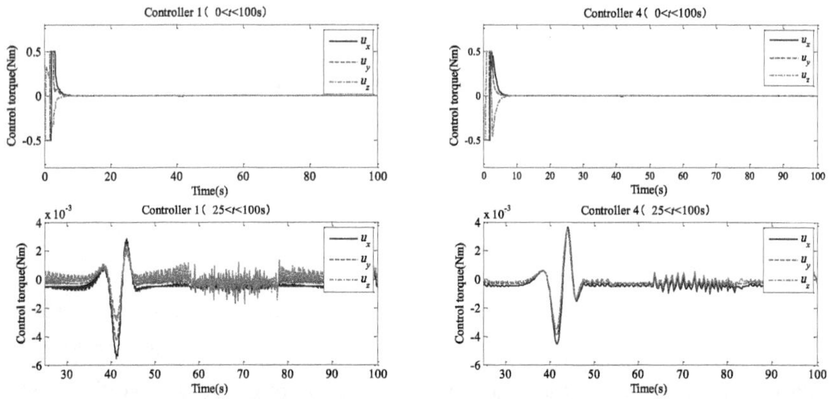

Figure 3.15 Corresponding control input comparison of controller 1 and controller 4 under C2.

C3:

For C3, the related matrices are:

$$A_c = \begin{bmatrix} 12.5471 & 16.5340 & 21.6925 & 44.2272 & -15.0062 & -175.4423 \\ 12.1279 & 14.3253 & 19.7875 & 76.3745 & -53.5849 & -160.7416 \\ 10.3196 & 12.8711 & 16.3752 & 67.0458 & -10.9690 & -182.7185 \\ 16.0719 & 19.1721 & 24.6902 & 50.1752 & -20.7552 & -191.6033 \\ 14.2781 & 17.7139 & 22.3539 & 75.6599 & -52.6564 & -173.2024 \\ 12.0525 & 14.6662 & 19.3230 & 64.5405 & -15.1092 & -182.3891 \end{bmatrix}$$

$$B_c = 1.0 \times 10^3 \times \begin{bmatrix} 1.9538 & 1.9311 & 2.4458 & 6.7315 & 6.7411 & 7.4210 \\ 1.5805 & 2.0863 & 2.2798 & 5.8757 & 6.5506 & 6.8177 \\ 1.4217 & 1.6168 & 2.3384 & 5.1042 & 5.3806 & 6.3180 \\ 1.4065 & 1.5835 & 2.1407 & 6.8813 & 7.1237 & 7.8611 \\ 1.1861 & 1.5269 & 1.9343 & 6.1055 & 6.5951 & 7.1151 \\ 0.9902 & 1.1952 & 1.7404 & 5.1468 & 5.4372 & 6.1765 \end{bmatrix}$$

$$K = \begin{bmatrix} -0.0454 & 0.0299 & 0.0250 & -22.3618 & 14.2496 & 11.8856 \\ 0.0268 & -0.0467 & 0.0200 & 12.7950 & -22.8580 & 9.5961 \\ 0.0184 & 0.0166 & -0.0452 & 8.8389 & 7.9534 & -22.0277 \end{bmatrix}$$

Then, aiming at C3, simulations are made with respect to controller 2 and controller 4, the corresponding attitude angle, angular velocity and control input are shown in Figs. 3.16–3.18. From which one can see that both controller 2 and controller 4 can make the first attitude stabilization

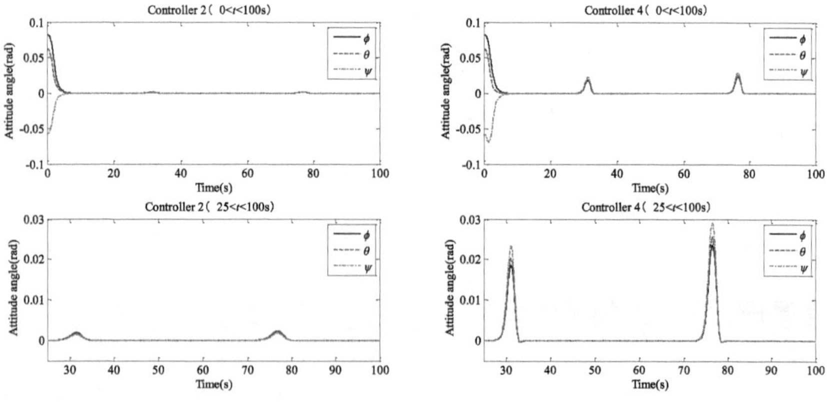

Figure 3.16 Corresponding attitude angle comparison of controller 2 and controller 4 under C3.

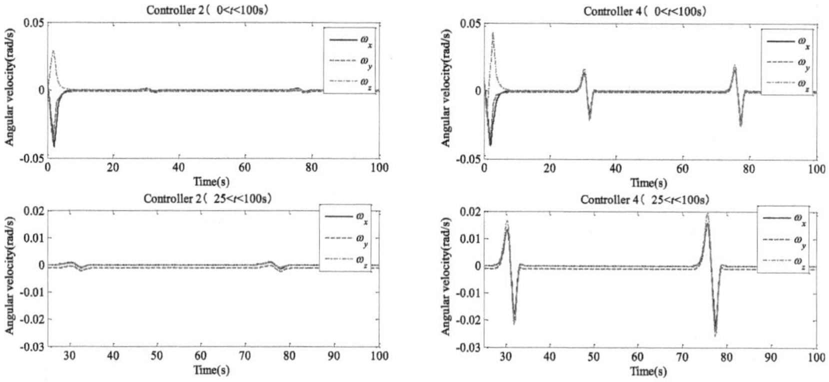

Figure 3.17 Corresponding angular velocity comparison of controller 2 and controller 4 under C3.

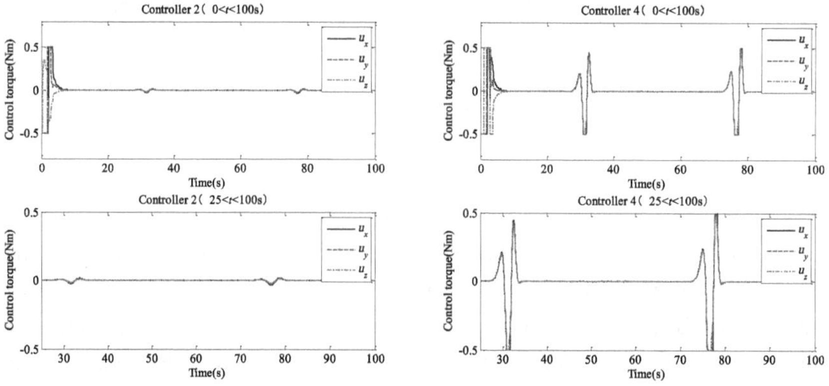

Figure 3.18 Corresponding control input comparison of controller 2 and controller 4 under C3.

of spacecraft at about 8 seconds, but the error peak will appear at about 30 seconds, and the time interval of two adjacent error peak is about 45 seconds. However, the attitude angle error peak corresponding to controller 2 is about 2×10^{-3} rad, while the attitude angle error peak corresponding to controller 4 is close to 0.03 rad, 15 times of controller 2, which indicates that under C3, the multiplicative gain perturbation of controller affects spacecraft attitude system controlled by controller 4 very much, which cannot be neglected, and controller 2 shows excellent nonfragility which illustrates the good performance of the proposed method described in Theorem 3.2.

C4:

For C4, the related matrices are:

$$A_c = \begin{bmatrix} 12.3163 & 16.1691 & 19.8890 & 35.7472 & -17.0792 & -157.2639 \\ 11.9087 & 13.9755 & 18.1298 & 68.7761 & -55.6821 & -144.0428 \\ 10.1353 & 12.5685 & 14.9258 & 60.3596 & -12.5629 & -168.3803 \\ 15.8880 & 18.9231 & 22.9831 & 42.4895 & -23.4993 & -174.3079 \\ 14.1036 & 17.4787 & 20.8106 & 68.8934 & -55.2929 & -157.5897 \\ 11.9108 & 14.4655 & 18.0147 & 58.7539 & -17.1364 & -169.3451 \end{bmatrix}$$

$$B_c = 1.0 \times 10^3 \times \begin{bmatrix} 1.9757 & 1.8886 & 2.4540 & 6.6700 & 6.6194 & 7.0999 \\ 1.6038 & 2.0499 & 2.2877 & 5.8205 & 6.4379 & 6.5229 \\ 1.4426 & 1.5858 & 2.3461 & 5.0580 & 5.2840 & 6.0608 \\ 1.4322 & 1.5473 & 2.1596 & 6.8486 & 7.0416 & 7.5760 \\ 1.2108 & 1.4953 & 1.9511 & 6.0754 & 6.5195 & 6.8570 \\ 1.0110 & 1.1689 & 1.7548 & 5.1223 & 5.3735 & 5.9577 \end{bmatrix}$$

$$K = \begin{bmatrix} -0.0455 & 0.0299 & 0.0250 & -22.6501 & 14.4392 & 12.0448 \\ 0.0269 & -0.0467 & 0.0201 & 12.9683 & -23.1526 & 9.7248 \\ 0.0184 & 0.0166 & -0.0452 & 8.9602 & 8.0640 & -22.3108 \end{bmatrix}$$

Finally, we compare the simulation results of controller 3 and controller 4 under C4, and the indices are still attitude angle, angular velocity and control input, which are shown in Figs. 3.19–3.21. From which one can see that both controller 3 and controller 4 can make the first attitude stabilization of spacecraft at about 8 seconds, but the error peak will

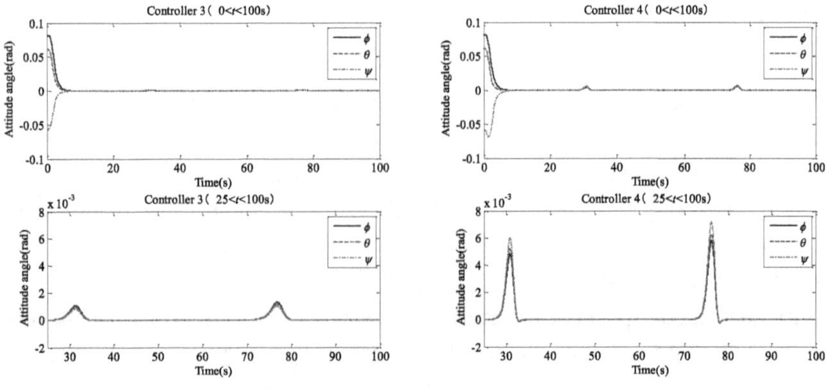

Figure 3.19 Corresponding attitude angle comparison of controller 3 and controller 4 under C4.

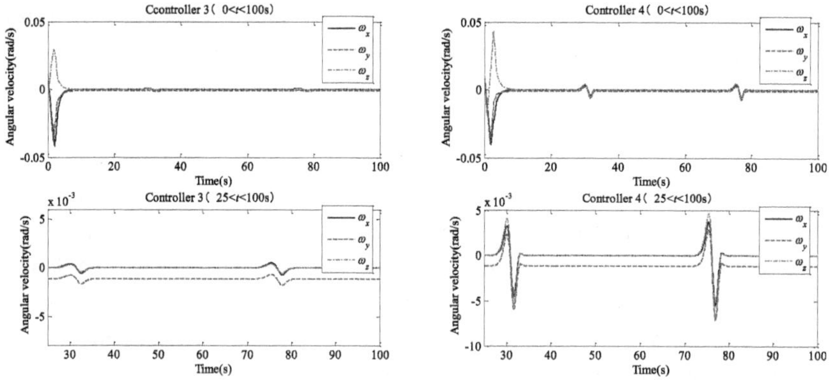

Figure 3.20 Corresponding angular velocity comparison of controller 3 and controller 4 under C4.

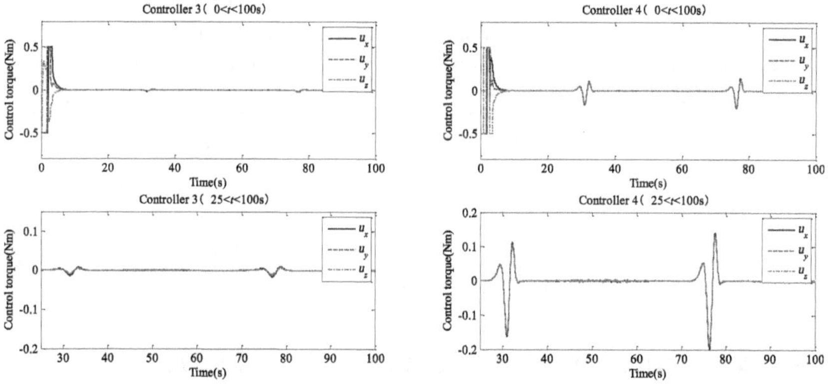

Figure 3.21 Corresponding control input comparison of controller 3 and controller 4 under C4.

appear at about 30 seconds, and the time interval of two adjacent error peak is about 45 seconds. However, the attitude angle error peak corresponding to controller 3 is about 1×10^{-3} rad, while the attitude angle error peak corresponding to controller 4 is close to 7×10^{-3} rad, 7 times of controller 3, and the upper and lower error peak difference of angular velocity is nearly 0.01 rad s^{-1}, which indicates that under C4, the multiplicative gain perturbation of controller affects spacecraft attitude system controlled by controller 4 very much, which cannot be neglected, and controller 3 can still guarantee the small range of attitude angle and angular velocity error, which illustrates the good performance of the proposed method described in Theorem 3.2.

There are the error peaks and the time intervals of two adjacent error peaks throughout the simulation examples, because there is periodic perturbation in the controller with the maximum at certain time. According to the simulation results from C1 to C4, we choose five indices of the first error peak time t_1, the time interval of two adjacent error peak Δt, the first stabilization time t_s, attitude angle error peak of DOFNFC e_{m1}, and that of SOFHC e_{m2} to analyze the impact of $\boldsymbol{M}_3$, $\boldsymbol{F}_3$ and $\boldsymbol{N}_3$ on the final state of spacecraft attitude control system. Based on the above simulation results, one can obtain the corresponding five indices shown in Table 3.2.

Regard C1 as reference group, C2, C3 and C4 as comparison groups, and some conclusions can be drawn:

1. The value of k only affects the first error peak time t_1 and the time interval of two adjacent error peak Δt, the smaller k is, that is, the more slowly multiplicative gain perturbation varies, the later t_1 is, and the longer Δt is.
2. The value of $\boldsymbol{M}_3$ ($\boldsymbol{N}_3$) only affects the attitude error peak, the bigger $\boldsymbol{M}_3$ ($\boldsymbol{N}_3$) is, the bigger multiplicative gain perturbation is, the bigger the attitude error peak is. The small variation of gain perturbation may lead to considerable influence on controller performance.
3. Under the condition of bounded multiplicative gain perturbation, DOFNFC can guarantee the small range of attitude angle and angular velocity error, but SOFHC is very sensitive to the nonlinear perturbation of controller gain.

3.4.3 Simulation results under coexisting additive and multiplicative perturbations

In this section, the effectiveness of the proposed controller under coexisting additive and multiplicative perturbations is illustrated through simulations of a postcapture spacecraft. For this example, the initial state is chosen as $\boldsymbol{x}(0) = [0.15\ \text{rad},\ 0.12\ \text{rad},\ -0.12\ \text{rad},\ -0.01\ \text{rad s}^{-1},\ -0.01\ \text{rad s}^{-1},\ 0.01\ \text{rad s}^{-1}]$, and $u_{m1} = u_{m2} = u_{m3} = 0.25$ Nm. Other simulation parameters are identical to the initial set.

Table 3.2 Five indices corresponding to C1–C4.

Set	t_1/s	Δt/s	t_s/s	e_{m1}/rad	e_{m2}/rad
C1	30	45	8	5×10^{-4}	5×10^{-4}
C2	40	60	8	6×10^{-4}	6×10^{-4}
C3	30	45	8	2×10^{-3}	3×10^{-2}
C4	30	45	8	1×10^{-3}	7×10^{-3}

Using Theorem 3.4 yields the related matrices as follows

$$A_c = 1.0 \times 10^5 \times \begin{bmatrix} -1.3796 & -1.3795 & -1.3795 & -1.3808 & -1.3785 & -1.3792 \\ 1.7386 & 1.7386 & 1.7386 & 1.7402 & 1.7372 & 1.7386 \\ 5.7049 & 5.7050 & 5.7049 & 5.7077 & 5.7048 & 5.7029 \\ -4.8823 & -4.8824 & -4.8824 & -4.8854 & -4.8813 & -4.8811 \\ -2.4164 & -2.4165 & -2.4165 & -2.4171 & -2.4174 & -2.4153 \\ 1.2325 & 1.2325 & 1.2325 & 1.2330 & 1.2327 & 1.2318 \end{bmatrix}$$

$$B_c = 1.0 \times 10^5 \times \begin{bmatrix} 5.9965 & 2.3255 & 2.7830 & 3.4702 & 3.4682 & 4.1112 \\ -1.5654 & -1.9538 & -2.5815 & -2.9829 & -2.9895 & -3.7898 \\ -2.2822 & -8.0965 & -9.4302 & -11.1343 & -12.0370 & -13.9969 \\ -0.1151 & 6.7194 & 8.0197 & 9.2608 & 10.0250 & 11.8544 \\ -0.6580 & 3.1702 & 3.7773 & 4.3578 & 4.7383 & 5.5937 \\ -1.3758 & -2.1647 & -2.5683 & -2.9716 & -3.2048 & -3.7725 \end{bmatrix}$$

$$K = \begin{bmatrix} -3026.4 & -3026.4 & -3026.4 & -3022.8 & -3032.6 & -3023.9 \\ -207.74 & -207.74 & -207.74 & -213.05 & -198.88 & -211.29 \\ 3257.8 & 3257.8 & 3257.8 & 3259.5 & 3255.1 & 3258.8 \end{bmatrix}$$

Applying the controller gives the time responses of attitude angle and angular velocity as shown in Figs. 3.22 and 3.23, respectively. It is observed that the attitude angle and angular velocity converge to within a small bound in 15 seconds. After 15 seconds, the accuracy of the attitude

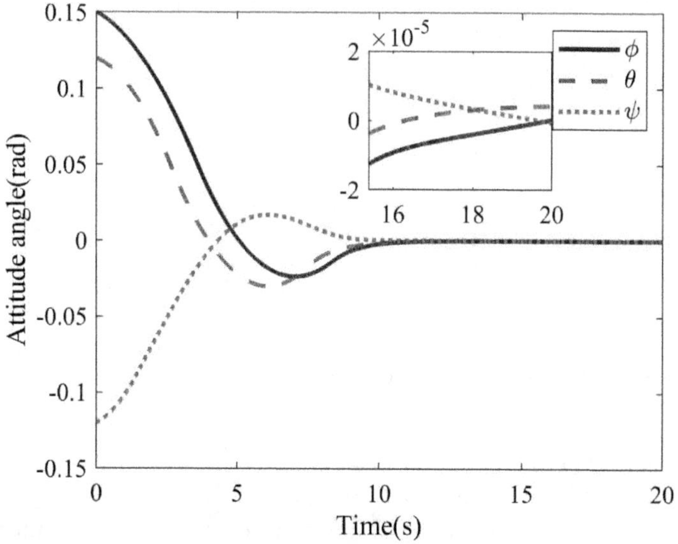

Figure 3.22 Time response of attitude angle.

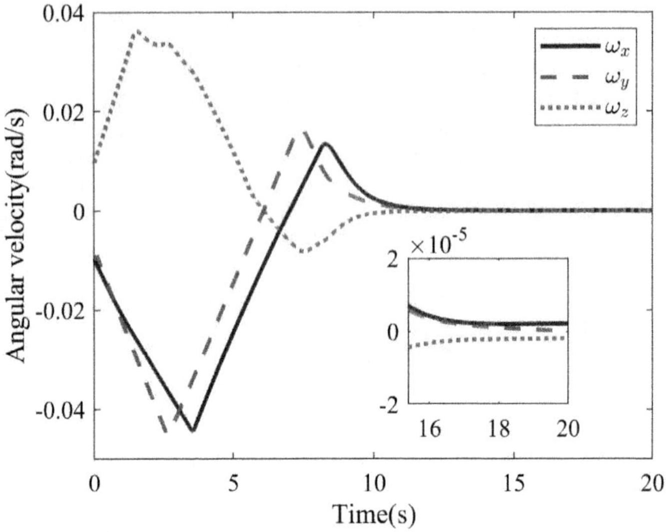

Figure 3.23 Time response of angular velocity.

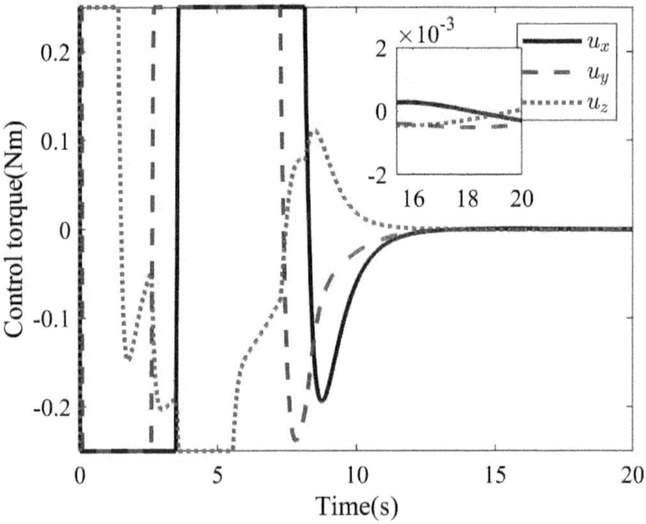

Figure 3.24 Time response of control torque.

angle is approximately 2×10^{-5}rad and that of angular velocity is approximately 1×10^{-5}rad s^{-1} in steady-state error. The control torque is plotted in Fig. 3.24, from which one can see that it is no more than 0.25 Nm. It is concluded that the designed nonfragile controller can stabilize the attitude control system rapidly in the presence of coexisting additive and multiplicative perturbations.

3.5 Conclusions

In this chapter, the dynamic output feedback nonfragile controller has been introduced to solve the spacecraft attitude stabilization problem when nonlinear perturbations such as model parameter uncertainty and controller gain perturbation, are considered. With respect to additive perturbation and multiplicative perturbation of the controller, this chapter proposes the corresponding algorithms respectively, which can guarantee that the closed-loop system satisfies H_∞ performance constraint, control input constraint and quadratic stability. Under the limited and continuous control input, the dynamic output feedback nonfragile controller can attain attitude stabilization of the spacecraft quickly. Meanwhile, when the controller gain perturbation is not taken into account, the standard output feedback H_∞ controller is designed for comparison, which can make attitude stabilization of spacecraft quickly without considering nonlinear perturbation of controller. Once the nonlinear perturbation of controller is considered, this controller will lead to spacecraft rotation under the condition of additive gain perturbation, and with increasing time, the rotation frequency will increase, leading to a sharp decline in system performance. The impact of multiplicative gain perturbation is closely related to perturbation value and the gain perturbation will make the system show a cyclical error peak. The standard output feedback H_∞ controller is highly fragile to nonlinear perturbation of controller while the proposed dynamic output feedback nonfragile controller shows excellent robustness and nonfragility.

References

[1] Pukdeboon C, Kumam P. Robust optimal sliding mode control for spacecraft position and attitude maneuvers. Aerospace Science and Technology 2015;43:329–42.
[2] Liu H, Guo L, Zhang Y. An anti-disturbance PD control scheme for attitude control and stabilization of flexible spacecrafts. Nonlinear Dynamics 2012;67(3):2081–8.
[3] Liu C, Sun Z, Shi K, et al. Robust non-fragile state feedback attitude control for uncertain spacecraft with input saturation. Proceedings of the Institution of Mechanical Engineers, Part G: Journal of Aerospace Engineering 2016;. Available from: http://doi.org/10.1177/0954410016679194.
[4] Tiwari PM, Janardhanan S, Nabi M. Rigid spacecraft attitude control using adaptive integral second order sliding mode. Aerospace Science and Technology 2015;42:50–7.

[5] Wie B, Plescia CT. Attitude stabilization of flexible spacecraft during stationkeeping maneuvers. Journal of Guidance Control & Dynamics 1984;7(7):430–6.
[6] Wie B, Lehner JA, Plescia CT. Roll/yaw control of a flexible spacecraft using skewed bias momentum wheels. Journal of Guidance, Control, and Dynamics 1985;8(4):447–53.
[7] Wie B, Barba PM. Quaternion feedback for spacecraft large angle maneuvers. Journal of Guidance, Control, and Dynamics 1985;8(3):360–5.
[8] Jovan DBoškovic, Li SM, Mehra RK. Robust adaptive variable structure control of spacecraft under control input saturation. Journal of Guidance Control & Dynamics 2001;24(1):14–22.
[9] Hu Q, Ma G, Xie L. Robust and adaptive variable structure output feedback control of uncertain systems with input nonlinearity. Automatica 2008;44(2):552–9.
[10] Tayebi A. Unit quaternion-based output feedback for the attitude tracking problem. IEEE Transactions on Automatic Control 2008;53(6):1516–20.
[11] Caccavale F, Villani L. Output feedback control for attitude tracking. Systems & Control Letters 1999;38(2):91–8.
[12] Hu J, Zhang H. Output feedback control for rigid-body attitude with constant disturbances. International Journal of Control 2015;88(3):602–12.
[13] Nicosia S, Tomei E. Nonlinear observer and output feedback attitude control of spacecraft. IEEE Transactions on Aerospace and Electronic Systems 1992;28(4):970–7.
[14] Singh SN, Zhang R. Adaptive output feedback control of spacecraft with flexible appendages by modeling error compensation. Acta Astronautica 2004;54(4):229–43.
[15] Wu S, Wen S. Robust H_∞ output feedback control for attitude stabilization of a flexible spacecraft. Nonlinear Dynamics 2016;84(1):405–12.
[16] Hu Q, Li B, Zhang A. Robust finite-time control allocation in spacecraft attitude stabilization under actuator misalignment. Nonlinear Dynamics 2013;73(1–2):53–71.
[17] Lee D, Vukovich G. Robust adaptive terminal sliding mode control on SE(3) for autonomous spacecraft rendezvous and docking. Nonlinear Dynamics 2016;83(4):2263–79.
[18] Elmali H, Olgac N. Robust output tracking control of nonlinear MIMO systems via sliding mode technique. Automatica 1992;28(1):145–51.
[19] Chang YC. Robust tracking control for nonlinear MIMO systems via fuzzy approaches. Automatica 2000;36(10):1535–45.
[20] Wei Y, Peng X, Qiu J. Robust and non-fragile static output feedback control for continuous-time semi-Markovian jump systems. Transactions of the Institute of Measurement & Control 2016;38(9):1136–50.
[21] He S, Xu H. Non-fragile finite-time filter design for time-delayed Markovian jumping systems via T–S fuzzy model approach. Nonlinear Dynamics 2015;80(3):1159–71.
[22] Sivaranjani K, Rakkiyappan R, Lakshmanan S, et al. Robust non-fragile control for offshore steel jacket platform with nonlinear perturbations. Nonlinear Dynamics 2015;81(4):2043–57.
[23] Huang SJ, Yang GH. Non-fragile H_∞ dynamic output feedback control for uncertain Takagi–Sugeno fuzzy systems with time-varying delay. International Journal of Systems Science 2016;47(12):2954–64.
[24] Yang X, Gao H, Shi P, et al. Robust H_∞ control for a class of uncertain mechanical systems. International Journal of Control 2010;83(7):1303–24.
[25] Cao YY, Lin Z. Robust stability analysis and fuzzy-scheduling control for nonlinear systems subject to actuator saturation. IEEE Transactions on Fuzzy Systems 2003;11(1):57–67.

CHAPTER 4

Observer-based fault tolerant delayed control

4.1 Introduction

The trend of the present generation of spacecraft is towards smaller vehicles capable of high-precision pointing and better robustness toward external disturbances and various perturbations. Future space missions demand the development of effective attitude control methods to ensure rapid and accurate response to various attitude maneuvering commands. However, some problems are frequently encountered in practical applications, such as modeling uncertainties, external disturbances, time-varying input delays, actuator faults, and actuator saturation, which may result in performance degradation and otherwise affect the performance of spacecraft. A significant challenge arises when these issues are considered simultaneously. Although much literature has addressed the attitude control problem of the spacecraft, and there are many sophisticated control methods, including variable structure control [1,2], adaptive control [3,4], feedback control [5], sampled-data control [6,7], and optimal control [8–10], etc., or syntheses of these methods, very few have considered these issues concurrently.

Broadly speaking, the model parameters of a spacecraft cannot be exactly determined, and a spacecraft attitude system is always subjected to various uncertainties. Further, the dynamics of a spacecraft is often time-varying and highly nonlinear in the presence of uncertainties due to environmental disturbances [11]. Parameter uncertainty in spacecraft systems has been taken into account in much literature, such as [12–14], etc., which demonstrates that it is necessary to consider modeling uncertainties to achieve high-precision and stable performance. Some literature has considered fault tolerant control in [15–17], since the actuator faults can lead to severe performance deterioration, or even instability of control systems. A nonregressor-based indirect approach was presented in [15] to achieve attitude tracking control of spacecraft in the presence of modeling uncertainties, external disturbances, actuator failures and limited resources, which showed good performance against these disturbances or perturbations. However, this work did not address the influence of time-varying

Spacecraft Attitude Control. DOI: https://doi.org/10.1016/B978-0-323-99005-9.00004-3

input delay, thus this controller may become ineffective when input delay exists. In [16], an adaptive sliding mode fault-tolerant attitude tracking control scheme was developed for flexible spacecraft with partial loss of actuator effectiveness, where a neural network was employed to account for system uncertainties and an online updating law was used to estimate the upper bound of actuator faults. To further address the actuator saturation problem, a modified fault tolerant control law was also presented. In [17], an adaptive control algorithm was used to estimate the disturbance and the boundary of actuator faults and saturation, where no prior information of the fault was needed. Most of the aforementioned references are state feedback methods and assume that full-state information is available for spacecraft control. Thus it is vital to address the fault tolerant control problem with only output measurements, and a fault estimation scheme can be incorporated directly into the fault tolerant controller design. Methods based on this idea have been developed in recent years, for example, sliding mode observer-based fault tolerant control scheme [18,19], and adaptive observer-based fault tolerant control scheme [20,21]. In addition, in several practical applications, there is an unavoidable time delay within the control system due to delays in measurements or actuators. A spacecraft attitude control system designed based on feedback schemes is usually robust to a small amount of time delay. However, if the time delay increases due to subsystem component failure or other factors, the long-term effect of this large time delay may induce undesirable behavior of controlled attitude motion or even lead to instability of the controlled system designed based on no delay [22,23]. Refs. [24] and [25] proposed an interesting fault estimation mechanism based on intermediate estimator but did not consider the problem of actuator saturation or contain state estimation information explicitly in the designed controller. The actuator types for attitude control of spacecraft include mechanical and electrical components, thrusters, and environmental force actuators. Thus actuator saturation is another critical issue that must be coped with in fault tolerant control of spacecraft attitude system as pointed out in [16]. Also, in reality, the output of spacecraft actuators is constrained. Therefore, if an unknown actuator fault occurs, an uncorrected attitude control system could produce inappropriate commands causing actuators to saturate rendering any further control effort ineffective.

Thus it is worth pointing out that, still now, solving the observer-based fault tolerant control problem considering model parameter uncertainty, external disturbances, time-varying input delay, actuator faults, and actuator saturation is an open problem. Each of these issues on its own can severely degrade control performance or even cause system instability, and it will be much more difficult to satisfy the control objective when all of these issues are considered simultaneously.

Motivated by the preceding discussions, this chapter presents an observer-based fault tolerant control scheme to achieve high-precision attitude stabilization and control of spacecraft with explicit consideration of model parameter uncertainty, time-varying delay input, external disturbances, actuator faults and saturation. The design method of this controller is based on linear matrix inequalities (LMIs), and LMIs today have become a highly effective technique in the field of stability and control with advantages of global optimal solution and numerical reliability, multiple objective design, and mature software packages. Specifically, many optimization problems involving LMIs have solvable global optimal solutions and can be solved numerically efficiently and reliably even when their sizes are very large. It is very convenient to handle multiple objective control systems design, and the solution to the multiobjective design problem is straightforward. To solve problems related with LMIs, there have been a few mature software packages, for example, YALMIP, CVX, and the MATLAB LMI toolbox, which have been found to be very efficient. The main contributions of this work can be summarized as four points. First, the proposed observer-based fault tolerant control scheme can stabilize attitude angle and angular velocity to the vicinity of the expected states with high precision, in spite of simultaneous model parameter uncertainty, time-varying delay input, external disturbances, actuator faults, and saturation. Second, an explicit bound of the estimation errors of attitude angle and angular velocity is obtained by introducing a new variable to construct a new state vector. Third, the proposed controller contains the estimation information of time-varying delay and fault signal, and thus explicitly accounts for time-varying delay and fault signal in the plant. Fourth, by exploiting the online adaptive estimation of the actuator fault, the proposed control method can achieve the goal of fault tolerant control without use of any fault detection and isolation mechanism to determine the fault information.

This chapter is organized as follows. The next section briefly describes the attitude dynamics of spacecraft with model parameter uncertainty, input delay, and actuator fault, and the corresponding observer-based control problem is formulated. Then, the design method of an observer-based fault tolerant controller based on LMIs is proposed by introducing a new variable, and stability analysis of the closed-loop system is performed via a Lyapunov approach. It is followed by simulations of the observer-based fault tolerant controller and prediction-based sampled-data H_∞ controller, which illustrate the superiority of the proposed approach. Finally, some conclusions are drawn.

4.2 Problem formulation

4.2.1 Attitude system description

The spacecraft is assumed to be a rigid body with actuators that provide torques about three mutually perpendicular axes. The three axes define a body-fixed frame. The attitude dynamics equation is

$$\boldsymbol{I}_b\dot{\boldsymbol{\omega}} + \boldsymbol{\omega} \times (\boldsymbol{I}_b\boldsymbol{\omega}) = \boldsymbol{T}_c + \boldsymbol{T}_g + \boldsymbol{T}_d \tag{4.1}$$

where $\boldsymbol{I}_b$ and $\boldsymbol{\omega}$ denote the inertia matrix and the angular velocity, respectively, $\boldsymbol{T}_c$, $\boldsymbol{T}_g$, and $\boldsymbol{T}_d$ denote the control input torque, the gravitational torque, and the external disturbance torque, respectively.

Provided the body-fixed frame is chosen to be a principal-axis frame, then Eq. (4.1) can be converted into the form of components

$$\begin{cases} I_x\ddot{\phi} + 4(I_y - I_z)\omega_0^2\phi + (I_y - I_z - I_x)\omega_0\dot{\psi} = T_{cx} + T_{dx} \\ I_y\ddot{\theta} + 3\omega_0^2(I_x - I_z)\theta = T_{cy} + T_{dy} \\ I_z\ddot{\psi} + (I_y - I_x)\omega_0^2\psi + (I_x + I_z - I_y)\omega_0\dot{\phi} = T_{cz} + T_{dz} \end{cases} \tag{4.2}$$

where I_x, I_y, and I_z denote the three components of inertia matrix; ω_x, ω_y, and ω_z denote the three components of angular velocity; T_{cx}, T_{cy}, and T_{cz} denote the three components of control input torque; T_{dx}, T_{dy}, and T_{dz} denote the three components of external disturbance torque;ω_0 represents the orbital angular velocity of spacecraft; Euler angles ϕ, θ and ψ are roll, pitch, and yaw attitude angle, respectively, and a typical sequence of three successive rotations is yaw→pitch→roll.

Let

$$\boldsymbol{x} = \begin{bmatrix}\phi & \theta & \psi & \dot{\varphi} & \dot{\theta} & \dot{\psi}\end{bmatrix}^T, \boldsymbol{y} = \begin{bmatrix}\phi & \theta & \psi & \omega_x & \omega_y & \omega_z\end{bmatrix}^T,$$
$$\boldsymbol{u} = \begin{bmatrix}T_{cx} & T_{cy} & T_{cz}\end{bmatrix}^T, \ \boldsymbol{w} = \begin{bmatrix}T_{dx} & T_{dy} & T_{dz}\end{bmatrix}^T$$

Then, Eq. (4.2) can be written as

$$\begin{cases} \dot{\boldsymbol{x}}(t) = \boldsymbol{A}\boldsymbol{x}(t) + \boldsymbol{B}_1\boldsymbol{u}(t) + \boldsymbol{B}_2\boldsymbol{w}(t) \\ \boldsymbol{y}(t) = \boldsymbol{C}_1\boldsymbol{x}(t) + \boldsymbol{D}_1 \end{cases} \tag{4.3}$$

where,

$$A=\begin{bmatrix} 0 & 0 & 0 & 1 & 0 & 0 \\ 0 & 0 & 0 & 0 & 1 & 0 \\ 0 & 0 & 0 & 0 & 0 & 1 \\ A_{41} & 0 & 0 & 0 & 0 & -\omega_0 I_x^{-1}(I_y-I_x-I_z) \\ 0 & A_{52} & 0 & 0 & 0 & 0 \\ 0 & 0 & A_{63} & \omega_0 I_z^{-1}(I_y-I_x-I_z) & 0 & 0 \end{bmatrix}$$

$$A_{41}=-4\omega_0^2 I_x^{-1}(I_y-I_z),\ A_{52}=-3\omega_0^2 I_y^{-1}(I_x-I_z),\ A_{63}=-\omega_0^2 I_z^{-1}(I_y-I_x)$$

$$\boldsymbol{B}_1=\boldsymbol{B}_2=\begin{bmatrix}\boldsymbol{0}_{3\times 3} & diag(I_x^{-1},I_y^{-1},I_z^{-1})\end{bmatrix}^T$$

$$\boldsymbol{D}_1=\begin{bmatrix}\boldsymbol{0}_{1\times 4} & -\omega_0 & 0\end{bmatrix}^T,\ \boldsymbol{C}_1=\begin{bmatrix}\boldsymbol{I}_{3\times 3} & \boldsymbol{0}_{3\times 3}\\ \boldsymbol{B} & \boldsymbol{I}_{3\times 3}\end{bmatrix},\ \boldsymbol{B}=\begin{bmatrix}0 & 0 & -\omega_0\\ 0 & 0 & 0\\ \omega_0 & 0 & 0\end{bmatrix}$$

Assumption 4.1: The external disturbance $\boldsymbol{w}(t)$ is assumed to be bounded, that is, its Euclidean norm satisfies $\|\boldsymbol{w}(t)\|\le\kappa_1$, where $\kappa_1>0$.

Remark 4.1: Assumption 4.1 is reasonable as the external disturbances primarily include magnetic torque, solar radiation, and aerodynamic torque, which are all bounded [26,27].

In addition, the following important points should be noted.

1. Model parameter uncertainty

Due to the parameter perturbations, modeling errors, and environmental changes, model uncertainty widely exists in physical systems. If model uncertainty is not considered during control system design, the controller obtained is highly likely to fail in practical situations. To analytically study the stabilization of uncertain fault tolerant attitude control system, we will assume that the admissible parameter uncertainties have a norm bounded uncertainty form in order to have a sufficient condition for exponential stability, which is the most adopted form in robust stability analysis [28]. Considering parameter uncertainty, $\boldsymbol{A}$ is transformed into $\boldsymbol{A}+\Delta\boldsymbol{A}$, where $\Delta\boldsymbol{A}$ denotes model parameter uncertainty and satisfies matching condition, that is,

$$\Delta\boldsymbol{A}(t)=\boldsymbol{M}_1\boldsymbol{F}_1(t)\boldsymbol{N}_1,\ \|\boldsymbol{F}_1(t)\|\le 1,\ \forall t$$

where $\boldsymbol{M}_1$ and $\boldsymbol{N}_1$ are known constant real matrices of appropriate dimensions and $\boldsymbol{F}_1(t)$ is a Lipschitz measurable matrix function.

2. Input delay

In a control system, input delay is due to many factors, such as the reaction time of actuators and sensors, faults of actuators and sensors, signal sampling, and transmission, etc. The effects of these time-delay phenomena will all be reflected in the control input. Time delay input will present a barrier to carry out required performances and even destroy the stability of the closed-loop system. Here, $\tau(t)$ denotes time-varying delay, and we assume that it satisfies $l_1 \leq \tau(t) \leq l_2, \dot{\tau}(t) \leq \vartheta$, where l_1, l_2 and ϑ are constant scalars. As is well known, an input delayed time will usually cause performance degradation and even instability.

3. Actuator faults

Since the performance of an attitude control system can be severely impaired by improper actuator actions, actuator fault has been considered to be one of the most critical challenges to be solved. Taking this issue into account, let $\boldsymbol{E}$ with full-column rank denote the distribution matrix of fault signal $\boldsymbol{f}(t)$ appearing in the input. Specifically, it represents the process fault if $\boldsymbol{E} \neq \boldsymbol{B}_1$, and it also represents the actuator fault if $\boldsymbol{E} = \boldsymbol{B}_1$.

4. Actuator saturation

Actuator saturation is a ubiquitous phenomenon due to the constraints of physical characteristics in real engineering applications, which will often have a negative impact on the stability and performance of an attitude control system. When the spacecraft performs an attitude maneuver, if the theoretical control input is larger than the saturation value of actual control input, then the actual control input $sat(\boldsymbol{u})$ should be described as

$$sat(u_{i=1,2,3}(t)) = sign(u_i(t))min\{|u_i(t)|, u_{mi}\} \tag{4.4}$$

where u_{mi} is the upper boundary of control input that the actuator can provide.

Assumption 4.2: There exists a positive scalar κ_2, such that the first derivative of the fault signal $\boldsymbol{f}(t)$ satisfies $\|\dot{\boldsymbol{f}}(t)\| \leq \kappa_2$.

Remark 4.2: Assumption 4.2 is common in the existing recognition techniques on observer design [29,30]. It should be noted that the norm of the fault signal $\boldsymbol{f}(t)$ can be unknown. Thus it is more general than the fault-tolerant sliding-mode-observer method in [31], which requires preliminary knowledge of the bound of the fault.

Assumption 4.3: $rank(\boldsymbol{B}_1, \boldsymbol{E}) = rank(\boldsymbol{B}_1)$, there exists a matrix $\boldsymbol{B}_1^*$ such that $(\boldsymbol{I} - \boldsymbol{B}_1\boldsymbol{B}_1^*)\boldsymbol{E} = 0$.

Remark 4.3: Assumption 4.3 is general in [32], which means that the faults that appear in the actuator can be compensated by the control input.

Considering the above factors, the new linear form of spacecraft attitude control system can be expressed as

$$\begin{cases} \dot{\boldsymbol{x}}(t) = (\boldsymbol{A}+\Delta\boldsymbol{A})\boldsymbol{x}(t) + \boldsymbol{B}_1\boldsymbol{u}(t,t-\tau(t)) + \boldsymbol{E}\boldsymbol{f}(t) + \boldsymbol{B}_2\boldsymbol{w}(t) \\ \boldsymbol{y}(t) = \boldsymbol{C}_1\boldsymbol{x}(t) + \boldsymbol{D}_1 \end{cases} \tag{4.5}$$

4.2.2 Control objective

The objective of this chapter is to construct an observer-based fault tolerant controller such that the states of the resulting closed-loop system are uniformly ultimately bounded in the presence of model parameter uncertainty, external disturbances, time-varying input delay, actuator faults, and actuator saturation. It is worth pointing out that the attitude information and fault signal are required to be observed simultaneously.

4.3 Observer-based fault tolerant control scheme

4.3.1 Intermediate observer design

In this section, to estimate the attitude information and fault signal simultaneously such that the fault tolerant control scheme is designed based on the estimation, it is necessary to construct an auxiliary variable. It should be noted that this auxiliary variable can combine the attitude information and fault signal together so as to bring convenience. The new variable is described as

$$\boldsymbol{\xi}(t) = \boldsymbol{f}(t) - \eta\boldsymbol{E}^T\boldsymbol{x}(t) \tag{4.6}$$

$$\dot{\boldsymbol{\xi}}(t) = \dot{\boldsymbol{f}}(t) - \eta\boldsymbol{E}^T\left[(\boldsymbol{A}+\Delta\boldsymbol{A})\boldsymbol{x}(t) + \boldsymbol{B}_1\boldsymbol{u}(t,t-\tau(t)) + \boldsymbol{E}\boldsymbol{\xi}(t) + \eta\boldsymbol{E}\boldsymbol{E}^T\boldsymbol{x}(t) + \boldsymbol{B}_2\boldsymbol{w}(t)\right] \tag{4.7}$$

where η is a scalar variable to be chosen.

The proposed observer is based on

$$\dot{\hat{x}}(t) = A\hat{x}(t) + B_1 u(t,t-\tau(t)) + E\hat{f}(t) + L(y(t) - \hat{y}(t)) \tag{4.8}$$

$$\dot{\hat{\xi}}(t) = -\eta E^T \left[A\hat{x}(t) + B_1 u(t,t-\tau(t)) + E\hat{\xi}(t) + \eta E E^T \hat{x}(t) \right] \tag{4.9}$$

$$\hat{y}(t) = C_1 \hat{x}(t) + D_1 \tag{4.10}$$

$$\hat{f}(t) = \hat{\xi}(t) + \eta E^T \hat{x}(t) \tag{4.11}$$

where $\hat{x}(t)$, $\hat{\xi}(t)$, $\hat{y}(t)$, $\hat{f}(t)$ are the estimates of $x(t)$, $\xi(t)$, $y(t)$ and $f(t)$, respectively, and L is the observer gain matrix.

4.3.2 Delayed controller design

Based on the above estimation, the controller is designed as

$$u(t,t-\tau(t)) = B_1^*(B_1 B_1^*)^* \left[-B_1 K\hat{x}(t) + A_d \hat{x}(t-\tau(t)) - E\hat{f}(t) \right] \tag{4.12}$$

where K is chosen such that $A - B_1 K$ is Hurwitz, which can be guaranteed by the condition of (20) below. $\hat{x}(t-\tau(t))$ means the time profiles of states, which is zero when $t < \tau(t)$.

Remark 4.4: Here, the matrices B_1^* and $(B_1 B_1^*)^*$ are the Moore–Penrose pseudoinverses of B_1 and $B_1 B_1^*$, respectively. It follows that $B_1 B_1^* B_1 = B_1; B_1^* B_1 B_1^* = B_1^*; (B_1 B_1^*)^T = B_1 B_1^*; (B_1^* B_1)^T = B_1^* B_1$.

Denote $e_x(t) = x(t) - \hat{x}(t)$, $e_\xi(t) = \xi(t) - \hat{\xi}(t)$, $e_f(t) = f(t) - \hat{f}(t)$ and $e_x(t-\tau(t)) = x(t-\tau(t)) - \hat{x}(t-\tau(t))$, where $x(t-\tau(t))$ denotes the influence of input delay in terms of attitude information, substituting Eq. (4.12) into the first term of Eq. (4.5) and combining Eqs. (4.8)–(4.11) yields the overall closed-loop system as follows

$$\begin{aligned} \dot{x}(t) &= (A - B_1 K)x(t) + B_1 K e_x(t) + \Delta A x(t) + A_d x(t-\tau(t)) \\ &\quad - A_d e_x(t-\tau(t)) + E e_\xi(t) + \eta E E^T e_x(t) + B_2 w(t) \end{aligned} \tag{4.13}$$

$$\dot{e}_x(t) = (A - LC_1)e_x(t) + \Delta A x(t) + E e_\xi(t) + \eta E E^T e_x(t) + B_2 w(t) \tag{4.14}$$

$$\dot{e}_\xi(t) = \dot{f}(t) - \eta E^T \left[A e_x(t) + \Delta A x(t) + E e_\xi(t) + \eta E E^T e_x(t) + B_2 w(t) \right] \tag{4.15}$$

Thus the objective of this chapter is to seek the controller gain matrix $\boldsymbol{K}$ and observer gain matrix $\boldsymbol{L}$ such that the states of the resulting closed-loop system represented by Eqs. (4.13)–(4.15) are uniformly ultimately bounded.

4.3.3 Control solution

Before the main results are introduced, we first introduce the following essential preliminary lemmas.

Lemma 4.1: [33] Let $\overline{\boldsymbol{H}}, \overline{\boldsymbol{E}}$ and $\overline{\boldsymbol{F}}(t)$ be real matrices of appropriate dimensions with $\left\|\overline{\boldsymbol{F}}(t)\right\| \leq 1$, then for any scalar $\overline{\xi} > 0$,

$$\overline{\boldsymbol{H}}\overline{\boldsymbol{F}}(t)\overline{\boldsymbol{E}} + \overline{\boldsymbol{E}}^T\overline{\boldsymbol{F}}(t)^T\overline{\boldsymbol{H}}^T \leq \overline{\xi}^{-1}\overline{\boldsymbol{H}}\overline{\boldsymbol{H}}^T + \overline{\xi}\overline{\boldsymbol{E}}^T\overline{\boldsymbol{E}}$$

Lemma 4.2: (Schur complement lemma) Let the partitioned matrix

$$\boldsymbol{A} = \begin{bmatrix} \boldsymbol{A}_{11} & \boldsymbol{A}_{12} \\ * & \boldsymbol{A}_{22} \end{bmatrix}$$

be symmetric. Then

$$\boldsymbol{A} < 0 \Leftrightarrow \boldsymbol{A}_{11} < 0, \boldsymbol{A}_{22} - \boldsymbol{A}_{12}^T\boldsymbol{A}_{11}^{-1}\boldsymbol{A}_{12} < 0 \Leftrightarrow \boldsymbol{A}_{22} < 0, \boldsymbol{A}_{11} - \boldsymbol{A}_{12}\boldsymbol{A}_{22}^{-1}\boldsymbol{A}_{12}^T < 0$$

Theorem 4.1: The states of the closed-loop system represented by Eqs. (4.13)–(4.15) are uniformly ultimately bounded, and converge to a small residual set, if for symmetric positive definite matrix $\widehat{\boldsymbol{P}}$, scalars $\delta > 0, \vartheta > 0, \eta > 0, \xi_i > 0$, $i = 1, 2, \ldots, 14$, given matrices $\boldsymbol{M} = [\underset{\sim}{\boldsymbol{M}}_1^T \quad \underset{\sim}{\boldsymbol{M}}_2^T \quad \underset{\sim}{\boldsymbol{M}}_3^T \quad \underset{\sim}{\boldsymbol{M}}_4^T \quad \underset{\sim}{\boldsymbol{M}}_5^T \quad \underset{\sim}{\boldsymbol{M}}_6^T]^T$ and $\mathbf{N} = [\underset{\sim}{\boldsymbol{N}}_1^T \quad \underset{\sim}{\boldsymbol{N}}_2^T \quad \underset{\sim}{\boldsymbol{N}}_3^T \quad \underset{\sim}{\boldsymbol{N}}_4^T \quad \underset{\sim}{\boldsymbol{N}}_5^T \quad \underset{\sim}{\boldsymbol{N}}_6^T]^T$, there exist symmetric positive definite matrices $\boldsymbol{P}_i, i = 1, 2, 3, 4, 5$ and matrices $\boldsymbol{H}$ and $\boldsymbol{K}$, such that the following LMIs hold:

$$(\boldsymbol{A} - \boldsymbol{B}_1\boldsymbol{K})^T\widehat{\boldsymbol{P}} + \widehat{\boldsymbol{P}}(\boldsymbol{A} - \boldsymbol{B}_1\boldsymbol{K}) < 0 \tag{4.16}$$

$$\begin{bmatrix} \boldsymbol{\Theta}_{11} & \boldsymbol{\Theta}_{12} \\ * & \boldsymbol{\Theta}_{22} \end{bmatrix} < 0 \tag{4.17}$$

where,

$$\boldsymbol{\Theta}_{11}=\begin{bmatrix} \overline{\boldsymbol{\Theta}}_{11} & \overline{\boldsymbol{\Theta}}_{12} & \boldsymbol{P}_1\boldsymbol{A}_d & \overline{\boldsymbol{\Theta}}_{14} & \boldsymbol{P}_1\boldsymbol{E} & -\underset{\sim}{\boldsymbol{N}}_1 & \underset{\sim}{\boldsymbol{M}}_1 & \underset{\sim}{\boldsymbol{N}}_1 \\ * & \overline{\boldsymbol{\Theta}}_{22} & \underset{\sim}{\boldsymbol{M}}_3^{\mathrm{T}} & \overline{\boldsymbol{\Theta}}_{24} & \overline{\boldsymbol{\Theta}}_{25} & \overline{\boldsymbol{\Theta}}_{26} & \underset{\sim}{\boldsymbol{M}}_2 & \underset{\sim}{\boldsymbol{N}}_2 \\ * & * & 0 & \overline{\boldsymbol{\Theta}}_{34} & 0 & -\underset{\sim}{\boldsymbol{N}}_3 & \underset{\sim}{\boldsymbol{M}}_3 & \underset{\sim}{\boldsymbol{N}}_3 \\ * & * & * & \overline{\boldsymbol{\Theta}}_{44} & \overline{\boldsymbol{\Theta}}_{45} & \overline{\boldsymbol{\Theta}}_{46} & \underset{\sim}{\boldsymbol{M}}_4 & \underset{\sim}{\boldsymbol{N}}_4 \\ * & * & * & * & \overline{\boldsymbol{\Theta}}_{55} & -\underset{\sim}{\boldsymbol{N}}_5 & \underset{\sim}{\boldsymbol{M}}_5 & \underset{\sim}{\boldsymbol{N}}_5 \\ * & * & * & * & * & \overline{\boldsymbol{\Theta}}_{66} & \underset{\sim}{\boldsymbol{M}}_6 & \underset{\sim}{\boldsymbol{N}}_6 \\ * & * & * & * & * & * & -\dfrac{\delta\boldsymbol{P}_2}{l_2} & 0 \\ * & * & * & * & * & * & * & -\dfrac{\delta\boldsymbol{P}_2}{l_2-l_1} \end{bmatrix}$$

and where,

$$\overline{\boldsymbol{\Theta}}_{11}=(\boldsymbol{A}-\boldsymbol{B}_1\boldsymbol{K})^T\boldsymbol{P}_1+\boldsymbol{P}_1(\boldsymbol{A}-\boldsymbol{B}_1\boldsymbol{K})+(2l_2-l_1)\lambda_{\mathrm{mdA}}(\delta\boldsymbol{P}_2)$$

$$\overline{\boldsymbol{\Theta}}_{12}=\boldsymbol{P}_1\boldsymbol{B}_1\boldsymbol{K}+\eta\boldsymbol{P}_1\boldsymbol{E}\boldsymbol{E}^T+\underset{\sim}{\boldsymbol{M}}_1$$

$$\overline{\boldsymbol{\Theta}}_{14}=-\boldsymbol{P}_1\boldsymbol{A}_d-\underset{\sim}{\boldsymbol{M}}_1+\underset{\sim}{\boldsymbol{N}}_1$$

$$\begin{aligned}\overline{\boldsymbol{\Theta}}_{22}=&\ \boldsymbol{A}^T\boldsymbol{P}_2-\boldsymbol{C}_1^T\boldsymbol{H}^T+\boldsymbol{P}_2\boldsymbol{A}-\boldsymbol{H}\boldsymbol{C}_1+\eta\boldsymbol{P}_2\boldsymbol{E}\boldsymbol{E}^T+\eta\boldsymbol{E}\boldsymbol{E}^T\boldsymbol{P}_2\\ &+\boldsymbol{P}_4+\boldsymbol{P}_5+(2l_2-l_1)\eta\delta(\boldsymbol{A}^T\boldsymbol{P}_2-\boldsymbol{C}_1^T\boldsymbol{H}^T)\boldsymbol{E}\boldsymbol{E}^T\\ &+(2l_2-l_1)\eta\delta\boldsymbol{E}\boldsymbol{E}^T(\boldsymbol{P}_2\boldsymbol{A}-\boldsymbol{H}\boldsymbol{C}_1)\\ &+(2l_2-l_1)\eta^2\boldsymbol{E}\boldsymbol{E}^T\boldsymbol{E}\boldsymbol{E}^T+\underset{\sim}{\boldsymbol{M}}_2+\underset{\sim}{\boldsymbol{M}}_2^T\end{aligned}$$

$$\overline{\boldsymbol{\Theta}}_{24}=\underset{\sim}{\boldsymbol{M}}_4^T-\underset{\sim}{\boldsymbol{M}}_2+\underset{\sim}{\boldsymbol{N}}_2$$

$$\begin{aligned}\overline{\boldsymbol{\Theta}}_{25}=&\ (2l_2-l_1)\delta(\boldsymbol{A}^T\boldsymbol{P}_2-\boldsymbol{C}_1^T\boldsymbol{H}^T)\boldsymbol{E}+\boldsymbol{P}_2\boldsymbol{E}-\eta\boldsymbol{A}^T\boldsymbol{E}\boldsymbol{P}_3\\ &+(2l_2-l_1)\eta\boldsymbol{E}\boldsymbol{E}^T(\delta\boldsymbol{P}_2)\boldsymbol{E}-\eta^2\boldsymbol{E}\boldsymbol{E}^T\boldsymbol{E}\boldsymbol{P}_3+\underset{\sim}{\boldsymbol{M}}_5^T\end{aligned}$$

$$\overline{\boldsymbol{\Theta}}_{26}=\underset{\sim}{\boldsymbol{M}}_6^T-\underset{\sim}{\boldsymbol{N}}_2$$

$$\overline{\boldsymbol{\Theta}}_{34}=-\underset{\sim}{\boldsymbol{M}}_3+\underset{\sim}{\boldsymbol{N}}_3$$

$$\overline{\boldsymbol{\Theta}}_{44}=-(1-\vartheta)\boldsymbol{P}_5-\underset{\sim}{\boldsymbol{M}}_4-\underset{\sim}{\boldsymbol{M}}_4^T+\underset{\sim}{\boldsymbol{N}}_4+\underset{\sim}{\boldsymbol{N}}_4^T$$

$$\overline{\boldsymbol{\Theta}}_{45}=-\underset{\sim}{\boldsymbol{M}}_5^T+\underset{\sim}{\boldsymbol{N}}_5^T$$

$$\overline{\boldsymbol{\Theta}}_{46}=-\underset{\sim}{\boldsymbol{M}}_6^T+\underset{\sim}{\boldsymbol{N}}_6^T-\underset{\sim}{\boldsymbol{N}}_4$$

$$\overline{\boldsymbol{\Theta}}_{55}=(2l_2-l_1)\boldsymbol{E}^T(\delta\boldsymbol{P}_2)\boldsymbol{E}-\eta(\boldsymbol{P}_3\boldsymbol{E}^T\boldsymbol{E}+\boldsymbol{E}^T\boldsymbol{E}\boldsymbol{P}_3)$$

$$\overline{\boldsymbol{\Theta}}_{66}=-\boldsymbol{P}_4-\underset{\sim}{\boldsymbol{N}}_6-\underset{\sim}{\boldsymbol{N}}_6^T$$

and

$$\boldsymbol{\Theta}_{12} = \begin{bmatrix} \boldsymbol{\Theta}_{121} & \boldsymbol{\Theta}_{122} & \boldsymbol{\Theta}_{123} \end{bmatrix}$$

where,

$$\boldsymbol{\Theta}_{121} = \begin{bmatrix} \boldsymbol{P}_1\boldsymbol{M}_1 & \boldsymbol{N}_1^T & \boldsymbol{P}_1\boldsymbol{B}_2 & 0 & 0 & 0 \\ 0 & 0 & 0 & \boldsymbol{P}_2\boldsymbol{M}_1 & \boldsymbol{P}_2\boldsymbol{B}_2 & \delta\boldsymbol{H}_0 \\ 0 & 0 & 0 & 0 & 0 & 0 \\ 0 & 0 & 0 & 0 & 0 & 0 \\ 0 & 0 & 0 & 0 & 0 & 0 \\ 0 & 0 & 0 & 0 & 0 & 0 \\ 0 & 0 & 0 & 0 & 0 & 0 \\ 0 & 0 & 0 & 0 & 0 & 0 \end{bmatrix}$$

$$\boldsymbol{\Theta}_{122} = \begin{bmatrix} 0 & 0 & 0 & 0 \\ \delta\boldsymbol{H}_0\boldsymbol{M}_1 & \delta\boldsymbol{H}_0\boldsymbol{B}_2 & \boldsymbol{E}\boldsymbol{E}^T(\delta\boldsymbol{P}_2)\boldsymbol{M}_1 & \boldsymbol{E}\boldsymbol{E}^T(\delta\boldsymbol{P}_2)\boldsymbol{B}_2 \\ 0 & 0 & 0 & 0 \\ 0 & 0 & 0 & 0 \\ 0 & 0 & 0 & 0 \\ 0 & 0 & 0 & 0 \\ 0 & 0 & 0 & 0 \\ 0 & 0 & 0 & 0 \end{bmatrix}$$

$$\boldsymbol{\Theta}_{123} = \begin{bmatrix} 0 & 0 & 0 & 0 & 0 \\ 0 & 0 & 0 & 0 & 0 \\ 0 & 0 & 0 & 0 & 0 \\ 0 & 0 & 0 & 0 & 0 \\ \boldsymbol{E}^T(\delta\boldsymbol{P}_2)\boldsymbol{M}_1 & \boldsymbol{E}^T(\delta\boldsymbol{P}_2)\boldsymbol{B}_2 & \boldsymbol{P}_3 & \boldsymbol{P}_3\boldsymbol{E}^T\boldsymbol{M}_1 & \boldsymbol{P}_3\boldsymbol{E}^T\boldsymbol{B}_2 \\ 0 & 0 & 0 & 0 & 0 \\ 0 & 0 & 0 & 0 & 0 \\ 0 & 0 & 0 & 0 & 0 \end{bmatrix}$$

and where $\boldsymbol{H}_0 = \boldsymbol{A}^T\boldsymbol{P}_2 - \boldsymbol{C}_1^T\boldsymbol{H}^T$;
and

$$\boldsymbol{\Theta}_{22} = diag\left(\begin{bmatrix} \boldsymbol{\Theta}_{221} & \boldsymbol{\Theta}_{222} & \boldsymbol{\Theta}_{223} & \boldsymbol{\Theta}_{224} \end{bmatrix}\right)$$

where,

$$\boldsymbol{\Theta}_{221} = \begin{bmatrix} -\xi_1\boldsymbol{I} & -\dfrac{\boldsymbol{I}}{\xi_1 + \xi_3 + \xi_6 + (2l_2\text{-}l_1)(\xi_8 + \xi_{10}^{-1} + \xi_{11}^{-1} + \xi_{12}^{-1})} \end{bmatrix}$$

$$\boldsymbol{\Theta}_{222}=\begin{bmatrix}-\xi_2\boldsymbol{I} & -\xi_3\boldsymbol{I} & -\xi_4\boldsymbol{I} & -\dfrac{\delta\boldsymbol{P}_2}{2l_2-l_1} & -\dfrac{\xi_8\boldsymbol{I}}{2l_2-l_1} & -\dfrac{\xi_9\boldsymbol{I}}{2l_2-l_1}\end{bmatrix}$$

$$\boldsymbol{\Theta}_{223}=\begin{bmatrix}-\dfrac{\boldsymbol{I}}{(2l_2-l_1)\xi_{11}\eta^2} & -\dfrac{\xi_{14}\boldsymbol{I}}{(2l_2-l_1)\eta^2} & -\dfrac{\boldsymbol{I}}{(2l_2-l_1)\xi_{10}}\end{bmatrix}$$

$$\boldsymbol{\Theta}_{224}=\begin{bmatrix}-\dfrac{\xi_{13}\boldsymbol{I}}{2l_2-l_1} & -\xi_5\boldsymbol{I} & -\dfrac{\xi_6\boldsymbol{I}}{\eta^2} & -\dfrac{\xi_7\boldsymbol{I}}{\eta^2}\end{bmatrix}$$

When LMIs (4.16)–(4.17) are feasible, one can obtain the controller gain matrix $\boldsymbol{K}$ and the observer gain matrix $\boldsymbol{L}=\boldsymbol{P}_2^{-1}\boldsymbol{H}$.

Proof: As mentioned before, the objective of this chapter is to seek the controller gain matrix and observer gain matrix such that the states of the resulting closed-loop system represented by Eqs. (4.13)–(4.15) are uniformly ultimately bounded in the presence of time-varying input delay and other constraints. However, these states are specified on different channels of the closed-loop system. Thus we should establish a common Lyapunov function including all these states and eliminating the effect of time-varying input delay and other constraints, such that observer-based fault tolerant controller design amounts to solving a system of LMIs.

Choose a Lyapunov function candidate as

$$\begin{aligned}V(t)&=\boldsymbol{x}(t)^T\boldsymbol{P}_1\boldsymbol{x}(t)+\boldsymbol{e}_x^T(t)\boldsymbol{P}_2\boldsymbol{e}_x(t)+\boldsymbol{e}_\xi^T(t)\boldsymbol{P}_3\boldsymbol{e}_\xi(t)\\&\quad+\int_{t-l_2}^{t}\boldsymbol{e}_x^T(s)\boldsymbol{P}_4\boldsymbol{e}_x(s)ds+\int_{t-\tau(t)}^{t}\boldsymbol{e}_x^T(s)\boldsymbol{P}_5\boldsymbol{e}_x(s)ds\\&\quad+\int_{-l_2}^{0}\int_{t+\theta}^{t}\dot{\boldsymbol{e}}_x^T(s)(\delta\boldsymbol{P}_2)\dot{\boldsymbol{e}}_x(s)dsd\theta+\int_{-l_2}^{-l_1}\int_{t+\theta}^{t}\dot{\boldsymbol{e}}_x^T(s)(\delta\boldsymbol{P}_2)\dot{\boldsymbol{e}}_x(s)dsd\theta\end{aligned}$$

Taking the derivative of $V(t)$ with respect to time yields

$$\begin{aligned}\dot{V}(t)&=\dot{\boldsymbol{x}}(t)^T\boldsymbol{P}_1\boldsymbol{x}(t)+\boldsymbol{x}(t)^T\boldsymbol{P}_1\dot{\boldsymbol{x}}(t)+\dot{\boldsymbol{e}}_x(t)^T\boldsymbol{P}_2\boldsymbol{e}_x(t)\\&\quad+\boldsymbol{e}_x(t)^T\boldsymbol{P}_2\dot{\boldsymbol{e}}_x(t)+\dot{\boldsymbol{e}}_\xi(t)^T\boldsymbol{P}_3\boldsymbol{e}_\xi(t)+\boldsymbol{e}_\xi(t)^T\boldsymbol{P}_3\dot{\boldsymbol{e}}_\xi(t)\\&\quad+\boldsymbol{e}_x^T(s)\boldsymbol{P}_4\boldsymbol{e}_x(s)|_{t-l_2}^{t}+\boldsymbol{e}_x^T(t)\boldsymbol{P}_5\boldsymbol{e}_x(t)-(1-\dot{\tau}(t))\boldsymbol{e}_x^T(t-\tau(t))\boldsymbol{P}_5\boldsymbol{e}_x(t-\tau(t))\\&\quad+l_2\dot{\boldsymbol{e}}_x^T(t)(\delta\boldsymbol{P}_2)\dot{\boldsymbol{e}}_x(t)-\int_{t-l_2}^{t}\dot{\boldsymbol{e}}_x^T(s)(\delta\boldsymbol{P}_2)\dot{\boldsymbol{e}}_x(s)ds+(l_2-l_1)\dot{\boldsymbol{e}}_x^T(t)(\delta\boldsymbol{P}_2)\dot{\boldsymbol{e}}_x(t)\\&\quad-\int_{t-l_2}^{t-l_1}\dot{\boldsymbol{e}}_x^T(s)(\delta\boldsymbol{P}_2)\dot{\boldsymbol{e}}_x(s)ds\end{aligned}$$

In view of Eqs. (4.13)–(4.15), it yields

$$
\begin{aligned}
\dot{V}(t) = {} & \boldsymbol{x}(t)^T\left[(\boldsymbol{A}-\boldsymbol{B}_1\boldsymbol{K})^T\boldsymbol{P}_1+\boldsymbol{P}_1(\boldsymbol{A}-\boldsymbol{B}_1\boldsymbol{K})\right]\boldsymbol{x}(t)+2\boldsymbol{x}(t)^T\boldsymbol{P}_1\boldsymbol{B}_1\boldsymbol{K}\boldsymbol{e}_x(t)\\
& +2\boldsymbol{x}(t)^T\boldsymbol{P}_1\Delta\boldsymbol{A}\boldsymbol{x}(t)+2\boldsymbol{x}(t)^T\boldsymbol{P}_1\boldsymbol{A}_d\boldsymbol{x}(t-\tau(t))\\
& -2\boldsymbol{x}(t)^T\boldsymbol{P}_1\boldsymbol{A}_d\boldsymbol{e}_x(t-\tau(t))+2\boldsymbol{x}(t)^T\boldsymbol{P}_1\boldsymbol{E}\boldsymbol{e}_\xi(t)\\
& +2\eta\boldsymbol{x}(t)^T\boldsymbol{P}_1\boldsymbol{E}\boldsymbol{E}^T\boldsymbol{e}_x(t)+2\boldsymbol{x}(t)^T\boldsymbol{P}_1\boldsymbol{B}_2\boldsymbol{w}(t)\\
& +\boldsymbol{e}_x(t)^T\left[(\boldsymbol{A}-\boldsymbol{L}\boldsymbol{C}_1)^T\boldsymbol{P}_2+\boldsymbol{P}_2(\boldsymbol{A}-\boldsymbol{L}\boldsymbol{C}_1)\right]\boldsymbol{e}_x(t)\\
& +2\boldsymbol{e}_x(t)^T\boldsymbol{P}_2\Delta\boldsymbol{A}\boldsymbol{x}(t)+2\boldsymbol{e}_x(t)^T\boldsymbol{P}_2\boldsymbol{E}\boldsymbol{e}_\xi(t)\\
& +\boldsymbol{e}_x(t)^T(\eta\boldsymbol{P}_2\boldsymbol{E}\boldsymbol{E}^T+\eta\boldsymbol{E}\boldsymbol{E}^T\boldsymbol{P}_2)\boldsymbol{e}_x(t)\\
& +2\boldsymbol{e}_x(t)^T\boldsymbol{P}_2\boldsymbol{B}_2\boldsymbol{w}(t)+2\boldsymbol{e}_\xi(t)^T\boldsymbol{P}_3\dot{\boldsymbol{f}}(t)\\
& -\eta\boldsymbol{e}_\xi(t)^T(\boldsymbol{P}_3\boldsymbol{E}^T\boldsymbol{E}+\boldsymbol{E}^T\boldsymbol{E}\boldsymbol{P}_3)\boldsymbol{e}_\xi(t)\\
& -2\eta\boldsymbol{e}_\xi(t)^T\boldsymbol{P}_3\boldsymbol{E}^T\left[\boldsymbol{A}\boldsymbol{e}_x(t)+\Delta\boldsymbol{A}\boldsymbol{x}(t)+\eta\boldsymbol{E}\boldsymbol{E}^T\boldsymbol{e}_x(t)+\boldsymbol{B}_2\boldsymbol{w}(t)\right]\\
& +\boldsymbol{e}_x^T(t)\boldsymbol{P}_4\boldsymbol{e}_x(t)-\boldsymbol{e}_x^T(t-l_2)\boldsymbol{P}_4\boldsymbol{e}_x(t-l_2)\\
& +\boldsymbol{e}_x^T(t)\boldsymbol{P}_5\boldsymbol{e}_x(t)-(1-\dot{\tau}(t))\boldsymbol{e}_x^T(t-\tau(t))\boldsymbol{P}_5\boldsymbol{e}_x(t-\tau(t))\\
& +l_2\dot{\boldsymbol{e}}_x^T(t)(\delta\boldsymbol{P}_2)\dot{\boldsymbol{e}}_x(t)-\int_{t-l_2}^{t}\dot{\boldsymbol{e}}_x^T(s)(\delta\boldsymbol{P}_2)\dot{\boldsymbol{e}}_x(s)ds\\
& +(l_2-l_1)\dot{\boldsymbol{e}}_x^T(t)(\delta\boldsymbol{P}_2)\dot{\boldsymbol{e}}_x(t)-\int_{t-l_2}^{t-l_1}\dot{\boldsymbol{e}}_x^T(s)(\delta\boldsymbol{P}_2)\dot{\boldsymbol{e}}_x(s)ds
\end{aligned}
$$

Define $\chi(t)=[\boldsymbol{x}(t)^T\ \ \boldsymbol{e}_x(t)^T\ \ \boldsymbol{x}(t-\tau(t))^T\ \ \boldsymbol{e}_x(t-\tau(t))^T\ \ \boldsymbol{e}_\xi(t)^T\ \ \boldsymbol{e}_x(t-l_2)^T]^T$, $\boldsymbol{M}=[\underset{\sim}{\boldsymbol{M}}_1^T\ \ \underset{\sim}{\boldsymbol{M}}_2^T\ \ \underset{\sim}{\boldsymbol{M}}_3^T\ \ \underset{\sim}{\boldsymbol{M}}_4^T\ \ \underset{\sim}{\boldsymbol{M}}_5^T\ \ \underset{\sim}{\boldsymbol{M}}_6^T]^T$, and $\boldsymbol{N}=[\underset{\sim}{\boldsymbol{N}}_1^T\ \ \underset{\sim}{\boldsymbol{N}}_2^T\ \ \underset{\sim}{\boldsymbol{N}}_3^T\ \ \underset{\sim}{\boldsymbol{N}}_4^T\ \ \underset{\sim}{\boldsymbol{N}}_5^T\ \ \underset{\sim}{\boldsymbol{N}}_6^T]^T$, where $\underset{\sim}{\boldsymbol{M}}_i, \underset{\sim}{\boldsymbol{N}}_i, i=1,2,3,4,5,6$ denote the components of $\boldsymbol{M}$ and $\boldsymbol{N}$, respectively.

According to Lemma 4.1 and the given constraint conditions, one has

$$\begin{aligned}
\dot{V}(t) \leq\ & \boldsymbol{x}(t)^T\left[(\boldsymbol{A}-\boldsymbol{B}_1\boldsymbol{K})^T\boldsymbol{P}_1+\boldsymbol{P}_1(\boldsymbol{A}-\boldsymbol{B}_1\boldsymbol{K})\right]\boldsymbol{x}(t)+2\boldsymbol{x}(t)^T\boldsymbol{P}_1\boldsymbol{B}_1\boldsymbol{K}\boldsymbol{e}_x(t)\\
&+\xi_1^{-1}\boldsymbol{x}(t)^T\boldsymbol{P}_1\boldsymbol{M}_1\boldsymbol{M}_1^T\boldsymbol{P}_1\boldsymbol{x}(t)+\xi_1\boldsymbol{x}(t)^T\boldsymbol{N}_1^T\boldsymbol{N}_1\boldsymbol{x}(t)\\
&+2\boldsymbol{x}(t)^T\boldsymbol{P}_1\boldsymbol{A}_d\boldsymbol{x}(t-\tau(t))-2\boldsymbol{x}(t)^T\boldsymbol{P}_1\boldsymbol{A}_d\boldsymbol{e}_x(t-\tau(t))\\
&+2\boldsymbol{x}(t)^T\boldsymbol{P}_1\boldsymbol{E}\boldsymbol{e}_\xi(t)+2\eta\boldsymbol{x}(t)^T\boldsymbol{P}_1\boldsymbol{E}\boldsymbol{E}^T\boldsymbol{e}_x(t)+\xi_2^{-1}\boldsymbol{x}(t)^T\boldsymbol{P}_1\boldsymbol{B}_2\boldsymbol{B}_2^T\boldsymbol{P}_1\boldsymbol{x}(t)\\
&+\xi_2\kappa_1^2+\boldsymbol{e}_x(t)^T\left[(\boldsymbol{A}-\boldsymbol{L}\boldsymbol{C}_1)^T\boldsymbol{P}_2+\boldsymbol{P}_2(\boldsymbol{A}-\boldsymbol{L}\boldsymbol{C}_1)\right]\boldsymbol{e}_x(t)\\
&+\xi_3^{-1}\boldsymbol{e}_x(t)^T\boldsymbol{P}_2\boldsymbol{M}_1\boldsymbol{M}_1^T\boldsymbol{P}_2\boldsymbol{e}_x(t)+\xi_3\boldsymbol{x}(t)^T\boldsymbol{N}_1^T\boldsymbol{N}_1\boldsymbol{x}(t)\\
&+2\boldsymbol{e}_x(t)^T\boldsymbol{P}_2\boldsymbol{E}\boldsymbol{e}_\xi(t)+\boldsymbol{e}_x(t)^T(\eta\boldsymbol{P}_2\boldsymbol{E}\boldsymbol{E}^T+\eta\boldsymbol{E}\boldsymbol{E}^T\boldsymbol{P}_2)\boldsymbol{e}_x(t)\\
&+\xi_4^{-1}\boldsymbol{e}_x(t)^T\boldsymbol{P}_2\boldsymbol{B}_2\boldsymbol{B}_2^T\boldsymbol{P}_2\boldsymbol{e}_x(t)+\xi_4\kappa_1^2+\xi_5^{-1}\boldsymbol{e}_\xi(t)^T\boldsymbol{P}_3\boldsymbol{P}_3\boldsymbol{e}_\xi(t)\\
&+\xi_5\kappa_2^2-\eta\boldsymbol{e}_\xi(t)^T(\boldsymbol{P}_3\boldsymbol{E}^T\boldsymbol{E}+\boldsymbol{E}^T\boldsymbol{E}\boldsymbol{P}_3)\boldsymbol{e}_\xi(t)-2\eta\boldsymbol{e}_\xi(t)^T\boldsymbol{P}_3\boldsymbol{E}^T\boldsymbol{A}\boldsymbol{e}_x(t)\\
&-2\eta^2\boldsymbol{e}_\xi(t)^T\boldsymbol{P}_3\boldsymbol{E}^T\boldsymbol{E}\boldsymbol{E}^T\boldsymbol{e}_x(t)+\xi_6^{-1}\eta^2\boldsymbol{e}_\xi(t)^T\boldsymbol{P}_3\boldsymbol{E}^T\boldsymbol{M}_1\boldsymbol{M}_1^T\boldsymbol{E}\boldsymbol{P}_3\boldsymbol{e}_\xi(t)\\
&+\xi_6\boldsymbol{x}(t)^T\boldsymbol{N}_1^T\boldsymbol{N}_1\boldsymbol{x}(t)+\xi_7^{-1}\eta^2\boldsymbol{e}_\xi(t)^T\boldsymbol{P}_3\boldsymbol{E}^T\boldsymbol{B}_2\boldsymbol{B}_2{}^T\boldsymbol{E}\boldsymbol{P}_3\boldsymbol{e}_\xi(t)\\
&+\xi_7\kappa_1^2+\boldsymbol{e}_x^T(t)\boldsymbol{P}_4\boldsymbol{e}_x(t)-\boldsymbol{e}_x^T(t-l_2)\boldsymbol{P}_4\boldsymbol{e}_x(t-l_2)\\
&+\boldsymbol{e}_x^T(t)\boldsymbol{P}_5\boldsymbol{e}_x(t)-(1-\vartheta)\boldsymbol{e}_x^T(t-\tau(t))\boldsymbol{P}_5\boldsymbol{e}_x(t-\tau(t))\\
&+l_2\dot{\boldsymbol{e}}_x^T(t)(\delta\boldsymbol{P}_2)\dot{\boldsymbol{e}}_x(t)-\int_{t-\tau(t)}^{t}\dot{\boldsymbol{e}}_x^T(s)(\delta\boldsymbol{P}_2)\dot{\boldsymbol{e}}_x(s)ds\\
&+(l_2-l_1)\dot{\boldsymbol{e}}_x^T(t)(\delta\boldsymbol{P}_2)\dot{\boldsymbol{e}}_x(t)-\int_{t-l_2}^{t-\tau(t)}\dot{\boldsymbol{e}}_x^T(s)(\delta\boldsymbol{P}_2)\dot{\boldsymbol{e}}_x(s)ds\\
&+2\boldsymbol{\chi}(t)^T\boldsymbol{M}\left[\boldsymbol{e}_x(t)-\boldsymbol{e}_x(t-\tau(t))-\int_{t-\tau(t)}^{t}\dot{\boldsymbol{e}}_x(s)ds\right]\\
&+2\boldsymbol{\chi}(t)^T\boldsymbol{N}\left[\boldsymbol{e}_x(t-\tau(t))-\boldsymbol{e}_x(t-l_2)-\int_{t-l_2}^{t-\tau(t)}\dot{\boldsymbol{e}}_x(s)ds\right]
\end{aligned}$$

Since $l_1 \leq \tau(t) \leq l_2$, it can be derived that

$$\begin{cases} l_2\boldsymbol{\chi}(t)^T\boldsymbol{M}(\delta\boldsymbol{P}_2)^{-1}\boldsymbol{M}^T\boldsymbol{\chi}(t) \geq \tau(t)\boldsymbol{\chi}(t)^T\boldsymbol{M}(\delta\boldsymbol{P}_2)^{-1}\boldsymbol{M}^T\boldsymbol{\chi}(t)\\ (l_2-l_1)\boldsymbol{\chi}(t)^T\boldsymbol{N}(\delta\boldsymbol{P}_2)^{-1}\boldsymbol{N}^T\boldsymbol{\chi}(t) \geq (l_2-\tau(t))\boldsymbol{\chi}(t)^T\boldsymbol{N}(\delta\boldsymbol{P}_2)^{-1}\boldsymbol{N}^T\boldsymbol{\chi}(t)\end{cases}$$

Then,

$$\begin{aligned}
&-\int_{t-\tau(t)}^{t}(\dot{\boldsymbol{e}}_x^T(s)(\delta\boldsymbol{P}_2)\dot{\boldsymbol{e}}_x(s)+2\boldsymbol{\chi}(t)^T\boldsymbol{M}\dot{\boldsymbol{e}}_x(s))ds\\
&=-\int_{t-\tau(t)}^{t}\left[\boldsymbol{\chi}(t)^T\boldsymbol{M}+\dot{\boldsymbol{e}}_x(s)^T(\delta\boldsymbol{P}_2)\right](\delta\boldsymbol{P}_2)^{-1}\left[\boldsymbol{M}^T\boldsymbol{\chi}(t)+(\delta\boldsymbol{P}_2)\dot{\boldsymbol{e}}_x(s)\right]ds\\
&\quad+\tau(t)\boldsymbol{\chi}(t)^T\boldsymbol{M}(\delta\boldsymbol{P}_2)^{-1}\boldsymbol{M}^T\boldsymbol{\chi}(t)\\
&\leq-\int_{t-\tau(t)}^{t}\left[\boldsymbol{\chi}(t)^T\boldsymbol{M}+\dot{\boldsymbol{e}}_x(s)^T(\delta\boldsymbol{P}_2)\right](\delta\boldsymbol{P}_2)^{-1}\left[\boldsymbol{M}^T\boldsymbol{\chi}(t)+(\delta\boldsymbol{P}_2)\dot{\boldsymbol{e}}_x(s)\right]ds\\
&\quad+l_2\boldsymbol{\chi}(t)^T\boldsymbol{M}(\delta\boldsymbol{P}_2)^{-1}\boldsymbol{M}^T\boldsymbol{\chi}(t)
\end{aligned}$$

and

$$
\begin{aligned}
&-\int_{t-l_2}^{t-\tau(t)}(\dot{e}_x^T(s)(\delta P_2)\dot{e}_x(s)+2\chi(t)^T N\dot{e}_x(s))ds \\
&=-\int_{t-l_2}^{t-\tau(t)}\left[\chi(t)^T N+\dot{e}_x(s)^T(\delta P_2)\right](\delta P_2)^{-1}\left[N^T\chi(t)+(\delta P_2)\dot{e}_x(s)\right]ds \\
&\quad+(l_2-\tau(t))\chi(t)^T N(\delta P_2)^{-1}N^T\chi(t) \\
&\leq-\int_{t-l_2}^{t-\tau(t)}\left[\chi(t)^T N+\dot{e}_x(s)^T(\delta P_2)\right](\delta P_2)^{-1}\left[N^T\chi(t)+(\delta P_2)\dot{e}_x(s)\right]ds \\
&\quad+(l_2-l_1)\chi(t)^T N(\delta P_2)^{-1}N^T\chi(t)
\end{aligned}
$$

According to Eq. (4.14), one knows

$$
\begin{aligned}
&\dot{e}_x^T(t)(\delta P_2)\dot{e}_x(t)=e_x(t)^T(A-LC_1)^T(\delta P_2)(A-LC_1)e_x(t) \\
&\quad+2e_x(t)^T(A-LC_1)^T(\delta P_2)\Delta A x(t)+2e_x(t)^T(A-LC_1)^T(\delta P_2)Ee_\xi(t) \\
&\quad+\eta e_x(t)^T\left[(A-LC_1)^T(\delta P_2)EE^T+EE^T(\delta P_2)(A-LC_1)\right]e_x(t) \\
&\quad+2e_x(t)^T(A-LC_1)^T(\delta P_2)B_2 w(t)+x(t)^T\Delta A^T(\delta P_2)\Delta A x(t) \\
&\quad+2x(t)^T\Delta A^T(\delta P_2)Ee_\xi(t)+2\eta x(t)^T\Delta A^T(\delta P_2)EE^T e_x(t) \\
&\quad+2x(t)^T\Delta A^T(\delta P_2)B_2 w(t)+e_\xi(t)^T E^T(\delta P_2)Ee_\xi(t) \\
&\quad+2\eta e_\xi(t)^T E^T(\delta P_2)EE^T e_x(t)+2e_\xi(t)^T E^T(\delta P_2)B_2 w(t) \\
&\quad+\eta^2 e_x(t)^T EE^T EE^T e_x(t)+2\eta e_x(t)^T EE^T(\delta P_2)B_2 w(t) \\
&\quad+w(t)^T B_2^T(\delta P_2)B_2 w(t)
\end{aligned}
$$

With Lemma 4.1 and the given constraint conditions, one has

$$
\begin{aligned}
&\dot{e}_x^T(t)(\delta P_2)\dot{e}_x(t)\leq e_x(t)^T(A-LC_1)^T(\delta P_2)(A-LC_1)e_x(t) \\
&\quad+\xi_8^{-1}e_x(t)^T(A-LC_1)^T(\delta P_2)M_1M_1^T(\delta P_2)(A-LC_1)e_x(t) \\
&\quad+\xi_8 x(t)^T N_1^T N_1 x(t)+2e_x(t)^T(A-LC_1)^T(\delta P_2)Ee_\xi(t) \\
&\quad+\eta e_x(t)^T\left[(A-LC_1)^T(\delta P_2)EE^T+EE^T(\delta P_2)(A-LC_1)\right]e_x(t) \\
&\quad+\xi_9^{-1}e_x(t)^T(A-LC_1)^T(\delta P_2)B_2B_2^T(\delta P_2)(A-LC_1)e_x(t)+\xi_9\kappa_1^2 \\
&\quad+x(t)^T\Delta A^T(\delta P_2)\Delta A x(t)+\xi_{10}^{-1}x(t)^T N_1^T N_1 x(t) \\
&\quad+\xi_{10}e_\xi(t)^T E^T(\delta P_2)M_1M_1^T(\delta P_2)Ee_\xi(t)+\xi_{11}^{-1}x(t)^T N_1^T N_1 x(t) \\
&\quad+\xi_{11}\eta^2 e_x(t)^T EE^T(\delta P_2)M_1M_1^T(\delta P_2)EE^T e_x(t)+\xi_{12}^{-1}x(t)^T N_1^T N_1 x(t) \\
&\quad+\xi_{12}w(t)^T B_2^T(\delta P_2)M_1M_1^T(\delta P_2)B_2 w(t)+e_\xi(t)^T E^T(\delta P_2)Ee_\xi(t) \\
&\quad+2\eta e_\xi(t)^T E^T(\delta P_2)EE^T e_x(t)+\xi_{13}^{-1}e_\xi(t)^T E^T(\delta P_2)B_2B_2^T(\delta P_2)Ee_\xi(t)+\xi_{13}\kappa_1^2 \\
&\quad+\eta^2 e_x(t)^T EE^T EE^T e_x(t)+\xi_{14}^{-1}\eta^2 e_x(t)^T EE^T(\delta P_2)B_2B_2^T(\delta P_2)EE^T e_x(t) \\
&\quad+\xi_{14}\kappa_1^2+w(t)^T B_2^T(\delta P_2)B_2 w(t)
\end{aligned}
$$

Therefore, the first derivative of the Lyapunov function becomes

$$\dot{V}(t)\leq \boldsymbol{\chi}(t)^T\begin{bmatrix} \boldsymbol{\Pi}_{11} & \boldsymbol{\Pi}_{12} & \boldsymbol{P}_1\boldsymbol{A}_d & -\boldsymbol{P}_1\boldsymbol{A}_d & \boldsymbol{P}_1\boldsymbol{E} & 0 \\ * & \boldsymbol{\Pi}_{22} & 0 & 0 & \boldsymbol{\Pi}_{25} & 0 \\ * & * & 0 & 0 & 0 & 0 \\ * & * & * & -(1-\vartheta)\boldsymbol{P}_5 & 0 & 0 \\ * & * & * & * & \boldsymbol{\Pi}_{55} & 0 \\ * & * & * & * & * & -\boldsymbol{P}_4 \end{bmatrix}\boldsymbol{\chi}(t)$$
$$+\xi_2\kappa_1^2+\xi_4\kappa_1^2+\xi_5\kappa_2^2+\xi_7\kappa_1^2+l_2\dot{\boldsymbol{e}}_x^T(t)(\delta\boldsymbol{P}_2)\dot{\boldsymbol{e}}_x(t)+(l_2-l_1)\dot{\boldsymbol{e}}_x^T(t)(\delta\boldsymbol{P}_2)\dot{\boldsymbol{e}}_x(t)$$
$$+2\boldsymbol{\chi}(t)^T\boldsymbol{M}[\boldsymbol{e}_x(t)-\boldsymbol{e}_x(t-\tau(t))]+2\boldsymbol{\chi}(t)^T\boldsymbol{N}[\boldsymbol{e}_x(t-\tau(t))-\boldsymbol{e}_x(t-l_2)]$$
$$-\int_{t-\tau(t)}^{t}\left[\boldsymbol{\chi}(t)^T\boldsymbol{M}+\dot{\boldsymbol{e}}_x(s)(\delta\boldsymbol{P}_2)\right](\delta\boldsymbol{P}_2)^{-1}\left[\boldsymbol{M}^T\boldsymbol{\chi}(t)+(\delta\boldsymbol{P}_2)^T\dot{\boldsymbol{e}}_x(s)^T\right]ds$$
$$-\int_{t-l_2}^{t-\tau(t)}\left[\boldsymbol{\chi}(t)^T\boldsymbol{N}+\dot{\boldsymbol{e}}_x(s)(\delta\boldsymbol{P}_2)\right](\delta\boldsymbol{P}_2)^{-1}\left[\boldsymbol{N}^T\boldsymbol{\chi}(t)+(\delta\boldsymbol{P}_2)^T\dot{\boldsymbol{e}}_x(s)^T\right]ds$$
$$+l_2\boldsymbol{\chi}(t)^T\boldsymbol{M}(\delta\boldsymbol{P}_2)^{-1}\boldsymbol{M}^T\boldsymbol{\chi}(t)+(l_2-l_1)\boldsymbol{\chi}(t)^T\boldsymbol{N}(\delta\boldsymbol{P}_2)^{-1}\boldsymbol{N}^T\boldsymbol{\chi}(t)$$

where,

$$\begin{aligned}
\boldsymbol{\Pi}_{11} &= (\boldsymbol{A}-\boldsymbol{B}_1\boldsymbol{K})^T\boldsymbol{P}_1+\boldsymbol{P}_1(\boldsymbol{A}-\boldsymbol{B}_1\boldsymbol{K})+\xi_1^{-1}\boldsymbol{P}_1\boldsymbol{M}_1\boldsymbol{M}_1^T\boldsymbol{P}_1 \\
&\quad +\xi_1\boldsymbol{N}_1^T\boldsymbol{N}_1+\xi_2^{-1}\boldsymbol{P}_1\boldsymbol{B}_2\boldsymbol{B}_2^T\boldsymbol{P}_1+\xi_3\boldsymbol{N}_1^T\boldsymbol{N}_1+\xi_6\boldsymbol{N}_1^T\boldsymbol{N} \\
\boldsymbol{\Pi}_{12} &= \boldsymbol{P}_1\boldsymbol{B}_1\boldsymbol{K}+\eta\boldsymbol{P}_1\boldsymbol{E}\boldsymbol{E}^T \\
\boldsymbol{\Pi}_{22} &= (\boldsymbol{A}-\boldsymbol{L}\boldsymbol{C}_1)^T\boldsymbol{P}_2+\boldsymbol{P}_2(\boldsymbol{A}-\boldsymbol{L}\boldsymbol{C}_1)+\xi_3^{-1}\boldsymbol{P}_2\boldsymbol{M}_1\boldsymbol{M}_1^T\boldsymbol{P}_2 \\
&\quad +\eta\boldsymbol{P}_2\boldsymbol{E}\boldsymbol{E}^T+\eta\boldsymbol{E}\boldsymbol{E}^T\boldsymbol{P}_2+\xi_4^{-1}\boldsymbol{P}_2\boldsymbol{B}_2\boldsymbol{B}_2^T\boldsymbol{P}_2+\boldsymbol{P}_4+\boldsymbol{P}_5 \\
\boldsymbol{\Pi}_{25} &= \boldsymbol{P}_2\boldsymbol{E}-\eta\boldsymbol{A}^T\boldsymbol{E}\boldsymbol{P}_3-\eta^2\boldsymbol{E}\boldsymbol{E}^T\boldsymbol{E}\boldsymbol{P}_3 \\
\boldsymbol{\Pi}_{55} &= -\eta(\boldsymbol{P}_3\boldsymbol{E}^T\boldsymbol{E}+\boldsymbol{E}^T\boldsymbol{E}\boldsymbol{P}_3)+\xi_6^{-1}\eta^2\boldsymbol{P}_3\boldsymbol{E}^T\boldsymbol{M}_1\boldsymbol{M}_1^T\boldsymbol{E}\boldsymbol{P}_3 \\
&\quad \xi_5^{-1}\boldsymbol{P}_3\boldsymbol{P}_3+\xi_7^{-1}\eta^2\boldsymbol{P}_3\boldsymbol{E}^T\boldsymbol{B}_2\boldsymbol{B}_2{}^T\boldsymbol{E}\boldsymbol{P}_3
\end{aligned}$$

that is,

$$\dot{V}(t)\leq \boldsymbol{\chi}(t)^T\begin{bmatrix} \tilde{\boldsymbol{\Pi}}_{11} & \boldsymbol{\Pi}_{12} & \boldsymbol{P}_1\boldsymbol{A}_d & -\boldsymbol{P}_1\boldsymbol{A}_d & \boldsymbol{P}_1\boldsymbol{E} & 0 \\ * & \tilde{\boldsymbol{\Pi}}_{22} & 0 & 0 & \tilde{\boldsymbol{\Pi}}_{25} & 0 \\ * & * & 0 & 0 & 0 & 0 \\ * & * & * & -(1-\vartheta)\boldsymbol{P}_5 & 0 & 0 \\ * & * & * & * & \tilde{\boldsymbol{\Pi}}_{55} & 0 \\ * & * & * & * & * & -\boldsymbol{P}_4 \end{bmatrix}\boldsymbol{\chi}(t)$$
$$+\xi_2\kappa_1^2+\xi_4\kappa_1^2+\xi_5\kappa_2^2+\xi_7\kappa_1^2+(2l_2\text{-}l_1)(\xi_9+\xi_{13}+\xi_{14})\kappa_1^2$$
$$+2\boldsymbol{\chi}(t)^T\boldsymbol{M}[\boldsymbol{e}_x(t)-\boldsymbol{e}_x(t-\tau(t))]+2\boldsymbol{\chi}(t)^T\boldsymbol{N}[\boldsymbol{e}_x(t-\tau(t))-\boldsymbol{e}_x(t-l_2)]$$
$$+l_2\boldsymbol{\chi}(t)^T\boldsymbol{M}(\delta\boldsymbol{P}_2)^{-1}\boldsymbol{M}^T\boldsymbol{\chi}(t)+(l_2-l_1)\boldsymbol{\chi}(t)^T\boldsymbol{N}(\delta\boldsymbol{P}_2)^{-1}\boldsymbol{N}^T\boldsymbol{\chi}(t)$$
$$+(2l_2-l_1)\boldsymbol{w}(t)^T\boldsymbol{B}_2^T\left(\xi_{12}(\delta\boldsymbol{P}_2)\boldsymbol{M}_1\boldsymbol{M}_1^T(\delta\boldsymbol{P}_2)+(\delta\boldsymbol{P}_2)\right)\boldsymbol{B}_2\boldsymbol{w}(t)$$

where

$$\begin{aligned}
\tilde{\boldsymbol{\Pi}}_{11} &= \boldsymbol{\Pi}_{11} + (2l_2 - l_1)(\xi_8 + \xi_{10}^{-1} + \xi_{11}^{-1} + \xi_{12}^{-1})\boldsymbol{N}_1^T\boldsymbol{N}_1 + (2l_2 - l_1)\lambda_{\mathrm{mdA}}(\delta\boldsymbol{P}_2) \\
\tilde{\boldsymbol{\Pi}}_{22} &= \boldsymbol{\Pi}_{22} + (2l_2 - l_1)(\boldsymbol{A}-\boldsymbol{L}\boldsymbol{C}_1)^T(\delta\boldsymbol{P}_2)(\boldsymbol{A} - \boldsymbol{L}\boldsymbol{C}_1) \\
&\quad + (2l_2 - l_1)\xi_8^{-1}(\boldsymbol{A}-\boldsymbol{L}\boldsymbol{C}_1)^T(\delta\boldsymbol{P}_2)\boldsymbol{M}_1\boldsymbol{M}_1^T(\delta\boldsymbol{P}_2)(\boldsymbol{A} - \boldsymbol{L}\boldsymbol{C}_1) \\
&\quad + (2l_2 - l_1)\eta(\boldsymbol{A}-\boldsymbol{L}\boldsymbol{C}_1)^T(\delta\boldsymbol{P}_2)\boldsymbol{E}\boldsymbol{E}^T + (2l_2 - l_1)\eta\boldsymbol{E}\boldsymbol{E}^T(\delta\boldsymbol{P}_2)(\boldsymbol{A} - \boldsymbol{L}\boldsymbol{C}_1) \\
&\quad + (2l_2 - l_1)\xi_9^{-1}(\boldsymbol{A}-\boldsymbol{L}\boldsymbol{C}_1)^T(\delta\boldsymbol{P}_2)\boldsymbol{B}_2\boldsymbol{B}_2^T(\delta\boldsymbol{P}_2)(\boldsymbol{A} - \boldsymbol{L}\boldsymbol{C}_1) \\
&\quad + (2l_2 - l_1)\xi_{11}\eta^2\boldsymbol{E}\boldsymbol{E}^T(\delta\boldsymbol{P}_2)\boldsymbol{M}_1\boldsymbol{M}_1^T(\delta\boldsymbol{P}_2)\boldsymbol{E}\boldsymbol{E}^T + (2l_2 - l_1)\eta^2\boldsymbol{E}\boldsymbol{E}^T\boldsymbol{E}\boldsymbol{E}^T \\
&\quad + (2l_2 - l_1)\xi_{14}^{-1}\eta^2\boldsymbol{E}\boldsymbol{E}^T(\delta\boldsymbol{P}_2)\boldsymbol{B}_2\boldsymbol{B}_2^T(\delta\boldsymbol{P}_2)\boldsymbol{E}\boldsymbol{E}^T \\
\tilde{\boldsymbol{\Pi}}_{25} &= \boldsymbol{\Pi}_{25} + (2l_2 - l_1)(\boldsymbol{A}-\boldsymbol{L}\boldsymbol{C}_1)^T(\delta\boldsymbol{P}_2)\boldsymbol{E}+(2l_2 - l_1)\eta\boldsymbol{E}\boldsymbol{E}^T(\delta\boldsymbol{P}_2)\boldsymbol{E} \\
\tilde{\boldsymbol{\Pi}}_{55} &= \boldsymbol{\Pi}_{55} + (2l_2 - l_1)\xi_{10}\boldsymbol{E}^T(\delta\boldsymbol{P}_2)\boldsymbol{M}_1\boldsymbol{M}_1^T(\delta\boldsymbol{P}_2)\boldsymbol{E} + (2l_2 - l_1)\boldsymbol{E}^T(\delta\boldsymbol{P}_2)\boldsymbol{E} \\
&\quad + (2l_2 - l_1)\xi_{13}^{-1}\boldsymbol{E}^T(\delta\boldsymbol{P}_2)\boldsymbol{B}_2\boldsymbol{B}_2^T(\delta\boldsymbol{P}_2)\boldsymbol{E}
\end{aligned}$$

where λ_{mdA} denotes $\lambda_{\max}(\Delta\boldsymbol{A}^T\Delta\boldsymbol{A})$, which can be calculated before simulation.

Then,

$$\begin{aligned}
\dot{V}(t) \leq\, &\boldsymbol{\chi}(t)^T\left\{\boldsymbol{\Theta}_N + l_2\boldsymbol{M}(\delta\boldsymbol{P}_2)^{-1}\boldsymbol{M}^T + (l_2 - l_1)\boldsymbol{N}(\delta\boldsymbol{P}_2)^{-1}\boldsymbol{N}^T\right\}\boldsymbol{\chi}(t) \\
&+ \xi_2\kappa_1^2 + \xi_4\kappa_1^2 + \xi_5\kappa_2^2 + \xi_7\kappa_1^2 + (2l_2 - l_1)(\xi_9 + \xi_{13} + \xi_{14})\kappa_1^2 \\
&+ (2l_2 - l_1)\boldsymbol{w}(t)^T\boldsymbol{B}_2^T\left(\xi_{12}(\delta\boldsymbol{P}_2)\boldsymbol{M}_1\boldsymbol{M}_1^T(\delta\boldsymbol{P}_2) + (\delta\boldsymbol{P}_2)\right)\boldsymbol{B}_2\boldsymbol{w}(t)
\end{aligned}$$

where,

$$\boldsymbol{\Theta}_N = \begin{bmatrix} \boldsymbol{\Theta}_{N11} & \boldsymbol{\Theta}_{N12} \\ * & \boldsymbol{\Theta}_{N22} \end{bmatrix},$$

and where,

$$\boldsymbol{\Theta}_{N11} = \begin{bmatrix} \tilde{\boldsymbol{\Pi}}_{11} & \boldsymbol{\Pi}_{12}+\underset{\sim}{\boldsymbol{M}}_1 & \boldsymbol{P}_1\boldsymbol{A}_d \\ * & \tilde{\boldsymbol{\Pi}}_{22} + \underset{\sim}{\boldsymbol{M}}_2 + \underset{\sim}{\boldsymbol{M}}_2^T & \underset{\sim}{\boldsymbol{M}}_3^T \\ * & * & 0 \end{bmatrix},$$

$$\boldsymbol{\Theta}_{N12} = \begin{bmatrix} -\boldsymbol{P}_1\boldsymbol{A}_d - \underset{\sim}{\boldsymbol{M}}_1 + \underset{\sim}{\boldsymbol{N}}_1 & \boldsymbol{P}_1\boldsymbol{E} & -\underset{\sim}{\boldsymbol{N}}_1 \\ \underset{\sim}{\boldsymbol{M}}_4^T - \underset{\sim}{\boldsymbol{M}}_2 + \underset{\sim}{\boldsymbol{N}}_2 & \tilde{\boldsymbol{\Pi}}_{25} + \underset{\sim}{\boldsymbol{M}}_5^T & \underset{\sim}{\boldsymbol{M}}_6^T - \underset{\sim}{\boldsymbol{N}}_2 \\ -\underset{\sim}{\boldsymbol{M}}_3 + \underset{\sim}{\boldsymbol{N}}_3 & 0 & -\underset{\sim}{\boldsymbol{N}}_3 \end{bmatrix},$$

$$\boldsymbol{\Theta}_{N22} = \begin{bmatrix} -(1-\vartheta)\boldsymbol{P}_5 - \underset{\sim}{\boldsymbol{M}}_4 - \underset{\sim}{\boldsymbol{M}}_4^T + \underset{\sim}{\boldsymbol{N}}_4 + \underset{\sim}{\boldsymbol{N}}_4^T & -\underset{\sim}{\boldsymbol{M}}_5^T + \underset{\sim}{\boldsymbol{N}}_5^T & -\underset{\sim}{\boldsymbol{M}}_6^T + \underset{\sim}{\boldsymbol{N}}_6^T - \underset{\sim}{\boldsymbol{N}}_4 \\ * & \tilde{\boldsymbol{\Pi}}_{55} & -\underset{\sim}{\boldsymbol{N}}_5 \\ * & * & -\boldsymbol{P}_4 - \underset{\sim}{\boldsymbol{N}}_6 - \underset{\sim}{\boldsymbol{N}}_6^T \end{bmatrix}$$

Denote $\boldsymbol{\Theta}_P = -\{\boldsymbol{\Theta}_N + l_2\boldsymbol{M}(\delta\boldsymbol{P}_2)^{-1}\boldsymbol{M}^T + (l_2 - l_1)\boldsymbol{N}(\delta\boldsymbol{P}_2)^{-1}\boldsymbol{N}^T\}$, $\boldsymbol{M}_0 = (2l_2 - l_1)\boldsymbol{B}_2^T(\xi_{12}(\delta\boldsymbol{P}_2)\boldsymbol{M}_1\boldsymbol{M}_1^T(\delta\boldsymbol{P}_2) + (\delta\boldsymbol{P}_2))\boldsymbol{B}_2$, one knows

$$\begin{aligned}\dot{V}(t) \le & -\chi(t)^T\boldsymbol{\Theta}_P\chi(t) + \xi_2\kappa_1^2 + \xi_4\kappa_1^2 + \xi_5\kappa_2^2 + \xi_7\kappa_1^2 \\ & + (2l_2 - l_1)(\xi_9 + \xi_{13} + \xi_{14})\kappa_1^2 + \lambda_{\max}(\boldsymbol{M}_0)\kappa_1^2\end{aligned}$$

Let $\gamma = \xi_2\kappa_1^2 + \xi_4\kappa_1^2 + \xi_5\kappa_2^2 + \xi_7\kappa_1^2 + (2l_2 - l_1)(\xi_9 + \xi_{13} + \xi_{14})\kappa_1^2 + \lambda_{\max}(\boldsymbol{M}_0)\kappa_1^2$, it can be concluded that if $\boldsymbol{\Theta}_P > 0$, then

$$\dot{V}(t) \le -\lambda_{\min}(\boldsymbol{\Theta}_P)\|\chi\|^2 + \gamma$$

Furthermore, define a set S described by

$$S = \left\{\chi(t)\middle|\lambda_{\min}(\boldsymbol{\Theta}_P)\|\chi\|^2 > \gamma\right\}$$

Let $\overline{S}$ be the complement of S. It follows that $\dot{V}(t) < 0$, $\chi(t) \in S$, and the state $\chi(t)$ will converge asymptotically to $\overline{S}$. With **Lemma 4.2**, $\boldsymbol{\Theta}_P > 0$ equals the following inequality:

$$\begin{bmatrix} \boldsymbol{\Theta}_N & \boldsymbol{M} & \boldsymbol{N} \\ * & -\dfrac{\delta\boldsymbol{P}_2}{l_2} & 0 \\ * & * & -\dfrac{\delta\boldsymbol{P}_2}{l_2 - l_1} \end{bmatrix} < 0$$

According to Lemma 4.2, and defining $\boldsymbol{H} = \boldsymbol{P}_2\boldsymbol{L}$, one can obtain

$$\begin{bmatrix} \boldsymbol{\Theta}_{11} & \boldsymbol{\Theta}_{12} \\ * & \boldsymbol{\Theta}_{22} \end{bmatrix} < 0$$

where,

$$\boldsymbol{\Theta}_{11} = \begin{bmatrix} \overline{\boldsymbol{\Theta}}_{11} & \overline{\boldsymbol{\Theta}}_{12} & \boldsymbol{P}_1\boldsymbol{A}_d & \overline{\boldsymbol{\Theta}}_{14} & \boldsymbol{P}_1\boldsymbol{E} & -\underset{\sim}{\boldsymbol{N}}_1 & \underset{\sim}{\boldsymbol{M}}_1 & \underset{\sim}{\boldsymbol{N}}_1 \\ * & \overline{\boldsymbol{\Theta}}_{22} & \underset{\sim}{\boldsymbol{M}}_3^T & \overline{\boldsymbol{\Theta}}_{24} & \overline{\boldsymbol{\Theta}}_{25} & \overline{\boldsymbol{\Theta}}_{26} & \underset{\sim}{\boldsymbol{M}}_2 & \underset{\sim}{\boldsymbol{N}}_2 \\ * & * & 0 & \overline{\boldsymbol{\Theta}}_{34} & 0 & -\underset{\sim}{\boldsymbol{N}}_3 & \underset{\sim}{\boldsymbol{M}}_3 & \underset{\sim}{\boldsymbol{N}}_3 \\ * & * & * & \overline{\boldsymbol{\Theta}}_{44} & \overline{\boldsymbol{\Theta}}_{45} & \overline{\boldsymbol{\Theta}}_{46} & \underset{\sim}{\boldsymbol{M}}_4 & \underset{\sim}{\boldsymbol{N}}_4 \\ * & * & * & * & \overline{\boldsymbol{\Theta}}_{55} & -\underset{\sim}{\boldsymbol{N}}_5 & \underset{\sim}{\boldsymbol{M}}_5 & \underset{\sim}{\boldsymbol{N}}_5 \\ * & * & * & * & * & \overline{\boldsymbol{\Theta}}_{66} & \underset{\sim}{\boldsymbol{M}}_6 & \underset{\sim}{\boldsymbol{N}}_6 \\ * & * & * & * & * & * & -\dfrac{\delta\boldsymbol{P}_2}{l_2} & 0 \\ * & * & * & * & * & * & * & -\dfrac{\delta\boldsymbol{P}_2}{l_2 - l_1} \end{bmatrix}$$

and where,

$$\overline{\boldsymbol{\Theta}}_{11} = (\boldsymbol{A}-\boldsymbol{B}_1\boldsymbol{K})^T\boldsymbol{P}_1 + \boldsymbol{P}_1(\boldsymbol{A}-\boldsymbol{B}_1\boldsymbol{K}) + (2l_2 - l_1)\lambda_{\text{mdA}}(\delta\boldsymbol{P}_2)$$

$$\overline{\boldsymbol{\Theta}}_{12} = \boldsymbol{P}_1\boldsymbol{B}_1\boldsymbol{K}+\eta\boldsymbol{P}_1\boldsymbol{E}\boldsymbol{E}^T+\underset{\sim}{\boldsymbol{M}}_1$$

$$\overline{\boldsymbol{\Theta}}_{14} = -\boldsymbol{P}_1\boldsymbol{A}_d - \underset{\sim}{\boldsymbol{M}}_1 + \underset{\sim}{\boldsymbol{N}}_1$$

$$\begin{aligned}\overline{\boldsymbol{\Theta}}_{22} = {} & \boldsymbol{A}^T\boldsymbol{P}_2 - \boldsymbol{C}_1^T\boldsymbol{H}^T + \boldsymbol{P}_2\boldsymbol{A} - \boldsymbol{H}\boldsymbol{C}_1 + \eta\boldsymbol{P}_2\boldsymbol{E}\boldsymbol{E}^T + \eta\boldsymbol{E}\boldsymbol{E}^T\boldsymbol{P}_2 \\ & +\boldsymbol{P}_4+\boldsymbol{P}_5 + (2l_2 - l_1)\eta\delta(\boldsymbol{A}^T\boldsymbol{P}_2 - \boldsymbol{C}_1^T\boldsymbol{H}^T)\boldsymbol{E}\boldsymbol{E}^T \\ & + (2l_2 - l_1)\eta\delta\boldsymbol{E}\boldsymbol{E}^T(\boldsymbol{P}_2\boldsymbol{A} - \boldsymbol{H}\boldsymbol{C}_1) \\ & + (2l_2 - l_1)\eta^2\boldsymbol{E}\boldsymbol{E}^T\boldsymbol{E}\boldsymbol{E}^T + \underset{\sim}{\boldsymbol{M}}_2 + \underset{\sim}{\boldsymbol{M}}_2^T\end{aligned}$$

$$\overline{\boldsymbol{\Theta}}_{24} = \underset{\sim}{\boldsymbol{M}}_4^T - \underset{\sim}{\boldsymbol{M}}_2 + \underset{\sim}{\boldsymbol{N}}_2$$

$$\begin{aligned}\overline{\boldsymbol{\Theta}}_{25} = {} & (2l_2 - l_1)\delta(\boldsymbol{A}^T\boldsymbol{P}_2 - \boldsymbol{C}_1^T\boldsymbol{H}^T)\boldsymbol{E}+\boldsymbol{P}_2\boldsymbol{E} - \eta\boldsymbol{A}^T\boldsymbol{E}\boldsymbol{P}_3 \\ & +(2l_2 - l_1)\eta\boldsymbol{E}\boldsymbol{E}^T(\delta\boldsymbol{P}_2)\boldsymbol{E} - \eta^2\boldsymbol{E}\boldsymbol{E}^T\boldsymbol{E}\boldsymbol{P}_3 + \underset{\sim}{\boldsymbol{M}}_5^T\end{aligned}$$

$$\overline{\boldsymbol{\Theta}}_{26} = \underset{\sim}{\boldsymbol{M}}_6^T - \underset{\sim}{\boldsymbol{N}}_2$$

$$\overline{\boldsymbol{\Theta}}_{34} = -\underset{\sim}{\boldsymbol{M}}_3 + \underset{\sim}{\boldsymbol{N}}_3$$

$$\overline{\boldsymbol{\Theta}}_{44} = -(1-\vartheta)\boldsymbol{P}_5 - \underset{\sim}{\boldsymbol{M}}_4 - \underset{\sim}{\boldsymbol{M}}^T{}_4 + \underset{\sim}{\boldsymbol{N}}_4 + \underset{\sim}{\boldsymbol{N}}_4^T$$

$$\overline{\boldsymbol{\Theta}}_{45} = -\underset{\sim}{\boldsymbol{M}}_5^T + \underset{\sim}{\boldsymbol{N}}_5^T$$

$$\overline{\boldsymbol{\Theta}}_{46} = -\underset{\sim}{\boldsymbol{M}}_6^T + \underset{\sim}{\boldsymbol{N}}_6^T - \underset{\sim}{\boldsymbol{N}}_4$$

$$\overline{\boldsymbol{\Theta}}_{55} = (2l_2 - l_1)\boldsymbol{E}^T(\delta\boldsymbol{P}_2)\boldsymbol{E} - \eta(\boldsymbol{P}_3\boldsymbol{E}^T\boldsymbol{E}+\boldsymbol{E}^T\boldsymbol{E}\boldsymbol{P}_3)$$

$$\overline{\boldsymbol{\Theta}}_{66} = -\boldsymbol{P}_4 - \underset{\sim}{\boldsymbol{N}}_6 - \underset{\sim}{\boldsymbol{N}}_6^T$$

and

$$\boldsymbol{\Theta}_{12} = \begin{bmatrix}\boldsymbol{\Theta}_{121} & \boldsymbol{\Theta}_{122} & \boldsymbol{\Theta}_{123}\end{bmatrix}$$

where,

$$\boldsymbol{\Theta}_{121} = \begin{bmatrix} \boldsymbol{P}_1\boldsymbol{M}_1 & \boldsymbol{N}_1^T & \boldsymbol{P}_1\boldsymbol{B}_2 & 0 & 0 & 0 \\ 0 & 0 & 0 & \boldsymbol{P}_2\boldsymbol{M}_1 & \boldsymbol{P}_2\boldsymbol{B}_2 & \delta\boldsymbol{H}_0 \\ 0 & 0 & 0 & 0 & 0 & 0 \\ 0 & 0 & 0 & 0 & 0 & 0 \\ 0 & 0 & 0 & 0 & 0 & 0 \\ 0 & 0 & 0 & 0 & 0 & 0 \\ 0 & 0 & 0 & 0 & 0 & 0 \\ 0 & 0 & 0 & 0 & 0 & 0 \end{bmatrix}$$

$$\boldsymbol{\Theta}_{122}=\begin{bmatrix} 0 & 0 & 0 & 0 \\ \delta\boldsymbol{H}_0\boldsymbol{M}_1 & \delta\boldsymbol{H}_0\boldsymbol{B}_2 & \boldsymbol{E}\boldsymbol{E}^T(\delta\boldsymbol{P}_2)\boldsymbol{M}_1 & \boldsymbol{E}\boldsymbol{E}^T(\delta\boldsymbol{P}_2)\boldsymbol{B}_2 \\ 0 & 0 & 0 & 0 \\ 0 & 0 & 0 & 0 \\ 0 & 0 & 0 & 0 \\ 0 & 0 & 0 & 0 \\ 0 & 0 & 0 & 0 \\ 0 & 0 & 0 & 0 \end{bmatrix}$$

$$\boldsymbol{\Theta}_{123}=\begin{bmatrix} 0 & 0 & 0 & 0 & 0 \\ 0 & 0 & 0 & 0 & 0 \\ 0 & 0 & 0 & 0 & 0 \\ 0 & 0 & 0 & 0 & 0 \\ \boldsymbol{E}^T(\delta\boldsymbol{P}_2)\boldsymbol{M}_1 & \boldsymbol{E}^T(\delta\boldsymbol{P}_2)\boldsymbol{B}_2 & \boldsymbol{P}_3 & \boldsymbol{P}_3\boldsymbol{E}^T\boldsymbol{M}_1 & \boldsymbol{P}_3\boldsymbol{E}^T\boldsymbol{B}_2 \\ 0 & 0 & 0 & 0 & 0 \\ 0 & 0 & 0 & 0 & 0 \\ 0 & 0 & 0 & 0 & 0 \end{bmatrix}$$

and where $\boldsymbol{H}_0=\boldsymbol{A}^T\boldsymbol{P}_2-\boldsymbol{C}_1^T\boldsymbol{H}^T$;

and

$$\boldsymbol{\Theta}_{22}=diag\left(\left[\boldsymbol{\Theta}_{221} \quad \boldsymbol{\Theta}_{222} \quad \boldsymbol{\Theta}_{223} \quad \boldsymbol{\Theta}_{224}\right]\right)$$

where,

$$\boldsymbol{\Theta}_{221}=\begin{bmatrix} -\xi_1\boldsymbol{I} & -\dfrac{\boldsymbol{I}}{\xi_1+\xi_3+\xi_6+(2l_2\text{-}l_1)(\xi_8+\xi_{10}^{-1}+\xi_{11}^{-1}+\xi_{12}^{-1})} \end{bmatrix}$$

$$\boldsymbol{\Theta}_{222}=\begin{bmatrix} -\xi_2\boldsymbol{I} & -\xi_3\boldsymbol{I} & -\xi_4\boldsymbol{I} & -\dfrac{\delta\boldsymbol{P}_2}{2l_2-l_1} & -\dfrac{\xi_8\boldsymbol{I}}{2l_2-l_1} & -\dfrac{\xi_9\boldsymbol{I}}{2l_2-l_1} \end{bmatrix}$$

$$\boldsymbol{\Theta}_{223}=\begin{bmatrix} -\dfrac{\boldsymbol{I}}{(2l_2-l_1)\xi_{11}\eta^2} & -\dfrac{\xi_{14}\boldsymbol{I}}{(2l_2-l_1)\eta^2} & -\dfrac{\boldsymbol{I}}{(2l_2-l_1)\xi_{10}} \end{bmatrix}$$

$$\boldsymbol{\Theta}_{224}=\begin{bmatrix} -\dfrac{\xi_{13}\boldsymbol{I}}{2l_2-l_1} & -\xi_5\boldsymbol{I} & -\dfrac{\xi_6\boldsymbol{I}}{\eta^2} & -\dfrac{\xi_7\boldsymbol{I}}{\eta^2} \end{bmatrix}$$

This completes the proof of Theorem 4.1.

Remark 4.5: Before solving inequality (4.17), inequality (4.16) should be solved to obtain the controller gain $\boldsymbol{K}$. The observer gain $\boldsymbol{L}$ can be then obtained by solving inequality (4.17). It should be noted that the feasibility of the LMIs can be improved by adjusting the aforementioned parameters, that is, $\delta, \vartheta, \eta, \xi_{i=1,2,\ldots,14}$, $\boldsymbol{M}$,$\boldsymbol{N}$ and $\widehat{\boldsymbol{P}}$.

Remark 4.6: Since the auxiliary variable $\boldsymbol{\xi}(t)$ can combine the attitude information and fault signal together, it is necessary to introduce this variable to estimate the attitude information and fault signal simultaneously such that the fault tolerant control scheme is designed based on the estimation. If we choose $\boldsymbol{e}_f(t)$ instead of $\boldsymbol{e}_\xi(t)$ as one of the components of $\boldsymbol{\chi}(t)$, the derived inequality will include the observer gain matrix $\boldsymbol{L}$, so that the solution cannot be obtained via the LMI method. This is also the reason why we introduce the new variable $\boldsymbol{\xi}(t)$. As long as γ is designed with possibly small value, that is, scalars $\xi_{i=2,4,5,7,9,13,14}$, κ_1, κ_2 are preset with possibly small values, the states of the closed-loop system will converge asymptotically to the vicinity of the expected states.

4.4 Simulation test

In this section, the effectiveness of the observer-based fault tolerant controller is illustrated with simulations of a spacecraft attitude control system. In this example, the moments of inertia are taken as $I_x = 20$ kg m^2, $I_y = 18$ kg m^2, $I_z = 15$ kg m^2, and the orbit is a 300 km circular orbit. Choose the initial state as $\boldsymbol{x}(0) = [0.08$ rad, 0.06 rad, -0.06 rad, -0.01 rad s^{-1}, -0.01 rad s^{-1}, 0.01 rad s$^{-1}]^T$, the maximum control torque is assumed to be $u_{m1} = u_{m2} = u_{m3} = 0.25$ Nm. This constraint on control torque is considered in the simulation.

The external disturbances used in the simulation are taken to be of the following form and ω_0 is the orbital frequency.

$$\boldsymbol{w}(t) = 5 \times 10^{-4} \times \begin{bmatrix} \sin(100\omega_0\pi t) \\ \cos(100\omega_0\pi t) \\ \cos(100\omega_0\pi t+\pi/3) \end{bmatrix} \text{Nm}$$

Meanwhile, choose

$$\begin{aligned}
&\boldsymbol{M}_1 = 0.1 \times [0.8 \quad 1.1 \quad 1.3 \quad 1.5 \quad 1.6 \quad 1.8]^T, \\
&\boldsymbol{N}_1 = 0.4 \times [0.1 \quad -0.2 \quad 0.3 \quad -0.4 \quad -0.2 \quad 1],\ \boldsymbol{A}_d = 0.3\boldsymbol{A}, \\
&\boldsymbol{F}_1(t) = \sin(100\omega_0 t),\ \xi_{i=1,2,\cdots,14} = 0.01,\ l_1 = 0.2,\ l_2 = 0.8, \\
&\vartheta = 0.3,\ \eta = 0.6,\ \delta = 0.1,\ \widehat{\boldsymbol{P}} = 30 \times diag[2\ 1\ 3\ 4\ 2\ 5], \\
&\underset{\sim}{\boldsymbol{M}}_1 = \boldsymbol{I}_{6\times 6},\ \underset{\sim}{\boldsymbol{M}}_2 = -\boldsymbol{I}_{6\times 6},\ \underset{\sim}{\boldsymbol{M}}_3 = \boldsymbol{0}_{6\times 6}, \\
&\underset{\sim}{\boldsymbol{M}}_4 = \boldsymbol{0}_{6\times 6},\ \underset{\sim}{\boldsymbol{M}}_5 = \boldsymbol{0}_{3\times 6},\ \underset{\sim}{\boldsymbol{M}}_6 = \boldsymbol{0}_{6\times 6}, \\
&\underset{\sim}{\boldsymbol{N}}_1 = -\boldsymbol{I}_{6\times 6},\ \underset{\sim}{\boldsymbol{N}}_2 = \boldsymbol{0}_{6\times 6},\ \underset{\sim}{\boldsymbol{N}}_3 = \boldsymbol{I}_{6\times 6}, \\
&\underset{\sim}{\boldsymbol{N}}_4 = \boldsymbol{0}_{6\times 6},\ \underset{\sim}{\boldsymbol{N}}_5 = \boldsymbol{0}_{3\times 6},\ \underset{\sim}{\boldsymbol{N}}_6 = -\boldsymbol{I}_{6\times 6}
\end{aligned}$$

Table 4.1 The form of $\bar{f}(t)$.

t/s	[0,10]	(10,45]	(45,60]	(60,75]	(75,100]
$\bar{f}(t)$/Nm	0	$\sin(0.1\pi t)$	1	$\cos(0.1\pi t)$	0

Here we assume the time-varying delay $\tau(t) = 0.5 + 0.3\sin(t)$. The fault signal $f(t)$ is assumed as $f(t) = [0.005\ 0.006\ 0.007]^T\bar{f}(t)$, and the form of $\bar{f}(t)$ is shown in Table 4.1.

The actuator fault is assumed to occur.

For comparison, two cases are studied in this section. In case A, the proposed observer-based fault-tolerant controller (OBFTC) is applied for attitude stabilization. In case B, the prediction-based sampled-data H_∞ controller (PBSDHC) introduced in [6] is employed. All simulation parameters are identical in the two cases.

4.4.1 Simulation results using the proposed controller

With inequality (4.16), the controller gain matrix $\boldsymbol{K}$ is

$$\boldsymbol{K} = \begin{bmatrix} 10 & 0 & 0 & 14.8924 & 0 & -0.0040 \\ 0 & 9 & 0 & 0 & 26.8064 & 0 \\ 0 & 0 & 9 & -0.0054 & 0 & 8.9355 \end{bmatrix}$$

The related matrices of observer gain are

$$\boldsymbol{P}_2 = \begin{bmatrix} 8.1598 & 0.8885 & 0.4163 & -0.4860 & -0.7006 & -1.0940 \\ 0.8885 & 5.8727 & 0.0783 & -0.5072 & -0.6677 & -0.9680 \\ 0.4163 & 0.0783 & 4.9509 & -0.5388 & -0.6642 & -0.9073 \\ -0.4860 & -0.5072 & -0.5388 & 3.5754 & -0.7168 & -0.8067 \\ -0.7006 & -0.6677 & -0.6642 & -0.7168 & 3.3828 & -0.7980 \\ -1.0940 & -0.9680 & -0.9073 & -0.8067 & -0.7980 & 3.1160 \end{bmatrix}$$

$$\boldsymbol{H} = \begin{bmatrix} 189.6127 & 146.0682 & 145.5760 & 114.5379 & 106.2898 & 104.2901 \\ 148.7442 & 190.7270 & 147.4262 & 108.9514 & 112.9326 & 105.5656 \\ 149.9955 & 149.1831 & 191.5207 & 109.6295 & 108.2711 & 111.5415 \\ 258.1546 & 257.6289 & 257.3930 & 260.7559 & 216.8100 & 214.0882 \\ 262.0670 & 261.5982 & 261.3527 & 223.3369 & 263.7536 & 219.0877 \\ 268.3320 & 268.0118 & 267.8330 & 231.2780 & 229.8245 & 268.8084 \end{bmatrix}$$

The eigenvalues of $\boldsymbol{P}_2$ are 0.9455, 4.0418, 4.2154, 5.1248, 5.7001, 9.0301, so $\boldsymbol{P}_2 > 0$. Then, one has

$$\boldsymbol{L} = \boldsymbol{P}_2^{-1}\boldsymbol{H}$$
$$= \begin{bmatrix} 86.9244 & 80.8629 & 81.3354 & 69.4814 & 68.4086 & 68.4609 \\ 119.4023 & 127.7316 & 120.4159 & 100.8053 & 101.5928 & 100.4794 \\ 147.9544 & 148.4821 & 157.3979 & 124.4829 & 124.1568 & 124.9074 \\ 255.1709 & 255.7887 & 256.3896 & 228.4791 & 217.7149 & 216.7422 \\ 281.1579 & 281.8420 & 282.4782 & 240.8721 & 250.1478 & 238.8605 \\ 334.8721 & 335.7175 & 336.4679 & 287.0174 & 285.9129 & 295.1722 \end{bmatrix}$$

Applying the controller gain $\boldsymbol{K}$ and observer gain $\boldsymbol{L}$ to Eqs. (4.13)–(4.15) gives the time responses of attitude angle and angular velocity as shown in Figs. 4.1 and 4.2, respectively. It is observed that the attitude angle and angular velocity converge to a small bound with a settling time of approximately 16 seconds. The accuracy of attitude angle is less than 1.5×10^{-4} rad and that of angular velocity is approximately 3×10^{-5} rad s^{-1} in steady-state error no matter whether the fault signal varies or is constant. For the analysis of the effectiveness of the presented observer, Figs. 4.3 and 4.4 show time responses of observer errors of attitude angle and angular velocity, respectively. As one can see, the developed observer has estimated the actual states successfully, and the observer errors also converge to the residual region, as shown in Figs. 4.3 and 4.4. These convergences were obtained in the presence of model parameter uncertainty, external

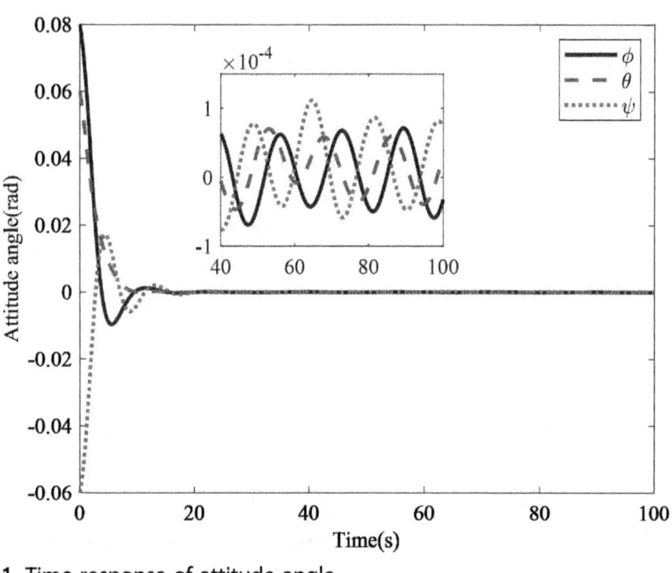

Figure 4.1 Time response of attitude angle.

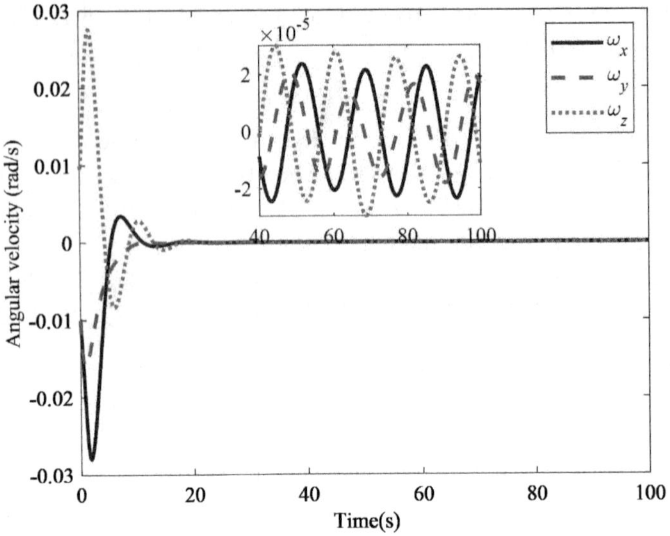

Figure 4.2 Time response of angular velocity.

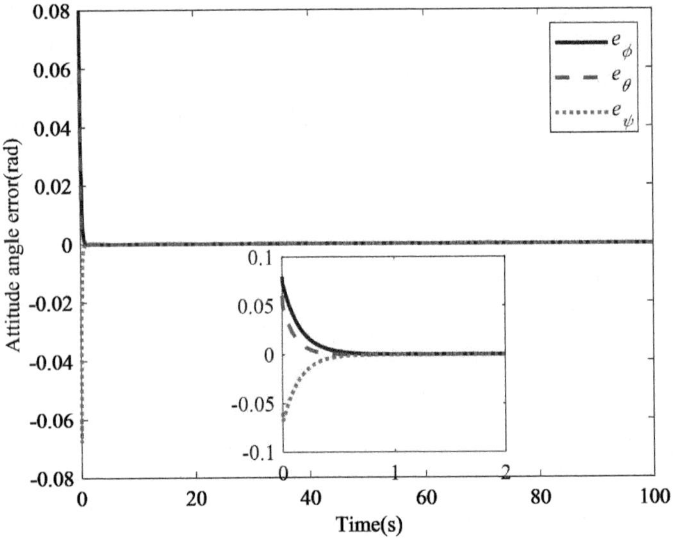

Figure 4.3 Time response of attitude angle error.

disturbances, actuator faults, and actuator saturation. Moreover, high precision of the observer is guaranteed, that is, $|e_\phi| < 9 \times 10^{-6}$rad, $|e_\theta| < 7.6 \times 10^{-6}$rad, $|e_\psi| < 7.2 \times 10^{-6}$rad, $|e_{\omega x}| < 6.3 \times 10^{-6}$rad s^{-1}, $|e_{\omega y}| < 1 \times 10^{-5}$rad s^{-1}, $|e_{\omega z}| < 2 \times 10^{-5}$rad s^{-1}. The fault signal and its estimation are plotted in Fig. 4.5, from which one can see that the observed fault signal is nearly the same

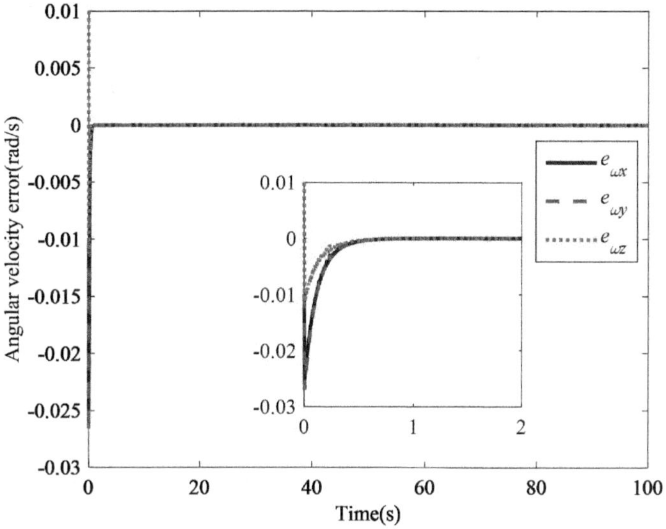

Figure 4.4 Time response of angular velocity error.

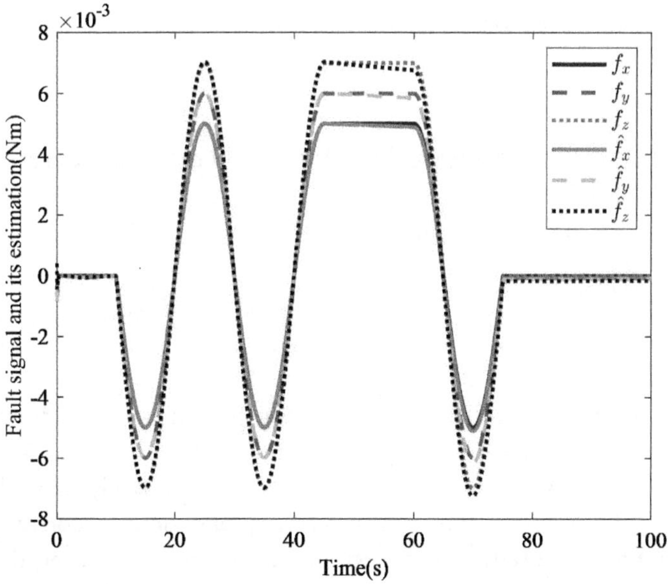

Figure 4.5 Fault signal and its estimation.

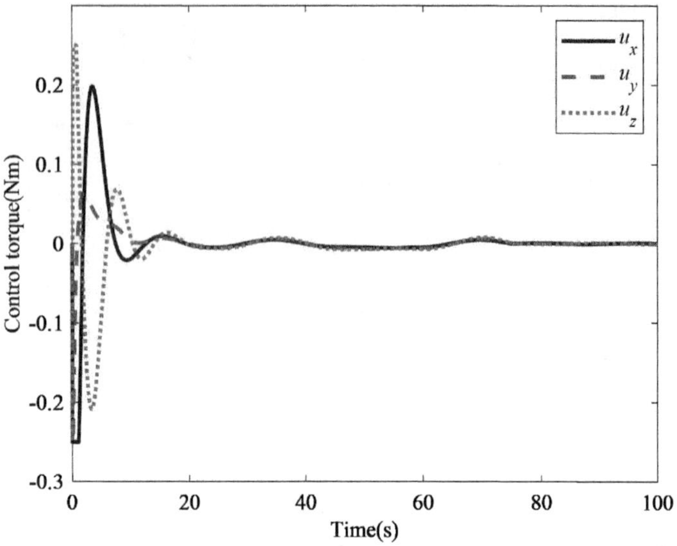

Figure 4.6 Time response of control torque.

as the real value. Based on these simulations, it is seen that both attitude information and fault signals are estimated with good accuracy. The control torque shown in Fig. 4.6 is no more than 0.25 Nm. Thus the simulation results verify the theoretical analysis and demonstrate the effectiveness of the proposed observer-based control strategy.

4.4.2 Simulation results using the prediction-based sampled-data H_∞ controller

In this case, the controller gain matrix $\boldsymbol{K}$ is

$$\boldsymbol{K} = \begin{bmatrix} -14.7353 & 0 & -0.0011 & -25.8891 & 0 & 0.0022 \\ 0 & -13.0087 & 0 & 0 & -23.0927 & 0 \\ -0.0123 & 0 & -9.2271 & 0.0023 & 0 & -17.5250 \end{bmatrix}$$

Applying this controller and considering the time-varying input delay and other factors, we obtain the time responses of attitude angle and angular velocity as shown in Figs. 4.7 and 4.8, respectively. It is observed that the attitude angle and angular velocity also quickly converge to a small bound with a settling time less than 20 seconds. However, the accuracy of the attitude angle is 1×10^{-3} rad s^{-1} and that of angular velocity

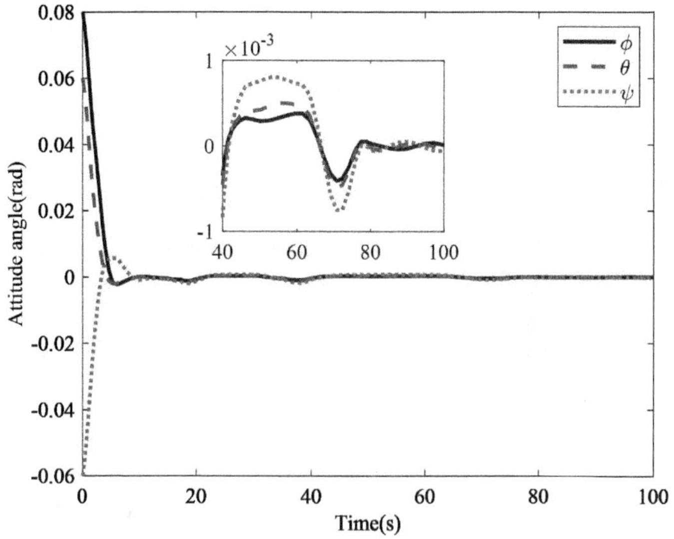

Figure 4.7 Time response of attitude angle using prediction-based sampled-data H_∞ controller.

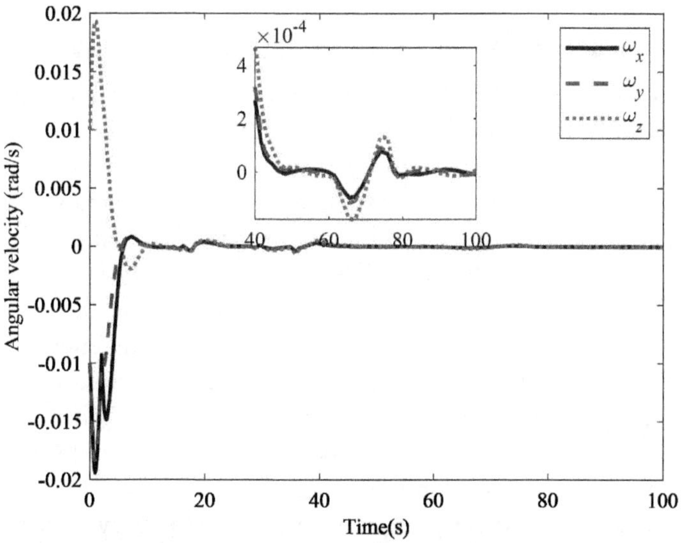

Figure 4.8 Time response of angular velocity using prediction-based sampled-data H_∞ controller.

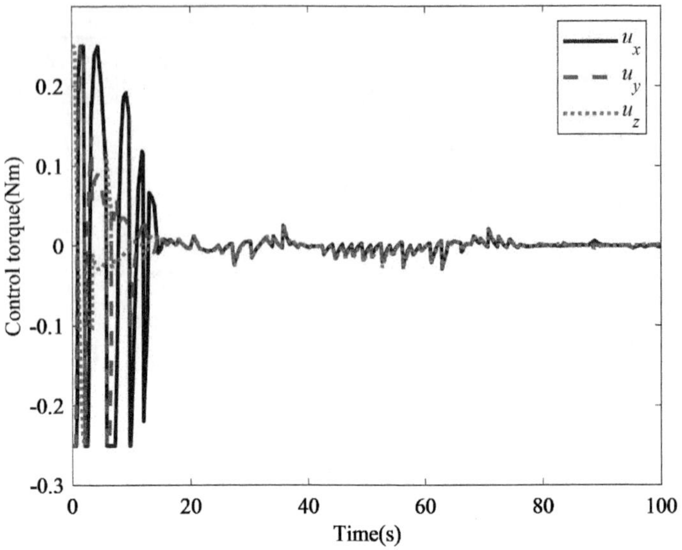

Figure 4.9 Time response of control torque using prediction-based sampled-data H_∞ controller.

is 5×10^{-4} rad s^{-1} in steady-state error when a fault signal exists. Fig. 4.9 shows the response curve of control torque using PBSDHC.

4.4.3 Comparison analysis using different controllers

Energy efficiency is an extremely important issue for achieving the target tasks and extending the mission operating life. Thus the optimization of energy consumption problem should also be taken into account for the spacecraft attitude control system design. To analyze the total energy consumption, the energy index function derived from the optimal cost function is defined by $E = 1/2 \int_0^T \|\boldsymbol{u}\| dt$, where T denotes the simulation time, and $T = 100$ seconds is chosen in the simulation. The bar visualization graphic and time response of the energy index of the two strategies with model parameter uncertainty, external disturbances, actuator faults, and actuator saturation are shown in Figs. 4.10 and 4.11, where the value of the bar in Fig. 4.10 denotes the corresponding energy consumption index in different intervals. The PBSDHC would evidently require much more energy to achieve the same mission compared with the scheme presented in this work. Other evaluation indices can also be seen in Table 4.2, from which one can

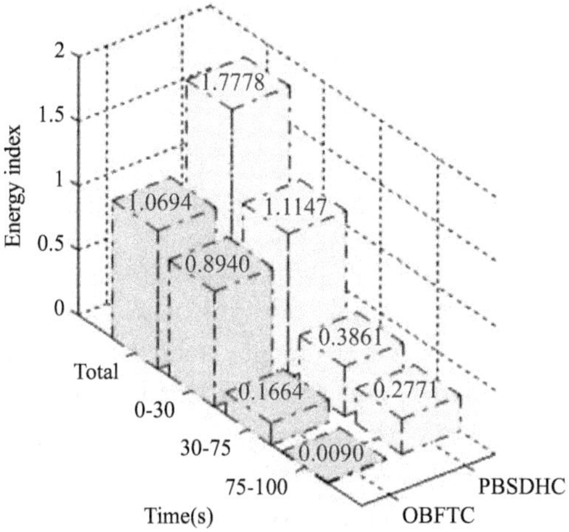

Figure 4.10 Bar visualization graphic of the energy index.

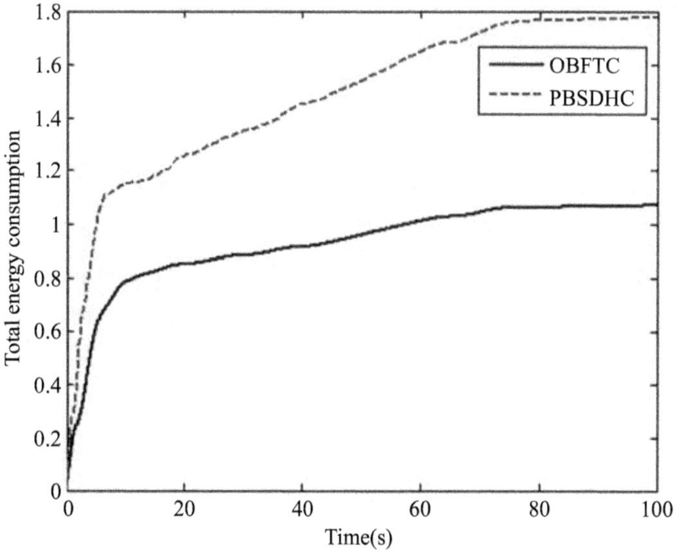

Figure 4.11 Time response of the total energy index.

Table 4.2 Other evaluation indices.

Control scheme	Accuracy of attitude angle (rad)	Accuracy of angular velocity (rad s^{-1})
Case A	1.5×10^{-4}	3×10^{-5}
Case B	1×10^{-3}	5×10^{-4}

see that the OBFTC is superior to the PBSDHC in the presence of model parameter uncertainty, external disturbances, actuator faults, and actuator saturation.

4.5 Conclusions

In this work, an observer-based fault tolerant control strategy is proposed for spacecraft attitude stabilization and control. A new auxiliary variable is introduced first in the observer design, which takes advantage of the spacecraft attitude dynamics and provides estimates of the attitude information and fault signals with uniformly ultimately bounded estimation errors. By using the observer-based fault tolerant controller, the states of the closed-loop system converge asymptotically to the vicinity of the expected states with very good attitude stabilization performance. More importantly, the effects of actuator fault and multiple unknown perturbations, such as external disturbances, model parameter uncertainty, and time-varying input delay, and actuator saturation are considered simultaneously in the design and analysis. Simulation results illustrate the superior performance of the proposed controller. It should also be pointed out that external disturbances are not considered explicitly in the design process of observer gain parameter, which remains a subject for future research.

References

[1] Vadali SR. Variable-structure control of spacecraft large-angle maneuvers. Journal of Guidance, Control, and Dynamics 1986;9(2):235–9.

[2] Boskovic JD, Li SM, Mehra RK. Robust adaptive variable structure control of spacecraft under control input saturation. Journal of Guidance, Control, and Dynamics 2001;24(1):14–22.

[3] Liu C, Sun Z, Ye D, Shi K. Robust adaptive variable structure tracking control for spacecraft chaotic attitude motion. IEEE Access 2018;6:3851–7.

[4] Egeland O, Godhavn JM. Passivity-based adaptive attitude control of a rigid spacecraft. IEEE Transactions on Automatic Control 1994;39(4):842–6.

[5] Liu C, Sun Z, Shi K, Wang F. Robust non-fragile state feedback attitude control for uncertain spacecraft with input saturation. Proceedings of the Institution of Mechanical Engineers, Part G: Journal of Aerospace Engineering 2018;232(2):246–59.

[6] Zhu B, Zhang Z, Zhou D, Ma J, Li S. Prediction-based sampled-data H_∞ controller design for attitude stabilization of a rigid spacecraft with disturbances. International Journal of Systems Science 2017;48(11):2356–67.

[7] Sakthivel R, Selvi S, Mathiyalagan K. Fault-tolerant sampled-data control of flexible spacecraft with probabilistic time delays. Nonlinear Dynamics 2015;79(3):1835–46.

[8] Krstic M, Tsiotras P. Inverse optimal stabilization of a rigid spacecraft. IEEE Transactions on Automatic Control 1999;44(5):1042–9.

[9] Schaub H, Junkins JL, Robinett RD. New penalty functions and optimal control formulation for spacecraft attitude control problems. Journal of Guidance, Control, and Dynamics 1997;20(3):428–34.
[10] Park Y. Robust and optimal attitude control of spacecraft with disturbances. International Journal of Systems Science 2015;46(7):1222–33.
[11] Liu C, Sun Z, Shi K, Wang F. Robust dynamic output feedback control for attitude stabilization of spacecraft with nonlinear perturbations. Aerospace Science and Technology 2017;64:102–21.
[12] Xia Y, Zhu Z, Fu M, Wang S. Attitude tracking of rigid spacecraft with bounded disturbances. IEEE Transactions on Industrial Electronics 2011;58(2):647–59.
[13] Xing GQ, Parvez SA. Nonlinear attitude state tracking control for spacecraft. Journal of Guidance, Control, and Dynamics 2001;24(3):624–6.
[14] Thakur D, Srikant S, Akella MR. Adaptive attitude-tracking control of spacecraft with uncertain time-varying inertia parameters. Journal of Guidance, Control, and Dynamics 2015;38(1):41–52.
[15] Cai W, Liao XH, Song YD. Indirect robust adaptive fault-tolerant control for attitude tracking of spacecraft. Journal of Guidance Control, and Dynamics 2008;31 (5):1456–63.
[16] Xiao B, Hu Q, Zhang Y. Adaptive sliding mode fault tolerant attitude tracking control for flexible spacecraft under actuator saturation. IEEE Transactions on Control Systems Technology 2012;20(6):1605–12.
[17] Zhang F, Jin L, Xu S. Fault tolerant attitude control for spacecraft with SGCMGs under actuator partial failure and actuator saturation. Acta Astronautica 2017;132:303–11.
[18] Jiang B, Shi P, Mao Z. Sliding mode observer-based fault estimation for nonlinear networked control systems. Circuits, Systems, and Signal Processing 2011;30 (1):1–16.
[19] Yin S, Yang H, Kaynak O. Sliding mode observer-based FTC for Markovian jump systems with actuator and sensor faults. IEEE Transactions on Automatic Control 2017;62(7):3551–8.
[20] Chen F, Jiang R, Zhang K, Jiang B, Tao G. Robust backstepping sliding-mode control and observer-based fault estimation for a quadrotor UAV. IEEE Transactions on Industrial Electronics 2016;63(8):5044–56.
[21] Jiang B, Xu D, Shi P, Lim CC. Adaptive neural observer-based backstepping fault tolerant control for near space vehicle under control effector damage. IET Control Theory & Applications 2014;8(9):658–66.
[22] Ailon A, Segev R, Arogeti S. A simple velocity-free controller for attitude regulation of a spacecraft with delayed feedback. IEEE Transactions on Automatic Control 2004;49(1):125–30.
[23] Samiei E, Butcher EA, Sanyal AK, Paz R. Attitude stabilization of rigid spacecraft with minimal attitude coordinates and unknown time-varying delay. Aerospace Science and Technology 2015;46:412–21.
[24] Zhu J, Yang G. Fault accommodation for linear systems with time-varying delay. International Journal of Systems Science 2017;48(2):316–23.
[25] Zhu J, Yang G, Wang H, Wang F. Fault estimation for a class of nonlinear systems based on intermediate estimator. IEEE Transactions on Automatic Control 2016;61 (9):2518–24.
[26] Liu M, Zhang L, Shi P, Karimi HR. Robust control of stochastic systems against bounded disturbances with application to flight control. IEEE Transactions on Industrial Electronics 2014;61(3):1504–15.
[27] Ding S, Zheng W. Nonsmooth attitude stabilization of a flexible spacecraft. IEEE Transactions on Aerospace and Electronic Systems 2014;50(2):1163–81.

[28] Mahmoud M, Jiang J, Zhang Y. Active fault tolerant control systems: stochastic analysis and synthesis. Springer-Verlag Berlin Heidelberg; 2003. p. 128–9.
[29] Huang S, Yang G. Fault tolerant controller design for t–s fuzzy systems with time-varying delay and actuator faults: a k-step fault-estimation approach. IEEE Transactions on Fuzzy Systems 2014;22(6):1526–40.
[30] You F, Li H, Wang F, Guan S. Robust fast adaptive fault estimation for systems with time-varying interval delay. Journal of the Franklin Institute 2015;352(12):5486–513.
[31] Shi P, Liu M, Zhang L. Fault-tolerant sliding-mode-observer synthesis of Markovian jump systems using quantized measurements. IEEE Transactions on Industrial Electronics 2015;62(9):5910–18.
[32] Tabatabaeipour SM, Bak T. Robust observer-based fault estimation and accommodation of discrete-time piecewise linear systems. Journal of the Franklin Institute 2014;51(1):277–95.
[33] Liu C, Ye D, Shi K, Sun Z. Robust high-precision attitude control for flexible spacecraft with improved mixed H_2/H_∞ control strategy under poles assignment constraint. Acta Astronautica 2017;136:166–75.

CHAPTER 5

Observer-based fault tolerant nonfragile control

5.1 Introduction

Many future spacecraft missions will be required to achieve high-precision pointing, rapid attitude maneuvers and stabilization in the presence of external environmental disturbances, performance constraints, imperfections of internal components, measurement errors, and actuator saturation, and even failure. In actual system operation, due to word length limits, conversion accuracy, truncation error in numerical operation and other reasons, controller gain perturbations often appear. It should also be mentioned that actuator structural perturbations, whose influence can also be regarded as controller perturbations, often occur during maneuvers, and may result in unsatisfactory performance or even instability. These challenges have attracted considerable interest in the literature, and several specific control approaches have been developed in response, such as sliding mode control [1,2], adaptive control [3,4], output feedback control [5,6], optimal control [7,8], and robust control or their integrations [9,10], etc. Although these methods have demonstrated satisfactory results for some situations, there are few actual results available that deal with these issues concurrently.

Generally speaking, the dynamics of a spacecraft is often time-varying and highly nonlinear in the presence of uncertainties due to environmental disturbances [11]. In addition, the uncertainty in the inertia matrix of the spacecraft also results in model parameter uncertainty. Thus much literature has taken model parameter uncertainty into consideration in spacecraft systems, and has shown that it is necessary to account for model parameter uncertainty to achieve highly precise and stable performance [12,13]. In contrast to most conventional controllers designed for the fault-free case, which does not take the possibility of fault occurrence into consideration, the fault tolerant controller can guarantee desirable stability and performance properties even in the event of component fault/failure. Although various fault tolerant control methods have been considered by

Spacecraft Attitude Control. DOI: https://doi.org/10.1016/B978-0-323-99005-9.00005-5

many researchers, adaptive control, sliding mode control, and control allocation are the three most widely used schemes for the design of spacecraft attitude fault tolerant controllers [14]. The work in [15] showed that adaptive control can be successfully used to achieve attitude tracking maneuvers for a rigid spacecraft with actuator failures. In addition, uncertain inertia parameters, external disturbances and input constraint were also addressed. However, this work did not address the influence of controller perturbations, so that this controller could become ineffective in the face of controller perturbation. In [16], an adaptive fault tolerant controller based on variable structure control was proposed to address the attitude tracking control problem of spacecraft in the presence of external disturbances, inexact knowledge of fault information, unknown actuator failure and input saturation, and this controller was capable of controlling the transient response of the closed-loop system based on a parameter that was adaptively adjusted. As sliding mode control has advantages of rapid response, insensitivity to parameters uncertainty and external disturbances, a variety of sliding mode control approaches have been developed and applied to spacecraft attitude fault tolerant control system design [17]. For example, two fault tolerant control laws including continuous sliding mode control and nonsingular terminal sliding mode control were developed in [18], which can guarantee global asymptotic convergence of the attitude control error in the presence of unknown external perturbations. In [19], the attitude tracking problem for an over-actuated spacecraft considering actuator faults, imprecise fault estimation and external disturbances was addressed, where the actuator fault was tackled directly without reconfiguring the controller. If full-state information is unavailable for spacecraft attitude, it is necessary to address the fault tolerant control problem using only output measurements. In such a case, the problem of fault estimation and measurement errors should also be considered. In recent years, sliding mode observer-based fault tolerant control schemes [20] and adaptive observer-based fault tolerant control schemes [21,22] have been developed. Also as is well known, H_∞ control can guarantee robust stability in the presence of parameter uncertainties and external disturbances, and it has been exploited in many papers [23,24]. In addition, controller perturbation often occurs in initial phase or decay phase of control system operation, and it has been taken into account in much of the literature, such as [25] and [26]. The approaches involved in controller design considering controller perturbation are the so-called nonfragile control methods. Refs. [6,27] have investigated nonfragile control strategies in terms of additive perturbation and multiplicative perturbation for spacecraft attitude control system in detail, but they did not consider actuator faults or measurement errors. Furthermore, the approach in [27] will be ineffective if attitude information cannot be obtained exactly, as in the case of sensor failure or other factors. Refs. [28,29] proposed a fault estimation mechanism based on intermediate estimator, but did not consider the problem of actuator saturation or controller perturbation in the designed controller, nor the H_∞ performance constraint. Actuators for attitude control of spacecraft include mechanical and electrical components, thrusters and environmental force actuators, and actuator saturation is another critical issue that must be dealt with in fault tolerant control of spacecraft attitude system as pointed out in [30]. If an unknown actuator fault occurs, the attitude control system uncorrected could produce inappropriate commands, causing actuators to saturate rendering any further control effort ineffective.

It is thus clear that still now, the observer-based fault tolerant control problem considering model parameter uncertainty, measurement errors, external disturbances, controller perturbation, and actuator fault and saturation is an open problem. On their own, each of these issues can severely degrade control performance or even cause system instability, and it is much more difficult to satisfy the control objective when all of these issues are considered simultaneously.

Motivated by the preceding discussions, this chapter presents a novel stochastically intermediate observer-based fault tolerant nonfragile H_∞ control scheme to achieve high-precision attitude stabilization of spacecraft with explicit consideration of H_∞ performance, quadratic stability, model parameter uncertainty, measurement errors, external disturbances, controller perturbation, and actuator fault and saturation. The main contributions of this work can be summarized as follows. First, a stochastically intermediate variable is introduced, which contains the Bernoulli random variable, and its expectation is exploited for observer construction. In other words, this approach can tolerate stochastic failure. Second, the observer-based fault tolerant control scheme developed can simultaneously achieve attitude stabilization and estimate the attitude information and fault signals with high precision and fast response, in spite of simultaneous existence of those issues mentioned previously. Third, the design process for the controller and observer contains information on controller perturbation, so that the developed controller is nonfragile compared with traditional controllers. Fourth, by exploiting online adaptive estimation of the attitude information and fault signals, the proposed control strategy can achieve the goal of fault tolerant control without use of any prior fault information or measured attitude information. Fifth, compared with some existing methods, the established approach requires neither equation constraints nor observer matching conditions and can greatly reduce energy consumption, thus it is less conservative and more economical.

The remainder of this chapter is outlined as follows. Section 5.2 presents the attitude dynamics of spacecraft with model parameter uncertainty, measurement errors and actuator fault, and the stochastically intermediate observer-based control problem is formulated. Section 5.3 develops the design method of the observer-based controller in terms of additive and multiplicative controller perturbation, and stability analysis of the closed-loop system is performed via a Lyapunov approach. Numerical simulations in terms of the above two cases are performed and comparisons are made with the existing state feedback nonfragile controller in Section 5.4. Finally, conclusions are drawn.

5.2 Problem formulation

5.2.1 Attitude system description

The spacecraft is assumed to be a rigid body with actuators that provide torques about three mutually perpendicular axes. The three axes define a body-fixed frame. The attitude dynamics equation is

$$\boldsymbol{I}_b\dot{\boldsymbol{\omega}} + \boldsymbol{\omega} \times (\boldsymbol{I}_b\boldsymbol{\omega}) = \boldsymbol{T}_c + \boldsymbol{T}_g + \boldsymbol{T}_d \tag{5.1}$$

where $\boldsymbol{I}_b$ and $\boldsymbol{\omega}$ denote the inertia matrix and the angular velocity, respectively, and $\boldsymbol{T}_c$, $\boldsymbol{T}_g$, and $\boldsymbol{T}_d$ denote the control input torque, the gravitational torque and the external disturbance torque, respectively.

With the body-fixed frame chosen to be a principal-axis frame, substituting the expression of gravitational torque into Eq. (5.1) allows it to be converted into component form

$$\begin{cases} I_x\ddot{\phi} + 4(I_y - I_z)\omega_0^2\phi + (I_y - I_z - I_x)\omega_0\dot{\psi} = T_{cx} + T_{dx} \\ I_y\ddot{\theta} + 3\omega_0^2(I_x - I_z)\theta = T_{cy} + T_{dy} \\ I_z\ddot{\psi} + (I_y - I_x)\omega_0^2\psi + (I_x + I_z - I_y)\omega_0\dot{\phi} = T_{cz} + T_{dz} \end{cases} \tag{5.2}$$

where I_x, I_y and I_z denote the three components of inertia matrix; ϕ, θ and ψ denote the three components of attitude angle; T_{cx}, T_{cy} and T_{cz} denote the three components of control input torque; T_{dx}, T_{dy} and T_{dz}

denote the three components of external disturbance torque; ω_0 represents the orbital angular velocity of the spacecraft; Euler angles ϕ, θ and ψ are roll, pitch and yaw attitude angle, respectively, and a typical sequence of three successive rotations is yaw→pitch→roll.

Let

$$\boldsymbol{x} = \begin{bmatrix} \phi & \theta & \psi & \dot{\phi} & \dot{\theta} & \dot{\psi} \end{bmatrix}^T, \boldsymbol{y} = \begin{bmatrix} \phi & \theta & \psi & \omega_x & \omega_y & \omega_z \end{bmatrix}^T,$$

$$\boldsymbol{z} = \begin{bmatrix} \phi & \theta & \psi & \dot{\phi} & \dot{\theta} & \dot{\psi} \end{bmatrix}^T, \boldsymbol{u} = \begin{bmatrix} T_{cx} & T_{cy} & T_{cz} \end{bmatrix}^T, \boldsymbol{w} = \begin{bmatrix} T_{dx} & T_{dy} & T_{dz} \end{bmatrix}^T,$$

where $\boldsymbol{y}$ is the measured output and $\boldsymbol{z}$ is the controlled output of the system.

Then, Eq. (5.2) can be written as

$$\begin{cases} \dot{\boldsymbol{x}}(t) = \boldsymbol{A}\boldsymbol{x}(t) + \boldsymbol{B}_1\boldsymbol{u}(t) + \boldsymbol{B}_2\boldsymbol{w}(t) \\ \boldsymbol{y}(t) = \boldsymbol{C}_1\boldsymbol{x}(t) + \boldsymbol{D}_1 \\ \boldsymbol{z}(t) = \boldsymbol{C}_2\boldsymbol{x}(t) \end{cases} \tag{5.3}$$

where,

$$\boldsymbol{A} = \begin{bmatrix} 0 & 0 & 0 & 1 & 0 & 0 \\ 0 & 0 & 0 & 0 & 1 & 0 \\ 0 & 0 & 0 & 0 & 0 & 1 \\ A_{41} & 0 & 0 & 0 & 0 & -\omega_0 I_x^{-1}(I_y - I_x - I_z) \\ 0 & A_{52} & 0 & 0 & 0 & 0 \\ 0 & 0 & A_{63} & \omega_0 I_z^{-1}(I_y - I_x - I_z) & 0 & 0 \end{bmatrix}$$

$$A_{41} = -4\omega_0^2 I_x^{-1}(I_y - I_z),\ A_{52} = -3\omega_0^2 I_y^{-1}(I_x - I_z),\ A_{63} = -\omega_0^2 I_z^{-1}(I_y - I_x)$$

$$\boldsymbol{B}_1 = \boldsymbol{B}_2 = \begin{bmatrix} \boldsymbol{0}_{3\times3} & diag(I_x^{-1}, I_y^{-1}, I_z^{-1}) \end{bmatrix}^T,\ \boldsymbol{D}_1 = \begin{bmatrix} \boldsymbol{0}_{1\times4} & -\omega_0 & 0 \end{bmatrix}^T,$$

$$\boldsymbol{C}_1 = \begin{bmatrix} \boldsymbol{I}_{3\times3} & \boldsymbol{0}_{3\times3} \\ \boldsymbol{B} & \boldsymbol{I}_{3\times3} \end{bmatrix},\ \boldsymbol{B} = \begin{bmatrix} 0 & 0 & -\omega_0 \\ 0 & 0 & 0 \\ \omega_0 & 0 & 0 \end{bmatrix}, \boldsymbol{C}_2 = \boldsymbol{I}_{6\times6}$$

In addition, the following important points should be noted.

1. Model parameter uncertainty

Due to parameter perturbations, modeling errors, and environmental changes, model uncertainty is prevalent in physical systems. If model uncertainty is not considered during control system design, the controller obtained is highly likely to fail in real situations. The admissible parameter uncertainty is often assumed to be norm-bounded, which is the most adopted form in robust stability analysis [31]. Taking into account parameter uncertainty, $\boldsymbol{A}$ will be transformed into

$A + \Delta A$, where ΔA denotes model parameter uncertainty, and satisfies a matching condition, that is,

$$\Delta A(t) = M_1 F_1(t) N_1, \ \|F_1(t)\| \leq 1, \ \forall t$$

where M_1 and N_1 are known constant real matrices of appropriate dimension, which characterize how the uncertain parameter in F_1(t) enters the nominal matrix A, and F_1(t) is an unknown matrix function with Lebesgue measurable elements. It is worth mentioning that the form of ΔA is easily obtained using matrix factorization approach.

2. Controller gain perturbation

   A growing error and certain unknown actuator features can lead to the problem of actuator perturbation or nominal controller perturbation, which is named here as controller gain perturbation. Taking the actuator perturbation to be Ξ and the nominal controller gain to be K, adding Ξ to the control system produces the actual controller gain matrix $K(I+\Xi)$, that is,

$$K(I + \Xi) = K + K\Xi = K + \Delta K \tag{5.4}$$

   ΔK is the controller gain perturbation generated by the actuator uncertainty or other factors, and two types are considered as follows:

   When ΔK is an additive perturbation, it satisfies

$$\Delta K = M_2 F_2(t) N_2, \ \|F_2(t)\| \leq 1 \tag{5.5}$$

   When ΔK is a multiplicative perturbation, it satisfies

$$\Delta K = M_3 F_3(t) N_3 K, \ \|F_3(t)\| \leq 1 \tag{5.6}$$

   The definitions of M_2, N_2,M_3, N_3, $F_2(t)$ and $F_3(t)$ are similar to ΔA.

3. Measurement errors

   Due to the unexpected variations in the environment, limitations of measurement technique, and aging of measuring equipment, there may appear measurement errors $v(t) \in L_2[0, \infty)$ between the system outputs and the measurement outputs, especially for electronic measurements [32].

4. Actuator faults

   Since the performance of attitude control system can be severely impaired by improper actuator actions, actuator faults must be considered to be one of the most critical challenges to be solved. To take this issue into account, let E with full-column rank denote the distribution matrix of fault signal $f(t)$ appearing in the input, with $\dot{f}(t) \in L_2[0, \infty)$. Specifically, it represents the process faults if $E \neq B_1$, and it also represents actuator faults if $E = B_1$.

5. Actuator saturation

Actuator saturation is a ubiquitous phenomenon due to the constraints of physical characteristics in real engineering applications, which will often have a negative impact on the stability and performance of an attitude control system. When the spacecraft performs an attitude maneuver, if the theoretical control input is larger than the saturation value of the actual control input, the actual control input $sat(\boldsymbol{u})$ should be described as

$$sat(u_{i=1,2,3}(t)) = sign(u_i(t))min\{|u_i(t)|, u_{mi}\} \tag{5.7}$$

where u_{mi} is the upper bound of control input that the actuator can provide.

Some assumptions and remarks are given for the convenience of subsequent derivation.

Assumption 5.1: The external disturbance $\boldsymbol{w}(t)$ is assumed to be bounded, that is, $\boldsymbol{w}(t)$ belongs to $L_2[0, \infty)$.

Remark 5.1: Assumption 5.1 is reasonable. The main reason is that the external disturbances primarily include magnetic torque, solar radiation and aerodynamic torque, which are all bounded [33].

Remark 5.2: This chapter does not define the boundary of measurement errors, and is therefore more general compared with [34] and [32] where the boundary is defined and should be known respectively.

Remark 5.3: The norm of the fault signal $\boldsymbol{f}(t)$ can be unknown. Thus it is more general than the fault-tolerant sliding-mode-observer method in [35], which requires preliminary knowledge of the bound of the fault.

Assumption 5.2: $rank(\boldsymbol{B}_1, \boldsymbol{E}) = rank(\boldsymbol{B}_1)$, and there exists a matrix $\boldsymbol{B}_1^*$ such that $(\boldsymbol{I} - \boldsymbol{B}_1\boldsymbol{B}_1^*)\boldsymbol{E}=0$.

Remark 5.4: Assumption 5.2 is general in [36], which means the faults which appear in the actuator can be compensated by the control input. In addition, the matrix $\boldsymbol{B}_1^*$ is the pseudoinverse of $\boldsymbol{B}_1$, which follows that $\boldsymbol{B}_1\boldsymbol{B}_1^*\boldsymbol{E} = \boldsymbol{E}$.

Considering the above factors, the spacecraft attitude control system can now be expressed as:

$$\begin{cases} \dot{\boldsymbol{x}}(t) = (\boldsymbol{A}+\Delta\boldsymbol{A})\boldsymbol{x}(t) + \boldsymbol{B}_1\boldsymbol{u}(t) + \boldsymbol{E}\boldsymbol{f}(t) + \boldsymbol{B}_2\boldsymbol{w}(t) \\ \boldsymbol{y}(t) = \boldsymbol{C}_1\boldsymbol{x}(t) + \boldsymbol{D}_1 + \boldsymbol{v}(t) \\ \boldsymbol{z}(t) = \boldsymbol{C}_2\boldsymbol{x}(t) \end{cases} \tag{5.8}$$

5.2.2 Stochastically intermediate observer design

In this chapter, an intermediate observer with stochastic failure is constructed for system (5.8). To introduce the observer, a stochastically intermediate variable is introduced first:

$$\bar{\boldsymbol{\xi}}(t) = \boldsymbol{f}(t) - \varpi(t)\boldsymbol{\Omega}\boldsymbol{x}(t) \tag{5.9}$$

Here, $\boldsymbol{\Omega}$ is designed as follow:

$$\boldsymbol{\Omega} = \eta \boldsymbol{E}^T \tag{5.10}$$

where η is a scalar.

The variable $\varpi(t)$ is a Bernoulli random variable and described as

$$\varpi(t) = \begin{cases} 1, & \textit{If success of } \boldsymbol{\Omega}x(t) \textit{ occurs} \\ 0, & \textit{If failure of } \boldsymbol{\Omega}x(t) \textit{ occurs} \end{cases} \tag{5.11}$$

One can write $\varpi(t) \sim \text{Ber}(\varpi)$, where ϖ is the probability of success, and the random variable $\varpi(t)$ has the following distribution

$$P(\varpi(t) = 1) = \varpi, \; P(\varpi(t) = 0) = 1 - \varpi \tag{5.12}$$

Letting $\boldsymbol{\xi}(t)$ be the expectation of $\bar{\boldsymbol{\xi}}(t)$, one can conclude that

$$\boldsymbol{\xi}(t) = \boldsymbol{f}(t) - \varpi\eta\boldsymbol{E}^T\boldsymbol{x}(t) \tag{5.13}$$

$$\begin{aligned}\dot{\boldsymbol{\xi}}(t) &= \dot{\boldsymbol{f}}(t) - \varpi\eta\boldsymbol{E}^T\left[(\boldsymbol{A}+\Delta\boldsymbol{A})\boldsymbol{x}(t) + \boldsymbol{B}_1\boldsymbol{u}(t) + \boldsymbol{E}\boldsymbol{f}(t) + \boldsymbol{B}_2\boldsymbol{w}(t)\right] \\ &= \dot{\boldsymbol{f}}(t) - \varpi\eta\boldsymbol{E}^T\left[(\boldsymbol{A}+\Delta\boldsymbol{A})\boldsymbol{x}(t) + \boldsymbol{B}_1\boldsymbol{u}(t) + \boldsymbol{E}\boldsymbol{\xi}(t) + \varpi\eta\boldsymbol{E}\boldsymbol{E}^T\boldsymbol{x}(t) + \boldsymbol{B}_2\boldsymbol{w}(t)\right]\end{aligned} \tag{5.14}$$

Then, the intermediate observer is based on

$$\dot{\hat{\boldsymbol{x}}}(t) = \boldsymbol{A}\hat{\boldsymbol{x}}(t) + \boldsymbol{B}_1\boldsymbol{u}(t) + \boldsymbol{E}\hat{\boldsymbol{f}}(t) + \boldsymbol{L}(\boldsymbol{y}(t) - \hat{\boldsymbol{y}}(t)) \tag{5.15}$$

$$\dot{\hat{\boldsymbol{\xi}}}(t) = -\varpi\eta\boldsymbol{E}^T\left[\boldsymbol{A}\hat{\boldsymbol{x}}(t) + \boldsymbol{B}_1\boldsymbol{u}(t) + \boldsymbol{E}\hat{\boldsymbol{\xi}}(t) + \varpi\eta\boldsymbol{E}\boldsymbol{E}^T\hat{\boldsymbol{x}}(t)\right] \tag{5.16}$$

$$\hat{\boldsymbol{y}}(t) = \boldsymbol{C}_1\hat{\boldsymbol{x}}(t) + \boldsymbol{D}_1 \tag{5.17}$$

$$\hat{\boldsymbol{f}}(t) = \hat{\boldsymbol{\xi}}(t) + \varpi\eta\boldsymbol{E}^T\hat{\boldsymbol{x}}(t) \tag{5.18}$$

where $\hat{\boldsymbol{x}}(t)$, $\hat{\boldsymbol{\xi}}(t)$, $\hat{\boldsymbol{y}}(t)$, $\hat{\boldsymbol{f}}(t)$ are estimates of $\boldsymbol{x}(t)$, $\boldsymbol{\xi}(t)$, $\boldsymbol{y}(t)$ and $\boldsymbol{f}(t)$, respectively, and $\boldsymbol{L}$ is the observer gain matrix.

5.2.3 Nonfragile controller design

Based on the above estimation, the controller is designed as

$$\boldsymbol{u}(t) = -(\boldsymbol{K}+\Delta\boldsymbol{K})\hat{\boldsymbol{x}}(t) - \boldsymbol{B}_1^*\boldsymbol{E}\hat{\boldsymbol{f}}(t) \tag{5.19}$$

where, the term $\boldsymbol{B}_1^*\boldsymbol{E}\hat{\boldsymbol{f}}(t)$ is added to the input to compensate for the effect of the fault signal; $\boldsymbol{K}$ is the controller gain matrix to be determined and $\hat{\boldsymbol{x}}(t)$ is the estimation of attitude information used for feedback control.

Denoting $\boldsymbol{e}_x(t) = \boldsymbol{x}(t) - \hat{\boldsymbol{x}}(t), \boldsymbol{e}_\xi(t) = \boldsymbol{\xi}(t) - \hat{\boldsymbol{\xi}}(t)$, and substituting Eq. (5.19) into the first equation of Eq. (5.8) yields

$$\begin{aligned}
\dot{\boldsymbol{x}}(t) &= (\boldsymbol{A}+\Delta\boldsymbol{A})\boldsymbol{x}(t) + \left[-\boldsymbol{B}_1(\boldsymbol{K}+\Delta\boldsymbol{K})\hat{\boldsymbol{x}}(t) - \boldsymbol{E}\hat{\boldsymbol{f}}(t)\right] + \boldsymbol{E}\boldsymbol{f}(t) + \boldsymbol{B}_2\boldsymbol{w}(t) \\
&= (\boldsymbol{A} - \boldsymbol{B}_1\boldsymbol{K})\boldsymbol{x}(t) + \boldsymbol{B}_1\boldsymbol{K}\boldsymbol{e}_x(t) + \Delta\boldsymbol{A}\boldsymbol{x}(t) - \boldsymbol{B}_1\Delta\boldsymbol{K}\boldsymbol{x}(t) \\
&\quad + \boldsymbol{B}_1\Delta\boldsymbol{K}\boldsymbol{e}_x(t) + \boldsymbol{E}\boldsymbol{e}_f(t) + \boldsymbol{B}_2\boldsymbol{w}(t) \\
&= (\boldsymbol{A} - \boldsymbol{B}_1\boldsymbol{K})\boldsymbol{x}(t) + \boldsymbol{B}_1\boldsymbol{K}\boldsymbol{e}_x(t) + \Delta\boldsymbol{A}\boldsymbol{x}(t) - \boldsymbol{B}_1\Delta\boldsymbol{K}\boldsymbol{x}(t) \\
&\quad + \boldsymbol{B}_1\Delta\boldsymbol{K}\boldsymbol{e}_x(t) + \boldsymbol{E}\boldsymbol{e}_\xi(t) + \varpi\eta\boldsymbol{E}\boldsymbol{E}^T\boldsymbol{e}_x(t) + \boldsymbol{B}_2\boldsymbol{w}(t)
\end{aligned} \tag{5.20}$$

Then, it can be concluded that

$$\begin{aligned}
\dot{\boldsymbol{e}}_x(t) &= (\boldsymbol{A}+\Delta\boldsymbol{A})\boldsymbol{x}(t) + \boldsymbol{E}\boldsymbol{f}(t) + \boldsymbol{B}_2\boldsymbol{w}(t) - \boldsymbol{A}\hat{\boldsymbol{x}}(t) - \boldsymbol{E}\hat{\boldsymbol{f}}(t) - \boldsymbol{L}(\boldsymbol{y}(t) - \hat{\boldsymbol{y}}(t)) \\
&= (\boldsymbol{A} - \boldsymbol{L}\boldsymbol{C}_1)\boldsymbol{e}_x(t) + \Delta\boldsymbol{A}\boldsymbol{x}(t) + \boldsymbol{E}\boldsymbol{e}_\xi(t) + \varpi\eta\boldsymbol{E}\boldsymbol{E}^T\boldsymbol{e}_x(t) + \boldsymbol{B}_2\boldsymbol{w}(t) - \boldsymbol{L}\boldsymbol{v}(t)
\end{aligned} \tag{5.21}$$

$$\begin{aligned}
\dot{\boldsymbol{e}}_\xi(t) &= \dot{\boldsymbol{f}}(t) - \varpi\eta\boldsymbol{E}^T\left[(\boldsymbol{A}+\Delta\boldsymbol{A})\boldsymbol{x}(t) + \boldsymbol{B}_1\boldsymbol{u}(t) + \boldsymbol{E}\boldsymbol{\xi}(t) + \varpi\eta\boldsymbol{E}\boldsymbol{E}^T\boldsymbol{x}(t)\right. \\
&\quad \left. + \boldsymbol{B}_2\boldsymbol{w}(t) - \boldsymbol{A}\hat{\boldsymbol{x}}(t) - \boldsymbol{B}_1\boldsymbol{u}(t) - \boldsymbol{E}\hat{\boldsymbol{\xi}}(t) - \varpi\eta\boldsymbol{E}\boldsymbol{E}^T\hat{\boldsymbol{x}}(t)\right] \\
&= \dot{\boldsymbol{f}}(t) - \varpi\eta\boldsymbol{E}^T\left[\boldsymbol{A}\boldsymbol{e}_x(t)+\Delta\boldsymbol{A}\boldsymbol{x}(t) + \boldsymbol{E}\boldsymbol{e}_\xi(t) + \varpi\eta\boldsymbol{E}\boldsymbol{E}^T\boldsymbol{e}_x(t) + \boldsymbol{B}_2\boldsymbol{w}(t)\right]
\end{aligned} \tag{5.22}$$

Remark 5.5: Compared with the traditional methods, the approach in this chapter is less conservative. For example, the methods in [37] and [38] have equation constraints, while the methods in [39] and [40] require an observer matching condition. In contrast, the scheme in this chapter requires neither equation constraints nor observer matching conditions. The introduced intermediate variable $\overline{\boldsymbol{\xi}}(t)$ is stochastic, for $\varpi(t)$ can take value 0 or 1 with finite probability. As the intermediate variable contains random variable $\varpi(t)$, its traditional derivative is meaningless, where the method in [34] cannot be applied directly. To tackle this difficulty, another intermediate variable based on the expectation of such a stochastically intermediate variable is defined, which is of central importance for construction of the observer.

5.2.4 Control objective

Thus the objective of this chapter is to seek the controller gain matrix $\boldsymbol{K}$ and observer gain matrix $\boldsymbol{L}$ such that the resulting closed-loop system represented by Eqs. (5.20)–(5.22) satisfies the following conditions:

1. The closed-loop system is quadratically stable (see Definition 5.1 below) while $\boldsymbol{w}(t) = 0, \boldsymbol{v}(t) = 0, \dot{\boldsymbol{f}}(t) = 0$.
2. For a given scalar $\gamma > 0$, $\left\| \mathbf{G}_{z\hat{w}}(s) \right\|_{\infty} < \gamma$ is guaranteed, where $\mathbf{G}_{z\hat{w}}(s)$ denotes the closed-loop transfer function from $\hat{\boldsymbol{w}}$ to $\boldsymbol{z}$, and $\hat{\boldsymbol{w}}$ is a constructed vector with norm $|\hat{w}| = \sqrt{w^2 + v^2 + \dot{f}^2}$.
3. The input torque cannot exceed its upper bound.

Definition 5.1: **quadratic stability** If there is a symmetric positive definite matrix $\boldsymbol{P} > 0$ and a positive constant α, for arbitrary uncertainty, the time derivative of Lyapunov function $V(\chi, t)$ satisfies

$$\dot{V}(\chi, t) \le -\alpha \left\| \chi(t) \right\|_2^2 \tag{5.23}$$

Here, $\chi(t)$ denotes the triple $\left(\boldsymbol{x}(t), \boldsymbol{e}_x(t), \boldsymbol{e}_\xi(t)\right)$. Then, the system represented by Eqs. (5.20)–(5.22) $(\boldsymbol{w}(t) = 0, \boldsymbol{v}(t) = 0, \dot{\boldsymbol{f}}(t) = 0)$ is said to be quadratically stable.

In short, with the designed controller and intermediate observer, the spacecraft attitude control system should achieve attitude stabilization and attitude information and actuator faults are estimated simultaneously in the presence of model parameter uncertainty, controller gain perturbation, external disturbances, actuator faults, actuator saturation and measurement errors.

5.3 Feasible solution for both cases

5.3.1 Some lemmas

Before developing the main results, we introduce the following preliminary lemmas.

Lemma 5.1: Let $\overline{\boldsymbol{H}}, \overline{\boldsymbol{E}}$ and $\overline{\boldsymbol{F}}(t)$ be the real matrices of appropriate dimensions with $\left\| \overline{\boldsymbol{F}}(t) \right\| \le 1$, then for any scalar $\overline{\xi} > 0$,

$$\overline{\boldsymbol{H}}\overline{\boldsymbol{F}}(t)\overline{\boldsymbol{E}} + \overline{\boldsymbol{E}}^T \overline{\boldsymbol{F}}(t)^T \overline{\boldsymbol{H}}^T \le \overline{\xi}^{-1} \overline{\boldsymbol{H}}\overline{\boldsymbol{H}}^T + \overline{\xi}\overline{\boldsymbol{E}}^T\overline{\boldsymbol{E}}$$

Lemma 5.2: (Schur complement lemma) Let the partitioned matrix

$$\boldsymbol{A}=\begin{bmatrix}\boldsymbol{A}_{11} & \boldsymbol{A}_{12}\\ * & \boldsymbol{A}_{22}\end{bmatrix}$$

be symmetric. Then

$$\boldsymbol{A}<0 \Leftrightarrow \boldsymbol{A}_{11}<0,\ \boldsymbol{A}_{22}-\boldsymbol{A}_{12}^T\boldsymbol{A}_{11}^{-1}\boldsymbol{A}_{12}<0 \Leftrightarrow \boldsymbol{A}_{22}<0,\ \boldsymbol{A}_{11}-\boldsymbol{A}_{12}\boldsymbol{A}_{22}^{-1}\boldsymbol{A}_{12}^T<0$$

5.3.2 Sufficient conditions under additive perturbation

Theorem 5.1 now gives the existence conditions of the controller gain matrix $\boldsymbol{K}$ and observer gain matrix $\boldsymbol{L}$ under additive perturbation.

Theorem 5.1: When the controller gain perturbation is in the form of Eq. (5.5), for given scalars $\varpi\in(0,1], \eta>0, \xi_i>0, i=1,2,\ldots,5$, and given symmetric positive definite matrix $\widehat{\boldsymbol{P}}$, then the system represented by Eqs. (5.20)–(5.22) is quadratically stable under controller (5.19), and $\boldsymbol{z}(t)$ satisfies H_∞ performance constraint if there exist symmetric positive definite matrices $\boldsymbol{P}_i, i=1,2,3$, matrices $\boldsymbol{H}$ and $\boldsymbol{K}$, such that the following linear matrix inequalities (LMIs) hold:

$$(\boldsymbol{A}-\boldsymbol{B}_1\boldsymbol{K})^T\widehat{\boldsymbol{P}}+\widehat{\boldsymbol{P}}(\boldsymbol{A}-\boldsymbol{B}_1\boldsymbol{K})<0 \tag{5.24}$$

$$\begin{bmatrix}\boldsymbol{\Theta}_{11} & \boldsymbol{\Theta}_{12}\\ * & \boldsymbol{\Theta}_{22}\end{bmatrix}<0 \tag{5.25}$$

where,

$$\boldsymbol{\Theta}_{11}=\begin{bmatrix}\tilde{\boldsymbol{\Pi}}_{11} & \boldsymbol{P}_1\boldsymbol{B}_1\boldsymbol{K}+\varpi\eta\boldsymbol{P}_1\boldsymbol{E}\boldsymbol{E}^T & \boldsymbol{P}_1\boldsymbol{E}\\ * & \tilde{\boldsymbol{\Pi}}_{22} & \tilde{\boldsymbol{\Pi}}_{23}\\ * & * & \tilde{\boldsymbol{\Pi}}_{33}\end{bmatrix},$$

$$\boldsymbol{\Theta}_{12}=\begin{bmatrix}\boldsymbol{P}_1\boldsymbol{B}_2 & 0 & 0 & \boldsymbol{P}_1\boldsymbol{M}_1 & \boldsymbol{N}_1^T & \boldsymbol{P}_1\boldsymbol{B}_1\boldsymbol{M}_2 & \boldsymbol{N}_2^T & \boldsymbol{C}_2^T & 0 & 0 & 0\\ \boldsymbol{P}_2\boldsymbol{B}_2 & -\boldsymbol{H} & 0 & 0 & 0 & 0 & 0 & 0 & \boldsymbol{N}_2^T & \boldsymbol{P}_2\boldsymbol{M}_1 & 0\\ -\varpi\eta\boldsymbol{P}_3\boldsymbol{E}^T\boldsymbol{B}_2 & 0 & \boldsymbol{P}_3 & 0 & 0 & 0 & 0 & 0 & 0 & 0 & \boldsymbol{P}_3\boldsymbol{E}^T\boldsymbol{M}_1\end{bmatrix}$$

$$\boldsymbol{\Theta}_{22}=blkdiag\left(-\gamma^2\boldsymbol{I}\ \ -\gamma^2\boldsymbol{I}\ \ -\gamma^2\boldsymbol{I}\ \ -\xi_1\boldsymbol{I}\ \ -(\xi_1+\xi_4+\varpi\eta\xi_5)^{-1}\boldsymbol{I}\ \ -(\xi_2^{-1}+\xi_3^{-1})^{-1}\boldsymbol{I}\ \ -\xi_2^{-1}\boldsymbol{I}\ \ -\boldsymbol{I}\ \ -\xi_3^{-1}\boldsymbol{I}\ \ -\xi_4\boldsymbol{I}\ \ -(\varpi\eta)^{-1}\xi_5\boldsymbol{I}\right)$$

and where,

$$\begin{aligned}
\tilde{\boldsymbol{\Pi}}_{11}&=(\boldsymbol{A}-\boldsymbol{B}_1\boldsymbol{K})^T\boldsymbol{P}_1+\boldsymbol{P}_1(\boldsymbol{A}-\boldsymbol{B}_1\boldsymbol{K})\\
\tilde{\boldsymbol{\Pi}}_{22}&=\boldsymbol{A}^T\boldsymbol{P}_2-\boldsymbol{C}_1^T\boldsymbol{H}^T+\boldsymbol{P}_2\boldsymbol{A}-\boldsymbol{H}\boldsymbol{C}_1+\varpi\eta\boldsymbol{P}_2\boldsymbol{E}\boldsymbol{E}^T+\varpi\eta\boldsymbol{E}\boldsymbol{E}^T\boldsymbol{P}_2\\
\tilde{\boldsymbol{\Pi}}_{23}&=\boldsymbol{P}_2\boldsymbol{E}-\varpi\eta\boldsymbol{A}^T\boldsymbol{E}\boldsymbol{P}_3-\varpi^2\eta^2\boldsymbol{E}\boldsymbol{E}^T\boldsymbol{E}\boldsymbol{P}_3\\
\tilde{\boldsymbol{\Pi}}_{33}&=-\varpi\eta(\boldsymbol{P}_3\boldsymbol{E}^T\boldsymbol{E}+\boldsymbol{E}^T\boldsymbol{E}\boldsymbol{P}_3)
\end{aligned}$$

When LMIs (5.24) and (5.25) are feasible, one can obtain the controller gain matrix $\boldsymbol{K}$ and the observer gain matrix $\boldsymbol{L} = \boldsymbol{P}_2^{-1}\boldsymbol{H}$.

Proof: First, one shows that the closed-loop system is quadratically stable while $\boldsymbol{w}(t) = 0, \boldsymbol{v}(t) = 0, \dot{\boldsymbol{f}}(t) = 0$.

Choose the Lyapunov function candidate as

$$V(t) = \boldsymbol{x}(t)^T\boldsymbol{P}_1\boldsymbol{x}(t) + \boldsymbol{e}_x^T(t)\boldsymbol{P}_2\boldsymbol{e}_x(t) + \boldsymbol{e}_\xi^T(t)\boldsymbol{P}_3\boldsymbol{e}_\xi(t)$$

Then

$$\begin{aligned}\dot{V}(t) &= \dot{\boldsymbol{x}}(t)^T\boldsymbol{P}_1\boldsymbol{x}(t) + \boldsymbol{x}(t)^T\boldsymbol{P}_1\dot{\boldsymbol{x}}(t) + \dot{\boldsymbol{e}}_x(t)^T\boldsymbol{P}_2\boldsymbol{e}_x(t)\\ &\quad + \boldsymbol{e}_x(t)^T\boldsymbol{P}_2\dot{\boldsymbol{e}}_x(t) + \dot{\boldsymbol{e}}_\xi(t)^T\boldsymbol{P}_3\boldsymbol{e}_\xi(t) + \boldsymbol{e}_\xi(t)^T\boldsymbol{P}_3\dot{\boldsymbol{e}}_\xi(t)\\ &= \boldsymbol{x}(t)^T\left[(\boldsymbol{A}-\boldsymbol{B}_1\boldsymbol{K})^T\boldsymbol{P}_1 + \boldsymbol{P}_1(\boldsymbol{A}-\boldsymbol{B}_1\boldsymbol{K})\right]\boldsymbol{x}(t)\\ &\quad + 2\boldsymbol{x}(t)^T\boldsymbol{P}_1\boldsymbol{B}_1\boldsymbol{K}\boldsymbol{e}_x(t) + 2\boldsymbol{x}(t)^T\boldsymbol{P}_1\Delta\boldsymbol{A}\boldsymbol{x}(t)\\ &\quad - 2\boldsymbol{x}(t)^T\boldsymbol{P}_1\boldsymbol{B}_1\Delta\boldsymbol{K}\boldsymbol{x}(t) + 2\boldsymbol{x}(t)^T\boldsymbol{P}_1\boldsymbol{B}_1\Delta\boldsymbol{K}\boldsymbol{e}_x(t)\\ &\quad + 2\boldsymbol{x}(t)^T\boldsymbol{P}_1\boldsymbol{E}\boldsymbol{e}_\xi(t) + 2\varpi\eta\boldsymbol{x}(t)^T\boldsymbol{P}_1\boldsymbol{E}\boldsymbol{E}^T\boldsymbol{e}_x(t) + 2\boldsymbol{x}(t)^T\boldsymbol{P}_1\boldsymbol{B}_2\boldsymbol{w}(t)\\ &\quad + \boldsymbol{e}_x(t)^T\left[(\boldsymbol{A}-\boldsymbol{L}\boldsymbol{C}_1)^T\boldsymbol{P}_2 + \boldsymbol{P}_2(\boldsymbol{A}-\boldsymbol{L}\boldsymbol{C}_1)\right]\boldsymbol{e}_x(t) + 2\boldsymbol{e}_x(t)^T\boldsymbol{P}_2\Delta\boldsymbol{A}\boldsymbol{x}(t)\\ &\quad + 2\boldsymbol{e}_x(t)^T\boldsymbol{P}_2\boldsymbol{E}\boldsymbol{e}_\xi(t) + \boldsymbol{e}_x(t)^T(\varpi\eta\boldsymbol{P}_2\boldsymbol{E}\boldsymbol{E}^T + \varpi\eta\boldsymbol{E}\boldsymbol{E}^T\boldsymbol{P}_2)\boldsymbol{e}_x(t)\\ &\quad + 2\boldsymbol{e}_x(t)^T\boldsymbol{P}_2\boldsymbol{B}_2\boldsymbol{w}(t) - 2\boldsymbol{e}_x(t)^T\boldsymbol{P}_2\boldsymbol{L}\boldsymbol{v}(t) + 2\boldsymbol{e}_\xi(t)^T\boldsymbol{P}_3\dot{\boldsymbol{f}}(t)\\ &\quad - \varpi\eta\boldsymbol{e}_\xi(t)^T(\boldsymbol{P}_3\boldsymbol{E}^T\boldsymbol{E}+\boldsymbol{E}^T\boldsymbol{E}\boldsymbol{P}_3)\boldsymbol{e}_\xi(t)\\ &\quad - 2\varpi\eta\boldsymbol{e}_\xi(t)^T\boldsymbol{P}_3\boldsymbol{E}^T\left[\boldsymbol{A}\boldsymbol{e}_x(t)+\Delta\boldsymbol{A}\boldsymbol{x}(t) + \varpi\eta\boldsymbol{E}\boldsymbol{E}^T\boldsymbol{e}_x(t) + \boldsymbol{B}_2\boldsymbol{w}(t)\right]\end{aligned} \tag{5.26}$$

Define $\chi(t) = [\,\boldsymbol{x}(t)^T \quad \boldsymbol{e}_x(t)^T \quad \boldsymbol{e}_\xi(t)^T\,]^T$, according to Lemma 5.1 and the given constraint conditions, one has

$$\begin{aligned}\dot{V}(t) \le{}& \boldsymbol{x}(t)^T\left[(\boldsymbol{A}-\boldsymbol{B}_1\boldsymbol{K})^T\boldsymbol{P}_1 + \boldsymbol{P}_1(\boldsymbol{A}-\boldsymbol{B}_1\boldsymbol{K})\right]\boldsymbol{x}(t) + 2\boldsymbol{x}(t)^T\boldsymbol{P}_1\boldsymbol{B}_1\boldsymbol{K}\boldsymbol{e}_x(t)\\ &+ \xi_1^{-1}\boldsymbol{x}(t)^T\boldsymbol{P}_1\boldsymbol{M}_1\boldsymbol{M}_1^T\boldsymbol{P}_1\boldsymbol{x}(t) + \xi_1\boldsymbol{x}(t)^T\boldsymbol{N}_1^T\boldsymbol{N}_1\boldsymbol{x}(t)\\ &+ \xi_2^{-1}\boldsymbol{x}(t)^T\boldsymbol{P}_1\boldsymbol{B}_1\boldsymbol{M}_2\boldsymbol{M}_2^T\boldsymbol{B}_1^T\boldsymbol{P}_1\boldsymbol{x}(t) + \xi_2\boldsymbol{x}(t)^T\boldsymbol{N}_2^T\boldsymbol{N}_2\boldsymbol{x}(t)\\ &+ \xi_3^{-1}\boldsymbol{x}(t)^T\boldsymbol{P}_1\boldsymbol{B}_1\boldsymbol{M}_2\boldsymbol{M}_2^T\boldsymbol{B}_1^T\boldsymbol{P}_1\boldsymbol{x}(t) + \xi_3\boldsymbol{e}_x(t)^T\boldsymbol{N}_2^T\boldsymbol{N}_2\boldsymbol{e}_x(t)\\ &+ 2\boldsymbol{x}(t)^T\boldsymbol{P}_1\boldsymbol{E}\boldsymbol{e}_\xi(t) + 2\varpi\eta\boldsymbol{x}(t)^T\boldsymbol{P}_1\boldsymbol{E}\boldsymbol{E}^T\boldsymbol{e}_x(t)\\ &+ \boldsymbol{e}_x(t)^T\left[(\boldsymbol{A}-\boldsymbol{L}\boldsymbol{C}_1)^T\boldsymbol{P}_2 + \boldsymbol{P}_2(\boldsymbol{A}-\boldsymbol{L}\boldsymbol{C}_1)\right]\boldsymbol{e}_x(t)\\ &+ \xi_4^{-1}\boldsymbol{e}_x(t)^T\boldsymbol{P}_2\boldsymbol{M}_1\boldsymbol{M}_1^T\boldsymbol{P}_2\boldsymbol{e}_x(t) + \xi_4\boldsymbol{x}(t)^T\boldsymbol{N}_1^T\boldsymbol{N}_1\boldsymbol{x}(t) + 2\boldsymbol{e}_x(t)^T\boldsymbol{P}_2\boldsymbol{E}\boldsymbol{e}_\xi(t)\\ &+ \boldsymbol{e}_x(t)^T(\varpi\eta\boldsymbol{P}_2\boldsymbol{E}\boldsymbol{E}^T + \varpi\eta\boldsymbol{E}\boldsymbol{E}^T\boldsymbol{P}_2)\boldsymbol{e}_x(t) - \boldsymbol{e}_\xi(t)^T(\varpi\eta\boldsymbol{P}_3\boldsymbol{E}^T\boldsymbol{E}+\varpi\eta\boldsymbol{E}^T\boldsymbol{E}\boldsymbol{P}_3)\boldsymbol{e}_\xi(t)\\ &- 2\varpi\eta\boldsymbol{e}_\xi(t)^T\boldsymbol{P}_3\boldsymbol{E}^T\boldsymbol{A}\boldsymbol{e}_x(t) - 2\varpi^2\eta^2\boldsymbol{e}_\xi(t)^T\boldsymbol{P}_3\boldsymbol{E}^T\boldsymbol{E}\boldsymbol{E}^T\boldsymbol{e}_x(t)\\ &+ \varpi\eta\xi_5^{-1}\boldsymbol{e}_\xi(t)^T\boldsymbol{P}_3\boldsymbol{E}^T\boldsymbol{M}_1\boldsymbol{M}_1^T\boldsymbol{E}\boldsymbol{P}_3\boldsymbol{e}_\xi(t)+\varpi\eta\xi_5\boldsymbol{x}(t)^T\boldsymbol{N}_1^T\boldsymbol{N}_1\boldsymbol{x}(t)\end{aligned}$$

Therefore, the first derivative of the Lyapunov function becomes

$$\dot{V}(t) \le \chi(t)^T \boldsymbol{\Theta}_N \chi(t)$$

where,

$$\boldsymbol{\Theta}_N = \begin{bmatrix} \boldsymbol{\Pi}_{11} & \boldsymbol{P}_1\boldsymbol{B}_1K + \varpi\eta\boldsymbol{P}_1\boldsymbol{E}\boldsymbol{E}^T & \boldsymbol{P}_1\boldsymbol{E} \\ * & \boldsymbol{\Pi}_{22} & \boldsymbol{\Pi}_{23} \\ * & * & \boldsymbol{\Pi}_{33} \end{bmatrix},$$

and where,

$$\begin{aligned} \boldsymbol{\Pi}_{11} &= (\boldsymbol{A} - \boldsymbol{B}_1\boldsymbol{K})^T\boldsymbol{P}_1 + \boldsymbol{P}_1(\boldsymbol{A} - \boldsymbol{B}_1\boldsymbol{K}) + \xi_1^{-1}\boldsymbol{P}_1\boldsymbol{M}_1\boldsymbol{M}_1^T\boldsymbol{P}_1 + \xi_1\boldsymbol{N}_1^T\boldsymbol{N}_1 \\ &\quad + \xi_2^{-1}\boldsymbol{P}_1\boldsymbol{B}_1\boldsymbol{M}_2\boldsymbol{M}_2^T\boldsymbol{B}_1^T\boldsymbol{P}_1 + \xi_2\boldsymbol{N}_2^T\boldsymbol{N}_2 \\ &\quad + \xi_3^{-1}\boldsymbol{P}_1\boldsymbol{B}_1\boldsymbol{M}_2\boldsymbol{M}_2^T\boldsymbol{B}_1^T\boldsymbol{P}_1 + \xi_4\boldsymbol{N}_1^T\boldsymbol{N}_1 + \varpi\eta\xi_5\boldsymbol{N}_1^T\boldsymbol{N}_1 \\ \boldsymbol{\Pi}_{22} &= (\boldsymbol{A} - \boldsymbol{L}\boldsymbol{C}_1)^T\boldsymbol{P}_2 + \boldsymbol{P}_2(\boldsymbol{A} - \boldsymbol{L}\boldsymbol{C}_1) + \xi_3\boldsymbol{N}_2^T\boldsymbol{N}_2 \\ &\quad + \xi_4^{-1}\boldsymbol{P}_2\boldsymbol{M}_1\boldsymbol{M}_1^T\boldsymbol{P}_2 + \varpi\eta\boldsymbol{P}_2\boldsymbol{E}\boldsymbol{E}^T + \varpi\eta\boldsymbol{E}\boldsymbol{E}^T\boldsymbol{P}_2 \\ \boldsymbol{\Pi}_{23} &= \boldsymbol{P}_2\boldsymbol{E} - \varpi\eta\boldsymbol{A}^T\boldsymbol{E}\boldsymbol{P}_3 - \varpi^2\eta^2\boldsymbol{E}\boldsymbol{E}^T\boldsymbol{E}\boldsymbol{P}_3 \\ \boldsymbol{\Pi}_{33} &= -\varpi\eta(\boldsymbol{P}_3\boldsymbol{E}^T\boldsymbol{E} + \boldsymbol{E}^T\boldsymbol{E}\boldsymbol{P}_3) + \varpi\eta\xi_5^{-1}\boldsymbol{P}_3\boldsymbol{E}^T\boldsymbol{M}_1\boldsymbol{M}_1^T\boldsymbol{E}\boldsymbol{P}_3 \end{aligned}$$

Considering inequality (5.25), with $\boldsymbol{H} = \boldsymbol{P}_2\boldsymbol{L}$, inequality (5.27) can be obtained.

$$\begin{bmatrix} \tilde{\boldsymbol{\Theta}}_{11} & \tilde{\boldsymbol{\Theta}}_{12} \\ * & \tilde{\boldsymbol{\Theta}}_{22} \end{bmatrix} < 0 \tag{5.27}$$

where,

$$\tilde{\boldsymbol{\Theta}}_{11} = \begin{bmatrix} \tilde{\boldsymbol{\Pi}}_{11} & \boldsymbol{P}_1\boldsymbol{B}_1\boldsymbol{K} + \varpi\eta\boldsymbol{P}_1\boldsymbol{E}\boldsymbol{E}^T & \boldsymbol{P}_1\boldsymbol{E} \\ * & \tilde{\boldsymbol{\Pi}}_{22} & \tilde{\boldsymbol{\Pi}}_{23} \\ * & * & \tilde{\boldsymbol{\Pi}}_{33} \end{bmatrix},$$

$$\tilde{\boldsymbol{\Theta}}_{12} = \begin{bmatrix} \boldsymbol{P}_1\boldsymbol{B}_2 & 0 & 0 & \boldsymbol{P}_1\boldsymbol{M}_1 & \boldsymbol{N}_1^T & \boldsymbol{P}_1\boldsymbol{B}_1\boldsymbol{M}_2 & \boldsymbol{N}_2^T & \boldsymbol{C}_2^T & 0 & 0 & 0 \\ \boldsymbol{P}_2\boldsymbol{B}_2 & -\boldsymbol{P}_2\boldsymbol{L} & 0 & 0 & 0 & 0 & 0 & 0 & \boldsymbol{N}_2^T & \boldsymbol{P}_2\boldsymbol{M}_1 & 0 \\ -\varpi\eta\boldsymbol{P}_3\boldsymbol{E}^T\boldsymbol{B}_2 & 0 & \boldsymbol{P}_3 & 0 & 0 & 0 & 0 & 0 & 0 & 0 & \boldsymbol{P}_3\boldsymbol{E}^T\boldsymbol{M}_1 \end{bmatrix}$$

$$\tilde{\boldsymbol{\Theta}}_{22} = blkdiag\left(-\gamma^2\boldsymbol{I} \;\; -\gamma^2\boldsymbol{I} \;\; -\gamma^2\boldsymbol{I} \;\; -\xi_1\boldsymbol{I} \;\; -(\xi_1 + \xi_4 + \varpi\eta\xi_5)^{-1}\boldsymbol{I} \;\; -(\xi_2^{-1} + \xi_3^{-1})^{-1}\boldsymbol{I} \;\; -\xi_2^{-1}\boldsymbol{I} \;\; -\boldsymbol{I} \;\; -\xi_3^{-1}\boldsymbol{I} \;\; -\xi_4\boldsymbol{I} \;\; -(\varpi\eta)^{-1}\xi_5\boldsymbol{I}\right)$$

and where,

$$\begin{aligned} \tilde{\boldsymbol{\Pi}}_{11} &= (\boldsymbol{A} - \boldsymbol{B}_1\boldsymbol{K})^T\boldsymbol{P}_1 + \boldsymbol{P}_1(\boldsymbol{A} - \boldsymbol{B}_1\boldsymbol{K}) \\ \tilde{\boldsymbol{\Pi}}_{22} &= (\boldsymbol{A} - \boldsymbol{L}\boldsymbol{C}_1)^T\boldsymbol{P}_2 + \boldsymbol{P}_2(\boldsymbol{A} - \boldsymbol{L}\boldsymbol{C}_1) + \varpi\eta\boldsymbol{P}_2\boldsymbol{E}\boldsymbol{E}^T + \varpi\eta\boldsymbol{E}\boldsymbol{E}^T\boldsymbol{P}_2 \\ \tilde{\boldsymbol{\Pi}}_{23} &= \boldsymbol{\Pi}_{23} \\ \tilde{\boldsymbol{\Pi}}_{33} &= -\varpi\eta(\boldsymbol{P}_3\boldsymbol{E}^T\boldsymbol{E} + \boldsymbol{E}^T\boldsymbol{E}\boldsymbol{P}_3) \end{aligned}$$

According to inequality (5.27), one knows $\boldsymbol{\Theta}_N < 0$, so that

$$\dot{V}(t) \leq \boldsymbol{\chi}(t)^T \boldsymbol{\Theta}_N \boldsymbol{\chi}(t) \leq \lambda_{\max}(\boldsymbol{\Theta}_N) \|\boldsymbol{\chi}\|^2 \tag{5.28}$$

Furthermore, let $\alpha = -\lambda_{\max}(\boldsymbol{\Theta}_N) > 0$, then,

$$\dot{V}(t) \leq -\alpha \|\boldsymbol{\chi}\|^2$$

that is, the closed-loop system is quadratically stable under the condition of controller (2.14).

Next, one shows the output $\boldsymbol{z}(t)$ satisfies H_∞ performance constraint.

To establish the L_2 [0, ∞) norm bound $\gamma^2 \|\hat{\boldsymbol{w}}(t)\|_2^2$, consider the following functional:

$$J = \int_0^\infty [\boldsymbol{z}(t)^T \boldsymbol{z}(t) - \gamma^2 \hat{\boldsymbol{w}}(t)^T \hat{\boldsymbol{w}}(t)] dt$$

As the closed-loop system has quadratic stability, for arbitrary nonzero $\hat{\boldsymbol{w}}(t) \in L_2[0, \infty)$, under zero initial conditions, one has

$$J = \int_0^\infty [\boldsymbol{z}(t)^T \boldsymbol{z}(t) - \gamma^2 \hat{\boldsymbol{w}}(t)^T \hat{\boldsymbol{w}}(t) + \dot{V}(\boldsymbol{x}(t))] dt - V(\infty) + V(0)$$

$$\leq \int_0^\infty \left\{ \begin{array}{l} \boldsymbol{x}(t)^T \boldsymbol{C}_2^T \boldsymbol{C}_2 \boldsymbol{x}(t) - \gamma^2 (\boldsymbol{w}(t)^T \boldsymbol{w}(t) + \boldsymbol{v}(t)^T \boldsymbol{v}(t) + \dot{\boldsymbol{f}}(t)^T \dot{\boldsymbol{f}}(t)) \\ + \boldsymbol{\chi}(t)^T \boldsymbol{\Theta}_N \boldsymbol{\chi}(t) + 2\boldsymbol{x}(t)^T \boldsymbol{P}_1 \boldsymbol{B}_2 \boldsymbol{w}(t) \\ + 2\boldsymbol{e}_x(t)^T \boldsymbol{P}_2 \boldsymbol{B}_2 \boldsymbol{w}(t) - 2\boldsymbol{e}_x(t)^T \boldsymbol{P}_2 \boldsymbol{L} \boldsymbol{v}(t) + 2\boldsymbol{e}_\xi(t)^T \boldsymbol{P}_3 \dot{\boldsymbol{f}}(t) - 2\varpi\eta \boldsymbol{e}_\xi(t)^T \boldsymbol{P}_3 \boldsymbol{E}^T \boldsymbol{B}_2 \boldsymbol{w}(t) \end{array} \right\} dt$$

$$\leq \int_0^\infty \begin{bmatrix} \boldsymbol{x}(t)^T & \boldsymbol{e}_x(t)^T & \boldsymbol{e}_\xi(t)^T & \boldsymbol{w}(t)^T & \boldsymbol{v}(t)^T & \dot{\boldsymbol{f}}(t)^T \end{bmatrix}$$

$$\begin{bmatrix} \boldsymbol{\Pi}_{11} + \boldsymbol{C}_2^T \boldsymbol{C}_2 & \boldsymbol{P}_1 \boldsymbol{B}_1 \boldsymbol{K} + \varpi\eta \boldsymbol{P}_1 \boldsymbol{E}\boldsymbol{E}^T & \boldsymbol{P}_1 \boldsymbol{E} & \boldsymbol{P}_1 \boldsymbol{B}_2 & 0 & 0 \\ * & \boldsymbol{\Pi}_{22} & \boldsymbol{\Pi}_{23} & \boldsymbol{P}_2 \boldsymbol{B}_2 & -\boldsymbol{P}_2 \boldsymbol{L} & 0 \\ * & * & \boldsymbol{\Pi}_{33} & -\varpi\eta \boldsymbol{P}_3 \boldsymbol{E}^T \boldsymbol{B}_2 & 0 & \boldsymbol{P}_3 \\ * & * & * & -\gamma^2 \boldsymbol{I} & 0 & 0 \\ * & * & * & * & -\gamma^2 \boldsymbol{I} & 0 \\ * & * & * & * & * & -\gamma^2 \boldsymbol{I} \end{bmatrix} \begin{bmatrix} \boldsymbol{x}(t) \\ \boldsymbol{e}_x(t) \\ \boldsymbol{e}_\xi(t) \\ \boldsymbol{w}(t) \\ \boldsymbol{v}(t) \\ \dot{\boldsymbol{f}}(t) \end{bmatrix} dt$$

According to Lemma 5.2 and inequality (5.27), $J < 0$ holds, that is, $\boldsymbol{z}$ (t) satisfies H_∞ performance constraint.

This completes the proof.

5.3.3 Sufficient conditions under multiplicative perturbation

Theorem 5.2 now gives the existence conditions of the controller gain matrix $\boldsymbol{K}$ and observer gain matrix $\boldsymbol{L}$ under multiplicative perturbation.

Theorem 5.2: When controller gain perturbation is in the form of Eq. (5.6), for given scalars $\varpi \in (0, 1], \eta > 0, \xi_i > 0, i = 1, 2, \ldots, 5,$ and given symmetric positive definite matrix $\widehat{\boldsymbol{P}}$, then the system represented by Eqs. (5.20)–(5.22) is quadratically stable under controller (5.19), and $\boldsymbol{z}(t)$ satisfies H_∞ performance constraint if there exist symmetric positive definite matrices $\boldsymbol{P}_i, i = 1, 2, 3$, matrices $\boldsymbol{H}$ and $\boldsymbol{K}$, such that the following LMIs hold:

$$(\boldsymbol{A}-\boldsymbol{B}_1\boldsymbol{K})^T\widehat{\boldsymbol{P}} + \widehat{\boldsymbol{P}}(\boldsymbol{A} - \boldsymbol{B}_1\boldsymbol{K}) < 0 \tag{5.29}$$

$$\begin{bmatrix} \boldsymbol{\Theta}_{11} & \boldsymbol{\Theta}_{12} \\ * & \boldsymbol{\Theta}_{22} \end{bmatrix} < 0 \tag{5.30}$$

where,

$$\boldsymbol{\Theta}_{11} = \begin{bmatrix} \tilde{\boldsymbol{\Pi}}_{11} & \boldsymbol{P}_1\boldsymbol{B}_1\boldsymbol{K}+\varpi\eta\boldsymbol{P}_1\boldsymbol{E}\boldsymbol{E}^T & \boldsymbol{P}_1\boldsymbol{E} \\ * & \tilde{\boldsymbol{\Pi}}_{22} & \tilde{\boldsymbol{\Pi}}_{23} \\ * & * & \tilde{\boldsymbol{\Pi}}_{33} \end{bmatrix}$$

$$\boldsymbol{\Theta}_{12} = \begin{bmatrix} \boldsymbol{P}_1\boldsymbol{B}_2 & 0 & 0 & \boldsymbol{P}_1\boldsymbol{M}_1 & \boldsymbol{N}_1^T & \boldsymbol{P}_1\boldsymbol{B}_1\boldsymbol{M}_3 & \boldsymbol{K}^T\boldsymbol{N}_3^T & \boldsymbol{C}_2^T & 0 & 0 & 0 \\ \boldsymbol{P}_2\boldsymbol{B}_2 & -\boldsymbol{H} & 0 & 0 & 0 & 0 & 0 & 0 & \boldsymbol{K}^T\boldsymbol{N}_3^T & \boldsymbol{P}_2\boldsymbol{M}_1 & 0 \\ -\varpi\eta\boldsymbol{P}_3\boldsymbol{E}^T\boldsymbol{B}_2 & 0 & \boldsymbol{P}_3 & 0 & 0 & 0 & 0 & 0 & 0 & 0 & \boldsymbol{P}_3\boldsymbol{E}^T\boldsymbol{M}_1 \end{bmatrix}$$

$$\boldsymbol{\Theta}_{22} = blkdiag\left(-\gamma^2\boldsymbol{I} \;\; -\gamma^2\boldsymbol{I} \;\; -\gamma^2\boldsymbol{I} \;\; -\xi_1\boldsymbol{I} \;\; -(\xi_1+\xi_4+\varpi\eta\xi_5)^{-1}\boldsymbol{I} \;\; -(\xi_2^{-1}+\xi_3^{-1})^{-1}\boldsymbol{I} \;\; -\xi_2^{-1}\boldsymbol{I} \;\; -\boldsymbol{I} \;\; -\xi_3^{-1}\boldsymbol{I} \;\; -\xi_4\boldsymbol{I} \;\; -(\varpi\eta)^{-1}\xi_5\boldsymbol{I}\right)$$

and where,

$$\begin{aligned}
\tilde{\boldsymbol{\Pi}}_{11} &= (\boldsymbol{A}-\boldsymbol{B}_1\boldsymbol{K})^T\boldsymbol{P}_1 + \boldsymbol{P}_1(\boldsymbol{A} - \boldsymbol{B}_1\boldsymbol{K}) \\
\tilde{\boldsymbol{\Pi}}_{22} &= \boldsymbol{A}^T\boldsymbol{P}_2 - \boldsymbol{C}_1^T\boldsymbol{H}^T + \boldsymbol{P}_2\boldsymbol{A} - \boldsymbol{H}\boldsymbol{C}_1 + \varpi\eta\boldsymbol{P}_2\boldsymbol{E}\boldsymbol{E}^T + \varpi\eta\boldsymbol{E}\boldsymbol{E}^T\boldsymbol{P}_2 \\
\tilde{\boldsymbol{\Pi}}_{23} &= \boldsymbol{P}_2\boldsymbol{E} - \varpi\eta\boldsymbol{A}^T\boldsymbol{E}\boldsymbol{P}_3 - \varpi^2\eta^2\boldsymbol{E}\boldsymbol{E}^T\boldsymbol{E}\boldsymbol{P}_3 \\
\tilde{\boldsymbol{\Pi}}_{33} &= -\varpi\eta(\boldsymbol{P}_3\boldsymbol{E}^T\boldsymbol{E}+\boldsymbol{E}^T\boldsymbol{E}\boldsymbol{P}_3)
\end{aligned}$$

When LMIs (5.29) and (5.30) are feasible, one can obtain the controller gain matrix $\boldsymbol{K}$ and the observer gain matrix $\boldsymbol{L} = \boldsymbol{P}_2^{-1}\boldsymbol{H}$.

Proof: First, one shows that the closed-loop system is quadratically stable while $\boldsymbol{w}(t) = 0, \boldsymbol{v}(t) = 0, \dot{\boldsymbol{f}}(t) = 0$.

Choose the Lyapunov function candidate as

$$V(t) = \boldsymbol{x}(t)^T\boldsymbol{P}_1\boldsymbol{x}(t) + \boldsymbol{e}_x^T(t)\boldsymbol{P}_2\boldsymbol{e}_x(t) + \boldsymbol{e}_\xi^T(t)\boldsymbol{P}_3\boldsymbol{e}_\xi(t)$$

Define $\chi(t) = [\,x(t)^T \quad e_x(t)^T \quad e_\xi(t)^T\,]^T$, according to Lemma 5.1 and the given constraint conditions, Eq. (5.26) gives

$$\begin{aligned}\dot{V}(t) \le\ & x(t)^T\left[(A-B_1K)^TP_1 + P_1(A-B_1K)\right]x(t) + 2x(t)^TP_1B_1Ke_x(t)\\&+\xi_1^{-1}x(t)^TP_1M_1M_1^TP_1x(t) + \xi_1x(t)^TN_1^TN_1x(t)\\&+\xi_2^{-1}x(t)^TP_1B_1M_3M_3^TB_1^TP_1x(t) + \xi_2x(t)^TK^TN_3^TN_3Kx(t)\\&+\xi_3^{-1}x(t)^TP_1B_1M_3M_3^TB_1^TP_1x(t) + \xi_3e_x(t)^TK^TN_3^TN_3Ke_x(t)\\&+2x(t)^TP_1Ee_\xi(t) + 2\varpi\eta x(t)^TP_1EE^Te_x(t) + e_x(t)^T\left[(A-LC_1)^TP_2\right.\\&\left.+P_2(A-LC_1)\right]e_x(t) + \xi_4^{-1}e_x(t)^TP_2M_1M_1^TP_2e_x(t)\\&+\xi_4x(t)N_1^TN_1x(t) + 2e_x(t)^TP_2Ee_\xi(t) + e_x(t)^T(\varpi\eta P_2EE^T\\&+\varpi\eta EE^TP_2)e_x(t) - e_\xi(t)^T(\varpi\eta P_3E^TE+\varpi\eta E^TEP_3)e_\xi(t)\\&-2\varpi\eta e_\xi(t)^TP_3E^TAe_x(t) - 2\varpi^2\eta^2e_\xi(t)^TP_3E^TEE^Te_x(t)\\&+\varpi\eta\xi_5^{-1}e_\xi(t)^TP_3E^TM_1M_1^TEP_3e_\xi(t)+\varpi\eta\xi_5x(t)^TN_1^TN_1x(t)\end{aligned}$$

Therefore, the first derivative of the Lyapunov function becomes

$$\dot{V}(t) \le \chi(t)^T\Theta_N\chi(t)$$

where,

$$\Theta_N = \begin{bmatrix}\Pi_{11} & P_1B_1K+\varpi\eta P_1EE^T & P_1E\\ * & \Pi_{22} & \Pi_{23}\\ * & * & \Pi_{33}\end{bmatrix},$$

and where,

$$\begin{aligned}\Pi_{11} &= (A-B_1K)^TP_1 + P_1(A-B_1K) + \xi_1^{-1}P_1M_1M_1^TP_1\\&\quad+\xi_1N_1^TN_1 + \xi_2^{-1}P_1B_1M_3M_3^TB_1^TP_1 + \xi_2K^TN_3^TN_3K\\&\quad+\xi_3^{-1}P_1B_1M_3M_3^TB_1^TP_1+\xi_4N_1^TN_1 + \varpi\eta\xi_5N_1^TN_1\\\Pi_{22} &= (A-LC_1)^TP_2 + P_2(A-LC_1) + \xi_3K^TN_3^TN_3K\\&\quad+\xi_4^{-1}P_2M_1M_1^TP_2 + \varpi\eta P_2EE^T + \varpi\eta EE^TP_2\\\Pi_{23} &= P_2E - \varpi\eta A^TEP_3 - \varpi^2\eta^2EE^TEP_3\\\Pi_{33} &= -\varpi\eta(P_3E^TE+E^TEP_3) + \varpi\eta\xi_5^{-1}P_3E^TM_1M_1^TEP_3\end{aligned}$$

Considering inequality (5.30), with $\boldsymbol{H} = \boldsymbol{P}_2\boldsymbol{L}$, inequality (5.31) can be obtained.

$$\begin{bmatrix} \tilde{\boldsymbol{\Theta}}_{11} & \tilde{\boldsymbol{\Theta}}_{12} \\ * & \tilde{\boldsymbol{\Theta}}_{22} \end{bmatrix} < 0 \tag{5.31}$$

where,

$$\tilde{\boldsymbol{\Theta}}_{11} = \begin{bmatrix} \tilde{\boldsymbol{\Pi}}_{11} & \boldsymbol{P}_1\boldsymbol{B}_1\boldsymbol{K}+\varpi\eta\boldsymbol{P}_1\boldsymbol{E}\boldsymbol{E}^T & \boldsymbol{P}_1\boldsymbol{E} \\ * & \tilde{\boldsymbol{\Pi}}_{22} & \tilde{\boldsymbol{\Pi}}_{23} \\ * & * & \tilde{\boldsymbol{\Pi}}_{33} \end{bmatrix},$$

$$\tilde{\boldsymbol{\Theta}}_{12} = \begin{bmatrix} \boldsymbol{P}_1\boldsymbol{B}_2 & 0 & 0 & \boldsymbol{P}_1\boldsymbol{M}_1 & \boldsymbol{N}_1^T & \boldsymbol{P}_1\boldsymbol{B}_1\boldsymbol{M}_3 & \boldsymbol{K}^T\boldsymbol{N}_3^T & \boldsymbol{C}_2^T & 0 & 0 & 0 \\ \boldsymbol{P}_2\boldsymbol{B}_2 & -\boldsymbol{P}_2\boldsymbol{L} & 0 & 0 & 0 & 0 & 0 & 0 & \boldsymbol{K}^T\boldsymbol{N}_3^T & \boldsymbol{P}_2\boldsymbol{M}_1 & 0 \\ -\varpi\eta\boldsymbol{P}_3\boldsymbol{E}^T\boldsymbol{B}_2 & 0 & \boldsymbol{P}_3 & 0 & 0 & 0 & 0 & 0 & 0 & 0 & \boldsymbol{P}_3\boldsymbol{E}^T\boldsymbol{M}_1 \end{bmatrix}$$

$$\tilde{\boldsymbol{\Theta}}_{22} = blkdiag\left(-\gamma^2\boldsymbol{I} \ \ -\gamma^2\boldsymbol{I} \ \ -\gamma^2\boldsymbol{I} \ \ -\xi_1\boldsymbol{I} \ \ -(\xi_1+\xi_4+\varpi\eta\xi_5)^{-1}\boldsymbol{I} \ \ -(\xi_2^{-1}+\xi_3^{-1})^{-1}\boldsymbol{I} \ \ -\xi_2^{-1}\boldsymbol{I} \ \ -\boldsymbol{I} \ \ -\xi_3^{-1}\boldsymbol{I} \ \ -\xi_4\boldsymbol{I} \ \ -(\varpi\eta)^{-1}\xi_5\boldsymbol{I}\right)$$

and where,

$$\begin{aligned} \tilde{\boldsymbol{\Pi}}_{11} &= (\boldsymbol{A}-\boldsymbol{B}_1\boldsymbol{K})^T\boldsymbol{P}_1 + \boldsymbol{P}_1(\boldsymbol{A}-\boldsymbol{B}_1\boldsymbol{K}) \\ \tilde{\boldsymbol{\Pi}}_{22} &= (\boldsymbol{A}-\boldsymbol{L}\boldsymbol{C}_1)^T\boldsymbol{P}_2 + \boldsymbol{P}_2(\boldsymbol{A}-\boldsymbol{L}\boldsymbol{C}_1) + \varpi\eta\boldsymbol{P}_2\boldsymbol{E}\boldsymbol{E}^T + \varpi\eta\boldsymbol{E}\boldsymbol{E}^T\boldsymbol{P}_2 \\ \tilde{\boldsymbol{\Pi}}_{23} &= \boldsymbol{\Pi}_{23} \\ \tilde{\boldsymbol{\Pi}}_{33} &= -\varpi\eta(\boldsymbol{P}_3\boldsymbol{E}^T\boldsymbol{E}+\boldsymbol{E}^T\boldsymbol{E}\boldsymbol{P}_3) \end{aligned}$$

According to inequality (5.31), one knows $\boldsymbol{\Theta}_N < 0$, then

$$\dot{V}(t) \le \chi(t)^T\boldsymbol{\Theta}_N\chi(t) \le \lambda_{\max}(\boldsymbol{\Theta}_N)\|\chi\|^2 \tag{5.32}$$

Furthermore, let $\alpha = -\lambda_{\max}(\boldsymbol{\Theta}_N) > 0$, then,

$$\dot{V}(t) \le -\alpha\|\chi\|^2$$

that is, the closed-loop system is quadratically stable under the condition of controller (5.19).

Next, one shows the output $\boldsymbol{z}(t)$ satisfies H_∞ performance constraint.

To establish the L_2 $[0, \infty)$ norm bound $\gamma^2\|\hat{\boldsymbol{w}}(t)\|_2^2$, consider the following functional:

$$J = \int_0^\infty [\boldsymbol{z}(t)^T\boldsymbol{z}(t) - \gamma^2\hat{\boldsymbol{w}}(t)^T\hat{\boldsymbol{w}}(t)]dt$$

As the closed-loop system has quadratic stability, for arbitrary nonzero $\hat{w}(t) \in L_2[0, \infty)$, under zero initial conditions,

$$J = \int_0^{\infty} [z(t)^T z(t) - \gamma^2 \hat{w}(t)^T \hat{w}(t) + \dot{V}(x(t))]dt - V(\infty) + V(0)$$

$$\leq \int_0^{\infty} \left\{ \begin{array}{l} x(t)^T C_2^T C_2 x(t) - \gamma^2 (w(t)^T w(t) + v(t)^T v(t) \\ + \dot{f}(t)^T \dot{f}(t)) + \chi(t)^T \Theta_{N1} \chi(t) + 2x(t)^T P_1 B_2 w(t) \\ + 2e_x(t)^T P_2 B_2 w(t) - 2e_x(t)^T P_2 L v(t) + 2e_\xi(t)^T P_3 \dot{f}(t) \\ - 2\varpi\eta e_\xi(t)^T P_3 E^T B_2 w(t) \end{array} \right\} dt$$

$$\leq \int_0^{\infty} \begin{bmatrix} x(t)^T & e_x(t)^T & e_\xi(t)^T & w(t)^T & v(t)^T & \dot{f}(t)^T \end{bmatrix}$$

$$\begin{bmatrix} \Pi_{11} + C_2^T C_2 & P_1 B_1 K + \varpi\eta P_1 E E^T & P_1 E & P_1 B_2 & 0 & 0 \\ * & \Pi_{22} & \Pi_{23} & P_2 B_2 & -P_2 L & 0 \\ * & * & \Pi_{33} & -\varpi\eta P_3 E^T B_2 & 0 & P_3 \\ * & * & * & -\gamma^2 I & 0 & 0 \\ * & * & * & * & -\gamma^2 I & 0 \\ * & * & * & * & * & -\gamma^2 I \end{bmatrix} \begin{bmatrix} x(t) \\ e_x(t) \\ e_\xi(t) \\ w(t) \\ v(t) \\ \dot{f}(t) \end{bmatrix} dt$$

According to Lemma 5.2 and inequality (5.27), $J < 0$ holds, that is, z (t) satisfies H_∞ performance constraint.

This completes the proof.

Remark 5.6: In Theorem 5.1 or Theorem 5.2, before solving inequality (5.25) or inequality (5.30), the controller gain matrix K should be chosen such that $A - B_1 K$ is Hurwitz with the conditions of (5.24) or (5.29). Then the observer gain L can be obtained by solving inequality (5.25) or inequality (5.30). It should be noted that the feasibility of the LMI can be improved by adjusting the aforementioned parameters, that is, ϖ, η, $\xi_{i=1,2,\ldots,5}$ and $\widehat{P}$.

Remark 5.7: The scalar γ can be regarded as an optimization variable to obtain a reduction in the H_∞ performance constraint bound. In this case, the minimum of H_∞ performance constraint bound with admissible controllers can be readily found by solving the following convex optimization problem: Minimize γ subject to the LMIs in Theorem 5.1 or Theorem 5.2.

5.4 Simulation test

In this section, the effectiveness of the observer-based fault tolerant controller is illustrated with simulations of a spacecraft attitude control system. For this example, the moments of inertia are taken to be I_x=20 kg m^2,

$I_y = 18\ \mathrm{kg\,m^2}$, $I_z = 15\ \mathrm{kg\,m^2}$, and the orbit is a 300 km circular orbit. The initial state is chosen as $\boldsymbol{x}(0) = [0.08\ \mathrm{rad},\ 0.06\ \mathrm{rad},\ -0.06\ \mathrm{rad},\ -0.01\ \mathrm{rad\,s^{-1}},\ -0.01\ \mathrm{rad\,s^{-1}},\ 0.01\ \mathrm{rad\,s^{-1}}]^T$, and the maximum control torque is assumed to be $u_{m1} = u_{m2} = u_{m3} = 0.25$ Nm. This constraint on control torque is considered in the simulation.

The external disturbances used in the simulation are supposed to be the following form and ω_0 is the orbital frequency.

$$\boldsymbol{w}(t) = 5 \times 10^{-4} \times \begin{bmatrix} \cos(\omega_0 t) \\ \cos(\omega_0 t + \pi/4) \\ \cos(\omega_0 t + \pi/3) \end{bmatrix} \mathrm{Nm}$$

The following parameters are also chosen:

$$\begin{aligned}
&\boldsymbol{M}_1 = 0.008 \times \begin{bmatrix} 0.8 & 1.1 & 1.3 & 1.5 & 1.6 & 1.8 \end{bmatrix}^T, \\
&\boldsymbol{N}_1 = 0.1 \times \begin{bmatrix} -0.1 & -0.2 & -0.3 & -0.4 & -0.2 & 1 \end{bmatrix}, \boldsymbol{F}_1(t) = \sin(100\omega_0 t), \\
&\boldsymbol{M}_2 = \begin{bmatrix} 0.01 & 0.01 & 0.01 \end{bmatrix}^T, \boldsymbol{N}_2 = \begin{bmatrix} 0.1 & 0.1 & 0.1 & 0.1 & 0.1 & 0.1 \end{bmatrix}, \\
&\boldsymbol{F}_2(t) = \sin(100\omega_0 t + \pi/4), \\
&\boldsymbol{M}_3 = \begin{bmatrix} 0.1 & 0.1 & 0.1 \end{bmatrix}^T, \boldsymbol{N}_3 = 10^{-3} \times \boldsymbol{1}_{6\times 3}, \\
&\boldsymbol{F}_3(t) = \begin{bmatrix} 0.3 & 0.4 & 0.5 & 0.3 & 0.5 & 0.4 \end{bmatrix} \cos(100\omega_0 t) \\
&\xi_1 = \xi_2 = \xi_3 = \xi_4 = \xi_5 = 0.02, \eta = 0.62, \widehat{\boldsymbol{P}} = 0.4 \times diag([1\ 2\ 1\ 2\ 1\ 2])
\end{aligned}$$

Here, we assume the fault signal $\boldsymbol{f}(t) = [0.005\ 0.006\ 0.007]^T \bar{f}(t)$, and the form of $\bar{f}(t)$ is shown in Table 5.1.

The measurement errors are assumed to be $\boldsymbol{v}(t) = 10^{-4} \times \begin{bmatrix} 4 & 5 & 6 & 0.2 & 0.2 & 0.2 \end{bmatrix}^T \sin(0.01\pi t)$, with units of rad for angle and $\mathrm{rad\,s^{-1}}$ for angular velocity. An actuator fault is assumed to occur, that is, $\boldsymbol{E} = \boldsymbol{B}_1$, and the probability of success is selected to be $\overline{\omega} = 0.5$.

For comparison, two cases are studied in this section. In case A, the proposed state feedback controller (SFC) and the fault-tolerant controller via stochastically intermediate observer (FTCSIO) are applied for spacecraft attitude stabilization with additive perturbation. In case B, the above two controllers are employed in spacecraft attitude stabilization with multiplicative perturbation. The state feedback controller is introduced in [27].

Table 5.1 The form of $\bar{f}(t)$.

t/s	[0,10]	(10,45]	(45,60]	(60,75]	(75,100]
$\bar{f}(t)$/Nm	0	$\sin(0.1\pi t)$	1	$\cos(0.1\pi t)$	0

5.4.1 Comparison analysis under additive perturbation

With inequality (5.24), the controller gain matrix $\boldsymbol{K}$ is

$$\boldsymbol{K}=\begin{bmatrix} 10 & 0 & 0 & 57.9437 & 0 & 2.0343 \\ 0 & 36 & 0 & 0 & 104.2987 & 0 \\ 0 & 0 & 7.5 & -1.5306 & 0 & 43.4578 \end{bmatrix}$$

Solving inequality (5.25) gives

$$\boldsymbol{P}_2=10^4\times\begin{bmatrix} 4.2832 & -1.2495 & -1.4792 & -0.0181 & 0.0084 & 0.0119 \\ -1.2495 & 3.4747 & -2.0285 & 0.0059 & -0.0287 & 0.0129 \\ -1.4792 & -2.0285 & 2.7907 & 0.0086 & 0.0126 & -0.0176 \\ -0.0181 & 0.0059 & 0.0086 & 0.0245 & -0.0110 & -0.0177 \\ 0.0084 & -0.0287 & 0.0126 & -0.0110 & 0.0426 & -0.0152 \\ 0.0119 & 0.0129 & -0.0176 & -0.0177 & -0.0152 & 0.0275 \end{bmatrix}$$

and

$$\boldsymbol{H}=10^4\times\begin{bmatrix} 8.3707 & 0.0378 & 0.0584 & 0.1539 & -0.1580 & -0.2124 \\ 0.0365 & 8.3967 & 0.0769 & -0.0012 & 0.2662 & -0.0481 \\ 0.0589 & 0.0766 & 8.4124 & -0.1081 & -0.1287 & 0.1762 \\ 0.1963 & -0.0333 & -0.0960 & 9.0663 & 0.3186 & 0.5209 \\ -0.1338 & 0.2474 & -0.1272 & 0.3314 & 8.7588 & 0.3172 \\ -0.1821 & -0.0788 & 0.1786 & 0.5297 & 0.3107 & 9.0684 \end{bmatrix}$$

Then, one has

$$\boldsymbol{L}=\boldsymbol{P}_2^{-1}\boldsymbol{H}$$
$$=10^4\times\begin{bmatrix} 0.0010 & 0.0013 & 0.0015 & 0.0037 & 0.0024 & 0.0036 \\ 0.0012 & 0.0019 & 0.0020 & 0.0040 & 0.0029 & 0.0041 \\ 0.0014 & 0.0021 & 0.0026 & 0.0056 & 0.0038 & 0.0058 \\ -0.0097 & 0.0116 & 0.0052 & 1.9983 & 1.2137 & 1.9604 \\ -0.0064 & 0.0081 & 0.0035 & 1.2158 & 0.7636 & 1.2066 \\ -0.0106 & 0.0116 & 0.0060 & 1.9614 & 1.2051 & 1.9631 \end{bmatrix}$$

Applying the controller gain $\boldsymbol{K}$ and observer gain $\boldsymbol{L}$ to Eqs. (5.20)–(5.22) gives the time responses of attitude angle and angular velocity as shown in Figs. 5.1 and 5.2 respectively. It is observed that the attitude angle and angular velocity converge to within a small bound. The attitude angle is approximately 1×10^{-3} rad and that of angular velocity is approximately 2×10^{-4} rad s^{-1} in steady-state error. For the analysis of the effectiveness of the presented observer, Figs. 5.3 and 5.4 show time responses of observation errors of attitude angle and angular velocity respectively. As can be seen, the

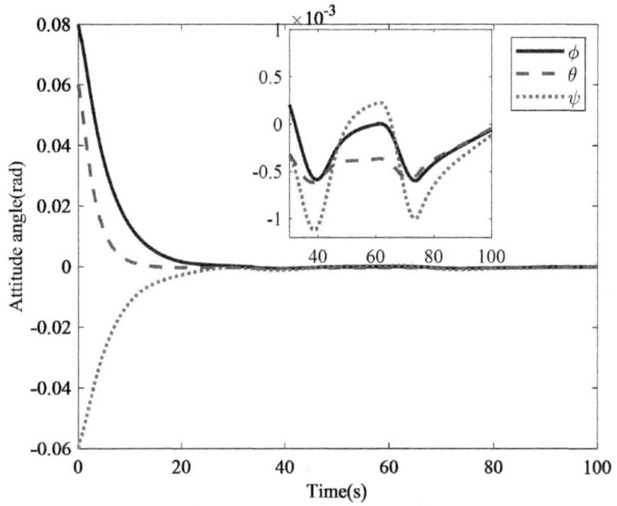

Figure 5.1 Time response of attitude angle with additive perturbation under fault-tolerant controller via stochastically intermediate observer.

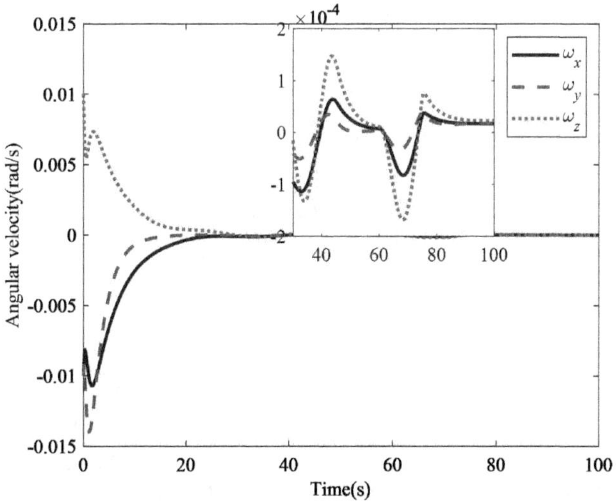

Figure 5.2 Time response of angular velocity with additive perturbation under fault-tolerant controller via stochastically intermediate observer.

estimated values converge to the real values rapidly and with high accuracy even when the conventional controller is employed. By applying the FTCSIO to the spacecraft attitude system, the observed attitude angle errors fall to within a small neighborhood containing zero (i.e., $|e_\phi| < 5 \times 10^{-4}$rad,

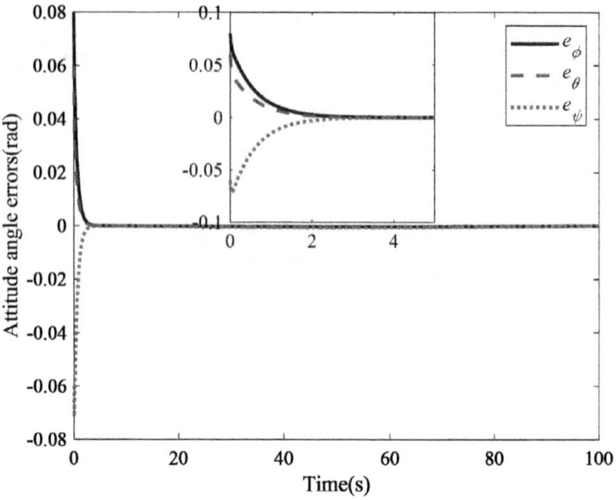

Figure 5.3 Time response of attitude angle errors with additive perturbation under fault-tolerant controller via stochastically intermediate observer.

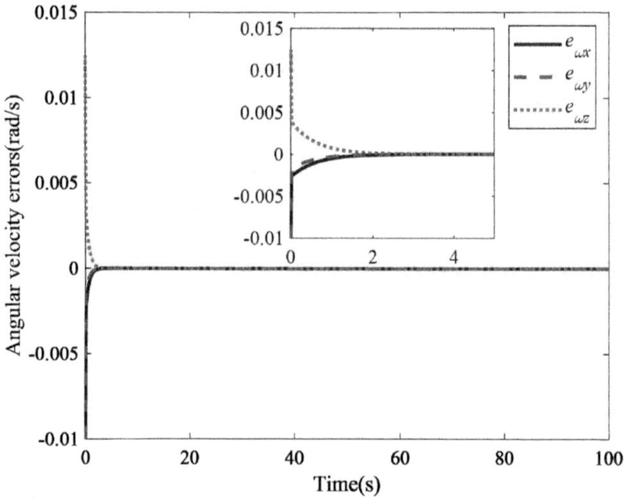

Figure 5.4 Time response of angular velocity errors with additive perturbation under fault-tolerant controller via stochastically intermediate observer.

$|e_\theta| < 3 \times 10^{-4}$ rad, $|e_\psi| < 6 \times 10^{-4}$ rad) within 3 seconds and the angular velocity errors reach $|e_{\omega x}| < 1 \times 10^{-4}$ rad s^{-1}, $|e_{\omega y}| < 0.7 \times 10^{-4}$ rad s^{-1} and $|e_{\omega z}| < 1.5 \times 10^{-4}$ rad s^{-1} at approximately 2 seconds. The fault signal and its stochastic estimation are plotted in Fig. 5.5, from which one can see that the observed fault signal is nearly the same as the real value. Based on these simulations, it is seen that both states and fault signals are estimated with good accuracy. The control torque shown in Fig. 5.6 is no more than 0.25 Nm. In order to further illustrate the influence of ϖ, the correlation between ϖ and η obtained by Theorem 5.1 is given in Fig. 5.7. It is observed that the curve is very steep when ϖ is near zero; otherwise, the

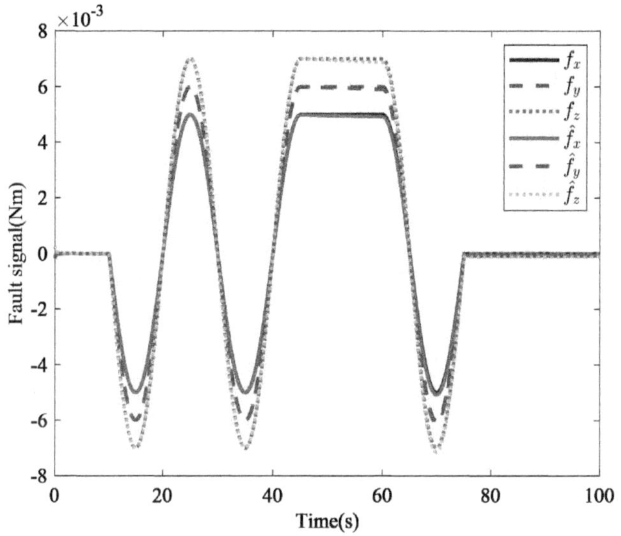

Figure 5.5 Fault signal and its stochastic estimation with additive perturbation under fault-tolerant controller via stochastically intermediate observer.

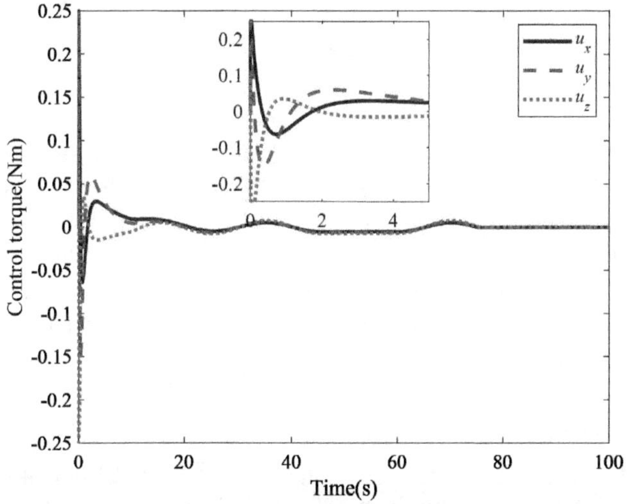

Figure 5.6 Time response of control torque with additive perturbation under fault-tolerant controller via stochastically intermediate observer.

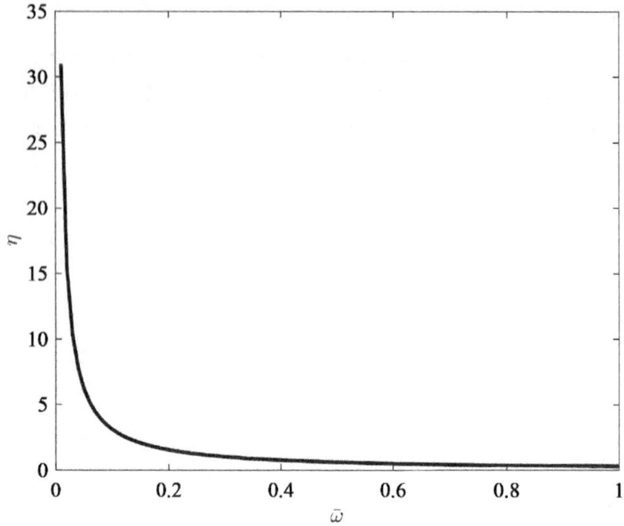

Figure 5.7 The correlation between ϖ and η.

curve is very flat. According to [29], larger ϖ will result in smaller η, which is consistent with the figure. Larger ϖ means greater effectiveness of $\eta E^T x(t)$, but the value of η cannot always increase. From the above analysis, it is clear that the parameter ϖ is of great importance.

When the state feedback controller is adopted, the controller gain matrix K is

$$K = \begin{bmatrix} -13.1786 & -1.1725 & -0.6418 & -26.6742 & -3.4368 & -2.5352 \\ -1.5258 & -13.0367 & -1.1603 & -3.3880 & -25.2173 & -2.4263 \\ -1.8498 & -1.7856 & -13.0968 & -2.2847 & -2.4131 & -24.0695 \end{bmatrix}$$

Applying this controller to the first term of Eq. (5.8) and considering the fault and additive perturbation, we obtain the time responses of attitude angle and angular velocity as shown in Figs. 5.8 and 5.9 respectively. It is observed that the attitude angle and angular velocity also converge to a small bound rapidly. The accuracy of attitude angle is less than 1×10^{-3} rad and that of angular velocity is less than 2×10^{-4} rad s^{-1} in steady-state error. Fig. 5.10 shows the response curve of control torque using the SFC.

Energy efficiency is an extremely important issue for achieving the target tasks and extending the mission operating life. Thus the optimization of energy consumption should be also taken into account for spacecraft attitude control system design. To analyze the total energy consumption,

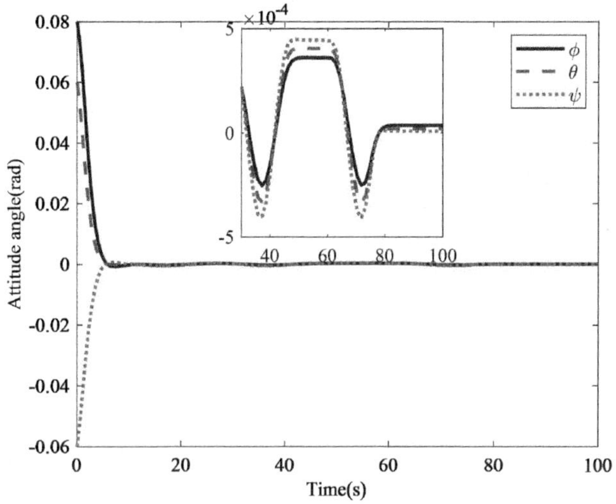

Figure 5.8 Time response of attitude angle with additive perturbation under state feedback controller.

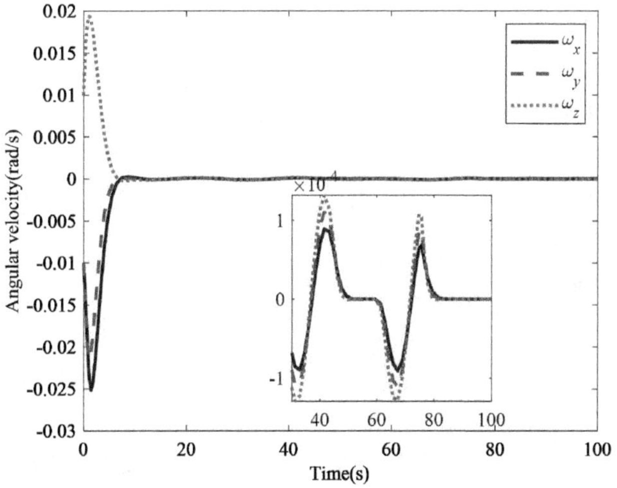

Figure 5.9 Time response of angular velocity with additive perturbation under state feedback controller.

an energy index function derived from the optimal cost function is defined by $E = 1/2\int_0^T \|\boldsymbol{u}\| dt$, where T denotes the simulation time, and $T = 100$ seconds is chosen in the simulation. The time responses of the energy consumption index at different time instants and the total energy

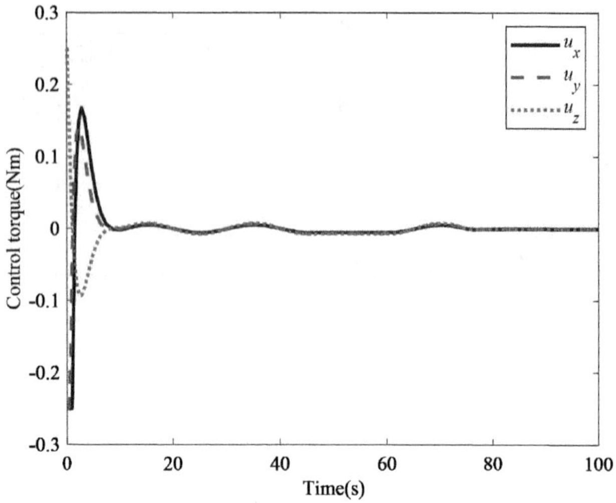

Figure 5.10 Time response of control torque with additive perturbation under state feedback controller.

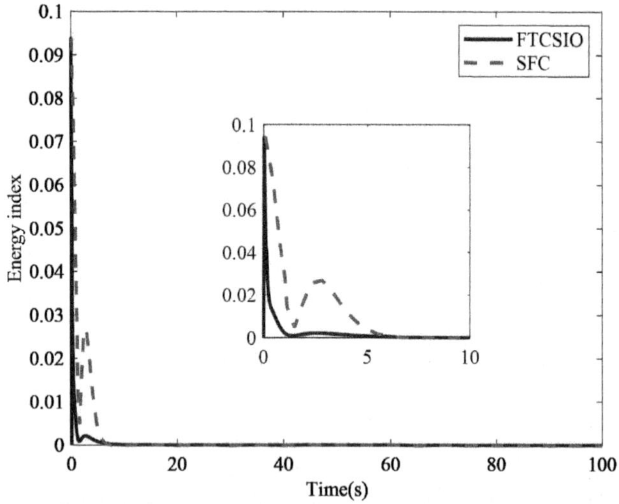

Figure 5.11 Time response of the energy consumption index at different time instants with additive perturbation.

consumption performance are shown in Figs. 5.11 and 5.12 respectively. Fig. 5.11 indicates that the state feedback controller would evidently require much more energy to achieve the same mission compared with the scheme presented in this work at any arbitrary time instant. Fig. 5.12

shows that the total energy consumption of the FTCSIO is only 0.02748 while that of the SFC is 0.1342. Thus the energy consumption is greatly reduced by 79.52% with the proposed control strategy. Although the SFC approach can achieve attitude stabilization in the presence of uncertainties or perturbations, it cannot obtain the fault information or attitude information, and results in higher energy consumption. From the perspective of energy consumption and unknown attitude information, it can be concluded that the proposed fault-tolerant controller in this chapter is superior to the state feedback controller. However, other evaluation indices can also be seen in Table 5.2, from which one can see that the state feedback controller is superior to the proposed fault-tolerant controller from the perspective of control accuracy.

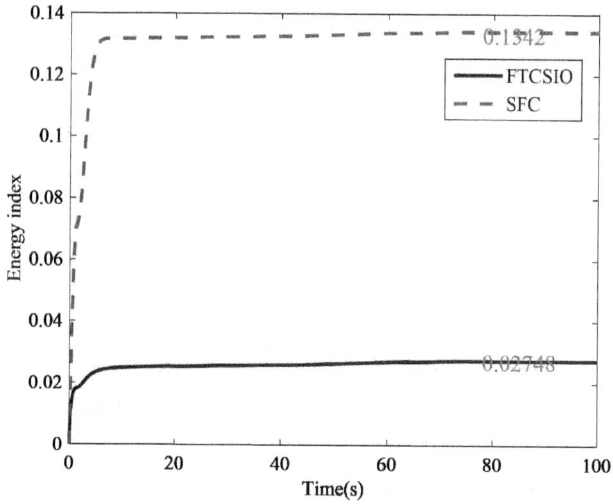

Figure 5.12 Time response of the total energy consumption performance comparisons with additive perturbation.

Table 5.2 Other evaluation indices with additive perturbation.

Control schemes	FTCSIO	SFC
Accuracy of attitude angle (rad)	1.2×10^{-3}	4.5×10^{-4}
Accuracy of angular velocity (rad s^{-1})	1.7×10^{-4}	1.4×10^{-4}
Overshoot of control torque (NM)	0.152	0.168

FTCSIO, fault-tolerant controller via stochastically intermediate observer; *SFC*, state feedback controller.

5.4.2 Comparison analysis under multiplicative perturbation

With inequality (5.29), the controller gain matrix $\boldsymbol{K}$ is

$$\boldsymbol{K}=\begin{bmatrix} 10 & 0 & 0 & 57.9437 & 0 & 2.0343 \\ 0 & 36 & 0 & 0 & 104.2987 & 0 \\ 0 & 0 & 7.5 & -1.5306 & 0 & 43.4578 \end{bmatrix}$$

Solving inequality (5.30) gives

$$\boldsymbol{P}_2=10^4\times\begin{bmatrix} 5.3800 & -1.5737 & -1.8675 & -0.0230 & 0.0108 & 0.0141 \\ -1.5737 & 4.3602 & -2.5602 & 0.0081 & -0.0358 & 0.0168 \\ -1.8675 & -2.5602 & 3.4998 & 0.0103 & 0.0159 & -0.0217 \\ -0.0230 & 0.0081 & 0.0103 & 0.0323 & -0.0150 & -0.0213 \\ 0.0108 & -0.0358 & 0.0159 & -0.0150 & 0.0540 & -0.0208 \\ 0.0141 & 0.0168 & -0.0217 & -0.0213 & -0.0208 & 0.0348 \end{bmatrix}$$

and

$$\boldsymbol{H}=10^5\times\begin{bmatrix} 1.0086 & 0.0085 & 0.0119 & 0.0396 & -0.0267 & -0.0348 \\ 0.0084 & 1.0147 & 0.0154 & -0.0062 & 0.0498 & -0.0150 \\ 0.0116 & 0.0156 & 1.0184 & -0.0195 & -0.0249 & 0.0363 \\ 0.0433 & -0.0092 & -0.0199 & 1.1254 & 0.0538 & 0.0849 \\ -0.0245 & 0.0479 & -0.0260 & 0.0555 & 1.0845 & 0.0560 \\ -0.0322 & -0.0181 & 0.0345 & 0.0851 & 0.0546 & 1.1339 \end{bmatrix}$$

Then, one has

$$\boldsymbol{L}=\boldsymbol{P}_2^{-1}\boldsymbol{H}$$
$$=10^4\times\begin{bmatrix} 0.0011 & 0.0014 & 0.0016 & 0.0040 & 0.0029 & 0.0041 \\ 0.0013 & 0.0021 & 0.0022 & 0.0044 & 0.0035 & 0.0047 \\ 0.0016 & 0.0023 & 0.0028 & 0.0061 & 0.0045 & 0.0065 \\ -0.0073 & 0.0118 & 0.0009 & 2.2352 & 1.4981 & 2.2814 \\ -0.0058 & 0.0093 & 0.0009 & 1.5028 & 1.0326 & 1.5490 \\ -0.0090 & 0.0122 & 0.0021 & 2.2756 & 1.5403 & 2.3620 \end{bmatrix}$$

Applying the controller gain $\boldsymbol{K}$ and observer gain $\boldsymbol{L}$ to Eqs. (5.20)–(5.22) gives the time responses of attitude angle and angular velocity as shown in Figs. 5.13 and 5.14 respectively, from which it is observed that the attitude angle and angular velocity converge to within a small bound. The accuracy of attitude angle is approximately 1×10^{-3} rad and that of angular velocity is approximately 2×10^{-4} rad s^{-1} in steady-state error. For the analysis of the effectiveness of the presented observer, Figs. 5.15 and 5.16 show time responses of observation errors of attitude angle and angular velocity

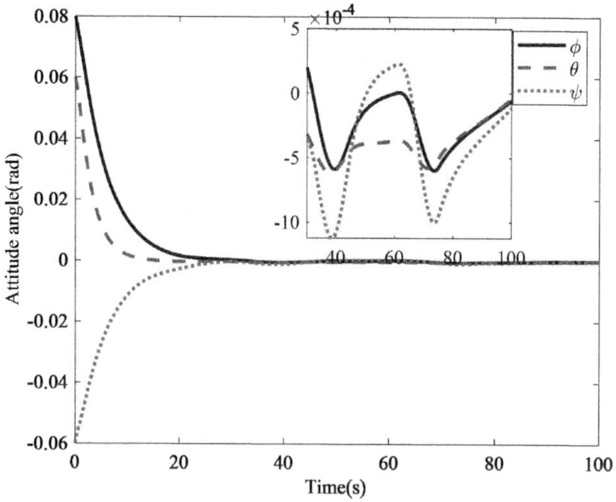

Figure 5.13 Time response of attitude angle with multiplicative perturbation under fault-tolerant controller via stochastically intermediate observer.

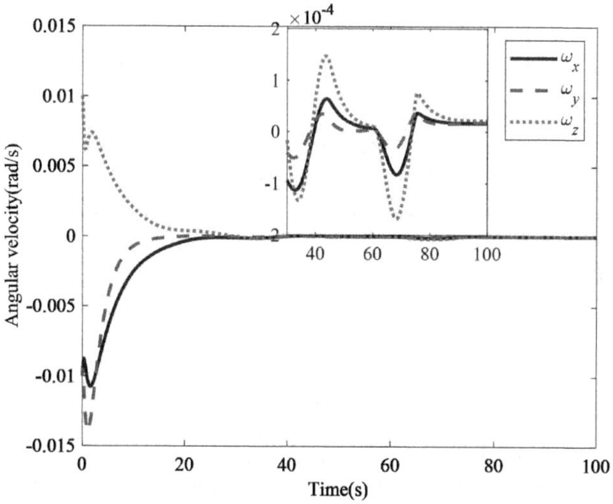

Figure 5.14 Time response of angular velocity with multiplicative perturbation under fault-tolerant controller via stochastically intermediate observer.

respectively. As can be seen, the estimated values converge to the real values rapidly and with high accuracy even when the conventional controller is employed. By applying the FTCSIO to the spacecraft attitude system, the observed attitude angle errors fall to within a small neighborhood containing

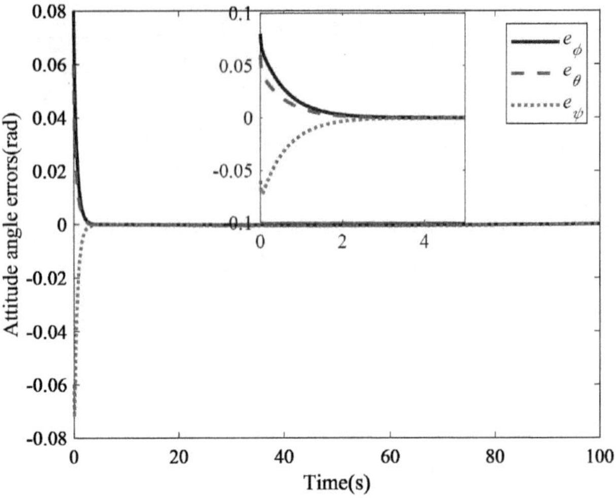

Figure 5.15 Time response of attitude angle errors with multiplicative perturbation under fault-tolerant controller via stochastically intermediate observer.

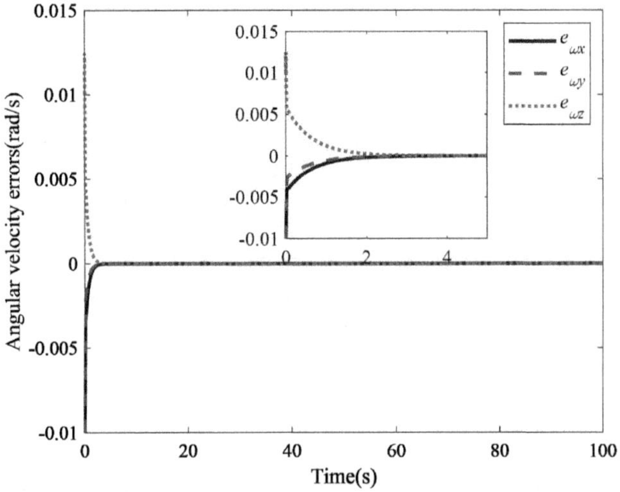

Figure 5.16 Time response of angular velocity errors with multiplicative perturbation under fault-tolerant controller via stochastically intermediate observer.

zero (i.e., $|e_\varphi| < 6 \times 10^{-4}$rad, $|e_\theta| < 3 \times 10^{-4}$rad, $|e_\psi| < 7 \times 10^{-4}$rad) within 3 s and the angular velocity errors reach $|e_{\omega x}| < 2 \times 10^{-4}$rad s^{-1}, $|e_{\omega y}| < 1.1 \times 10^{-4}$rad s^{-1} and $|e_{\omega z}| < 2.5 \times 10^{-4}$rad s^{-1} at approximately 2 seconds. The fault signal and its stochastic estimation are plotted in Fig. 5.17, from which one can see that the observed fault signal is nearly the same as the real value. Based on these simulations, it is seen that both states and fault signals are estimated with good accuracy. The control torque shown in Fig. 5.18 is no more than 0.25 Nm. To further illustrate the influence of ϖ, the correlation between ϖ and η obtained by Theorem 5.2 is the same as that in Fig. 5.7 and is omitted here. In this case, the parameter ϖ is of great importance.

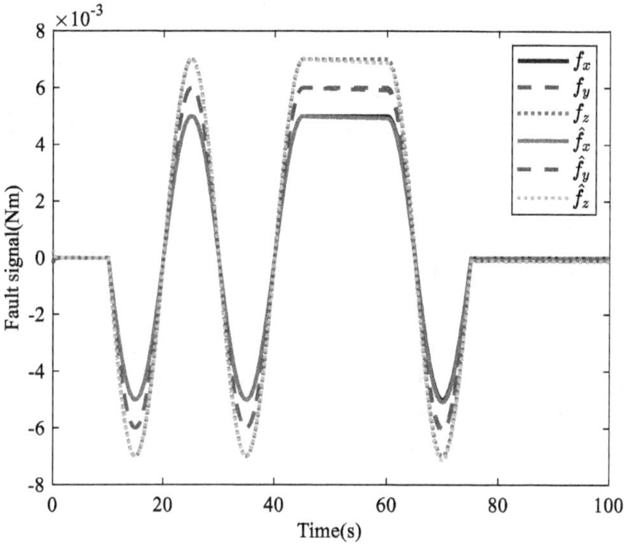

Figure 5.17 Fault signal and its stochastic estimation with multiplicative perturbation under fault-tolerant controller via stochastically intermediate observer.

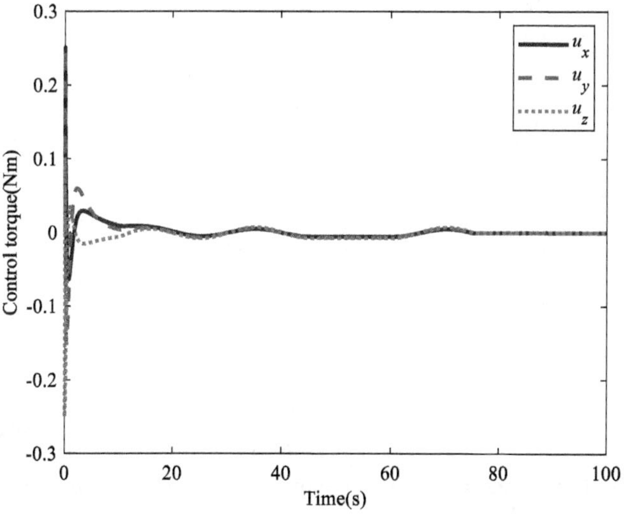

Figure 5.18 Time response of control torque with multiplicative perturbation under fault-tolerant controller via stochastically intermediate observer.

When the state feedback controller is adopted, the controller gain matrix $\boldsymbol{K}$ is

$$\boldsymbol{K} = \begin{bmatrix} -21.3485 & -10.7706 & -58.5235 & -45.0497 & -7.8576 & 17.8710 \\ 3.1250 & -38.2939 & -92.4675 & -25.0861 & -50.7369 & 5.0388 \\ 7.3569 & -4.8570 & -73.8583 & 3.5157 & -10.2769 & -89.5866 \end{bmatrix}$$

Applying this controller to the first term of Eq. (5.8) and considering the fault and multiplicative perturbation, we obtain the time responses of attitude angle and angular velocity as shown in Figs. 5.19 and 5.20 respectively. It is observed that the attitude angle and angular velocity also converge to a small bound quickly. The attitude angle is less than 1.5×10^{-4} rad and that of angular velocity is less than 4×10^{-5} rad s^{-1} in steady-state error. Fig. 5.21 shows the response curve of control torque using the SFC.

The time responses of the energy consumption index at different time instants and the total energy consumption performance are shown in Figs. 5.22 and 5.23 respectively. Fig. 5.22 indicates that the state feedback controller would evidently require much more energy to achieve the same mission compared with the scheme presented in this work at each time instant. Fig. 5.23 shows that the total energy consumption of the FTCSIO is only 0.02237 while that of the SFC is 0.4749. Thus the energy consumption is greatly reduced—by 95.29% with the proposed control strategy.

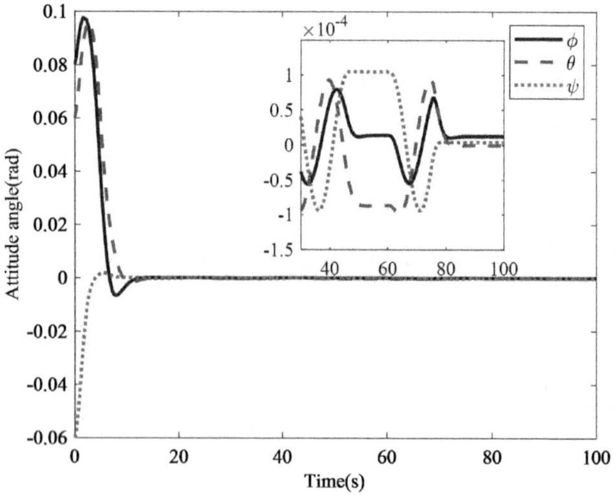

Figure 5.19 Time response of attitude angle with multiplicative perturbation under state feedback controller.

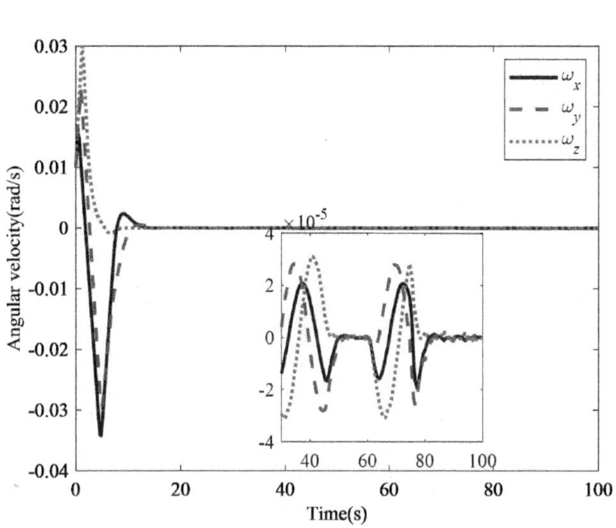

Figure 5.20 Time response of angular velocity with multiplicative perturbation under state feedback controller.

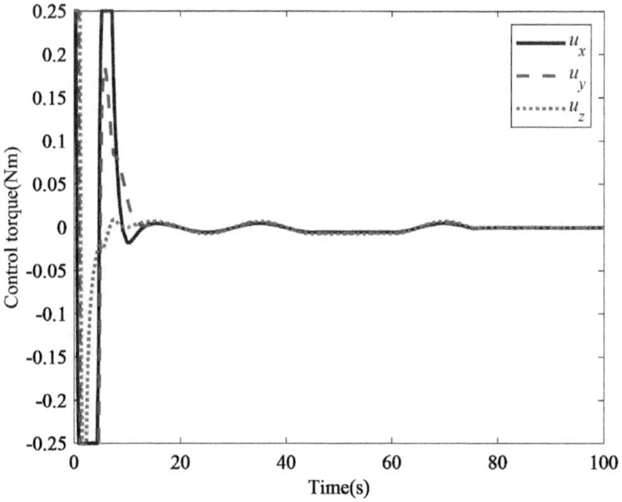

Figure 5.21 Time response of control torque with multiplicative perturbation under state feedback controller.

Although the SFC approach can achieve attitude stabilization in the presence of uncertainties or perturbations, it cannot obtain the fault information or attitude information, and results in higher energy consumption. It can be concluded here again, that according to the measures we use here, the proposed fault-tolerant controller in this chapter is superior to

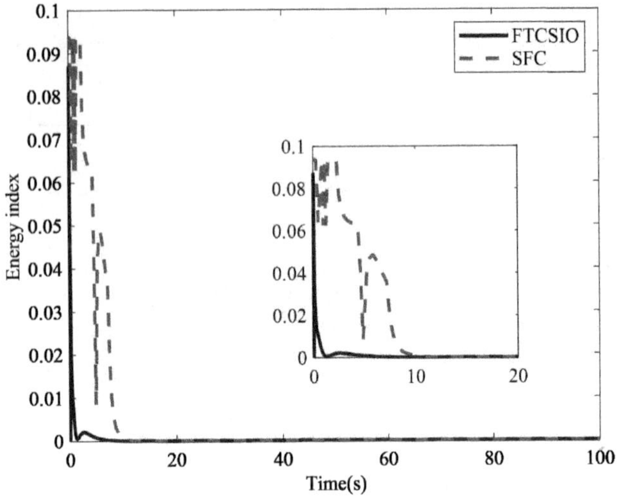

Figure 5.22 Time response of the energy consumption index at different time instants with multiplicative perturbation.

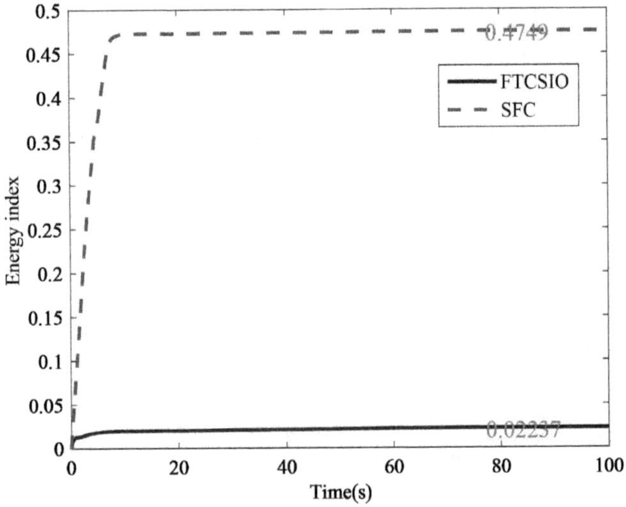

Figure 5.23 Time response of the total energy consumption performance comparisons with multiplicative perturbation.

the state feedback controller from the perspective of energy consumption and unknown attitude information. However, other evaluation indices can also be seen in Table 5.3, from which one can see that the state feedback controller is superior to the proposed fault-tolerant controller from the perspective of control accuracy.

Table 5.3 Other evaluation indices with multiplicative perturbation.

Control schemes	FTCSIO	SFC
Accuracy of attitude angle (rad)	1.1×10^{-3}	1.1×10^{-4}
Accuracy of angular velocity (rad s^{-1})	1.7×10^{-4}	3.1×10^{-5}
Overshoot of control torque (Nm)	0.15	0.25

FTCSIO, fault-tolerant controller via stochastically intermediate observer; *SFC*, state feedback controller.

5.5 Conclusions

In this chapter, a fault tolerant nonfragile H_∞ control scheme for spacecraft attitude control with controller perturbations is studied via a stochastically intermediate observer. Based on this intermediate observer, the proposed fault tolerant control approach can tolerate stochastic failure. Moreover, the states and fault signals of attitude control system can be estimated with high precision. By using the observer-based fault tolerant nonfragile H_∞ controller, the objective of attitude stabilization and control is achieved with very good performance. More importantly, the developed observer-based controller considers the constraints of H_∞ performance, quadratic stability, model parameter uncertainty, measurement errors, external disturbances, controller perturbation, and actuator fault and saturation simultaneously in the design and analysis. With respect to controller perturbation, two cases, of additive perturbation and multiplicative perturbation, are taken into account respectively in controller design and simulation analysis. Simulation results illustrate the advantage of the proposed method over the existing state feedback control method in terms of reduced energy consumption.

References

[1] Gui H, Vukovich G. Adaptive fault-tolerant spacecraft attitude control using a novel integral terminal sliding mode. International Journal of Robust and Nonlinear Control 2017;27(16):3174–96.

[2] Tiwari PM, Janardhanan S, Nabi M. Rigid spacecraft attitude control using adaptive integral second order sliding mode. Aerospace Science and Technology 2015;42:50–7.

[3] Han Y, Biggs JD, Cui N. Adaptive fault-tolerant control of spacecraft attitude dynamics with actuator failures. Journal of Guidance, Control, and Dynamics 2015;38(10):2033–42.

[4] De Ruiter AH. Adaptive spacecraft attitude control with actuator saturation. Journal of Guidance, Control, and Dynamics 2010;33(5):1692–6.

[5] Zou AM, Kumar KD, Hou ZG. Quaternion-based adaptive output feedback attitude control of spacecraft using Chebyshev neural networks. IEEE Transactions on Neural Networks 2010;21(9):1457–71.
[6] Liu C, Sun Z, Shi K, et al. Robust dynamic output feedback control for attitude stabilization of spacecraft with nonlinear perturbations. Aerospace Science and Technology 2017;64:102–21.
[7] Park Y. Robust and optimal attitude control of spacecraft with disturbances. International Journal of Systems Science 2015;46(7):1222–33.
[8] Sharma R, Tewari A. Optimal nonlinear tracking of spacecraft attitude maneuvers. IEEE Transactions on Control Systems Technology 2004;12(5):677–82.
[9] Wu Y, Han F, Zheng M, et al. Attitude control for on-orbit servicing spacecraft using hybrid actuator. Advances in Space Research 2018;61(6):1600–16.
[10] Wu Y, Han F, Zhang S, et al. Attitude agile maneuvering control for spacecraft equipped with hybrid actuators. Journal of Guidance, Control, and Dynamics 2018;41(3):803–12.
[11] Sakthivel R, Selvi S, Mathiyalagan K. Fault-tolerant sampled-data control of flexible spacecraft with probabilistic time delays. Nonlinear Dynamics 2015;79(3):1835–46.
[12] Xia Y, Zhu Z, Fu M, et al. Attitude tracking of rigid spacecraft with bounded disturbances. IEEE Transactions on Industrial Electronics 2011;58(2):647–59.
[13] Thakur D, Srikant S, Akella MR. Adaptive attitude-tracking control of spacecraft with uncertain time-varying inertia parameters. Journal of Guidance, Control, and Dynamics 2015;38(1):41–52.
[14] Yin S, Xiao B, Ding SX, et al. A review on recent development of spacecraft attitude fault tolerant control system. IEEE Transactions on Industrial Electronics 2016;63(5):3311–20.
[15] Cai W, Liao XH, Song YD. Indirect robust adaptive fault-tolerant control for attitude tracking of spacecraft. Journal of Guidance Control, and Dynamics 2008;31(5):1456–63.
[16] Bustan D, Sani SH, Pariz N. Adaptive fault-tolerant spacecraft attitude control design with transient response control. IEEE/ASME Transactions on Mechatronics 2014;19(4):1404–11.
[17] Xiao B, Hu Q, Wang D, et al. Attitude tracking control of rigid spacecraft with actuator misalignment and fault. IEEE Transactions on Control Systems Technology 2013;21(6):2360–6.
[18] Cao L, Chen X, Sheng T. Fault tolerant small satellite attitude control using adaptive non-singular terminal sliding mode. Advances in Space Research 2013;51(12):2374–93.
[19] Shen Q, Wang D, Zhu S, et al. Robust control allocation for spacecraft attitude tracking under actuator faults. IEEE Transactions on Control Systems Technology 2017;25(3):1068–75.
[20] Yin S, Yang H, Kaynak O. Sliding mode observer-based ftc for markovian jump systems with actuator and sensor faults. IEEE Transactions on Automatic Control 2017;62(7):3551–8.
[21] Chen F, Jiang R, Zhang K, et al. Robust backstepping sliding-mode control and observer-based fault estimation for a quadrotor UAV. IEEE Transactions on Industrial Electronics 2016;63(8):5044–656.
[22] Jiang B, Xu D, Shi P, et al. Adaptive neural observer-based backstepping fault tolerant control for near space vehicle under control effector damage. IET Control Theory & Applications 2014;8(9):658–66.
[23] Gao X, Teo KL, Duan GR. Robust H∞ control of spacecraft rendezvous on elliptical orbit. Journal of the Franklin Institute 2012;349(8):2515–29.

[24] Zhu HY, Wu HN, Wang JW. H∞ disturbance attenuation for nonlinear coupled parabolic PDE−ODE systems via fuzzy-model-based control approach. IEEE Transactions on Systems, Man, and Cybernetics: Systems 2016;47(8):1−12.
[25] Yang GH, Wang JL. Non-fragile H∞ control for linear systems with multiplicative controller gain variations. Automatica 2001;37(5):727−37.
[26] Kao Y, Xie J, Wang C, et al. A sliding mode approach to H∞ non-fragile observer-based control design for uncertain Markovian neutral-type stochastic systems. Automatica 2015;52:218−26.
[27] Liu C, Sun Z, Shi K, et al. Robust non-fragile state feedback attitude control for uncertain spacecraft with input saturation. Proceedings of the Institution of Mechanical Engineers, Part G: Journal of Aerospace Engineering 2018;232 (2):246−59.
[28] Zhu JW, Yang GH, Wang H, et al. Fault estimation for a class of nonlinear systems based on intermediate estimator. IEEE Transactions on Automatic Control 2016;61 (9):2518−24.
[29] Wang G, Yi C. Fault estimation for nonlinear systems by an intermediate estimator with stochastic failure. Nonlinear Dynamics 2017;89(2):1195−204.
[30] Xiao B, Hu Q, Zhang Y. Adaptive sliding mode fault tolerant attitude tracking control for flexible spacecraft under actuator saturation. IEEE Transactions on Control Systems Technology 2012;20(6):1605−12.
[31] Mahmoud M, Jiang J, Zhang Y. Active fault tolerant control systems: stochastic analysis and synthesis. Berlin Heidelberg: Springer-Verlag; 2003. p. 128−9.
[32] Wang H, Yang GH. Robust mixed l1/H_∞ filtering for affine fuzzy systems with measurement errors. IEEE Transactions on Cybernetics 2014;44(7):1100−10.
[33] Zhang A, Hu Q, Zhang Y. Observer-based attitude control for satellite under actuator fault. Journal of Guidance, Control, and Dynamics 2015;38(4):806−11.
[34] Zhu JW, Yang GH. Fault accommodation for uncertain linear systems with measurement errors. International Journal of Robust and Nonlinear Control 2017;27 (10):1841−54.
[35] Shi P, Liu M, Zhang L. Fault-tolerant sliding-mode-observer synthesis of Markovian jump systems using quantized measurements. IEEE Transactions on Industrial Electronics 2015;62(9):5910−18.
[36] Tabatabaeipour SM, Bak T. Robust observer-based fault estimation and accommodation of discrete-time piecewise linear systems. Journal of the Franklin Institute 2014;51(1):277−95.
[37] Zhang K, Jiang B, Shi P. Fast fault estimation and accommodation for dynamical systems. IET Control Theory & Applications 2009;3(2):189−99.
[38] Jiang B, Staroswiecki M, Cocquempot V. Fault accommodation for nonlinear dynamic systems. IEEE Transactions on Automatic Control 2006;51(9):1578−83.
[39] You F, Li H, Wang F, et al. Robust fast adaptive fault estimation for systems with time-varying interval delay. Journal of the Franklin Institute 2015;352 (12):5486−513.
[40] Huang SJ, Yang GH. Fault tolerant controller design for T-S fuzzy systems with time-varying delay and actuator fault: a k-step fault-estimation approach. IEEE Transactions on Fuzzy Systems 2014;22(6):1526−40.

CHAPTER 6

Disturbance observer-based control with input MRCs

6.1 Introduction

Attitude stabilization and control of rigid spacecraft have received increasing attention due to its wide applications in the field of space missions [1,2]. Most of these systems are in face of external disturbances, time-varying inertia uncertainties, and input magnitude and rate constraints (MRCs) resulting from physical limits of the onboard actuators. In particular, the inertia properties for a spacecraft with both rigid and nonrigid inertial components inevitably vary in practice due to many factors, for example, environmental disturbances, the uneven mass distribution and displacement of onboard payloads, and fuel depletion [3,4]. A challenging problem that arises in this domain is that the external disturbances and inertia uncertainties cannot be exactly known in advance, which further increases the complexity and difficulty in controller design. Future space missions demand that the new generation of spacecraft should be able to perform maneuvers with high precision and better robustness to external disturbances and inertia uncertainties. Thus it is necessary to develop effective control schemes for spacecraft attitude maneuvers in the presence of external disturbances, inertial uncertainties, and input MRCs.

For the attitude control of rigid spacecraft, various control approaches have been studied in much literature based on several inspiring approaches, such as sliding mode control [5,6], adaptive control [7], [8], feedback control [9,10], optimal control [11,12], disturbance observer-based control [13,14], and robust control or syntheses of these methods [15,16]. Some of these studies have addressed how to deal with inertial uncertainties; however, they generally assume that the nominal inertia has a known value, and the inertial uncertainties have a lower bound. For example, three nonsingular terminal sliding mode controllers were designed in [5] to deal with the issues of inertia uncertainties and external disturbances for attitude stabilization of rigid spacecraft. An integral

Spacecraft Attitude Control. DOI: https://doi.org/10.1016/B978-0-323-99005-9.00006-7

terminal sliding mode control method was studied in [6] to address the attitude tracking problem of a rigid spacecraft in the presence of uncertain moments of inertia, unknown disturbances, and sudden actuator faults. In [8], an adaptive feedback control law was presented for attitude tracking control problems in the face of external disturbances and uncertainties in the moments of inertia. In [10], a finite-time output feedback attitude tracking control law was proposed for rigid spacecraft based on finite-time observer and continuous finite-time control techniques. However, the inertial uncertainties in aforesaid references are regarded as parameter perturbations of the inertia matrix, and have lower bounds. In the event of larger inertial uncertainties, for example, the attitude control of combined spacecraft after rendezvous and docking, these methods are no longer applicable. To solve this problem, Nuthi and Subbarao proposed a computational adaptive optimal control law to estimate the unknown inertial parameters in [12], which showed successful regulation of angular velocities. However, this algorithm could only be implemented for the spacecraft operating in a relatively disturbance-free environment, that is, it cannot be applied to the spacecraft affected by unknown external disturbances. One way to overcome this problem is to estimate the disturbance for a control action to be taken, to compensate for the influence of disturbances [17]. Thus disturbance observer-based control has been widely studied and explored in practical applications for recent decades, due to its particular characteristics in dealing with external disturbances and model uncertainties [18].

In fact, from a practical viewpoint, another important problem that gains more attention is input constraints. As the onboard actuators for attitude control of spacecraft include mechanical and electrical components and the environmental force actuators, for example, thrusters, reaction wheels, and CMGs, which are easy to suffer from input constraints. Neglecting input constraints on both magnitude and dynamics can be source of undesirable or even catastrophic behavior for the closed-loop system. To cope with this critical issue, several methods have already been presented in the literature. For instance, a standard hyperbolic tangent function was used in [19] and [20], and a saturation function was used in [16] to solve the input constraints problem. Subsequently, an arc tangent function was adopted to model the constrained control input in [21], where the attitude reorientation problem of a satellite subject to external disturbances, inertial uncertainties, actuator faults, and input saturation was studied. The chapters mentioned above only considered the input magnitude constraint, however, the control constraint on input rate should also be taken into account for safe use of actuators [22,23]. The rate constraint is a dynamic nonlinearity, unlike the magnitude constraint which is a static nonlinearity. Physical, safety, or technological constraints show that the control actuators can neither provide unlimited magnitude signals nor unlimited speed of reaction. In [24], the problem of attitude tracking control of spacecraft subject to input MRCs was investigated based on the smooth hyperbolic tangent function, but with a large convergence error. To reduce the steady-state error and compensate for the influence of disturbances, a disturbance observer-based controller considering input MRCs was presented in [25], where the inertial uncertainties were neglected and only one-dimensional rotation was considered. It is worth noting that this method is also conservative, for it set the Lyapunov variable involved in estimation errors to be an identity matrix.

Motivated by the preceding discussions, this chapter presents a novel attitude control scheme without prior knowledge of the external disturbances and inertial uncertainties to achieve high-precision attitude stabilization and control of rigid spacecraft. The controller design is based on linear matrix inequalities (LMIs), which have really become a very effective technique in the field of stability and control with the advantages of global optimal solution and numerical reliability, multiple objective design, and mature software packages. The main contributions of this work are stated as follows. First, the attitude information and estimation errors of lumped disturbances can converge to the vicinity of the equilibrium state with high precision. Second, an explicit bound of attitude information and estimation errors of lumped disturbances is obtained in the process of Lyapunov analysis. Third, compared with the controller reported in [25], the proposed control law is less conservative for it doesn't restrict the Lyapunov variable involved in estimation errors and also extends the dimensions. Fourth, the developed control scheme is not only applicable to spacecraft subject to input MRCs, but also easily extendable to a more general class of second-order system with input MRCs.

The rest of this chapter is organized as follows. The next section briefly recalls the attitude dynamics of rigid spacecraft and converts it into a state-space form, where a lumped disturbance resulting from the external disturbances and inertia uncertainties is defined, and the corresponding control problem is formulated. Then, the main results of this work are presented, where the disturbance observer-based attitude controller for a rigid spacecraft with input MRCs is designed, and corresponding stability analysis is performed via a Lyapunov approach. It is followed by simulations of the proposed scheme, which demonstrate the effectiveness of the proposed approach. Finally, conclusions are drawn.

6.2 Problem formulation

6.2.1 Attitude system description

The spacecraft is assumed to be a rigid body with actuators that provide torques about three mutually perpendicular axes defining a body-fixed frame. The attitude dynamics equation is

$$\boldsymbol{I}_b\dot{\boldsymbol{\omega}} + \boldsymbol{\omega} \times (\boldsymbol{I}_b\boldsymbol{\omega}) = \boldsymbol{T}_c + \boldsymbol{T}_g + \boldsymbol{T}_d \tag{6.1}$$

where $\boldsymbol{I}_b$ and $\boldsymbol{\omega}$ denote the inertia matrix and the angular velocity, respectively, and $\boldsymbol{T}_c$, $\boldsymbol{T}_g$, and $\boldsymbol{T}_d$ denote the control input torque, the gravity-gradient torque and the external disturbance torque, respectively.

With the body-fixed frame chosen to be a principal-axis frame, substituting the expression of gravitational torque into Eq. (6.1) allows it to be converted into component form

$$\begin{cases} I_x\ddot{\phi} + 4(I_y - I_z)\omega_0^2\phi + (I_y - I_z - I_x)\omega_0\dot{\psi} = T_{cx} + T_{dx} \\ I_y\ddot{\theta} + 3\omega_0^2(I_x - I_z)\theta = T_{cy} + T_{dy} \\ I_z\ddot{\psi} + (I_y - I_x)\omega_0^2\psi + (I_x + I_z - I_y)\omega_0\dot{\phi} = T_{cz} + T_{dz} \end{cases} \tag{6.2}$$

where, T_{cx}, T_{cy} and T_{cz} denote the three components of control input torque; T_{dx}, T_{dy} and T_{dz} denote the three components of external disturbance torque.

Due to the parameter perturbations, modeling errors, and environmental changes, inertial uncertainties widely exist in physical systems. If inertial uncertainties are not considered for controller design, the controller obtained is likely to fail in practical situations. To study analytically the attitude stabilization of spacecraft, we will assume that the admissible parameter uncertainties have a norm-bounded form in order to have a sufficient conditions for exponential stability, which is the most adopted form in robust stability analysis.

Let

$$\boldsymbol{x} = \begin{bmatrix} \phi & \theta & \psi & \dot{\varphi} & \dot{\theta} & \dot{\psi} \end{bmatrix}^T, \ \boldsymbol{u} = \begin{bmatrix} T_{cx} & T_{cy} & T_{cz} \end{bmatrix}^T, \ \boldsymbol{w}_0 = \begin{bmatrix} T_{dx} & T_{dy} & T_{dz} \end{bmatrix}^T,$$

Then, Eq. (6.2) can be written as

$$\dot{\boldsymbol{x}}(t) = (\boldsymbol{A} + \Delta\boldsymbol{A})\boldsymbol{x}(t) + (\boldsymbol{B}_1 + \Delta\boldsymbol{B}_1)\boldsymbol{u}(t) + (\boldsymbol{B}_2 + \Delta\boldsymbol{B}_2)\boldsymbol{w}_0(t) \tag{6.3}$$

where $\Delta\boldsymbol{A}$, $\Delta\boldsymbol{B}_1$ and $\Delta\boldsymbol{B}_2$ denote matrices induced by inertia uncertainties, which are of the following forms:

$$\Delta\boldsymbol{A}=\begin{bmatrix}0&0&0&0&0&0\\0&0&0&0&0&0\\0&0&0&0&0&0\\\Delta A_{41}&0&0&0&0&\Delta A_{46}\\0&\Delta A_{52}&0&0&0&0\\0&0&\Delta A_{63}&\Delta A_{64}&0&0\end{bmatrix}$$

$$\Delta\boldsymbol{B}_1=\begin{bmatrix}0&0&0\\0&0&0\\0&0&0\\\Delta B_{11}&0&0\\0&\Delta B_{12}&0\\0&0&\Delta B_{13}\end{bmatrix},\ \Delta\boldsymbol{B}_2=\begin{bmatrix}0&0&0\\0&0&0\\0&0&0\\\Delta B_{21}&0&0\\0&\Delta B_{22}&0\\0&0&\Delta B_{23}\end{bmatrix}$$

and where ΔA_{41}, ΔA_{46}, ΔA_{52}, ΔA_{63} and ΔA_{64} denote the structure of $\Delta\boldsymbol{A}$; ΔB_{11}, ΔB_{12} and ΔB_{13} denote the structure of $\Delta\boldsymbol{B}_1$; ΔB_{21}, ΔB_{22} and ΔB_{23} denote the structure of $\Delta\boldsymbol{B}_2$. They can be respectively calculated by $\boldsymbol{A}$, $\boldsymbol{B}_1$ and $\boldsymbol{B}_2$ below.
and the coefficient matrices are denoted by

$$\boldsymbol{A}=\begin{bmatrix}0&0&0&1&0&0\\0&0&0&0&1&0\\0&0&0&0&0&1\\A_{41}&0&0&0&0&A_{46}\\0&A_{52}&0&0&0&0\\0&0&A_{63}&A_{64}&0&0\end{bmatrix}$$

where,

$$\begin{aligned}&A_{41}=-4\omega_0^2I_x^{-1}(I_y-I_z),\ A_{46}=-\omega_0I_x^{-1}(I_y-I_x-I_z),\\&A_{52}=-3\omega_0^2I_y^{-1}(I_x-I_z),\ A_{63}=-\omega_0^2I_z^{-1}(I_y-I_x),\\&A_{64}=\omega_0I_z^{-1}(I_y-I_x-I_z)\end{aligned}$$

and

$$\boldsymbol{B}_1=\boldsymbol{B}_2=\begin{bmatrix}\boldsymbol{0}_{3\times3} & diag(I_x^{-1},I_y^{-1},I_z^{-1})\end{bmatrix}^T$$

Therefore, one can conclude that $\Delta\boldsymbol{A}=\boldsymbol{A}(\boldsymbol{I}_b+\Delta\boldsymbol{I}_b)-\boldsymbol{A}(\boldsymbol{I}_b)$, $\Delta\boldsymbol{B}_1=\Delta\boldsymbol{B}_2=\boldsymbol{B}_1(\boldsymbol{I}_b+\Delta\boldsymbol{I}_b)-\boldsymbol{B}_1(\boldsymbol{I}_b)$, where $\boldsymbol{A}(\boldsymbol{I}_b+\Delta\boldsymbol{I}_b)$ and $\boldsymbol{B}_1(\boldsymbol{I}_b+\Delta\boldsymbol{I}_b)$

represent the coefficient matrices taking inertia uncertainties into account, $\boldsymbol{A}(\boldsymbol{I}_b)$ and $\boldsymbol{B}_1(\boldsymbol{I}_b)$ represent the nominal coefficient matrices. Henceforth $(\boldsymbol{I}_b)$ will be omitted for simplicity.

Thus Eq. (6.3) can be converted into the following form:

$$\dot{\boldsymbol{x}}(t) = \boldsymbol{A}\boldsymbol{x}(t) + \boldsymbol{B}_1\boldsymbol{u}(t) + \boldsymbol{B}_1\boldsymbol{w}(t) \tag{6.4}$$

where $\boldsymbol{w}(t) = \boldsymbol{w}_0(t) + \boldsymbol{B}_1^T(\boldsymbol{B}_1\boldsymbol{B}_1^T)^*(\Delta\boldsymbol{A}\boldsymbol{x}(t) + \Delta\boldsymbol{B}_1\boldsymbol{u}(t) + \Delta\boldsymbol{B}_1\boldsymbol{w}_0(t))$ is regarded as the lumped disturbance from the external disturbance and inertia uncertainties, and where the matrix $(\boldsymbol{B}_1\boldsymbol{B}_1^T)^*$ are the Moore–Penrose pseudoinverse of $\boldsymbol{B}_1\boldsymbol{B}_1^T$.

Assumption 6.1: The external disturbance torque $\boldsymbol{w}_0(t)$ and its first time derivative are assumed to be bounded.

Remark 6.1: The external disturbance primarily includes magnetic torque, solar radiation and aerodynamic torque, which are all continuous and bounded [26]. In addition, because the total control authority is limited, all three available control input torques should be continuous and bounded, which is reasonable in practice [27]. It is worth noting that the desired attitude motion represented by attitude angle and angular rate is also bounded and differentiable in practice [28]. Together, they result in the fact that the lumped disturbance torque $\boldsymbol{w}(t)$ and its first time derivative are bounded, that is, $\|\boldsymbol{w}(t)\| \le \kappa_1$, $\|\dot{\boldsymbol{w}}(t)\| \le \kappa_2$, where $\kappa_{1,2} > 0$.

6.2.2 Control objective

The objective of this chapter is to develop a disturbance observer-based controller such that the states of the resulting closed-loop system are uniformly ultimately bounded in the face of external disturbances, time-varying inertia uncertainties, and input MRCs.

6.3 Controller design and analysis

To achieve the objective, an auxiliary variable is first introduced as

$$\boldsymbol{z}(t) = \hat{\boldsymbol{w}}(t) - \boldsymbol{L}\boldsymbol{x}(t) \tag{6.5}$$

where, $\hat{\boldsymbol{w}}(t)$ is the estimate of $\boldsymbol{w}(t)$ and $\boldsymbol{L}$ is the disturbance observer gain matrix.

Then, to estimate the lumped disturbance, a disturbance observer is designed based on

$$\begin{cases} \dot{\boldsymbol{z}}(t) = -\boldsymbol{L}(\boldsymbol{A}\boldsymbol{x}(t) + \boldsymbol{B}_1\boldsymbol{u}(t) + \boldsymbol{B}_1\hat{\boldsymbol{w}}(t)) \\ \hat{\boldsymbol{w}}(t) = \boldsymbol{z}(t) + \boldsymbol{L}\boldsymbol{x}(t) \end{cases} \tag{6.6}$$

Based on the disturbance observer designed, the controller is designed as

$$\boldsymbol{u}(t) = -\boldsymbol{K}\boldsymbol{x}(t) - \hat{\boldsymbol{w}}(t) \tag{6.7}$$

where $\boldsymbol{K}$ is the controller gain matrix.

Denoting $\boldsymbol{e}(t) = \boldsymbol{w}(t) - \hat{\boldsymbol{w}}(t)$ and $\boldsymbol{\xi}(t) = [\,\boldsymbol{x}(t)^T \quad \boldsymbol{e}(t)^T\,]^T$, one has

$$\dot{\boldsymbol{\xi}}(t) = \boldsymbol{A}_\xi\boldsymbol{\xi}(t) + \boldsymbol{B}_\xi\dot{\boldsymbol{w}}(t) \tag{6.8}$$

where,

$$\boldsymbol{A}_\xi = \begin{bmatrix} \boldsymbol{A} - \boldsymbol{B}_1\boldsymbol{K} & \boldsymbol{B}_1 \\ 0 & -\boldsymbol{L}\boldsymbol{B}_1 \end{bmatrix},\ \boldsymbol{B}_\xi = \begin{bmatrix} 0 \\ \boldsymbol{I} \end{bmatrix}$$

Remark 6.2: As a consequence of the fact that $\boldsymbol{w}(t)$ includes uncertain magnitudes, the initial estimation error $\boldsymbol{e}(0)$ is not easy to be exactly known, and the initial state $\boldsymbol{x}(0)$ may be unavailable, but they should be norm-bounded in practice. Then, for any symmetric positive definite matrix $\tilde{\boldsymbol{P}}$ of appropriate dimensions, there exist two positive scalars κ_3 and κ_4 such that $\boldsymbol{e}(0)^T\tilde{\boldsymbol{P}}\boldsymbol{e}(0) \le \kappa_3^2\boldsymbol{1}_{1\times n}\tilde{\boldsymbol{P}}\boldsymbol{1}_{n\times 1}$ and $\boldsymbol{x}(0)^T\tilde{\boldsymbol{P}}\boldsymbol{x}(0) \le \kappa_4^2\boldsymbol{1}_{1\times m}\tilde{\boldsymbol{P}}\boldsymbol{1}_{m\times 1}$, where $\boldsymbol{1}_{i\times j}$ denotes an i by j matrix (i rows, j columns) whose elements are all one, n is the number of elements for $\boldsymbol{e}(0)$, and m is the number of elements for $\boldsymbol{x}(0)$.

6.3.1 Some lemmas

Before developing the main results, we introduce the following preliminary lemmas.

Lemma 6.1: [29] Let $\overline{\boldsymbol{H}}, \overline{\boldsymbol{E}}$ and $\overline{\boldsymbol{F}}(t)$ be real matrices of appropriate dimensions with $\|\overline{\boldsymbol{F}}(t)\| \le 1$, then for any scalar $\overline{\xi} > 0$, one has

$$\overline{\boldsymbol{H}}\overline{\boldsymbol{F}}(t)\overline{\boldsymbol{E}} + \overline{\boldsymbol{E}}^T\overline{\boldsymbol{F}}(t)^T\overline{\boldsymbol{H}}^T \le \overline{\xi}^{-1}\overline{\boldsymbol{H}}\overline{\boldsymbol{H}}^T + \overline{\xi}\overline{\boldsymbol{E}}^T\overline{\boldsymbol{E}}$$

Lemma 6.2: [30] Let function $V(t) \geq 0$ be a continuous function defined $\forall t \geq 0$ and $V(0)$ bounded, if the following inequality holds:

$$\dot{V}(t) \leq -c_1 V(t) + c_2$$

where $c_1 > 0, c_2$ are constants, then one can conclude that $V(t)$ is bounded.

Lemma 6.3: (Schur complement lemma) Let the partitioned matrix

$$\boldsymbol{A} = \begin{bmatrix} \boldsymbol{A}_{11} & \boldsymbol{A}_{12} \\ * & \boldsymbol{A}_{22} \end{bmatrix}$$

be symmetric. Then

$$\boldsymbol{A} < 0 \Leftrightarrow \boldsymbol{A}_{11} < 0,\ \boldsymbol{A}_{22} - \boldsymbol{A}_{12}^T \boldsymbol{A}_{11}^{-1} \boldsymbol{A}_{12} < 0 \Leftrightarrow \boldsymbol{A}_{22} < 0,\ \boldsymbol{A}_{11} - \boldsymbol{A}_{12} \boldsymbol{A}_{22}^{-1} \boldsymbol{A}_{12}^T < 0$$

6.3.2 Coexisting conditions for observer and controller gains

Theorem 6.1 now gives the coexisting conditions of the controller gain matrix $\boldsymbol{K}$ and disturbance observer gain matrix $\boldsymbol{L}$.

Theorem 6.1: The states of the closed-loop system represented by Eq. (6.8) are uniformly ultimately bounded, and the input MRCs, that is, $\|\boldsymbol{u}\| \leq \lambda_1$ and $\|\dot{\boldsymbol{u}}\| \leq \lambda_2$, are satisfied, if for positive scalars λ_1, λ_2, ξ_1, ξ_2, ξ_3, ξ_4, κ_1, κ_2, κ_3 and κ_4, there exist symmetric positive definite matrices $\boldsymbol{X}_1$, $\boldsymbol{X}_2$ and matrices $\boldsymbol{W}$ and $\boldsymbol{H}$, such that the following LMIs hold:

$$\begin{bmatrix} \boldsymbol{A}\boldsymbol{X}_1 - \boldsymbol{B}_1\boldsymbol{W} + \boldsymbol{X}_1\boldsymbol{A}^T - \boldsymbol{W}^T\boldsymbol{B}_1^T + \xi_2\boldsymbol{X}_1 & \boldsymbol{B}_1\boldsymbol{X}_2 & 0 \\ * & -\boldsymbol{H}^T - \boldsymbol{H} + \xi_2\boldsymbol{X}_2 & \boldsymbol{X}_2 \\ * & * & -\xi_1^{-1}\boldsymbol{I} \end{bmatrix} \leq 0 \tag{6.9}$$

$$\begin{bmatrix} \xi_2^{-1}\xi_1^{-1}\kappa_2^2 - \varpi & \kappa_4\boldsymbol{1}_{1\times 6} & \kappa_3\boldsymbol{1}_{1\times 3} \\ * & -\boldsymbol{X}_1 & 0 \\ * & * & -\boldsymbol{X}_2 \end{bmatrix} \leq 0 \tag{6.10}$$

$$\begin{bmatrix} -k_0\boldsymbol{X}_1 & 0 & 0 & \boldsymbol{W}^T \\ * & -k_0\boldsymbol{X}_2 & 0 & -\boldsymbol{X}_2 \\ * & * & -k_0\kappa_1^{-2}\boldsymbol{I} & \boldsymbol{I} \\ * & * & * & -\boldsymbol{I} \end{bmatrix} \leq 0 \tag{6.11}$$

$$\begin{bmatrix} -\frac{(1+\xi_3^{-1})}{(1+\xi_3)}k_0k_1\boldsymbol{X}_2 & \boldsymbol{X}_2\boldsymbol{B}_1^T & \boldsymbol{H}^T \\ * & -(1+\xi_4)^{-1}k_0^{-1}\boldsymbol{X}_1 & 0 \\ * & * & -(1+\xi_4^{-1})^{-1}\boldsymbol{I} \end{bmatrix} \leq 0 \tag{6.12}$$

$$\begin{bmatrix} -k_1\boldsymbol{X}_1 & \boldsymbol{X}_1\boldsymbol{A}^T - \boldsymbol{W}^T\boldsymbol{B}_1^T \\ * & -\boldsymbol{X}_1 \end{bmatrix} \leq 0 \tag{6.13}$$

where $k_0 = (\varpi + 1)^{-1}\lambda_1^2$ and $k_1 = (\varpi + 1)\lambda_2^2/((1 + \xi_3^{-1})\lambda_1^2\varpi)$.

When LMIs (6.9)–(6.13) are feasible, one can obtain the controller gain matrix $\boldsymbol{K}=\boldsymbol{W}\boldsymbol{X}_1^{-1}$ and the observer gain matrix $\boldsymbol{L}=\boldsymbol{H}(\boldsymbol{B}_1\boldsymbol{X}_2)^T[(\boldsymbol{B}_1\boldsymbol{X}_2)(\boldsymbol{B}_1\boldsymbol{X}_2)^T]^*$, where the matrix $[(\boldsymbol{B}_1\boldsymbol{X}_2)(\boldsymbol{B}_1\boldsymbol{X}_2)^T]^*$ is the Moore–Penrose pseudoinverse of $(\boldsymbol{B}_1\boldsymbol{X}_2)(\boldsymbol{B}_1\boldsymbol{X}_2)^T$.

6.3.3 Proof and analysis

First, one shows that the states of the closed-loop system are uniformly ultimately bounded.

Choose a Lyapunov function candidate as

$$V(t) = \boldsymbol{x}(t)^T\boldsymbol{P}_1\boldsymbol{x}(t) + \boldsymbol{e}^T(t)\boldsymbol{P}_2\boldsymbol{e}(t) \tag{6.14}$$

where $\boldsymbol{P}_1 = \boldsymbol{X}_1^{-1}$, $\boldsymbol{P}_2 = \boldsymbol{X}_2^{-1}$.

Taking the derivative of $V(t)$ with respect to time yields

$$\begin{aligned} \dot{V}(t) &= \dot{\boldsymbol{x}}(t)^T\boldsymbol{P}_1\boldsymbol{x}(t) + \boldsymbol{x}(t)^T\boldsymbol{P}_1\dot{\boldsymbol{x}}(t) + \dot{\boldsymbol{e}}(t)^T\boldsymbol{P}_2\boldsymbol{e}(t) + \boldsymbol{e}(t)^T\boldsymbol{P}_2\dot{\boldsymbol{e}}(t) \\ &= \boldsymbol{x}(t)^T\left[(\boldsymbol{A}-\boldsymbol{B}_1\boldsymbol{K})^T\boldsymbol{P}_1 + \boldsymbol{P}_1(\boldsymbol{A}-\boldsymbol{B}_1\boldsymbol{K})\right]\boldsymbol{x}(t) + 2\boldsymbol{x}(t)^T\boldsymbol{P}_1\boldsymbol{B}_1\boldsymbol{e}(t) \\ &\quad - \boldsymbol{e}(t)^T(\boldsymbol{B}_1^T\boldsymbol{L}^T\boldsymbol{P}_2 + \boldsymbol{P}_2\boldsymbol{L}\boldsymbol{B}_1)\boldsymbol{e}(t) + 2\boldsymbol{e}(t)^T\dot{\boldsymbol{w}}(t) \end{aligned}$$

With Lemma 6.1 and the given constraint conditions, one has

$$\begin{aligned} \dot{V}(t) &\leq \boldsymbol{x}(t)^T\left[(\boldsymbol{A}-\boldsymbol{B}_1\boldsymbol{K})^T\boldsymbol{P}_1 + \boldsymbol{P}_1(\boldsymbol{A}-\boldsymbol{B}_1\boldsymbol{K})\right]\boldsymbol{x}(t) + 2\boldsymbol{x}(t)^T\boldsymbol{P}_1\boldsymbol{B}_1\boldsymbol{e}(t) \\ &\quad - \boldsymbol{e}(t)^T(\boldsymbol{B}_1^T\boldsymbol{L}^T\boldsymbol{P}_2 + \boldsymbol{P}_2\boldsymbol{L}\boldsymbol{B}_1)\boldsymbol{e}(t) + \xi_1\boldsymbol{e}(t)^T\boldsymbol{e}(t) + \xi_1^{-1}\dot{\boldsymbol{w}}(t)^T\dot{\boldsymbol{w}}(t) \\ &\leq \boldsymbol{x}(t)^T\left[(\boldsymbol{A}-\boldsymbol{B}_1\boldsymbol{K})^T\boldsymbol{P}_1 + \boldsymbol{P}_1(\boldsymbol{A}-\boldsymbol{B}_1\boldsymbol{K})\right]\boldsymbol{x}(t) + 2\boldsymbol{x}(t)^T\boldsymbol{P}_1\boldsymbol{B}_1\boldsymbol{e}(t) \\ &\quad - \boldsymbol{e}(t)^T\left[(\boldsymbol{B}_1^T\boldsymbol{L}^T\boldsymbol{P}_2 + \boldsymbol{P}_2\boldsymbol{L}\boldsymbol{B}_1) - \xi_1\boldsymbol{I}\right]\boldsymbol{e}(t) + \xi_1^{-1}\kappa_2^2 \\ &= \boldsymbol{\xi}(t)^T\boldsymbol{\Theta}\boldsymbol{\xi}(t) + \xi_1^{-1}\kappa_2^2 \end{aligned}$$

where,

$$\boldsymbol{\Theta} = \begin{bmatrix} (\boldsymbol{A}-\boldsymbol{B}_1\boldsymbol{K})^T\boldsymbol{P}_1 + \boldsymbol{P}_1(\boldsymbol{A}-\boldsymbol{B}_1\boldsymbol{K}) & \boldsymbol{P}_1\boldsymbol{B}_1 \\ * & -(\boldsymbol{B}_1^T\boldsymbol{L}^T\boldsymbol{P}_2 + \boldsymbol{P}_2\boldsymbol{L}\boldsymbol{B}_1) + \xi_1\boldsymbol{I} \end{bmatrix}$$

Defining $W=KX_1$ and $H=LB_1X_2$, multiplied by diag$\{P_1, P_2, I\}$ and its transpose on both sides of inequality (6.9) yields inequality (6.15) as follows.

$$\begin{bmatrix} (A-B_1K)^TP_1+P_1(A-B_1K)+\xi_2P_1 & P_1B_1 & 0 \\ * & -B_1^TL^TP_2-P_2LB_1+\xi_2P_2 & I \\ * & * & -\xi_1^{-1}I \end{bmatrix} \le 0 \tag{6.15}$$

With Lemma 6.3, one has

$$\dot{V}(t) \le \xi(t)^T\Theta\xi(t)+\xi_1^{-1}\kappa_2^2 \le -\xi_2\xi(t)^T\begin{bmatrix} P_1 & 0 \\ * & P_2 \end{bmatrix}\xi(t)+\xi_1^{-1}\kappa_2^2 \tag{6.16}$$

that is,

$$\dot{V}(t)+\xi_2V(t) \le \xi_1^{-1}\kappa_2^2 \tag{6.17}$$

In fact, the uniform ultimate boundedness of system (6.8) is quite apparent by noticing the results of Lemma 6.2 and the definition of $V(t)$ in Eq. (6.14). For completeness, it is shown in the details below.

Multiplying (6.17) by $e^{\xi_2 t}$ yields

$$\frac{d}{dt}\left(V(t)e^{\xi_2 t}\right) \le \xi_1^{-1}\kappa_2^2e^{\xi_2 t} \tag{6.18}$$

Integrating (6.18) over $[0, t]$ leads to

$$0 \le V(t) \le V(0)e^{-\xi_2 t}-\xi_2^{-1}\xi_1^{-1}\kappa_2^2e^{-\xi_2 t}+\xi_2^{-1}\xi_1^{-1}\kappa_2^2 \le V(0)+\xi_2^{-1}\xi_1^{-1}\kappa_2^2 \tag{6.19}$$

where, $V(0)=x(0)^TP_1x(0)+e^T(0)P_2e(0)$.

Denoting $\overline{\sigma}_x=\sqrt{\kappa_2^2/(\xi_1\xi_2\lambda_{P_1\min})}$ and $\overline{\sigma}_e=\sqrt{\kappa_2^2/(\xi_1\xi_2\lambda_{P_2\min})}$ with $\lambda_{P_1\min}=\min_{\tau\in[0,t]}\lambda_{\min}(P_1(\tau))$ and $\lambda_{P_2\min}=\min_{\tau\in[0,t]}\lambda_{\min}(P_2(\tau))$, from Eq. (6.14), one has

$$\lambda_{P_1\min}\left\|x(t)\right\|^2 \le \lambda_{\min}(P_1(t))\left\|x(t)\right\|^2 \le x(t)^TP_1x(t) \le V(t) \tag{6.20}$$

$$\lambda_{P_2\min}\left\|e(t)\right\|^2 \le \lambda_{\min}(P_2(t))\left\|e(t)\right\|^2 \le e^T(t)P_2e(t) \le V(t) \tag{6.21}$$

Combining inequalities (6.19)–(6.21) yields

$$\left\|x(t)\right\| \le \sqrt{(V(0)e^{-\xi_2 t}-\xi_2^{-1}\xi_1^{-1}\kappa_2^2e^{-\xi_2 t}+\xi_2^{-1}\xi_1^{-1}\kappa_2^2)/\lambda_{P_1\min}} \tag{6.22}$$

$$\left\|e(t)\right\| \le \sqrt{(V(0)e^{-\xi_2 t}-\xi_2^{-1}\xi_1^{-1}\kappa_2^2e^{-\xi_2 t}+\xi_2^{-1}\xi_1^{-1}\kappa_2^2)/\lambda_{P_2\min}} \tag{6.23}$$

If it happens that $V(0)=\xi_2^{-1}\xi_1^{-1}\kappa_2^2$, then $\|\boldsymbol{x}(t)\|\leq\overline{\sigma}_x, \forall t\geq 0$.

If $V(0)\neq\xi_2^{-1}\xi_1^{-1}\kappa_2^2$, from inequality (6.22), one can conclude that for given scalar $\sigma_x>\overline{\sigma}_x$, there exists $\bar{t}$, such that for any $t>\bar{t}$, one has $\|\boldsymbol{x}(t)\|\leq\sigma_x$. Specifically, for given σ_x:

$$\sigma_x=\sqrt{(V(0)e^{-\xi_2\bar{t}}-\xi_2^{-1}\xi_1^{-1}\kappa_2^2e^{-\xi_2\bar{t}}+\xi_2^{-1}\xi_1^{-1}\kappa_2^2)/\lambda_{P_1\min}} \tag{6.24}$$

It can be obtained that

$$\bar{t}=-\frac{1}{\xi_2}\ln\left(\frac{\sigma_x^2\lambda_{P_1\min}-\xi_2^{-1}\xi_1^{-1}\kappa_2^2}{V(0)-\xi_2^{-1}\xi_1^{-1}\kappa_2^2}\right) \tag{6.25}$$

and

$$\lim_{t\to\infty}\left\|\boldsymbol{x}(t)\right\|=\overline{\sigma}_x \tag{6.26}$$

Similar conclusion about $\|\boldsymbol{e}(t)\|$ can be made, that is,

$$\lim_{t\to\infty}\left\|\boldsymbol{e}(t)\right\|=\overline{\sigma}_e \tag{6.27}$$

that is, the states of the closed-loop system are uniformly ultimately bounded under controller (6.7).

In addition, $\overline{\sigma}_x=\sqrt{\kappa_2^2/(\xi_1\xi_2\lambda_{P_1\min})}$ and $\overline{\sigma}_e=\sqrt{\kappa_2^2/(\xi_1\xi_2\lambda_{P_2\min})}$ imply that the ultimate bounds of the state and the estimation error can be reduced by choosing appropriate values of ξ_1 and ξ_2. Thus the design method is able to arbitrarily reduce those bounds through a proper choice of the values of ξ_1 and ξ_2.

According to the above analysis, one knows the state trajectory $\boldsymbol{x}(t)$ will not get out of a ball of arbitrarily specified radius σ_x. In addition, a value σ_{x0} can also be found such that starting the initial state from within the ball of radius σ_{x0} guarantees that the state will stay within the ball of radius σ_x thereafter. Similarly, the state trajectory $\boldsymbol{e}(t)$ will not get out of a ball of arbitrarily specified radius σ_e. In addition, a value σ_{e0} can also be found such that starting the initial state from within the ball of radius σ_{e0} guarantees that the state will stay within the ball of radius σ_e thereafter. The two balls of radii σ_x and σ_e, mean the two regions around the equilibrium points of $\boldsymbol{x}(t)$ and $\boldsymbol{e}(t)$, respectively. Thus the equilibrium state $\boldsymbol{\xi}(t)=0$ is said to be stable according to the basic concept of stability in [31]. This fact can be used to construct uncertainty regions around the nominal equilibrium point, and the acceptable region should guarantee the stability of the closed-loop system. In fact, the obtained sets defined by $\overline{\sigma}_x$ and $\overline{\sigma}_e$ are conservative regions, which is not surprising since the sets depend on the Lyapunov function $V(t)$. It is possible to beforehand obtain a certain degree of robustness for the sets defined by $\overline{\sigma}_x$ and $\overline{\sigma}_e$ by additionally specifying the regions defined by σ_x and σ_e larger than the actual sets. Hence, if a solution is obtained, the specified degree of robustness is achieved. However, if these regions are specified to be too large, it is possible that there is no solution to the corresponding LMI problem.

The next one shows the input magnitude constraint, that is, $\|\boldsymbol{u}\| \leq \lambda_1$, is satisfied.

According to Lemma 6.3, the inequality (6.10) is equivalent to

$$\xi_2^{-1}\xi_1^{-1}\kappa_2^2 + \kappa_4^2\mathbf{1}_{1\times 6}\boldsymbol{P}_1\mathbf{1}_{6\times 1} + \kappa_3^2\mathbf{1}_{1\times 3}\boldsymbol{P}_2\mathbf{1}_{3\times 1} \leq \overline{\omega} \tag{6.28}$$

For $\boldsymbol{e}(0)^T\boldsymbol{P}_2\boldsymbol{e}(0) \leq \kappa_3^2\mathbf{1}_{1\times 3}\boldsymbol{P}_2\mathbf{1}_{3\times 1}$ and $\boldsymbol{x}(0)^T\boldsymbol{P}_1\boldsymbol{x}(0) \leq \kappa_4^2\mathbf{1}_{1\times 6}\boldsymbol{P}_1\mathbf{1}_{6\times 1}$, it is easily obtained that

$$\xi_2^{-1}\xi_1^{-1}\kappa_2^2 + \boldsymbol{x}(0)^T\boldsymbol{P}_1\boldsymbol{x}(0) + \boldsymbol{e}(0)^T\boldsymbol{P}_2\boldsymbol{e}(0) \leq \overline{\omega} \tag{6.29}$$

Multiplying $\text{diag}\{\boldsymbol{P}_1, \boldsymbol{P}_2, \boldsymbol{I}, \boldsymbol{I}\}$ and its transpose on both sides of inequality (6.11) yields

$$\begin{bmatrix} -k_0\boldsymbol{P}_1 & 0 & 0 & \boldsymbol{K}^T \\ * & -k_0\boldsymbol{P}_2 & 0 & -\boldsymbol{I} \\ * & * & -k_0\kappa_1^{-2}\boldsymbol{I} & \boldsymbol{I} \\ * & * & * & -\boldsymbol{I} \end{bmatrix} \leq 0 \tag{6.30}$$

With Lemma 6.3, one has

$$\begin{bmatrix} \boldsymbol{K}^T \\ -\boldsymbol{I} \\ \boldsymbol{I} \end{bmatrix}\begin{bmatrix} \boldsymbol{K} & -\boldsymbol{I} & \boldsymbol{I} \end{bmatrix} \leq \begin{bmatrix} k_0\boldsymbol{P}_1 & 0 & 0 \\ * & k_0\boldsymbol{P}_2 & 0 \\ * & * & k_0\kappa_1^{-2}\boldsymbol{I} \end{bmatrix} \tag{6.31}$$

Then,

$$\begin{aligned}
\|\boldsymbol{u}(t)\|_2^2 &= \boldsymbol{u}(t)^T\boldsymbol{u}(t) \\
&= [-\boldsymbol{K}\boldsymbol{x}(t) - \hat{\boldsymbol{w}}(t)]^T[-\boldsymbol{K}\boldsymbol{x}(t) - \hat{\boldsymbol{w}}(t)] \\
&= [\boldsymbol{K}\boldsymbol{x}(t) + \boldsymbol{w}(t) - \boldsymbol{e}(t)]^T[\boldsymbol{K}\boldsymbol{x}(t) + \boldsymbol{w}(t) - \boldsymbol{e}(t)] \\
&= \begin{bmatrix} \boldsymbol{x}(t)^T & \boldsymbol{e}(t)^T & \boldsymbol{w}(t)^T \end{bmatrix}\begin{bmatrix} \boldsymbol{K}^T \\ -\boldsymbol{I} \\ \boldsymbol{I} \end{bmatrix}\begin{bmatrix} \boldsymbol{K} & -\boldsymbol{I} & \boldsymbol{I} \end{bmatrix}\begin{bmatrix} \boldsymbol{x}(t) \\ \boldsymbol{e}(t) \\ \boldsymbol{w}(t) \end{bmatrix} \\
&\leq \begin{bmatrix} \boldsymbol{x}(t)^T & \boldsymbol{e}(t)^T & \boldsymbol{w}(t)^T \end{bmatrix}\begin{bmatrix} k_0\boldsymbol{P}_1 & 0 & 0 \\ * & k_0\boldsymbol{P}_2 & 0 \\ * & * & k_0\kappa_1^{-2}\boldsymbol{I} \end{bmatrix}\begin{bmatrix} \boldsymbol{x}(t) \\ \boldsymbol{e}(t) \\ \boldsymbol{w}(t) \end{bmatrix} \\
&\leq k_0\left(\boldsymbol{x}(t)^T\boldsymbol{P}_1\boldsymbol{x}(t) + \boldsymbol{e}^T(t)\boldsymbol{P}_2\boldsymbol{e}(t) + \kappa_1^{-2}\boldsymbol{w}(t)^T\boldsymbol{w}(t)\right) \\
&\leq k_0(V(t) + 1)
\end{aligned} \tag{6.32}$$

According to inequalities (6.19) and (6.29), one knows

$$\left\| \boldsymbol{u}(t) \right\|_2^2 \leq k_0(\varpi + 1) = \lambda_1^2 \tag{6.33}$$

Finally one shows the input rate constraint, that is, $\|\dot{\boldsymbol{u}}\| \leq \lambda_2$, is satisfied.

With $\boldsymbol{H}=\boldsymbol{L}\boldsymbol{B}_1\boldsymbol{X}_2$, multiplying diag$\{\boldsymbol{P}_2, \boldsymbol{I}, \boldsymbol{I}\}$ and its transpose on both sides of inequality (6.12) yields

$$\begin{bmatrix} -\dfrac{(1+\xi_3^{-1})}{(1+\xi_3)}k_0k_1\boldsymbol{P}_2 & \boldsymbol{B}_1^T & \boldsymbol{B}_1^T\boldsymbol{L}^T \\ * & -(1+\xi_4)^{-1}k_0^{-1}\boldsymbol{X}_1 & 0 \\ * & * & -(1+\xi_4^{-1})^{-1}\boldsymbol{I} \end{bmatrix} \leq 0 \tag{6.34}$$

According to Lemma 6.3, the inequality (6.34) is equivalent to

$$(1+\xi_4^{-1})\boldsymbol{B}_1^T\boldsymbol{L}^T\boldsymbol{L}\boldsymbol{B}_1 + (1+\xi_4)k_0\boldsymbol{B}_1^T\boldsymbol{P}_1\boldsymbol{B}_1 \leq \frac{(1+\xi_3^{-1})}{(1+\xi_3)}k_0k_1\boldsymbol{P}_2 \tag{6.35}$$

Multiplying diag$\{\boldsymbol{P}_1, \boldsymbol{I}\}$ and its transpose on both sides of inequality (6.13) yields

$$\begin{bmatrix} -k_1\boldsymbol{P}_1 & \boldsymbol{A}^T - \boldsymbol{K}^T\boldsymbol{B}_1^T \\ * & -\boldsymbol{X}_1 \end{bmatrix} \leq 0 \tag{6.36}$$

With Lemma 6.3, one has

$$(\boldsymbol{A}-\boldsymbol{B}_1\boldsymbol{K})^T\boldsymbol{P}_1(\boldsymbol{A}-\boldsymbol{B}_1\boldsymbol{K}) \leq k_1\boldsymbol{P}_1 \tag{6.37}$$

Inequality (6.31) leads to

$$\boldsymbol{K}^T\boldsymbol{K} \leq k_0\boldsymbol{P}_1 \tag{6.38}$$

With inequalities (6.37) and (6.38), one has

$$(\boldsymbol{A}-\boldsymbol{B}_1\boldsymbol{K})^T\boldsymbol{K}^T\boldsymbol{K}(\boldsymbol{A}-\boldsymbol{B}_1\boldsymbol{K}) \leq k_0k_1\boldsymbol{P}_1 \tag{6.39}$$

With inequalities (6.35) and (6.38), one has

$$(1+\xi_4^{-1})\boldsymbol{B}_1^T\boldsymbol{L}^T\boldsymbol{L}\boldsymbol{B}_1 + (1+\xi_4)\boldsymbol{B}_1^T\boldsymbol{K}^T\boldsymbol{K}\boldsymbol{B}_1 \leq \frac{(1+\xi_3^{-1})}{(1+\xi_3)}k_0k_1\boldsymbol{P}_2 \tag{6.40}$$

Considering Eqs. (6.7) and (6.8), one has

$$\begin{aligned} \dot{\boldsymbol{u}}(t) &= -\boldsymbol{K}\dot{\boldsymbol{x}}(t) - \dot{\hat{\boldsymbol{w}}}(t) \\ &= -\boldsymbol{K}((\boldsymbol{A}-\boldsymbol{B}_1\boldsymbol{K})\boldsymbol{x}(t) + \boldsymbol{B}_1\boldsymbol{e}(t)) - \dot{\boldsymbol{w}}(t) - \boldsymbol{L}\boldsymbol{B}_1\boldsymbol{e}(t) + \dot{\boldsymbol{w}}(t) \\ &= -\boldsymbol{K}(\boldsymbol{A}-\boldsymbol{B}_1\boldsymbol{K})\boldsymbol{x}(t) - (\boldsymbol{K}\boldsymbol{B}_1 + \boldsymbol{L}\boldsymbol{B}_1)\boldsymbol{e}(t) \end{aligned} \tag{6.41}$$

Then,

$$\begin{aligned}
\|\dot{u}(t)\|_2^2 &= \dot{u}(t)^T\dot{u}(t)\\
&= [-K(A-B_1K)x(t)-(KB_1+LB_1)e(t)]^T[-K(A-B_1K)x(t)-(KB_1+LB_1)e(t)]\\
&= x(t)^T(A-B_1K)^TK^TK(A-B_1K)x(t)+2x(t)^T(A-B_1K)^TK^T(KB_1+LB_1)e(t)\\
&\quad+e(t)^T(KB_1+LB_1)^T(KB_1+LB_1)e(t)\\
&\le(1+\xi_3^{-1})x(t)^T(A-B_1K)^TK^TK(A-B_1K)x(t)\\
&\quad+(1+\xi_3)e(t)^T(KB_1+LB_1)^T(KB_1+LB_1)e(t)\le(1+\xi_3^{-1})k_0k_1x(t)^TP_1x(t)\\
&\quad+(1+\xi_3)e(t)^T\left((1+\xi_4^{-1})B_1^TL^TLB_1+(1+\xi_4)B_1^TK^TKB_1\right)e(t)\\
&\le(1+\xi_3^{-1})k_0k_1x(t)^TP_1x(t)+(1+\xi_3^{-1})k_0k_1e(t)^TP_2e(t)\\
&=(1+\xi_3^{-1})k_0k_1V(t)
\end{aligned} \tag{6.42}$$

According to inequalities (6.19) and (6.29), one knows

$$\|\dot{u}(t)\|_2^2<(1+\xi_3^{-1})k_0k_1\varpi=\lambda_2^2 \tag{6.43}$$

This completes the proof.

Remark 6.3: The inequality (6.9) in Theorem 6.1 can guarantee that the closed-loop system is uniformly, ultimately bounded, which implies that whatever the values of the initial state of the spacecraft and the estimation error of lumped disturbance are, the attitude information and the estimation error of lumped disturbance can converge to within a small bound containing zero with increasing time. The inequalities (6.10) and (6.11) can guarantee that the input magnitude constraint is satisfied. The attitude information is measured by attitude sensors and available in this chapter; and the initial state of the spacecraft only influences the input magnitude rather than the stability of the closed-loop system. However, to avoid using the initial state of the spacecraft in the determination of controller and observer gains, a positive scalar κ_4 is introduced. The other positive scalar κ_3 is introduced to avoid using the estimation error of lumped disturbance. Likewise, the inequalities (6.12) and (6.13) can guarantee that the input rate constraint is satisfied. In other words, there are a set of feasible solutions for the observer and controller gains that can drive the closed-loop system to be uniformly ultimately bounded, but the inequalities (6.10)–(6.13) decrease the range of feasible solutions for the observer and controller gains, they together result in the optimal feasible solutions for $\boldsymbol{K}$ and $\boldsymbol{L}$ that can guarantee the achievement of the objective in this chapter.

6.4 Simulation test

In this section, the effectiveness of the disturbance observer-based controller is illustrated with simulations of a spacecraft attitude control system. In this example, the nominal moments of inertia are taken as I_x=20 kg m^2, I_y=18 kg m^2, I_z=15 kg m^2, which is subject to uncertainties of moments of inertia, that is, $\Delta I_x = 0.15 I_x \sin(0.12\pi t)$, $\Delta I_y = 0.12 I_y \sin(0.12\pi t+\pi/4)$, $\Delta I_z = 0.1 I_z \sin(0.12\pi t+\pi/3)$, and the orbit is a 300 km circular orbit. Choose the initial states as $\boldsymbol{x}(0)$=[0.16 rad, 0.13 rad, −0.15 rad, − 0.02 rad s^{-1}, −0.02 rad s^{-1}, 0.02 rad s^{-1}]^T and $\boldsymbol{e}(0)$=[0.03 Nm, 0.05 Nm, −0.04 Nm]^T, the control torque is constrained by λ_1=2 Nm, λ_2=20 Nm s^{-1}. This constraint on control torque is considered in the simulation.

The environmental disturbance has been analyzed in [32] in detail, which includes gravity-gradient torque, aerodynamic torque, and Earth magnetic torque. As pointed out in this work, the sinusoid is the prototype of some periodic disturbances, and the environmental disturbances such as magnetic torque and gravity-gradient torque are cyclic essentially and they can be represented by sinusoids. In addition, the sinusoid function representing disturbance torques has also been used in many papers to test the robustness of proposed controllers, for example, [16,33]. Thus the external disturbance used in the simulation is taken to be of the following form and ω_0 is the orbital frequency.

$$\boldsymbol{w}_0(t) = 5\times 10^{-4}\times \begin{bmatrix} \sin(100\omega_0\pi t) \\ \cos(100\omega_0\pi t) \\ \cos(100\omega_0\pi t+\pi/3) \end{bmatrix} \text{Nm}$$

Meanwhile, according to the general maximum bounds for external disturbances, control torques and attitude information in practice, one can choose $\xi_1 = \xi_2 = \xi_3 = \xi_4 = 0.2$, $\kappa_1 = 0.07$, $\kappa_2 = 0.02$, $\kappa_3 = 0.06$, $\kappa_4 = 0.17$ and $\varpi = 0.5$, then we obtain $k_0 = 2.6667$ and $k_1 = 50$. It is worth noting that a proper choice of the values of ξ_1 and ξ_2 is able to reduce the ultimate bounds of the state and the estimation error, so that the performance requirement can be satisfied.

6.4.1 Nonzero angular rates

Solving LMIs (6.9)–(6.13) gives

$$X_1 = \begin{bmatrix} 24.3961 & -0.0231 & -0.1263 & -2.5390 & 0.0024 & -0.0053 \\ -0.0231 & 29.7780 & -0.0304 & 0.0024 & -3.1000 & 0.0032 \\ -0.1263 & -0.0304 & 42.1166 & 0.0315 & 0.0032 & -4.3867 \\ -2.5390 & 0.0024 & 0.0315 & 0.3294 & -0.0003 & -0.0015 \\ 0.0024 & -3.1000 & 0.0032 & -0.0003 & 0.4049 & -0.0004 \\ -0.0053 & 0.0032 & -4.3867 & -0.0015 & -0.0004 & 0.5791 \end{bmatrix}$$

$$X_2 = \begin{bmatrix} 0.5604 & 0.0016 & 0.0016 \\ 0.0016 & 0.5595 & 0.0016 \\ 0.0016 & 0.0016 & 0.5573 \end{bmatrix}$$

$$W = \begin{bmatrix} -3.8553 & 0.0042 & 0.0129 & 0.7105 & -0.0006 & 0.0011 \\ 0.0038 & -4.1912 & 0.0047 & -0.0006 & 0.7814 & -0.0007 \\ 0.0095 & 0.0039 & -4.8531 & -0.0035 & -0.0006 & 0.9243 \end{bmatrix}$$

$$H = \begin{bmatrix} 4.6212 & 0.0070 & 0.0072 \\ 0.0070 & 4.6183 & 0.0072 \\ 0.0072 & 0.0072 & 4.6105 \end{bmatrix}$$

Then, one can obtain the controller gain matrix and the disturbance observer gain matrix as follows

$$K = WX_1^{-1} = \begin{bmatrix} 0.3356 & 0.0000 & -0.0019 & 4.7431 & 0.0000 & 0.0033 \\ 0.0000 & 0.2965 & 0.0000 & -0.0000 & 4.2005 & 0.0001 \\ 0.0028 & 0.0000 & 0.2417 & 0.0039 & 0.0000 & 3.4266 \end{bmatrix}$$

$$L = H(B_1X_2)^T\left[(B_1X_2)(B_1X_2)^{\mathrm{T}}\right]^* = \begin{bmatrix} 0 & 0 & 0 & 164.9393 & -0.1984 & -0.1687 \\ 0 & 0 & 0 & -0.2204 & 148.5907 & -0.1704 \\ 0 & 0 & 0 & -0.2247 & -0.2043 & 124.0892 \end{bmatrix}$$

Applying the controller gain K and observer gain L to the closed-loop attitude control system of the rigid spacecraft expressed by (6.8) produces the simulation results of Figs. 6.1–6.6. Figs. 6.1 and 6.2 display the attitude angle and angular rate, respectively. It is observed that the attitude angle and angular rate both converge to a small bound containing zero with the accuracy less than 1.5×10^{-5} rad and 4×10^{-6} rad s^{-1} in steady-state error, respectively. To illustrate the effectiveness of the presented disturbance observer, Figs. 6.3 and 6.4 show the time responses of

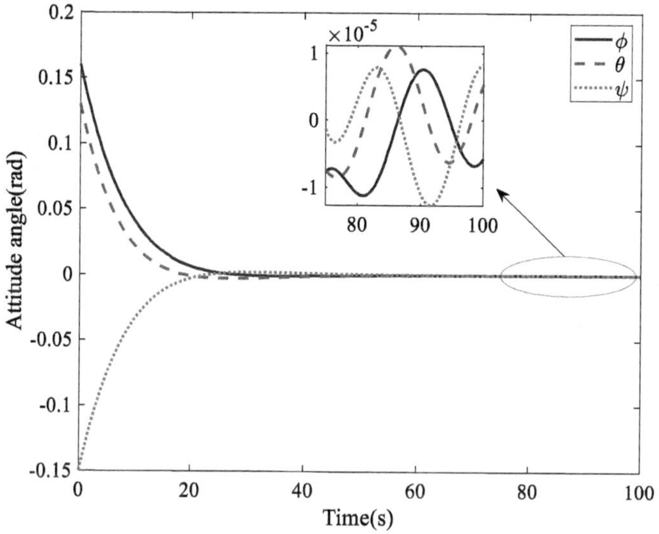

Figure 6.1 Time response of attitude angle.

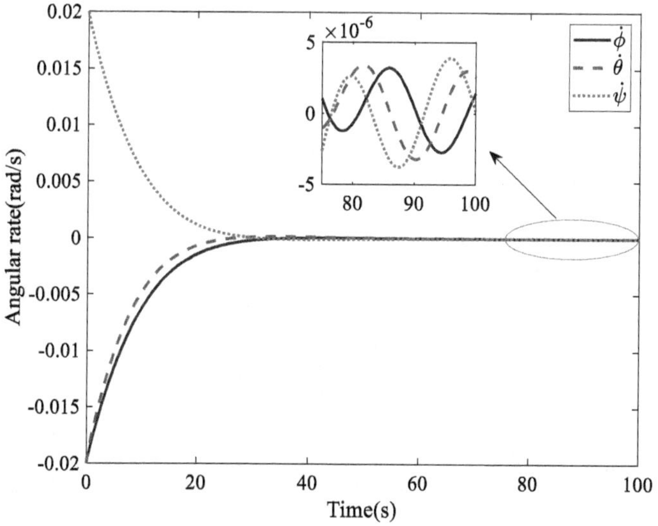

Figure 6.2 Time response of angular rate.

the lumped disturbance with its estimation and the estimation error, respectively. As one can see, the developed disturbance observer has successfully estimated the actual lumped disturbance within 1 second, and the estimation errors also converge to a small bound containing zero with the accuracy less than 2.5×10^{-5} Nm. The magnitude and rate of control

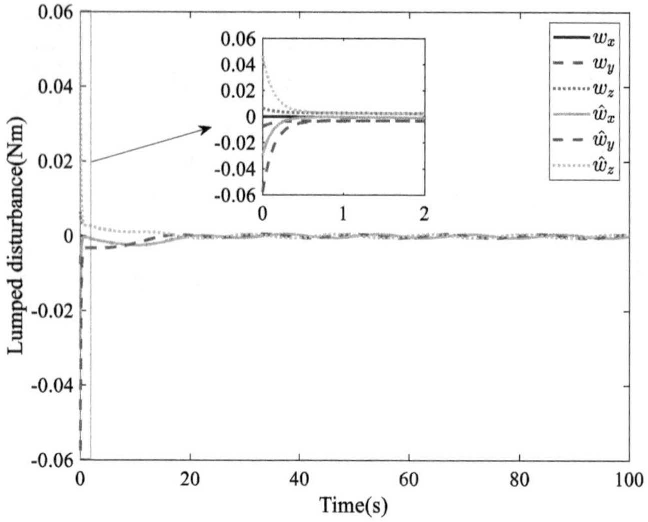

Figure 6.3 Lumped disturbance and its estimation.

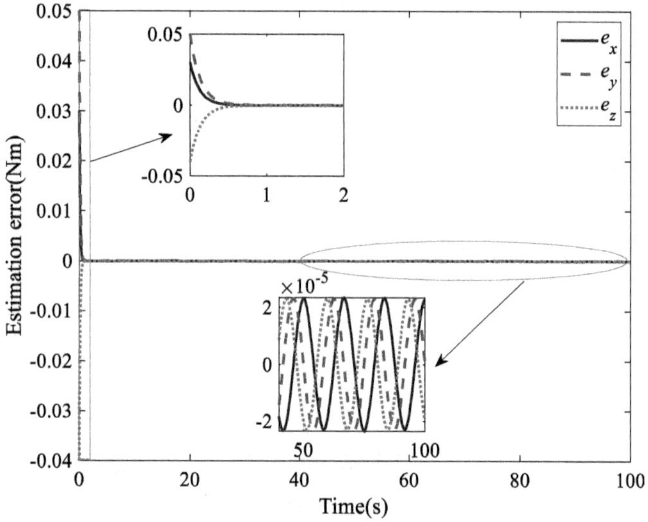

Figure 6.4 Estimation error.

torque are plotted in Figs. 6.5 and 6.6, from which one can see that its magnitude is less than 0.1 Nm with its rate less than 0.5 Nm s^{-1}. The simulation results thus verify the theoretical analysis and demonstrate the effectiveness of the presented control strategy via disturbance observer.

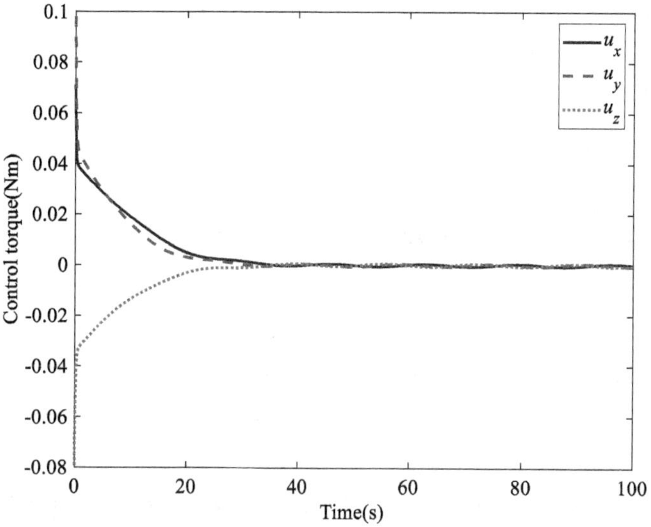

Figure 6.5 Time response of control torque magnitude.

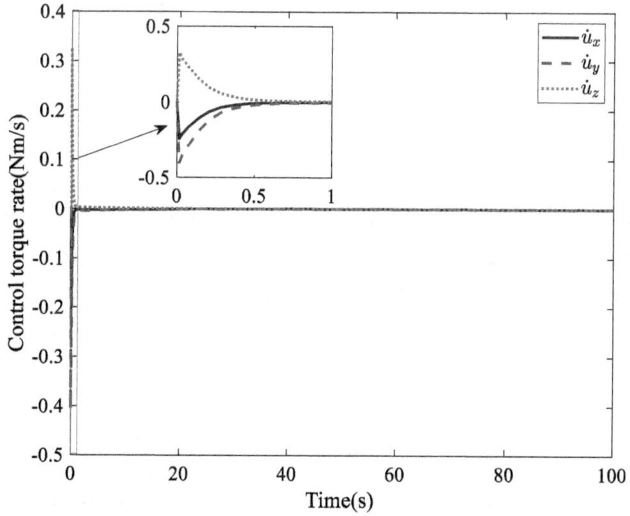

Figure 6.6 Time response of control torque rate.

6.4.2 Zero angular rates

To compare the performance of the controller scheme with different initial states and different input constraints, taking λ_2 for instance, another two comparative simulations are performed, that is, $\dot{\phi}(0) = \dot{\theta}(0) = \dot{\psi}(0) = 0$, $\lambda_2 = 20$ Nm s^{-1} and $\dot{\phi}(0) = \dot{\theta}(0) = \dot{\psi}(0) = 0$, $\lambda_2 = 2$ Nm s^{-1}, and other parameters are identical.

In the first case, that is, $\lambda_2 = 20$ Nm s^{-1}, the controller and the disturbance observer gain matrices are the same with Case A, that is,

$$\boldsymbol{K} = \begin{bmatrix} 0.3356 & 0.0000 & -0.0019 & 4.7431 & 0.0000 & 0.0033 \\ 0.0000 & 0.2965 & 0.0000 & -0.0000 & 4.2005 & 0.0001 \\ 0.0028 & 0.0000 & 0.2417 & 0.0039 & 0.0000 & 3.4266 \end{bmatrix}$$

$$\boldsymbol{L} = \begin{bmatrix} 0 & 0 & 0 & 164.9393 & -0.1984 & -0.1687 \\ 0 & 0 & 0 & -0.2204 & 148.5907 & -0.1704 \\ 0 & 0 & 0 & -0.2247 & -0.2043 & 124.0892 \end{bmatrix}$$

In the second case, that is, $\lambda_2 = 2$ Nm s^{-1}, such that $k_1 = 0.5$, and one can obtain the controller gain matrix and the disturbance observer gain matrix as follows:

$$\boldsymbol{K} = \begin{bmatrix} 0.3345 & 0.0001 & -0.0018 & 4.7179 & 0.0004 & 0.0036 \\ 0.0001 & 0.2971 & 0.0000 & 0.0004 & 4.1927 & 0.0004 \\ 0.0029 & 0.0001 & 0.2438 & 0.0043 & 0.0004 & 3.4365 \end{bmatrix}$$

$$\boldsymbol{L} = \begin{bmatrix} 0 & 0 & 0 & 20.1545 & -0.0203 & -0.0160 \\ 0 & 0 & 0 & -0.0225 & 18.2261 & -0.0176 \\ 0 & 0 & 0 & -0.0212 & -0.0211 & 15.2889 \end{bmatrix}$$

Applying the above two controllers and observers to the closed-loop system expressed by (6.8) produces the simulation results of Figs. 6.7–6.12. The time responses of attitude angle and angular rate are shown in Figs. 6.7 and 6.8, respectively. It is observed that the attitude angle and angular rate also quickly converge to a lower bound with a little longer settling time than subsection A. However, the accuracy for λ_2=20 Nm s^{-1} is much higher than for λ_2=2 Nm s^{-1}, with details shown in Table 6.1 below. The effectiveness of the presented disturbance observer is illustrated by Figs. 6.9 and 6.10, which show the response curves of the lumped disturbance with its estimation and the estimation error respectively. As seen, the estimation time for λ_2=20 Nm s^{-1} is shorter but with a higher estimation accuracy than for λ_2=2 Nm s^{-1}, with details shown in Table 6.1 below. The magnitude and rate of control torque are plotted in Figs. 6.11 and 6.12, from which one can see that both magnitudes of control torque are less than 0.05Nm with their rates less than 0.4 Nm s^{-1}, with details shown in Table 6.1 below. The simulation results thus verify the theoretical analysis and demonstrate the effectiveness of the presented control strategy via disturbance observer.

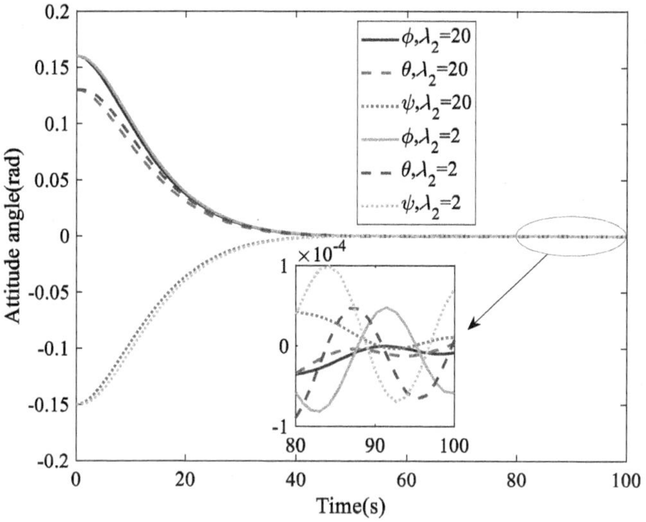

Figure 6.7 Time response of attitude angle with different initial conditions.

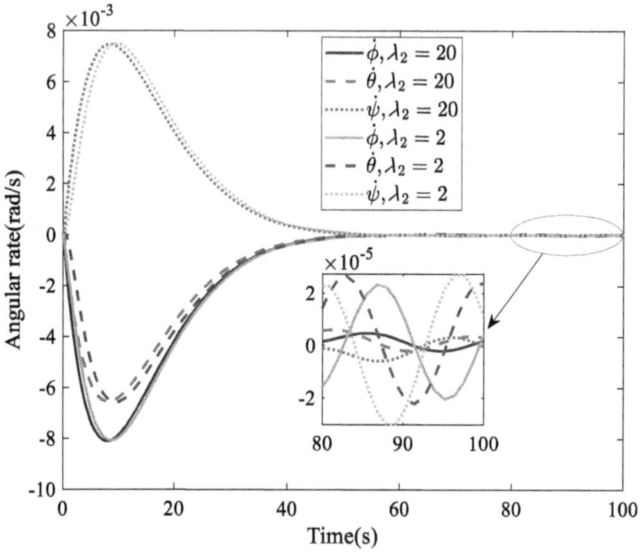

Figure 6.8 Time response of angular rate with different initial conditions.

6.4.3 Evaluation indices for the three conditions

To conduct a detailed analysis on the performance of the proposed control scheme, with respect to different initial states and different values of λ_2, some evaluation indices, that is, accuracy of attitude angle and angular

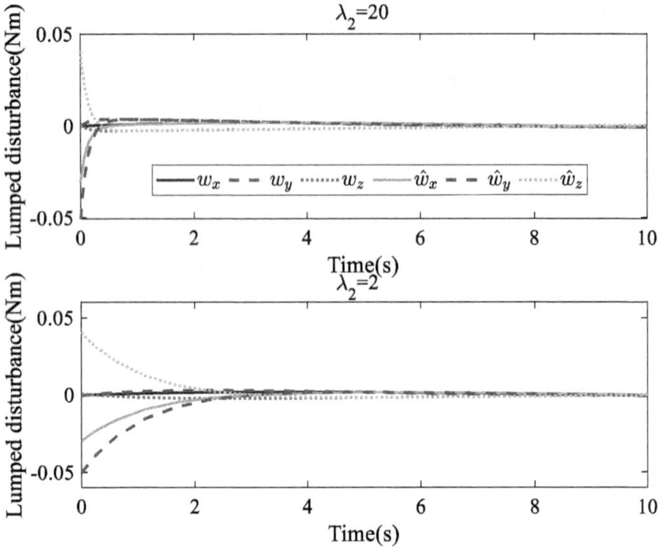

Figure 6.9 Lumped disturbance and its estimation with different initial conditions.

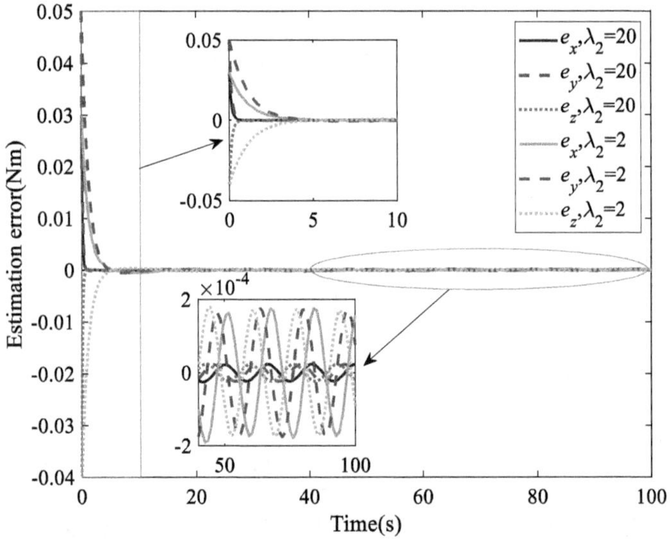

Figure 6.10 Estimation error with different initial conditions.

rate, estimation time, estimation accuracy, magnitude and rate of control torque, are illustrated in Table 6.1. It is obvious that the three groups of simulation results all support the effectiveness of the proposed control scheme based on disturbance observer, with only some differences in

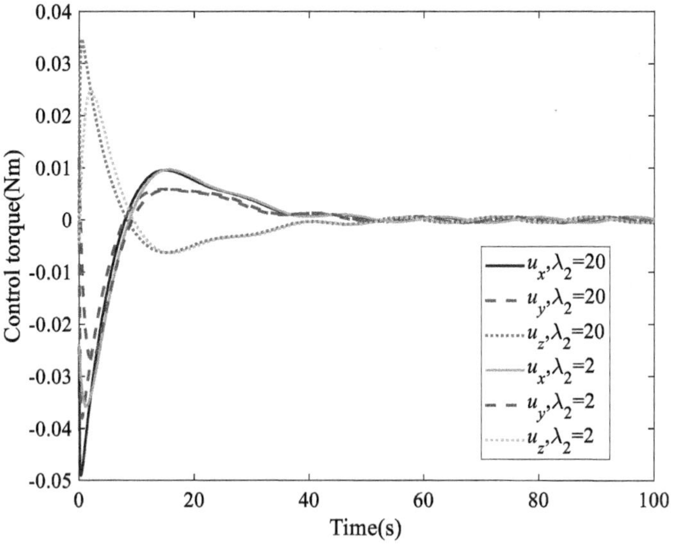

Figure 6.11 Time response of control torque magnitude with different initial conditions.

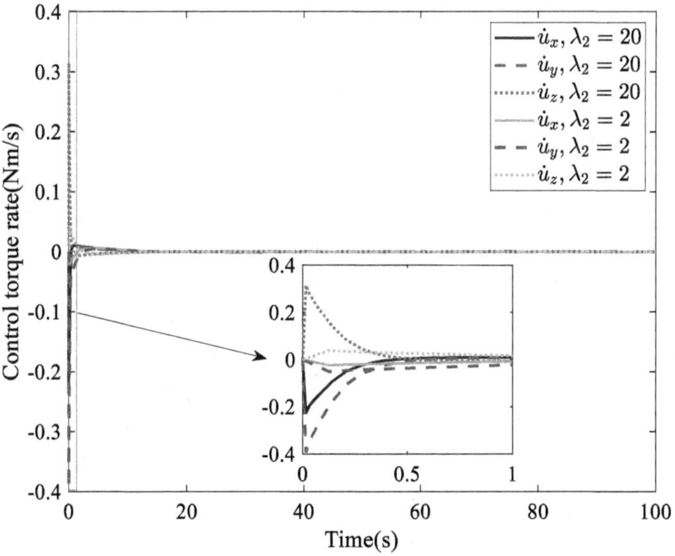

Figure 6.12 Time response of control torque rate with different initial conditions.

evaluation indices. For the nonzero angular rates and zero angular rates with identical settings, little difference of the evaluation indices has been generated; while for the zero angular rates with different values of λ_2, that is, λ_2=20 Nm s^{-1} and λ_2=2 Nm s^{-1} great difference of the evaluation

Table 6.1 Evaluation indices for the three conditions.

Evaluation indices	**Nonzero angular rates (λ_2=20)**	**Zero angular rates (λ_2=20)**	**Zero angular rates (λ_2=2)**
Accuracy of attitude angle (rad)	1.5×10^{-5}	5×10^{-5}	1×10^{-4}
Accuracy of angular rate (rad s^{-1})	4×10^{-6}	6×10^{-6}	3×10^{-5}
Estimation time (s)	1	1	5
Estimation accuracy (Nm)	2.5×10^{-5}	2.5×10^{-5}	2×10^{-4}
Control torque magnitude (Nm)	0.1	0.05	0.04
Control torque rate (Nm s^{-1})	0.41	0.39	0.05

indices has been generated. Thus it can be deduced that the value of λ_2 has great influence on control performance, while the initial states have little influence on control performance.

6.4.4 Parametric influence on control performance

To further analyze the influence of some parameters involved in the developed controller, this subsection will take varying κ_3 and κ_4 for examples, and other parameters are kept constant and identical. It is worth noting that the initial angular rates are nonzero values and $\lambda_2 = 20$ Nm s^{-1}. In this way, a wider family of results will be obtained to illustrate the effectiveness of the developed disturbance observer-based controller. Here, the parameter κ_3 is chosen to be 0.060, 0.062, 0.064, and 0.066 while the parameter κ_4 is chosen to be 0.160, 0.164, and 0.168, then six groups of simulation results are obtained as shown in Figs. 6.13–6.16. Figs. 6.13 and 6.14 display a family of attitude angles and angular rates, respectively. It is observed that all attitude angles and angular rates converge to a small bound containing zero with very small steady-state errors. It is noted that both families of attitude angles and angular rates differ little, which can be ignored. The magnitudes and rates of control torques are plotted in Figs. 6.15 and 6.16, from which one can see that the family of control torques and their first derivatives also differ little, which can be ignored. Thus it can be concluded that some parameters

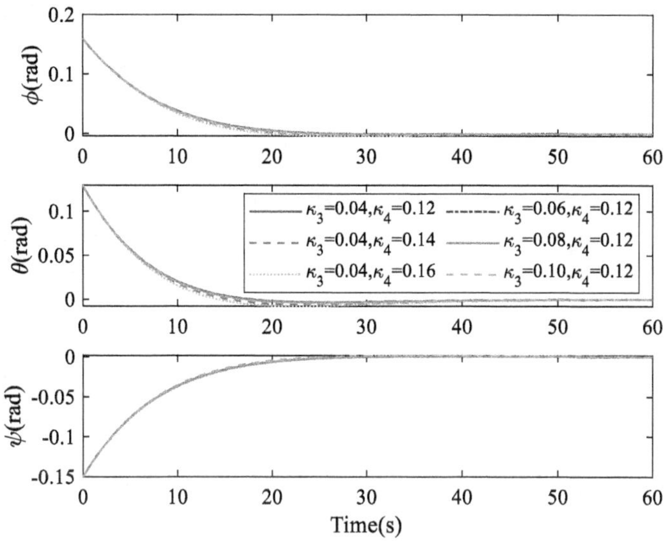

Figure 6.13 Attitude angle family.

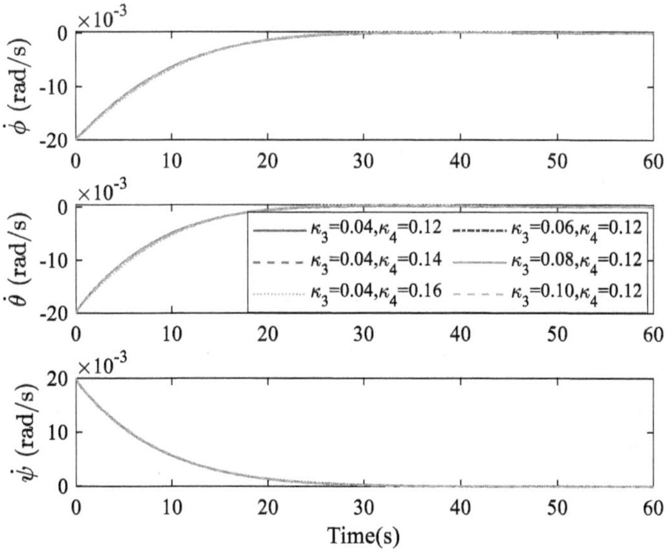

Figure 6.14 Angular rate family.

involved in the developed controller would have little influence on simulation results, and this also suggests that the closed-loop attitude control system are uniformly ultimately bounded with the presented control strategy, which also verifies the theoretical analysis.

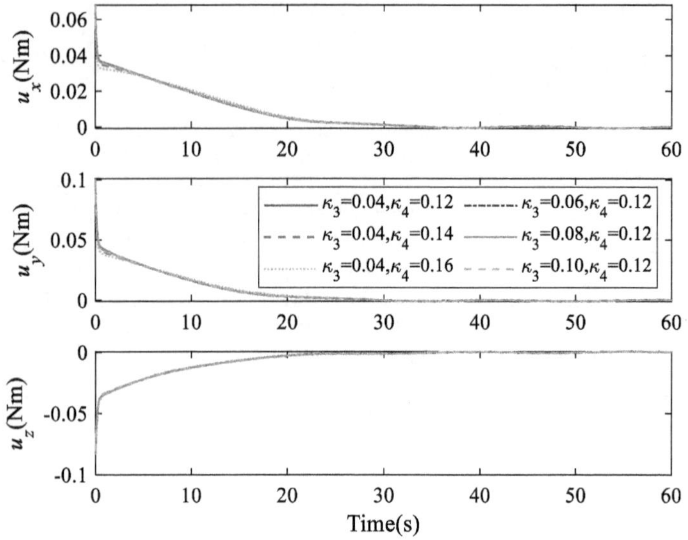

Figure 6.15 Control torque family.

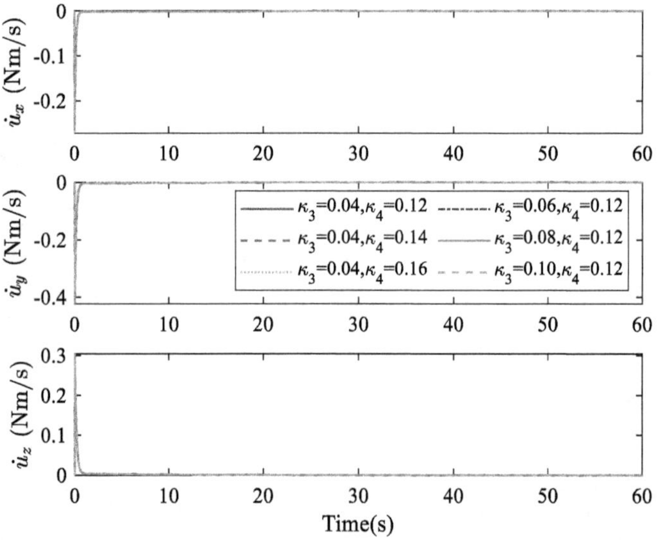

Figure 6.16 Control torque rate family.

6.5 Conclusions

In this work, a control strategy based on disturbance observer is proposed for attitude stabilization and control of rigid spacecraft with external disturbances, time-varying inertia uncertainties, and input MRCs. A new auxiliary variable is introduced first in disturbance observer design, and then a new closed-loop system consisting of attitude information and estimation errors of the lumped disturbance is established. By using the presented disturbance observer-based controller, the states of the closed-loop system converge asymptotically to the vicinity of the expected states with very good performance. More importantly, the effects of multiple unknown perturbations, such as external disturbances, inertia uncertainties, and input MRCs are considered simultaneously in the design and analysis. Simulation results illustrate the effectiveness and feasibility of the proposed controller, which also demonstrate that the input constraints have much greater influence than the initial states on control performance.

References

[1] Gui H, Vukovich G, Xu S. Attitude stabilization of a spacecraft with two parallel control moment gyroscopes. Journal of Guidance, Control, and Dynamics 2015;38(11):728–35.

[2] Liu C, Sun Z, Shi K, Wang F. Robust dynamic output feedback control for attitude stabilization of spacecraft with nonlinear perturbations. Aerospace Science and Technology 2017;64:102–21.

[3] Seo D, Akella MR. High-performance spacecraft adaptive attitude-tracking control through attracting-manifold design. Journal of Guidance, Control, and Dynamics 2008;31(4):884–91.

[4] Thakur D, Srikant S, Akella MR. Adaptive attitude-tracking control of spacecraft with uncertain time-varying inertia parameters. Journal of Guidance, Control, and Dynamics 2014;38(1):41–52.

[5] Eshghi S, Varatharajoo R. Nonsingular terminal sliding mode control technique for attitude tracking problem of a small satellite with combined energy and attitude control system (CEACS). Aerospace Science and Technology 2018;76:14–26.

[6] Gui H, Vukovich G. Adaptive fault-tolerant spacecraft attitude control using a novel integral terminal sliding mode. International Journal of Robust and Nonlinear Control 2017;27(16):3174–96.

[7] Liu C, Sun Z, Ye D, Shi K. Robust adaptive variable structure tracking control for spacecraft chaotic attitude motion. IEEE Access 2018;6:3851–7.

[8] Bai Y, Biggs JD, Zazzera FB, Cui N. Adaptive attitude tracking with active uncertainty rejection. Journal of Guidance, Control, and Dynamics 2018;41(2):550–8.

[9] Sun L. Constrained adaptive fault-tolerant attitude tracking control of rigid spacecraft. Advances in Space Research 2019;63(7):2229–38.

[10] Zou AM, Li W. Fixed-time output-feedback consensus tracking control for second-order multiagent systems. International Journal of Robust and Nonlinear Control 2019;29(13):4419–34.

[11] Park Y. Robust and optimal attitude control of spacecraft with disturbances. International Journal of Systems Science 2015;46(7):1222–33.

[12] Nuthi P, Subbarao K. Computational adaptive optimal control of spacecraft attitude dynamics with inertia-matrix identification. Journal of Guidance, Control, and Dynamics 2016;40(5):1258–62.

[13] Shi K, Liu C, Sun Z. Constrained fuel-free control for spacecraft electromagnetic docking in elliptical orbits. Acta Astronautica 2019;162:14–24.

[14] Sun L, Zheng Z. Disturbance-observer-based robust backstepping attitude stabilization of spacecraft under input saturation and measurement uncertainty. IEEE Transactions on Industrial Electronics 2017;64(10):7994–8002.

[15] Jin E, Sun Z. Robust controllers design with finite time convergence for rigid spacecraft attitude tracking control. Aerospace Science and Technology 2008;12(4):324–30.

[16] Liu C, Vukovich G, Sun Z, Shi K. Observer-based fault-tolerant attitude control for spacecraft with input delay. Journal of Guidance, Control, and Dynamics 2018; 41(9):2041–53.

[17] Liu C, Yue X, Shi K, Sun Z. Inertia-free attitude stabilization for flexible spacecraft with active vibration suppression. International Journal of Robust and Nonlinear Control 2019;29(18):6311–36.

[18] Qiao J, Zhang D, Zhu Y, Zhang P. Disturbance observer-based finite-time attitude maneuver control for micro satellite under actuator deviation fault. Aerospace Science and Technology 2018;82:262–71.

[19] Wallsgrove RJ, Akella MR. Globally stabilizing saturated attitude control in the presence of bounded unknown disturbances. Journal of Guidance, Control, and Dynamics 2005;28(5):957–63.

[20] Su Y, Zheng C. Globally asymptotic stabilization of spacecraft with simple saturated proportional-derivative control. Journal of Guidance, Control, and Dynamics 2011;34(6):1932–6.

[21] Hu Q, Shi Y, Shao X. Adaptive fault-tolerant attitude control for satellite reorientation under input saturation. Aerospace Science and Technology 2018;78:171–82.

[22] Zou AM, de Ruiter AH, Kumar KD. Disturbance observer-based attitude control for spacecraft with input MRS. IEEE Transactions on Aerospace and Electronic Systems 2018;55(1):384–96.

[23] Akella MR, Valdivia A, Kotamraju GR. Velocity-free attitude controllers subject to actuator magnitude and rate saturations. Journal of Guidance, Control, and Dynamics 2005;28(4):659–66.

[24] Zou AM, Kumar KD, Ruiter AH. Robust attitude tracking control of spacecraft under control input magnitude and rate saturations. International Journal of Robust and Nonlinear Control 2016;26(4):799–815.

[25] Liu Z, Liu J, Wang L. Disturbance observer based attitude control for flexible spacecraft with input magnitude and rate constraints. Aerospace Science and Technology 2018;72:486–92.

[26] Ding S, Zheng W. Nonsmooth attitude stabilization of a flexible spacecraft. IEEE Transactions on Aerospace and Electronic Systems 2014;50(2):1163–81.

[27] Bošković JD, Li SM, Mehra RK. Robust adaptive variable structure control of spacecraft under control input saturation. Journal of Guidance, Control, and Dynamics 2001;24(1):14–22.

[28] Boskovic JD, Li SM, Mehra RK. Robust tracking control design for spacecraft under control input saturation. Journal of Guidance, Control, and Dynamics 2004;27(4):627–33.

[29] Liu C, Ye D, Shi K, Sun Z. Robust high-precision attitude control for flexible spacecraft with improved mixed H_2/H_∞ control strategy under poles assignment constraint. Acta Astronautica 2017;136:166–75.

[30] Ge SS, Wang C. Adaptive neural control of uncertain MIMO nonlinear systems. IEEE Transactions on Neural Networks 2004;15(3):674–92.

[31] Slotine JJ, Li W. Applied nonlinear control. Englewood Cliffs, NJ: Prentice hall; 1991. p. 48.

[32] Yang CD, Sun YP. Mixed H_2/H_∞ state-feedback design for microsatellite attitude control. Control Engineering Practice 2002;10(9):951–70.

[33] Li Y, Ye D, Sun Z. Time efficient sliding mode controller based on bang-bang logic for satellite attitude control. Aerospace Science and Technology 2018;75:342–52.

CHAPTER 7

Improved mixed H_2/H_∞ control with poles assignment constraint

7.1 Introduction

Large flexible spacecraft systems have been developed for future applications in widespread communications, remote sensing, and related scientific researches in space [1]. Modern large spacecraft usually employ flexible appendages such as solar arrays and antennas. In rotational maneuvers of such spacecraft, elastic deformations in the flexible appendages often appear. Therefore it poses a challenging task for spacecraft designers to design the controller so as to provide high pointing precision while effectively suppressing the induced vibration. Besides, spacecraft are always subject to environmental or nonenvironmental disturbances, and the requirement for attitude stabilization performance becomes more rigorous. Using some actuators such as on-off thrusters or control moment gyroscopes to satisfy excellent accuracy requirement is a challenging task. It becomes much more complicated for flexible spacecraft where thruster firings or some parts of control moment gyroscopes can excite flexible modals resulting in poor accuracy or even attitude control instability [2]. However, when attitude control requires small control operations, reaction wheels or momentum wheels will be used. It can provide continuous control torques according to the desired torque profile for attitude stabilization. In such cases, thrusters or control moment gyroscopes should not be used any more, which will put forward higher requirements for the actuator.

During the past decades, considerable efforts have been made to study the attitude maneuvering and vibration suppression of flexible spacecraft and controller design theory has developed in a variety of directions. The adaptive control for rotational maneuvering and vibration suppression of an orbiting flexible spacecraft has been presented in [3–5]. However, these literatures only considered the pitch angle and its derivatives for feedback and flexible modes were not measured, which may cause parameter divergence and instability in the closed-loop system in front of unmodeled dynamics. Optimal control of flexible spacecraft has been considered in [6] and [7]. When we design such an attitude control law for spacecraft, it is desirable

Spacecraft Attitude Control. DOI: https://doi.org/10.1016/B978-0-323-99005-9.00007-9

that the attitude control law guarantees not only robustness with respect to disturbances but also optimality with respect to a performance index. However, the two objectives are conflicting factors and a trade-off between them is needed. Based on the Lyapunov function, a three-axis attitude tracking controller has been presented with robustness to parameter uncertainties and external disturbances for flexible spacecraft in [8]. The sliding mode fault tolerant control scheme has been developed in [9] in the presence of partial loss of actuator effectiveness fault and external disturbances. Despite several advantages, such as rapid response, low sensitivity to external perturbations and parameter variations, and low computational cost, sliding mode control has a chattering problem, which greatly hinders control precision. Based on the general sliding mode control, a new kind of sliding mode control technology named minimum sliding mode error feedback control (MSMEFC) has been proposed by Cao and Chen to tackle the problems of the uncertain disturbances [10–12]. MSMEFC has been used for fault tolerant control of spacecraft formation flying [10], fault tolerant control of small satellite attitude [11], and satellite attitude control and determination [12]. The integrating variable structure control and modal velocity feedback technique have been proposed for flexible spacecraft attitude control and vibration suppression in [13], which has provided great potential for flexible spacecraft control. To achieve the attitude synchronization for a group of flexible spacecraft and the induced vibrations suppression during formation maneuvers, the distributed cooperative control strategy based on the backstepping design and the neighbor-based design rule were proposed in [14]. However, new problems have been exposed in these integrated schemes, including complex algorithm structures and heavy computational burden. The fuzzy controller was designed in [15] for the attitude stabilization of the Republic of China Satellite (ROCSAT-1). Although this controller does not require gain settings and complicated computations, it is time consuming and cannot obtain satisfactory accuracy of the desired performance.

It is well known that H_2 control is often adopted to deal with transient performance while H_∞ control guarantees robust stability in the presence of parameter uncertainties and external disturbances. The mixed H_2/H_∞ controller can manage the trade-off between system performance and robustness, because it considers a mixed framework that can integrate optimal transient performance and robustness into a single controller. Consequently, it not only combines the merits of both H_2 optimal control and H_∞ robust control but can also achieve suboptimal control performance under a desired disturbance rejection constraint. The mixed H_2/H_∞ control problems have been widely investigated by many researchers [16–19]. In [16], the mixed H_2/H_∞ control performance has been investigated compared with other methods such as PD control, H_2 control, and H_∞ control for satellite attitude control system. In [17], the adaptive fuzzy mixed H_2/H_∞ control of nonlinear spacecraft systems was presented. Based on LMI approaches, Ref. [18] applied mixed H_2/H_∞ state feedback controller to the microsatellite control and verified its good performance to achieve a balanced compromise between H_2 and H_∞ performances. The mixed H_2/H_∞ controller with a parameter adaptive law was presented in [19] for spacecraft attitude control. However, the mixed H_2/H_∞ controllers mentioned above just made the Lyapunov matrix variable involved in H_2 performance constraint equal to that involved in H_∞ performance constraint, so that the multiobjective integrated problem became a convex optimization problem with great conservatives. Thus they can be called traditional mixed H_2/H_∞ controllers. To reduce conservativeness, the Lyapunov variables introduced in H_2 performance and H_∞ performance constraint should be different. In another way, one can also eliminate the coupling of Lyapunov matrix variables and system matrices by introducing a slack variable which provides an additional degree of freedom. In this way, one can obtain the improved mixed H_2/H_∞ controller. In addition, to achieve satisfactory transients in the process of designing controllers, one has to place the closed-loop poles in a suitable region of the complex plane. About 40 years ago, the pole allocation technique was proposed in [20] to allow controller designer to have direct control over the closed-loop system eigenvalues of flexible spacecraft. Recently, the poles placement theory in flexible spacecraft control has attracted great attention. In [21], the control of a large flexible platform in orbit based on pole placement was studied. The pole placement technique combined with optimal control was used in [22] for vibration suppression during maneuvering. However, to the best of our knowledge, the improved mixed H_2/H_∞ technique for flexible spacecraft attitude control and vibration suppression under poles assignment constraint has not been addressed, which is the focus of this chapter. The main contribution of this chapter is to design the robust high-precision attitude controller for flexible spacecraft with improved mixed H_2/H_∞ control strategy under poles assignment constraint. The proposed controller can simultaneously achieve attitude stabilization and vibration suppression. Actually, the robust mixed H_2/H_∞ control is one of the state feedback control methods. For state feedback control, all states of the system to be regulated were available to the controller, and it is complete feedback of system architecture information, while the static output feedback control is incomplete feedback of system architecture information. Thus we want to test the difference of feedback performance and choose the static output feedback controller to compare with the robust mixed H_2/H_∞ controller.

The remainder of this chapter is organized as follows. Section 7.2 establishes the state equations of flexible spacecraft dynamics and describes the purpose of this work. Section 7.3 proposes the design of robust H_2 and improved mixed H_2/H_∞ control under poles assignment constraint for flexible spacecraft system based on LMIs. Section 7.4 makes a detailed comparison analysis between improved mixed H_2/H_∞ controller under poles constraint and static output feedback controller in [23] by which this chapter is motivated. Besides, the simulation results using a traditional mixed H_2/H_∞ controller are also addressed. Finally, the concluding remarks and main references of this work are presented.

7.2 Problem formulation

7.2.1 Flexible spacecraft dynamics with two bending modes

This section introduces the attitude model of a flexible spacecraft, which is described as a rigid body with two flexible appendages such as solar panels. The dynamic models of the spacecraft are nonlinear and include rigid and flexible mode interaction. For flexible spacecraft with high-precision attitude control, the terminal three-dimensional rotations exhibit small values of the Euler angles, and the attitude system of the flexible spacecraft can be described by the following differential equations [24]

$$\begin{cases} \boldsymbol{J}\ddot{\boldsymbol{\theta}}+\boldsymbol{\Delta}_1\ddot{\boldsymbol{\eta}}_1+\boldsymbol{\Delta}_2\ddot{\boldsymbol{\eta}}_2=\boldsymbol{u}+\boldsymbol{d} \\ \ddot{\boldsymbol{\eta}}_1+2\xi_1\boldsymbol{\Omega}_1\dot{\boldsymbol{\eta}}_1+\boldsymbol{\Omega}_1^2\boldsymbol{\eta}_1+\boldsymbol{\Delta}_1^T\ddot{\boldsymbol{\theta}}=0 \\ \ddot{\boldsymbol{\eta}}_2+2\xi_2\boldsymbol{\Omega}_2\dot{\boldsymbol{\eta}}_2+\boldsymbol{\Omega}_2^2\boldsymbol{\eta}_2+\boldsymbol{\Delta}_2^T\ddot{\boldsymbol{\theta}}=0 \end{cases} \tag{7.1}$$

where $\boldsymbol{J}$ is the inertia matrix, $\boldsymbol{\theta}$ is the spacecraft attitude angle vector including yaw angle, pitch angle and roll angle, $\boldsymbol{\Delta}_i$ (i=1,2) is the coupling coefficient matrix of i-th solar panel, $\boldsymbol{\eta}_i$ (i=1,2) is the modal coordinate, ξ_i (i=1,2) and $\boldsymbol{\Omega}_i$ (i=1,2) are the modal damping ratio and the modal frequency matrix corresponding to $\boldsymbol{\eta}_i$, respectively, $\boldsymbol{u}$ is the control torque, and $\boldsymbol{d}$ represents the bounded disturbance torque.

Let $\hat{q}=\begin{bmatrix}\boldsymbol{\theta}^T & \boldsymbol{\eta}_1^T & \boldsymbol{\eta}_2^T\end{bmatrix}^T$, then Eq. (7.1) can be transformed into the following second-order matrix form:

$$\boldsymbol{R}\ddot{\hat{q}}+\boldsymbol{H}\dot{\hat{q}}+\boldsymbol{G}\hat{q}=\boldsymbol{L}_1\boldsymbol{u}+\boldsymbol{L}_2\boldsymbol{d} \tag{7.2}$$

where,

$$R = \begin{bmatrix} J & \Delta_1 & \Delta_2 \\ \Delta_1^T & I_{2\times 2} & 0_{2\times 2} \\ \Delta_2^T & 0_{2\times 2} & I_{2\times 2} \end{bmatrix}, H = \begin{bmatrix} 0_{3\times 3} & 0_{3\times 2} & 0_{3\times 2} \\ 0_{2\times 3} & 2\xi_1\Omega_1 & 0_{2\times 2} \\ 0_{2\times 3} & 0_{2\times 2} & 2\xi_2\Omega_2 \end{bmatrix},$$

$$G = \begin{bmatrix} 0_{3\times 3} & 0_{3\times 2} & 0_{3\times 2} \\ 0_{2\times 3} & \Omega_1^2 & 0_{2\times 2} \\ 0_{2\times 3} & 0_{2\times 2} & \Omega_2^2 \end{bmatrix}, L_1 = L_2 = \begin{bmatrix} I_{3\times 3} \\ 0_{2\times 2} \\ 0_{2\times 2} \end{bmatrix}$$

Choose state variable $x = \begin{bmatrix} \hat{q}^T & \dot{\hat{q}}^T \end{bmatrix}^T$, output variables $z_\infty = 10^{-3}\cdot R\cdot \ddot{\hat{q}}$, $z_2 = \hat{q}$, and Eq. (7.2) can be converted into the following state-space form:

$$\begin{cases} \dot{x} = Ax + B_1 u + B_2 d \\ z_\infty = C_1 x + D_1 u + D_2 d \\ z_2 = C_2 x \end{cases} \tag{7.3}$$

where $x = \begin{bmatrix} \theta^T & \eta_1^T & \eta_2^T & \dot{\theta}^T & \dot{\eta}_1^T & \dot{\eta}_2^T \end{bmatrix}^T$, $z_\infty = 10^{-3}\cdot R\cdot\begin{bmatrix} \ddot{\theta}^T & \ddot{\eta}_1^T & \ddot{\eta}_2^T \end{bmatrix}^T$, $z_2 = \begin{bmatrix} \theta^T & \eta_1^T & \eta_2^T \end{bmatrix}^T$, and the coefficient matrices are defined as follows:

$$A = \begin{bmatrix} 0 & I_{7\times 7} \\ -R^{-1}G & -R^{-1}H \end{bmatrix}, B_1 = \begin{bmatrix} 0_{7\times 3} \\ R^{-1}L_1 \end{bmatrix}, B_2 = \begin{bmatrix} 0_{7\times 3} \\ R^{-1}L_2 \end{bmatrix}$$
$$C_1 = 10^{-3}\times\begin{bmatrix} -G & -H \end{bmatrix}, C_2 = \begin{bmatrix} I_{7\times 7} & 0_{7\times 7} \end{bmatrix}, D_1 = 10^{-3}\times L_1, D_2 = 10^{-3}\times L_2$$

As to Eq. (7.3), design state feedback controller

$$u = Kx \tag{7.4}$$

Then, the closed-loop attitude control system of spacecraft becomes

$$\begin{cases} \dot{x} = (A+B_1K)x + B_2 d \\ z_\infty = (C_1 + D_1 K)x + D_2 d \\ z_2 = C_2 x \end{cases} \tag{7.5}$$

7.2.2 H_∞ and H_2 performance constraint

With some possibly small γ_∞ and c_2, which both denote the minimal attenuation level, the designed controller should satisfy the following H_∞ and H_2 performance constraints.

$$\left\| G_{z_\infty d} \right\|_\infty = \left\| (C_1+D_1K)(sI-(A+B_1K))^{-1}B_2+D_2 \right\|_\infty \le \gamma_\infty \tag{7.6}$$

$$\left\| G_{z_2 d} \right\|_2 = \left\| C_2(sI-(A+B_1K))^{-1}B_2 \right\|_2 \le \gamma_2 \tag{7.7}$$

where inequality (7.6) implies that the upper bound of H_∞ norm of the transfer function matrix from external disturbance to the output variable z_∞ is γ_∞, which can guarantee the external disturbance attenuation; inequality (7.7) implies that the upper bound of H_2 norm of the transfer function matrix from external disturbance to the output variable z_2 is γ_2, which can guarantee the good transient performance of system.

7.2.3 Poles assignment

Fast response can be achieved by placing the poles of the closed-loop system appropriately. Let D be a domain symmetric about the real axis on the complex plane, then a matrix $\boldsymbol{A}+\boldsymbol{B}_1\boldsymbol{K}\in\mathbb{R}^{n\times n}$ is said to be D-stable if all eigenvalues of $\boldsymbol{A}+\boldsymbol{B}_1\boldsymbol{K}$ satisfy $\lambda_i\in D, i=1,2,\ldots,n$. There are two special cases about the D domain on the complex plane.

Case 1: $D=H_{\alpha,\beta}=\{x+jy|-\beta<x<-\alpha\}$, which requires that all eigenvalues of the matrix $\boldsymbol{A}+\boldsymbol{B}_1\boldsymbol{K}$ are located in the strip region shown in Fig. 7.1.

Case 2: $D=D_{(q,r)}=\{x+jy\big|(x+q)^2+y^2<r^2\}$, which requires that all eigenvalues of the matrix $\boldsymbol{A}+\boldsymbol{B}_1\boldsymbol{K}$ are located in the disk region shown in Fig. 7.2.

In addition to the two special cases on the complex plane, LMI regions are more general. Let D be a region on the complex plane if there exist matrices $\boldsymbol{L}\in S^m$ and $\boldsymbol{M}\in\mathbb{R}^{m\times m}$ such that

$$D=\{s\big|s\in C,\boldsymbol{L}+s\boldsymbol{M}+\bar{s}\boldsymbol{M}^T<0\} \tag{7.8}$$

D is called an LMI region which is denoted by $D_{(L,M)}$.

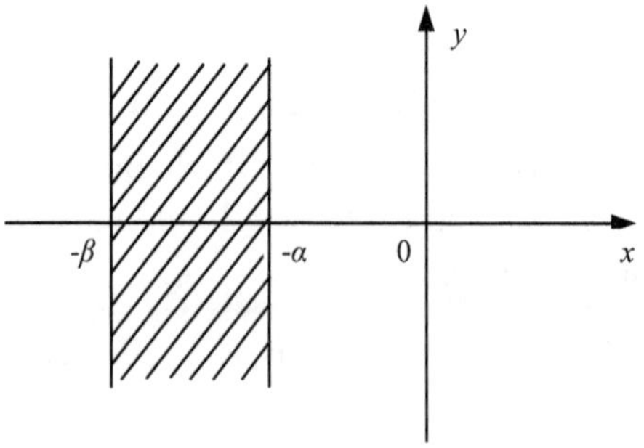

Figure 7.1 Strip region $H_{\alpha,\beta}$.

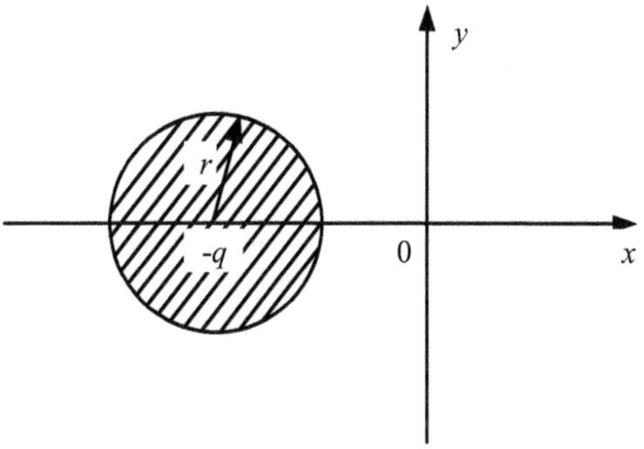

Figure 7.2 Disk region $D_{(q,r)}$.

The aforementioned two special cases are also LMI region virtually, because

$$\mathbf{H}_{\alpha,\beta} = \{x + jy | -\beta < x < -\alpha\} = \{s | -\beta < \mathrm{Re}(s) < -\alpha\}$$

$$= \left\{ s \middle| -\beta < \frac{1}{2}(s + \bar{s}) < -\alpha \right\} = \left\{ s \middle| \frac{1}{2}\mathrm{diag}(s, -s) + \frac{1}{2}\mathrm{diag}(\bar{s}, -\bar{s}) + \mathrm{diag}(\alpha, -\beta) < 0 \right\} \tag{7.9}$$

In this case, one has $\boldsymbol{L}$=diag(2α,-2β), $\boldsymbol{M}$=diag($1, -1$).

$$\mathrm{D}_{(q,r)} = \{x + jy | (x+q)^2 + y^2 < r^2\} = \{s | (s+q)(\bar{s}+q) < r^2\}$$

$$= \left\{ s \middle| \begin{bmatrix} -r & s+q \\ \bar{s}+q & -r \end{bmatrix} < 0 \right\} = \left\{ s \middle| \begin{bmatrix} -r & q \\ q & -r \end{bmatrix} + s\begin{bmatrix} 0 & 1 \\ 0 & 0 \end{bmatrix} + \bar{s}\begin{bmatrix} 0 & 0 \\ 1 & 0 \end{bmatrix} < 0 \right\} \tag{7.10}$$

In this case, one has $\boldsymbol{L}$=[$-r$, q; q, $-r$], $\boldsymbol{M}$=[0, 1; 0, 0].

7.2.4 Control objective

The objective of this chapter is to design an improved mixed H_2/H_∞ controller for flexible spacecraft under poles assignment constraint. It implies that under poles assignment constraint, for a given positive scalar γ_∞, to obtain the state feedback gain matrix $\boldsymbol{K}$, so that the spacecraft closed-loop attitude control system (7.5) is stable, let $\|G_{z_2 d}\|_2$ be the smallest with the constraint $\|\mathbf{G}_{z_\infty d}\|_\infty \le \gamma_\infty$.

7.3 Improved mixed H_2/H_∞ control law

7.3.1 Some lemmas

Before we prove our main results, we first introduce some essential preliminary lemmas.

Lemma 7.1: [25] According to bounded real lemma, for a given scalar γ_∞, the closed-loop system is stable and satisfies inequality (7.6), if there exists matrix $\boldsymbol{W}$ and symmetric positive definite matrix $\boldsymbol{X}$ such that

$$\begin{bmatrix} \boldsymbol{AX}+\boldsymbol{B}_1\boldsymbol{W}+(\boldsymbol{AX}+\boldsymbol{B}_1\boldsymbol{W})^T & \boldsymbol{B}_2 & (\boldsymbol{C}_1\boldsymbol{X}+\boldsymbol{D}_1\boldsymbol{W})^T \\ \boldsymbol{B}_2^T & -\gamma_\infty\boldsymbol{I} & \boldsymbol{D}_2^T \\ \boldsymbol{C}_1\boldsymbol{X}+\boldsymbol{D}_1\boldsymbol{W} & \boldsymbol{D}_2 & -\gamma_\infty\boldsymbol{I} \end{bmatrix} < 0 \tag{7.11}$$

Lemma 7.2: (Schur complement lemma) Let the partitioned matrix

$$\boldsymbol{A} = \begin{bmatrix} \boldsymbol{A}_{11} & \boldsymbol{A}_{12} \\ \boldsymbol{A}_{12}^T & \boldsymbol{A}_{22} \end{bmatrix}$$

be symmetric. Then

$$\boldsymbol{A}<0 \Leftrightarrow \boldsymbol{A}_{11}<0, \boldsymbol{A}_{22}-\boldsymbol{A}_{12}^T\boldsymbol{A}_{11}^{-1}\boldsymbol{A}_{12}<0 \Leftrightarrow \boldsymbol{A}_{22}<0, \boldsymbol{A}_{11}-\boldsymbol{A}_{12}\boldsymbol{A}_{22}^{-1}\boldsymbol{A}_{12}^T<0 \tag{7.12}$$

or

$$\boldsymbol{A}>0 \Leftrightarrow \boldsymbol{A}_{11}>0, \boldsymbol{A}_{22}-\boldsymbol{A}_{12}^T\boldsymbol{A}_{11}^{-1}\boldsymbol{A}_{12}>0 \Leftrightarrow \boldsymbol{A}_{22}>0, \boldsymbol{A}_{11}-\boldsymbol{A}_{12}\boldsymbol{A}_{22}^{-1}\boldsymbol{A}_{12}^T>0 \tag{7.13}$$

Lemma 7.3: Inequalities (7.6) and (7.7) are both satisfied if there exist matrix $\boldsymbol{W}$, symmetric matrix $\boldsymbol{Z}$, and symmetric positive definite matrix $\boldsymbol{X}$ such that

$$\begin{cases} \boldsymbol{AX}+\boldsymbol{B}_1\boldsymbol{W}+(\boldsymbol{AX}+\boldsymbol{B}_1\boldsymbol{W})^T+\boldsymbol{B}_2\boldsymbol{B}_2^T<0 \\ \begin{bmatrix} -\boldsymbol{Z} & \boldsymbol{C}_2\boldsymbol{X} \\ (\boldsymbol{C}_2\boldsymbol{X})^T & -\boldsymbol{X} \end{bmatrix}<0 \\ Trace(\boldsymbol{Z})<\rho \\ \begin{bmatrix} \boldsymbol{AX}+\boldsymbol{B}_1\boldsymbol{W}+(\boldsymbol{AX}+\boldsymbol{B}_1\boldsymbol{W})^T & \boldsymbol{B}_2 & (\boldsymbol{C}_1\boldsymbol{X}+\boldsymbol{D}_1\boldsymbol{W})^T \\ \boldsymbol{B}_2^T & -\gamma_\infty\boldsymbol{I} & \boldsymbol{D}_2^T \\ \boldsymbol{C}_1\boldsymbol{X}+\boldsymbol{D}_1\boldsymbol{W} & \boldsymbol{D}_2 & -\gamma_\infty\boldsymbol{I} \end{bmatrix}<0 \end{cases} \tag{7.14}$$

By minimizing $c_\infty\gamma_\infty + c_2\gamma_2^2$, where $\rho = \gamma_2^2$, the traditional mixed H_2/H_∞ controller can be constructed as $\boldsymbol{K} = \boldsymbol{W}\boldsymbol{X}^{-1}$.

Remark 7.1: : Lemma 7.3 gives a sufficient condition that guarantees the closed-loop system represented in Eq. (7.5) is robustly stable and satisfies H_2 and H_∞ norm bounds simultaneously. However, the mixed H_2/H_∞ controller designed in this way is conservative, for the Lyapunov variables introduced in H_2 and H_∞ performance constraint are the same, both represented by matrix W. That means the Lyapunov variable W responds to H_2 and H_∞ performance constraint at the same time.

Lemma 7.4: [26] (Reciprocal projection lemma) Let $\boldsymbol{P}$ be any given positive definite matrix. The following statements are equivalent:
1. LMI

$$\boldsymbol{\Psi}+\boldsymbol{S}+\boldsymbol{S}^T<0 \tag{7.15}$$

2. The LMI problem

$$\begin{bmatrix} \boldsymbol{\Psi}+\boldsymbol{P}-\boldsymbol{W}-\boldsymbol{W}^T & \boldsymbol{S}^T+\boldsymbol{W}^T \\ \boldsymbol{S}+\boldsymbol{W} & -\boldsymbol{P} \end{bmatrix}<0 \tag{7.16}$$

is feasible with respect to $\boldsymbol{W}$.

Lemma 7.5: [27] Let D be an LMI region on the complex plane with the form of Eq. (7.8), then matrix $\boldsymbol{A}$ is D-stable if there exists a symmetric positive definite matrix $\boldsymbol{P}$ such that

$$\boldsymbol{L}\otimes\boldsymbol{P}+\boldsymbol{M}\otimes(\boldsymbol{AP})+\boldsymbol{M}^T\otimes(\boldsymbol{AP})^T<0 \tag{7.17}$$

where $\otimes$ represents the Kronecker product.

7.3.2 H_2 control

Considering the system described by the first and third equations in Eq. (7.5), the H_2 performance can be described by Theorem 7.1.

Theorem 7.1: For a given scalar γ_2, the closed-loop system is stable and satisfies Eq. (7.7), if there exist matrices $\boldsymbol{N}_1=\boldsymbol{XB}_1\boldsymbol{K}$, $\boldsymbol{N}_2=\boldsymbol{Q}^T\boldsymbol{B}_1\boldsymbol{K}$, $\boldsymbol{N}_3=\boldsymbol{B}_1\boldsymbol{KQ}$, $\boldsymbol{Q}$, symmetric matrix $\boldsymbol{Z}$ and symmetric positive definite matrix $\boldsymbol{X}$, such that

(C_1): *Statement 1*
$\boldsymbol{A}+\boldsymbol{B}_1\boldsymbol{K}$is stable and $\left\|\boldsymbol{C}_2(s\boldsymbol{I}-(\boldsymbol{A}+\boldsymbol{B}_1\boldsymbol{K}))^{-1}\boldsymbol{B}_2\right\|_2\leq\gamma_2$.

(C_2): *Statement 2*

$$\begin{bmatrix} (\boldsymbol{XA}+\boldsymbol{N}_1)^T+(\boldsymbol{XA}+\boldsymbol{N}_1) & \boldsymbol{XB}_2 \\ \boldsymbol{B}_2^T\boldsymbol{X} & -\boldsymbol{I} \end{bmatrix}<0$$

$$\begin{bmatrix} \boldsymbol{X} & \boldsymbol{C}_2^T \\ \boldsymbol{C}_2 & \boldsymbol{Z} \end{bmatrix}>0$$

$Trace(\boldsymbol{Z})<\gamma_2^2$

(C_3): *Statement 3*

$$\begin{bmatrix} -\boldsymbol{Q}-\boldsymbol{Q}^T & (\boldsymbol{Q}^T\boldsymbol{A}+\boldsymbol{N}_2)+\boldsymbol{X} & \boldsymbol{Q}^T\boldsymbol{B}_2 & \boldsymbol{Q}^T \\ (\boldsymbol{Q}^T\boldsymbol{A}+\boldsymbol{N}_2)^T+\boldsymbol{X} & -\boldsymbol{X} & 0 & 0 \\ \boldsymbol{B}_2^T\boldsymbol{Q} & 0 & -\boldsymbol{I} & 0 \\ \boldsymbol{Q} & 0 & 0 & -\boldsymbol{X} \end{bmatrix}<0$$

$$\begin{bmatrix} \boldsymbol{X} & \boldsymbol{C}_2^T \\ \boldsymbol{C}_2 & \boldsymbol{Z} \end{bmatrix}>0$$

$Trace(\boldsymbol{Z})<\gamma_2^2$

(C_4): *Statement 4*

$$\begin{bmatrix} -\boldsymbol{Q}-\boldsymbol{Q}^T & (\boldsymbol{AQ}+\boldsymbol{N}_3)^T+\boldsymbol{X} & \boldsymbol{Q}^T\boldsymbol{C}_2^T & \boldsymbol{Q}^T \\ (\boldsymbol{AQ}+\boldsymbol{N}_3)+\boldsymbol{X} & -\boldsymbol{X} & 0 & 0 \\ \boldsymbol{C}_2\boldsymbol{Q} & 0 & -\boldsymbol{I} & 0 \\ \boldsymbol{Q} & 0 & 0 & -\boldsymbol{X} \end{bmatrix}<0$$

$$\begin{bmatrix} \boldsymbol{X} & \boldsymbol{B}_2 \\ \boldsymbol{B}_2^T & \boldsymbol{Z} \end{bmatrix}>0$$

$Trace(\boldsymbol{Z})<\gamma_2^2$

The above four statements are equivalent.

Proof:: The equivalence of (C_1) and (C_2) are standard, so we only need to prove the equivalence of (C_2), (C_3), and (C_4).

First, the equivalences of (C_2) and (C_3) are proved.

With Schur complement lemma, the first LMI in (C_2) can be written as

$$(\boldsymbol{XA}+\boldsymbol{N}_1)^T+(\boldsymbol{XA}+\boldsymbol{N}_1)+\gamma_2^{-1}\boldsymbol{XB}_2\boldsymbol{B}_2^T\boldsymbol{X}<0 \tag{7.18}$$

For $\boldsymbol{N}_1 = \boldsymbol{X}\boldsymbol{B}_1\boldsymbol{K}$, inequality (7.18) can be translated into

$$(\boldsymbol{A}+\boldsymbol{B}_1\boldsymbol{K})^T\boldsymbol{X} + \boldsymbol{X}(\boldsymbol{A}+\boldsymbol{B}_1\boldsymbol{K}) + \boldsymbol{X}\boldsymbol{B}_2\boldsymbol{B}_2^T\boldsymbol{X} < 0 \tag{7.19}$$

Let $\boldsymbol{Y} = \boldsymbol{X}^{-1}$, $\boldsymbol{Y}$ is also symmetric positive definite matrix, multiply $\boldsymbol{Y}$ and its transpose to both sides of inequality (7.19), and one can get

$$\boldsymbol{Y}^T(\boldsymbol{A}+\boldsymbol{B}_1\boldsymbol{K})^T\boldsymbol{X}\boldsymbol{Y} + \boldsymbol{Y}^T\boldsymbol{X}(\boldsymbol{A}+\boldsymbol{B}_1\boldsymbol{K})\boldsymbol{Y} + \boldsymbol{Y}^T\boldsymbol{X}\boldsymbol{B}_2\boldsymbol{B}_2^T\boldsymbol{X}\boldsymbol{Y} < 0 \tag{7.20}$$

With $\boldsymbol{X} = \boldsymbol{X}^T, \boldsymbol{Y} = \boldsymbol{Y}^T, \boldsymbol{Y} = \boldsymbol{X}^{-1}$, we have

$$\boldsymbol{Y}(\boldsymbol{A}+\boldsymbol{B}_1\boldsymbol{K})^T + (\boldsymbol{A}+\boldsymbol{B}_1\boldsymbol{K})\boldsymbol{Y} + \boldsymbol{B}_2\boldsymbol{B}_2^T < 0 \tag{7.21}$$

With Lemma 7.4, if $\boldsymbol{S} = \boldsymbol{Y}(\boldsymbol{A}+\boldsymbol{B}_1\boldsymbol{K})^T$, $\boldsymbol{\Psi} = \boldsymbol{B}_2\boldsymbol{B}_2^T$, inequality (7.21) is equivalent to inequality (7.22),

$$\begin{bmatrix} \boldsymbol{B}_2\boldsymbol{B}_2^T + \boldsymbol{P} - \boldsymbol{W} - \boldsymbol{W}^T & (\boldsymbol{A}+\boldsymbol{B}_1\boldsymbol{K})\boldsymbol{Y} + \boldsymbol{W}^T \\ \boldsymbol{Y}(\boldsymbol{A}+\boldsymbol{B}_1\boldsymbol{K})^T + \boldsymbol{W} & -\boldsymbol{P} \end{bmatrix} < 0 \tag{7.22}$$

With Schur complement lemma, inequality (7.22) can be converted into

$$\begin{bmatrix} \boldsymbol{P} - \boldsymbol{W} - \boldsymbol{W}^T & (\boldsymbol{A}+\boldsymbol{B}_1\boldsymbol{K})\boldsymbol{Y} + \boldsymbol{W}^T & \boldsymbol{B}_2 \\ \boldsymbol{Y}(\boldsymbol{A}+\boldsymbol{B}_1\boldsymbol{K})^T + \boldsymbol{W} & -\boldsymbol{P} & 0 \\ \boldsymbol{B}_2^T & 0 & -\boldsymbol{I} \end{bmatrix} < 0 \tag{7.23}$$

With the Schur complement lemma again, one can obtain

$$\begin{bmatrix} -\boldsymbol{W} - \boldsymbol{W}^T & (\boldsymbol{A}+\boldsymbol{B}_1\boldsymbol{K})\boldsymbol{Y} + \boldsymbol{W}^T & \boldsymbol{B}_2 & \boldsymbol{P} \\ \boldsymbol{Y}(\boldsymbol{A}+\boldsymbol{B}_1\boldsymbol{K})^T + \boldsymbol{W} & -\boldsymbol{P} & 0 & 0 \\ \boldsymbol{B}_2^T & 0 & -\boldsymbol{I} & 0 \\ \boldsymbol{P} & 0 & 0 & -\boldsymbol{P} \end{bmatrix} < 0 \tag{7.24}$$

Similarly, let $\boldsymbol{P} = \boldsymbol{X}^{-1}, \boldsymbol{Q} = \boldsymbol{W}^{-1}$, by performing the congruence transformation, that is, multiplied by a matrix $\tilde{\boldsymbol{M}}$ and its transpose on both sides of inequality (7.24), where

$$\tilde{\boldsymbol{M}} = \begin{bmatrix} \boldsymbol{Q} & 0 & 0 & 0 \\ 0 & \boldsymbol{X} & 0 & 0 \\ 0 & 0 & \boldsymbol{I} & 0 \\ 0 & 0 & 0 & \boldsymbol{X} \end{bmatrix} \tag{7.25}$$

Then

$$\tilde{\boldsymbol{M}}^T \begin{bmatrix} -\boldsymbol{W} - \boldsymbol{W}^T & (\boldsymbol{A}+\boldsymbol{B}_1\boldsymbol{K})\boldsymbol{Y} + \boldsymbol{W}^T & \boldsymbol{B}_2 & \boldsymbol{P} \\ \boldsymbol{Y}(\boldsymbol{A}+\boldsymbol{B}_1\boldsymbol{K})^T + \boldsymbol{W} & -\boldsymbol{P} & 0 & 0 \\ \boldsymbol{B}_2^T & 0 & -\boldsymbol{I} & 0 \\ \boldsymbol{P} & 0 & 0 & -\boldsymbol{P} \end{bmatrix} \tilde{\boldsymbol{M}} < 0 \tag{7.26}$$

One can get

$$\begin{bmatrix} -\boldsymbol{Q}-\boldsymbol{Q}^T & \boldsymbol{Q}^T(\boldsymbol{A}+\boldsymbol{B}_1\boldsymbol{K})+\boldsymbol{X} & \boldsymbol{Q}^T\boldsymbol{B}_2 & \boldsymbol{Q}^T \\ (\boldsymbol{A}+\boldsymbol{B}_1\boldsymbol{K})^T\boldsymbol{Q}+\boldsymbol{X} & -\boldsymbol{X} & 0 & 0 \\ \boldsymbol{B}_2^T\boldsymbol{Q} & 0 & -\boldsymbol{I} & 0 \\ \boldsymbol{Q} & 0 & 0 & -\boldsymbol{X} \end{bmatrix} < 0 \quad (7.27)$$

For $\boldsymbol{N}_2=\boldsymbol{Q}^T\boldsymbol{B}_1\boldsymbol{K}$, then,

$$\begin{bmatrix} -\boldsymbol{Q}-\boldsymbol{Q}^T & (\boldsymbol{Q}^T\boldsymbol{A}+\boldsymbol{N}_2)+\boldsymbol{X} & \boldsymbol{Q}^T\boldsymbol{B}_2 & \boldsymbol{Q}^T \\ (\boldsymbol{Q}^T\boldsymbol{A}+\boldsymbol{N}_2)^T+\boldsymbol{X} & -\boldsymbol{X} & 0 & 0 \\ \boldsymbol{B}_2^T\boldsymbol{Q} & 0 & -\boldsymbol{I} & 0 \\ \boldsymbol{Q} & 0 & 0 & -\boldsymbol{X} \end{bmatrix} < 0 \quad (7.28)$$

It can be seen that the first LMI in (C_3) is concluded, and the last two LMIs are the same as those in (C_2). Next, we will prove the equivalence of (C_3) and (C_4).

System $S_1((\boldsymbol{A}+\boldsymbol{B}_1\boldsymbol{K}),\boldsymbol{B}_2,\boldsymbol{C}_2)$

$$\begin{cases} \dot{\boldsymbol{x}}=(\boldsymbol{A}+\boldsymbol{B}_1\boldsymbol{K})x+\boldsymbol{B}_2\boldsymbol{d} \\ \boldsymbol{z}_2=\boldsymbol{C}_2\boldsymbol{x} \end{cases} \quad (7.29)$$

and system $S_2\left((\boldsymbol{A}+\boldsymbol{B}_1\boldsymbol{K})^T,\boldsymbol{C}_2^T,\boldsymbol{B}_2^T\right)$

$$\begin{cases} \dot{\boldsymbol{x}}=(\boldsymbol{A}+\boldsymbol{B}_1\boldsymbol{K})^T\boldsymbol{x}+\boldsymbol{C}_2^T\boldsymbol{d} \\ \boldsymbol{z}_2=\boldsymbol{B}_2^T\boldsymbol{x} \end{cases} \quad (7.30)$$

are dual systems. With transformation $((\boldsymbol{A}+\boldsymbol{B}_1\boldsymbol{K}),\boldsymbol{B}_2,\boldsymbol{C}_2)\rightarrow\left((\boldsymbol{A}+\boldsymbol{B}_1\boldsymbol{K})^T,\right.$ $\left.\boldsymbol{C}_2^T,\boldsymbol{B}_2^T\right)$, inequality (7.27) can be turned into

$$\begin{bmatrix} -\boldsymbol{Q}-\boldsymbol{Q}^T & \boldsymbol{Q}^T(\boldsymbol{A}+\boldsymbol{B}_1\boldsymbol{K})^T+\boldsymbol{X} & \boldsymbol{Q}^T\boldsymbol{C}_2^T & \boldsymbol{Q}^T \\ (\boldsymbol{A}+\boldsymbol{B}_1\boldsymbol{K})\boldsymbol{Q}+\boldsymbol{X} & -\boldsymbol{X} & 0 & 0 \\ \boldsymbol{C}_2\boldsymbol{Q} & 0 & -\boldsymbol{I} & 0 \\ \boldsymbol{Q} & 0 & 0 & -\boldsymbol{X} \end{bmatrix} < 0 \quad (7.31)$$

For $\boldsymbol{N}_3=\boldsymbol{B}_1\boldsymbol{K}\boldsymbol{Q}$, inequality (7.31) can yield

$$\begin{bmatrix} -\boldsymbol{Q}-\boldsymbol{Q}^T & (\boldsymbol{A}\boldsymbol{Q}+\boldsymbol{N}_3)^T+\boldsymbol{X} & \boldsymbol{Q}^T\boldsymbol{C}_2^T & \boldsymbol{Q}^T \\ (\boldsymbol{A}\boldsymbol{Q}+\boldsymbol{N}_3)+\boldsymbol{X} & -\boldsymbol{X} & 0 & 0 \\ \boldsymbol{C}_2\boldsymbol{Q} & 0 & -\boldsymbol{I} & 0 \\ \boldsymbol{Q} & 0 & 0 & -\boldsymbol{X} \end{bmatrix} < 0 \quad (7.32)$$

Similarly, the last two LMIs in (C_4) can be obtained with transformation $((\boldsymbol{A}+\boldsymbol{B}_1\boldsymbol{K}),\boldsymbol{B}_2,\boldsymbol{C}_2)\rightarrow\left((\boldsymbol{A}+\boldsymbol{B}_1\boldsymbol{K})^T,\boldsymbol{C}_2^T,\boldsymbol{B}_2^T\right)$.

7.3.3 Mixed H_2/H_∞ control

Based on Lemma 7.1, Lemma 7.5, and Theorem 7.1, one can generate the following Theorem 7.2 and Theorem 7.3 that can solve the mixed H_2/H_∞ problem under the pole assignment constraint. In fact, Theorem 7.2 is essentially equivalent to Theorem 7.3 based on the proof process of Theorem 7.1.

Theorem 7.2: For system (7.5), if there exist matrices $\boldsymbol{N}_2 = \boldsymbol{Q}^T\boldsymbol{B}_1\boldsymbol{K}$, $\boldsymbol{Q}$, $\boldsymbol{W}$, symmetric matrix $\boldsymbol{Z}$, and symmetric positive definite matrix $\boldsymbol{X}$, such that

$$\text{s.t.} \begin{bmatrix} -\boldsymbol{Q}-\boldsymbol{Q}^T & (\boldsymbol{Q}^T\boldsymbol{A}+\boldsymbol{N}_2)+\boldsymbol{X} & \boldsymbol{Q}^T\boldsymbol{B}_2 & \boldsymbol{Q}^T \\ (\boldsymbol{Q}^T\boldsymbol{A}+\boldsymbol{N}_2)^T+\boldsymbol{X} & -\boldsymbol{X} & 0 & 0 \\ \boldsymbol{B}_2^T\boldsymbol{Q} & 0 & -\boldsymbol{I} & 0 \\ \boldsymbol{Q} & 0 & 0 & -\boldsymbol{X} \end{bmatrix} < 0$$

$$\begin{bmatrix} \boldsymbol{X} & \boldsymbol{C}_2^T \\ \boldsymbol{C}_2 & \boldsymbol{Z} \end{bmatrix} > 0$$

$$Trace(\boldsymbol{Z}) < \rho$$

$$\begin{bmatrix} \boldsymbol{AX}+\boldsymbol{B}_1\boldsymbol{W}+(\boldsymbol{AX}+\boldsymbol{B}_1\boldsymbol{W})^T & \boldsymbol{B}_2 & (\boldsymbol{C}_1\boldsymbol{X}+\boldsymbol{D}_1\boldsymbol{W})^T \\ \boldsymbol{B}_2^T & -\gamma_\infty\boldsymbol{I} & \boldsymbol{D}_2^T \\ \boldsymbol{C}_1\boldsymbol{X}+\boldsymbol{D}_1\boldsymbol{W} & \boldsymbol{D}_2 & -\gamma_\infty\boldsymbol{I} \end{bmatrix} < 0$$

$$\boldsymbol{L}\otimes\boldsymbol{X}+\boldsymbol{M}\otimes(\boldsymbol{AX}+\boldsymbol{B}_1\boldsymbol{W})+\boldsymbol{M}^T\otimes(\boldsymbol{AX}+\boldsymbol{B}_1\boldsymbol{W})^T < 0$$

By minimizing $c_\infty\gamma_\infty + c_2\rho$, where $\rho = \gamma_2^2$, the state feedback gain matrix can be constructed as $\boldsymbol{K} = \boldsymbol{W}\boldsymbol{X}^{-1}$.

Proof: The first three LMIs in Theorem 7.2 correspond to Statement 3 in Theorem 7.1, which are about H_2 performance. The fourth LMI in Theorem 7.2 corresponds to Lemma 7.1, which is about H_∞ performance. Therefore we just need to prove the fifth LMI in Theorem 7.2. The proof process is as following:

With Lemma 7.5, when D is an LMI region on the complex plane with the form of Eq. (7.8), then the coefficient matrix $\boldsymbol{A}+\boldsymbol{B}_1\boldsymbol{K}$ of state equation represented in Eq. (7.5) is D-stable if there exists a symmetric positive definite matrix $\boldsymbol{X}$ such that

$$\boldsymbol{L}\otimes\boldsymbol{X}+\boldsymbol{M}\otimes((\boldsymbol{A}+\boldsymbol{B}_1\boldsymbol{K})\boldsymbol{X})+\boldsymbol{M}^T\otimes((\boldsymbol{A}+\boldsymbol{B}_1\boldsymbol{K})\boldsymbol{X})^T < 0$$

where $\otimes$ represents the Kronecker product.

That is,

$$L \otimes X + M \otimes (AX+B_1KX) + M^T \otimes (AX+B_1KX)^T < 0 \tag{7.33}$$

By defining $W=KX$, inequality (7.33) is turned into the fifth LMI in Theorem 7.2. This completes the proof.

Remark 7.2: The mixed H_2/H_∞ control problem can be converted into the above optimization problem from which one can see that the Lyapunov matrix variables $\mathbf{Q}$ and W are not the same. Theorem 7.2 gives a sufficient condition that guarantees the closed-loop system represented in Eq. (7.5) is robustly stable and satisfies H_2 and H_∞ norm bounds simultaneously. To reduce conservativeness, the Lyapunov variables introduced in H_2 and H_∞ performance constraint are different. That means the first Lyapunov variable $\mathbf{Q}$ responds to H_2 performance, and the second Lyapunov variable W responds to H_∞ performance.

Theorem 7.3: For system (7.5), if there exist matrices $N_3 = B_1KQ$, Q, W, R, symmetric matrix Z and symmetric positive definite matrix X, such that

$$\text{s.t.} \begin{bmatrix} -Q-Q^T & (AQ+N_3)^T+X & Q^TC_2^T & Q^T \\ (AQ+N_3)+X & -X & 0 & 0 \\ C_2Q & 0 & -I & 0 \\ Q & 0 & 0 & -X \end{bmatrix} < 0$$

$$\begin{bmatrix} X & B_2 \\ B_2^T & Z \end{bmatrix} > 0,$$

$$Trace(Z) < \rho,$$

$$\begin{bmatrix} AX+B_1W+(AX+B_1W)^T & B_2 & (C_1X+D_1W)^T \\ B_2^T & -\gamma_\infty I & D_2^T \\ C_1X+D_1W & D_2 & -\gamma_\infty I \end{bmatrix} < 0$$

$$L \otimes X + M \otimes (A^TX+R^T) + M^T \otimes (XA+R) < 0$$

By minimizing $c_\infty\gamma_\infty + c_2\rho$, where $\rho = \gamma_2^2$, the state feedback gain matrix can be constructed as $K = WX^{-1}$.

Proof: The first three LMIs in Theorem 7.3 correspond to *Statement 4* in Theorem 7.1, which are about H_2 performance. The fourth LMI in Theorem 7.3 corresponds to Lemma 7.1, which is about H_∞ performance. Therefore we just need to prove the fifth LMI in Theorem 7.3. The proof process is similar to the equivalence proof of (*C3*) and (*C4*) in Theorem 7.1.

System $S_1((\boldsymbol{A}+\boldsymbol{B}_1\boldsymbol{K}), \boldsymbol{B}_2, \boldsymbol{C}_2)$ represented by Eq. (7.29) and system $S_2((\boldsymbol{A}+\boldsymbol{B}_1\boldsymbol{K})^T, \boldsymbol{C}_2^T, \boldsymbol{B}_2^T)$ represented by Eq. (7.30) are dual systems. With transformation $((\boldsymbol{A}+\boldsymbol{B}_1\boldsymbol{K}), \boldsymbol{B}_2, \boldsymbol{C}_2) \rightarrow ((\boldsymbol{A}+\boldsymbol{B}_1\boldsymbol{K})^T, \boldsymbol{C}_2^T, \boldsymbol{B}_2^T)$, inequality (7.33) becomes

$$\boldsymbol{L} \otimes \boldsymbol{X} + \boldsymbol{M} \otimes (\boldsymbol{A}^T\boldsymbol{X}+\boldsymbol{K}^T\boldsymbol{B}_1^T\boldsymbol{X}) + \boldsymbol{M}^T \otimes (\boldsymbol{A}^T\boldsymbol{X}+\boldsymbol{K}^T\boldsymbol{B}_1^T\boldsymbol{X})^T < 0 \tag{7.34}$$

That is,

$$\boldsymbol{L} \otimes \boldsymbol{X} + \boldsymbol{M} \otimes (\boldsymbol{A}^T\boldsymbol{X}+\boldsymbol{K}^T\boldsymbol{B}_1^T\boldsymbol{X}) + \boldsymbol{M}^T \otimes (\boldsymbol{X}\boldsymbol{A}+\boldsymbol{X}\boldsymbol{B}_1\boldsymbol{K}) < 0 \tag{7.35}$$

By defining $\boldsymbol{R}=\boldsymbol{X}\boldsymbol{B}_1\boldsymbol{K}$, inequality (7.35) is turned into the fifth LMI in Theorem 3. This completes the proof.

Remark 7.3: The novel augmented LMI representations eliminate the coupling of Lyapunov matrix variables and system matrices by introducing slack variable $\boldsymbol{R}$ which provides additional degree of freedom. Additionally, the Lyapunov matrix variables $\boldsymbol{Q}$ and $\boldsymbol{W}$ are not the same. Theorem 7.3 gives a sufficient condition that guarantees the closed-loop system represented in Eq. (7.5) is robustly stable and satisfies H_2 and H_∞ norm bounds simultaneously. To reduce conservativeness, the Lyapunov variables introduced in H_2 and H_∞ performance constraint are different. That means the first Lyapunov variable $\boldsymbol{Q}$ responds to H_2 performance, and the second Lyapunov variable $\boldsymbol{W}$ responds to H_∞ performance.

7.4 Simulation test

To demonstrate the effective performance of the proposed method in this paper, one can choose the parameters of spacecraft and controllers similar to those in the Beginning of Life phase in [23], which are shown below.

The initial matrix

$$J = \begin{bmatrix} 74825 & -49 & -2348 \\ -49 & 127089 & -425 \\ -2348 & -425 & 99700 \end{bmatrix} \text{kg}\cdot\text{m}^2$$

The coupling coefficient matrices are given by

$$\boldsymbol{\Delta}_1 = \begin{bmatrix} 60.573 & 0.2415 \\ 10.205 & 210.630 \\ -295.675 & 11.6258 \end{bmatrix} \sqrt{\text{kg}}\cdot\text{m},\ \boldsymbol{\Delta}_2 = \begin{bmatrix} 42.8535 & 0.0001 \\ -0.0005 & 0.004 \\ 0.0001 & -44.058 \end{bmatrix} \sqrt{\text{kg}}\cdot\text{m}$$

The modal damping ratio $\xi_1 = \xi_2 = 0.015$, and the two modal frequency matrices are given by $\boldsymbol{\Omega}_1 = diag(0.02, 0.04)$, $\boldsymbol{\Omega}_2 = diag(0.15, 0.38)$. Meanwhile, external disturbance d is chosen as

$$\boldsymbol{d} = 5 \times 10^{-4} \begin{bmatrix} \sin(2\pi t/200) \\ \sin(2\pi t/200+\pi/4) \\ \sin(2\pi t/200+\pi/2) \end{bmatrix} \text{Nm}$$

We assume that the initial state $\boldsymbol{x}(0)=[0.3\times\pi/180,\ 0.2\times\pi/180,\ -0.3\times\pi/180,\ 1.2,\ 2.5,\ 0.12,\ 0.25,\ -0.03\times\pi/180,\ 0.02\times\pi/180,\ 0.02\times\pi/180,\ 0.15,\ 0.1,\ 0.015,\ 0.015]^T$, where the unit of attitude angle and angular velocity is rad and rad s^{-1}, respectively. The D domain required in poles assignment is defined as Fig. 7.3, where, q=0.5, r=3, α=0.

7.4.1 Simulation results using static output feedback controller

To correspond to [23], take roll axis for example and the numerical results using static output feedback controller are shown from Figs. 7.4 to 7.8. It is obvious that the attitude angle and angular velocity can converge to 0 second in 1500 seconds, and the control accuracy of attitude angle and

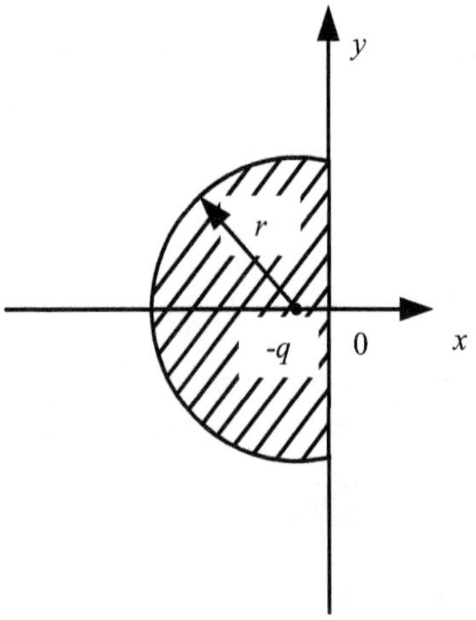

Figure 7.3 *D* domain in simulation.

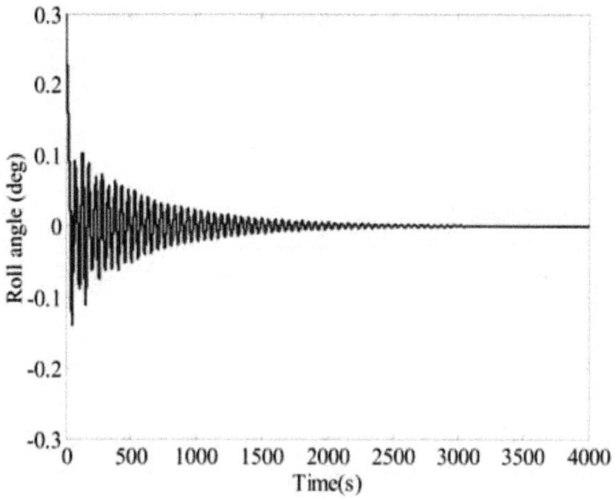

Figure 7.4 Roll angle using static output feedback controller.

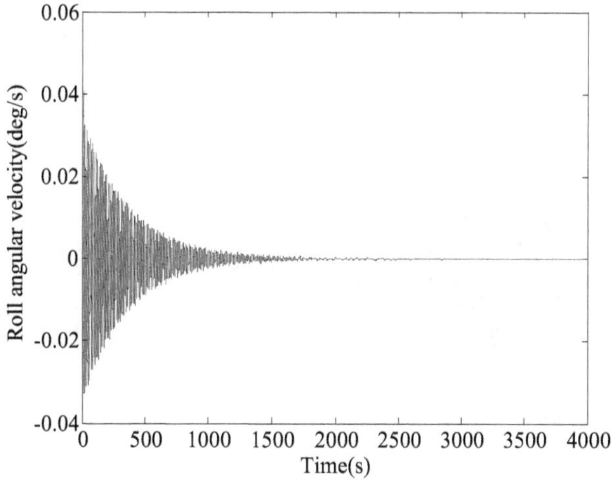

Figure 7.5 Roll angular velocity using static output feedback controller.

angular velocity is 0.005 degree and 0.001 degree second^{-1}, respectively. The modal coordinates of the two solar panels become smaller with increasing time. It suggests that the static output feedback controller can effectively suppress the vibration of solar panels. Specifically, the decreasing duration of the first modal coordinate from 3 to 0.1 requires 3×10^4 seconds and the maximum control torque is 6 Nm. For attitude stabilization of such a flexible spacecraft, a control moment gyroscope, instead of flywheel, should be used.

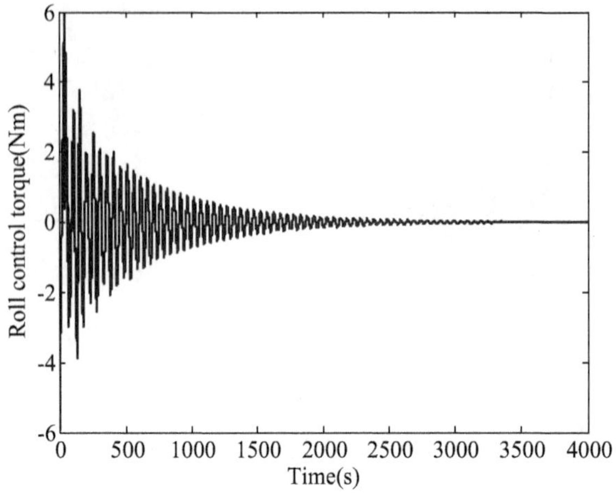

Figure 7.6 Roll control torque using static output feedback controller.

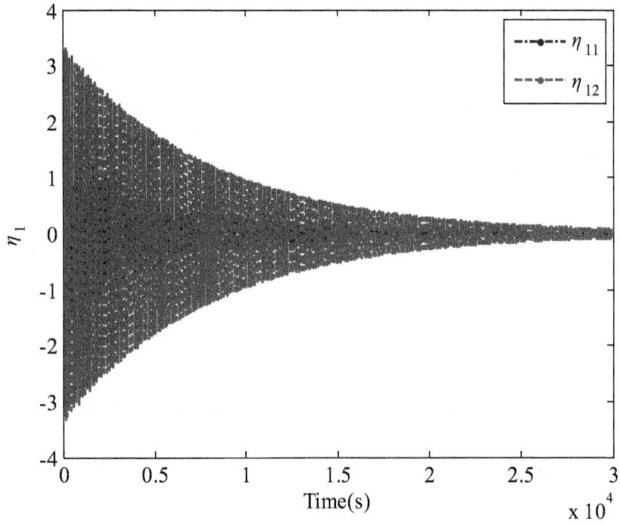

Figure 7.7 Modal coordinate of the first solar panel using static output feedback controller.

7.4.2 Simulation results using improved mixed H_2/H_∞ controller

To make comparisons, the proposed method presented in Theorem 7.2 is used. It can be obtained that the upper bounds of H_2 norm and H_∞

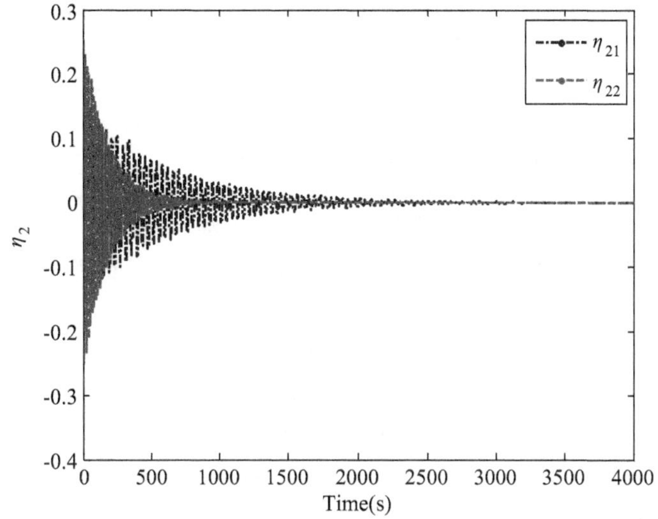

Figure 7.8 Modal coordinate of the second solar panel using static output feedback controller.

norm are γ_2=2 and γ_∞=0.002, respectively. The controller gain can be obtained as

$$\boldsymbol{K} = \begin{bmatrix} \boldsymbol{K}_1 & \boldsymbol{K}_2 & \boldsymbol{K}_3 & \boldsymbol{K}_4 \end{bmatrix}$$

$$\boldsymbol{K}_1 = \begin{bmatrix} -0.1750 & -7.3444 \times 10^{-10} & 1.8623 \times 10^{-7} \\ -6.9161 \times 10^{-10} & -0.1750 & 2.8234 \times 10^{-9} \\ 1.8692 \times 10^{-7} & 3.0000 \times 10^{-9} & -0.1750 \end{bmatrix},$$

$$\boldsymbol{K}_2 = \begin{bmatrix} -7.0290 \times 10^{-6} & 8.0328 \times 10^{-7} & -3.2502 \times 10^{-5} & -1.6883 \times 10^{-4} \\ -1.3362 \times 10^{-7} & -5.5011 \times 10^{-6} & -3.5155 \times 10^{-7} & -2.9618 \times 10^{-6} \\ 2.8781 \times 10^{-5} & -3.7877 \times 10^{-6} & 4.4221 \times 10^{-5} & 7.6340 \times 10^{-4} \end{bmatrix}$$

$$\boldsymbol{K}_3 = \begin{bmatrix} -0.8424 & -6.1726 \times 10^{-9} & 1.1647 \times 10^{-6} \\ -6.6313 \times 10^{-9} & -0.8424 & 2.2653 \times 10^{-8} \\ 1.1399 \times 10^{-6} & 2.0372 \times 10^{-8} & -0.8424 \end{bmatrix},$$

$$\boldsymbol{K}_2 = \begin{bmatrix} -3.5776 \times 10^{-5} & 4.9612 \times 10^{-6} & -1.2456 \times 10^{-4} & -0.0011 \\ -2.5918 \times 10^{-6} & -1.7672 \times 10^{-5} & -8.9623 \times 10^{-7} & -1.4878 \times 10^{-5} \\ 1.5059 \times 10^{-4} & -2.2134 \times 10^{-5} & 2.0275 \times 10^{-4} & 0.0047 \end{bmatrix}$$

Six open-loop poles are zero, which may influence the stability of the system; while the corresponding closed-loop poles are computed to be -0.4702, -0.4702, -0.4702, -0.3722, -0.3722, -0.3722, -0.0084 ± 0.4330i,

$-0.0024 \pm 0.1529i$, $-0.0035 \pm 0.0690i$, $-0.0009 \pm 0.0495i$, which are indeed in the LMI region D, as can be seen in Fig. 7.9.

The attitude angle, angular velocity, and corresponding control input including yaw, pitch and roll are shown from Figs. 7.10 to 7.12. Under the premise that the control accuracy of attitude angle is 0.005 degree, the convergence time of attitude angle is one-third of the original with approximately 500 seconds. When the control accuracy of angular velocity is 0.001 degree second^{-1}, the convergence time of angular velocity is

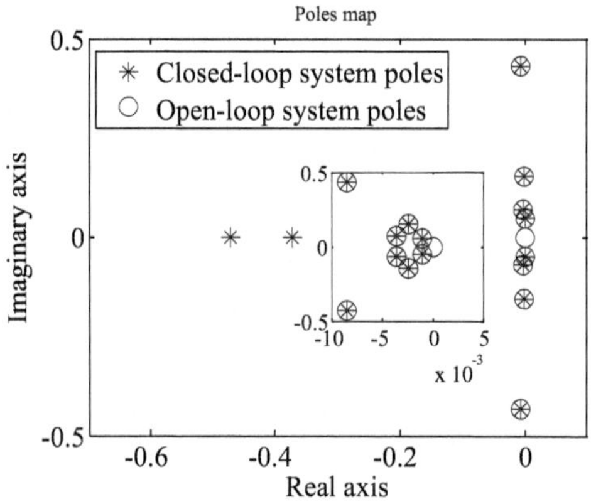

Figure 7.9 Poles assignment map.

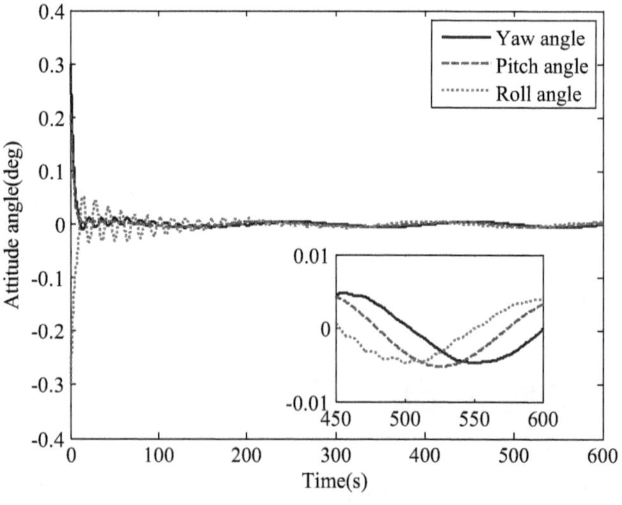

Figure 7.10 Attitude angle using an improved mixed H_2/H_∞ controller.

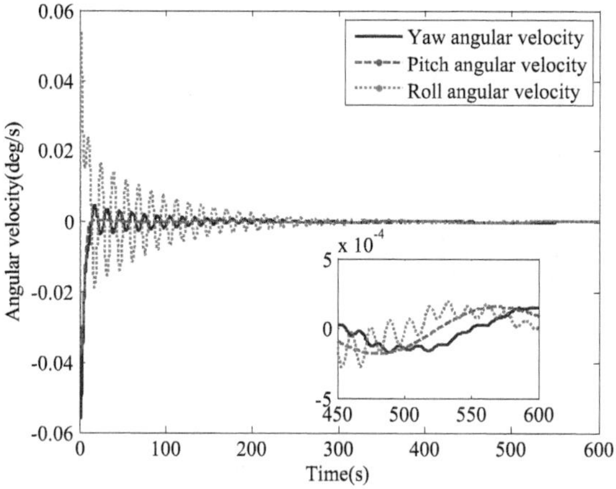

Figure 7.11 Angular velocity using an improved mixed H_2/H_∞ controller.

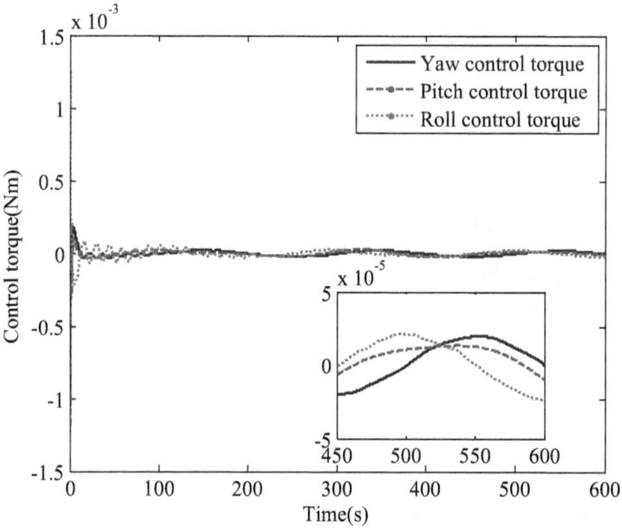

Figure 7.12 Control torque using an improved mixed H_2/H_∞ controller

approximately 135 seconds. Furthermore, the maximum control torque is much smaller than the original with no more than 0.0015 Nm. This phenomenon suggests that the actuator requirements have been reduced greatly, and flywheel control is sufficient for attitude stabilization of such a flexible spacecraft. In addition, the modal coordinates of the two solar panels are shown in Figs. 7.13 to 7.14. From the two figures, it can be observed that the modal coordinates become smaller over longer

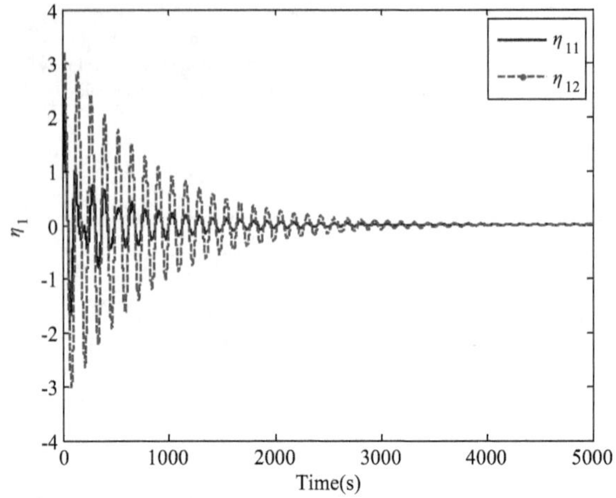

Figure 7.13 Modal coordinate of the first solar panel using an improved mixed H_2/H_∞ controller.

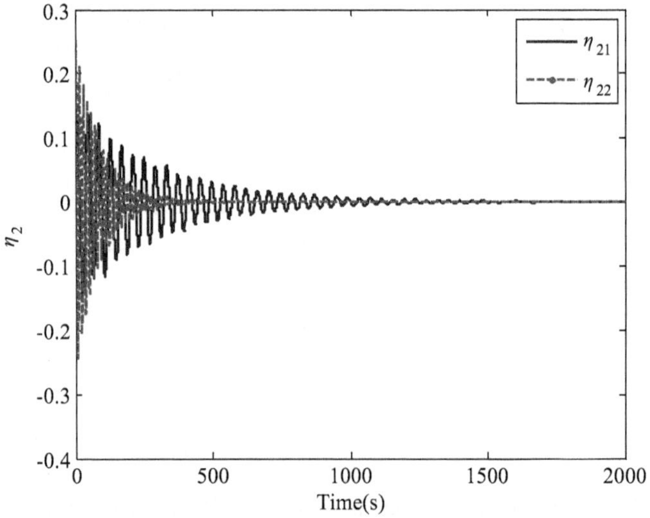

Figure 7.14 Modal coordinate of the second solar panel using an improved mixed H_2/H_∞ controller.

durations. This phenomenon illustrates that the improved mixed H_2/H_∞ controller with poles assignment can effectively suppress the vibration of the solar panels. Besides, the decreasing duration time of the first modal coordinate from 3 to 0.1 requires no more than 4000 seconds, which implies that it has reduced to two-fifteenths of the original.

7.4.3 Simulation results using a traditional mixed H_2/H_∞ controller

To verify the performance of the improved mixed H_2/H_∞ controller, the simulation results using the traditional mixed H_2/H_∞ controller presented in Lemma 7.3 are also performed. It can be obtained that the upper bounds of H_2 norm and H_∞ norm are γ_2=1.732 and γ_∞=0.001, respectively. The controller gain can be obtained as:

$$\boldsymbol{K}=\begin{bmatrix}\boldsymbol{K}_1 & \boldsymbol{K}_2 & \boldsymbol{K}_3 & \boldsymbol{K}_4\end{bmatrix}$$

$$\boldsymbol{K}_1=\begin{bmatrix} -0.1715 & -2.1269\times10^{-10} & 2.6639\times10^{-8} \\ -2.5006\times10^{-10} & -0.1715 & 6.4260\times10^{-10} \\ 2.4394\times10^{-8} & 4.5749\times10^{-10} & -0.1715 \end{bmatrix},$$

$$\boldsymbol{K}_2=\begin{bmatrix} -9.5626\times10^{-6} & 6.5073\times10^{-7} & -1.1823\times10^{-4} & -1.6770\times10^{-4} \\ -3.3416\times10^{-7} & -1.3316\times10^{-5} & -1.2101\times10^{-6} & -3.1607\times10^{-6} \\ 3.7219\times10^{-5} & -3.6442\times10^{-6} & 1.6196\times10^{-4} & 7.7487\times10^{-4} \end{bmatrix}$$

$$\boldsymbol{K}_3=\begin{bmatrix} -0.8309 & -1.5252\times10^{-9} & 1.8888\times10^{-7} \\ -1.7537\times10^{-9} & -0.8309 & 5.3471\times10^{-9} \\ 1.5926\times10^{-7} & 3.9582\times10^{-9} & -0.8309 \end{bmatrix},$$

$$\boldsymbol{K}_4=\begin{bmatrix} -5.8201\times10^{-5} & 8.7200\times10^{-6} & -5.8785\times10^{-4} & -8.3881\times10^{-4} \\ -8.9833\times10^{-6} & -6.7488\times10^{-5} & -3.2711\times10^{-6} & -4.2964\times10^{-6} \\ 2.2926\times10^{-4} & -4.0993\times10^{-5} & 8.5410\times10^{-4} & 0.0039 \end{bmatrix}$$

The attitude angle, angular velocity, and corresponding control input including yaw, pitch, and roll are shown from Figs. 7.15 to 7.17. The

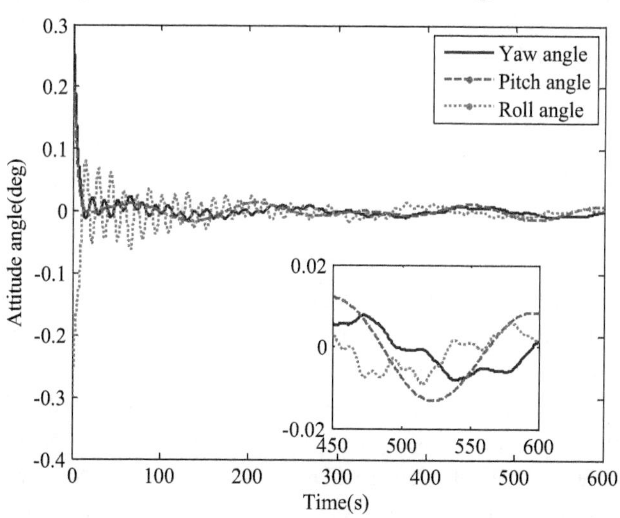

Figure 7.15 Attitude angle using traditional mixed H_2/H_∞ controller.

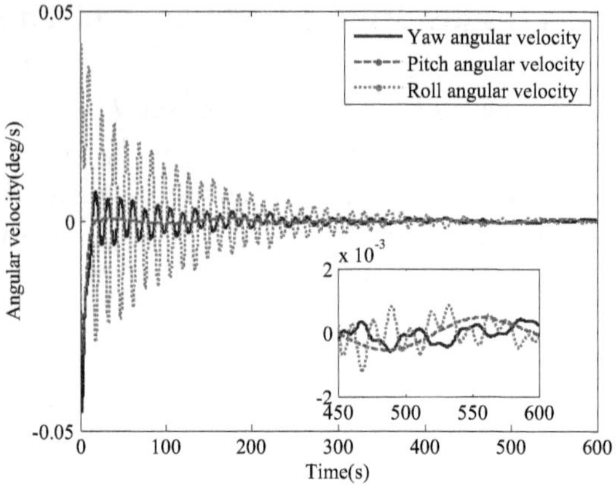

Figure 7.16 Angular velocity using traditional mixed H_2/H_∞ controller.

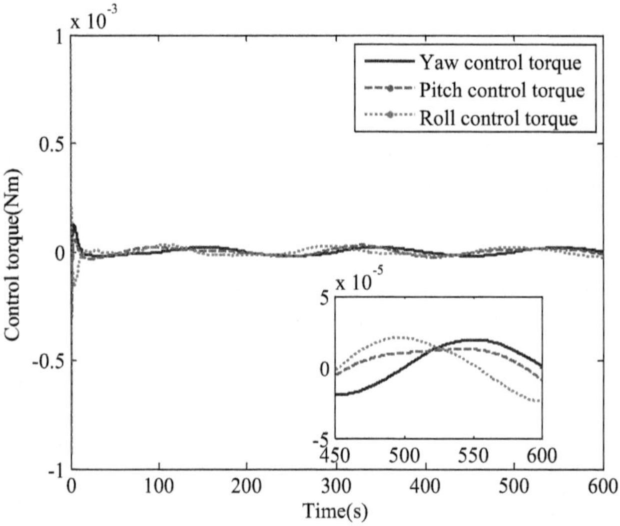

Figure 7.17 Control torque using traditional mixed H_2/H_∞ controller.

modal coordinates of the two solar panels are shown in Figs. 7.18 to 7.19. From the simulation figures, it is obvious that the modal coordinates become smaller with extended duration. It illustrates that the traditional mixed H_2/H_∞ controller can also effectively suppress the vibration of the solar panels. However, under the premise that the control accuracy of attitude angle is 0.005 degree, the robust stability of attitude system cannot be guaranteed. In this case, discussing the maximum of control torque

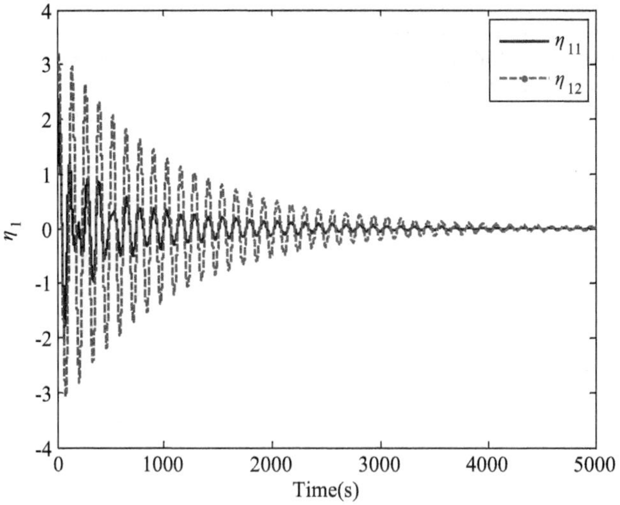

Figure 7.18 Modal coordinate of the first solar panel using traditional mixed H_2/H_∞ controller.

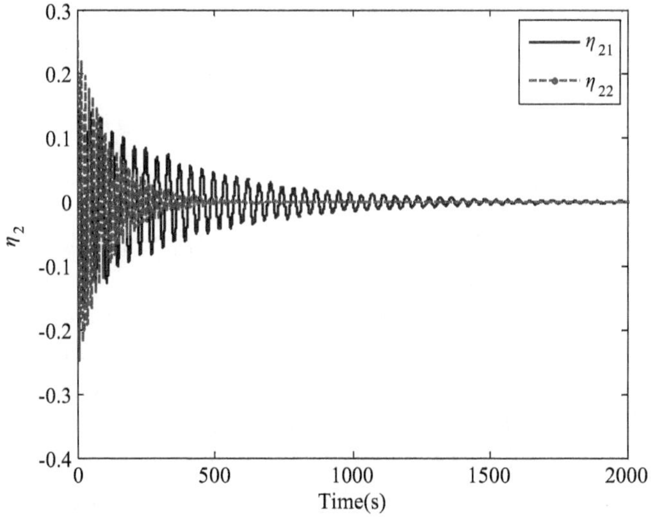

Figure 7.19 Modal coordinate of the second solar panel using traditional mixed H_2/H_∞ controller.

and other performance indices does not provide value. This also shows that the robust performance of improved mixed H_2/H_∞ controller under poles assignment constraint is much better.

Table 7.1 Comparison results.

Controller type	Stabilization time of the flexible spacecraft (s)	Maximum of control torque (Nm)	Decreasing duration time of the first modal coordinate (s)
Static output feedback controller	1500	6	3×10^4
Traditional mixed H_2/H_∞ controller	/	0.001	4000
Improved mixed H_2/H_∞ controller	500	0.0015	4000

7.4.4 Comparison analysis using different controllers

If we choose three comparison indices satisfying given accuracy including the stabilization time of the flexible spacecraft, maximum of control torque, and decreasing duration time of the first modal coordinate, the comparative results of the above three controllers can be illustrated in Table 7.1.

7.5 Conclusions

This chapter proposes the improved mixed H_2/H_∞ controller under poles assignment constraint which not only suppresses flexible vibration and external disturbances but also causes the spacecraft attitude to stabilize with high precision. The spacecraft attitude dynamics is firstly described and presented as a Lagrange-like model. Consequently, the improved mixed H_2/H_∞ controller under poles assignment constraint based on LMIs is designed to achieve attitude stabilization and vibration suppression in the presence of external disturbances. We confirm that the existing output feedback control method can achieve vibration suppression and robustness to external disturbance to obtain flexible spacecraft stabilization. However, we believe that the improved mixed H_2/H_∞ controller under poles assignment constraint presented in this chapter will lead to better results. Compared with the traditional mixed H_2/H_∞ controller, the improved controller shows better robustness. Besides, the improved controller can also reduce conservativeness by introducing different Lyapunov matrix variables in the H_2 and H_∞ performance constraint. The comparative results also indicate the greater advantage of the proposed method introduced in this chapter.

References

[1] Sakthivel R, Selvi S, Mathiyalagan K. Fault-tolerant sampled-data control of flexible spacecraft with probabilistic time delays. Nonlinear Dynamics 2015;79(3):1835–46.

[2] Song G, Agrawal BN. Vibration suppression of flexible spacecraft during attitude control. Acta Astronautica 2001;49:73–83.

[3] Singh SN, Zhang R. Adaptive output feedback control of spacecraft with flexible appendages by modeling error compensation. Acta Astronautica 2004;54(4):229–43.

[4] Lee KW, Singh SN. L_1 adaptive control of flexible spacecraft despite disturbances. Acta Astronautica 2012;80(80):24–35.

[5] Maganti GB, Singh SN. Simplified adaptive control of an orbiting flexible spacecraft. Acta Astronautica 2007;61(7–8):575–89.

[6] Zheng J, Banks SP, Alleyne H. Optimal attitude control for three-axis stabilized flexible spacecraft. Acta Astronautica 2005;56(5):519–28.

[7] Farrenkopf RL. Optimal open loop maneuver profiles for flexible spacecraft. Journal of Guidance Control & Dynamics 2012;2:272–80.

[8] Erdong J, Zhaowei S. Robust attitude tracking control of flexible spacecraft for achieving globally asymptotic stability. International Journal of Robust and Nonlinear Control 2009;19(11):1201–23.

[9] Xiao B, Hu Q, Zhang Y. Adaptive sliding mode fault tolerant attitude tracking control for flexible spacecraft under actuator saturation. IEEE Transactions on Control Systems Technology 2011;20(6):1605–12.

[10] Cao L, Chen X, Misra AK. Minimum sliding mode error feedback control for fault tolerant reconfigurable satellite formations with J2 perturbations. Acta Astronautica 2014;96:201–16.

[11] Cao L, Li X, Chen X, et al. Minimum sliding mode error feedback control for fault tolerant small satellite attitude control. Advances in Space Research 2014;53(2):309–24.

[12] Cao L, Chen X. Minimum sliding mode error feedback control and observer for satellite attitude control and determination. Proceedings of the Institution of Mechanical Engineers, Part G: Journal of Aerospace Engineering 2015;229(12):2242–57.

[13] Iyer A, Singh SN. Variable structure slewing control and vibration damping of flexible spacecraft. Acta Astronautica 1991;25(1):1–9.

[14] Du H, Li S. Attitude synchronization control for a group of flexible spacecraft. Automatica 2014;50(2):646–51.

[15] Cheng CH, Shu SL, Cheng PJ. Attitude control of a satellite using fuzzy controllers. Expert Systems with Applications 2009;36(3):6613–20.

[16] Won CH. Comparative study of various control methods for attitude control of a LEO satellite. Aerospace Science and Technology 1999;3(5):323–33.

[17] Chen BS, Wu CS, Jan YW. Adaptive fuzzy mixed H_2/H_∞ attitude control of spacecraft. IEEE Transactions on Aerospace and Electronic Systems 2000;36(4):1343–59.

[18] Yang CD, Sun YP. Mixed H_2/H_∞ state-feedback design for microsatellite attitude control. Control Engineering Practice 2002;10(9):951–70.

[19] Wu CS, Chen BS. Attitude control of spacecraft: mixed H_2/H_∞ approach. Journal of Guidance Control & Dynamics 2001;24(4):755–66.

[20] Tseng GT, Mahn RH. Flexible spacecraft control design using pole allocation technique. Journal of Guidance Control, and Dynamics 1978;1:279–81.
[21] Reddy AS SR, Bainum PM, Krishnal R, et al. Control of a large flexible platform in orbit. Journal of Guidance & Control 2012;4(6):642–9.
[22] Meirovitch L, Quinn RD. Maneuvering and vibration control of flexible spacecraft. Journal of Astronautical Sciences 1987;35:301–28.
[23] Wu S, Wen S. Robust H_∞ output feedback control for attitude stabilization of a flexible spacecraft. Nonlinear Dynamics 2016;84(1):405–12.
[24] Kida T, Yamaguchi I, Chida Y, et al. On-orbit robust control experiment of flexible spacecraft ETS-VI. Journal of Guidance, Control, and Dynamics 1997;20(5):865–72.
[25] Liu C, Sun Z, Shi K, et al. Mixed H_2/H_∞ control approach and its application in satellite attitude control system. International Journal of Engineering Research in Africa 2016;25:89–97.
[26] Apkarian P, Tuan HD, Bernussou J. Analysis, eigenstructure assignment and H_2 multichannel synthesis with enhanced LMI characterizations. Proceedings of the 39th IEEE Conference on Decision and Control, 2000: 1489–1494.
[27] Chilali M, Gahinet P. H_∞ design with pole placement constraints: an LMI approach. IEEE Transactions on Automatic Control 1996;41(3):358–67.

CHAPTER 8

Nonfragile H_∞ control with input constraints

8.1 Introduction

To minimize the launch costs of spacecraft, their weight is minimized, frequently leading to large, low weight flexible appendages and payloads such as deployable antennas and solar panels [1]. The interactions of elastic motions with spacecraft attitude together with increasingly stringent requirements on attitude performance require that active vibration control be implemented for spacecraft maneuvers and stabilization. In addition, spacecraft models on which control is based are often inaccurately known, and this is particularly the case for system uncertainties, that is, unstructured uncertainties resulting from deliberate neglect of actuator and sensor dynamics and higher-order structural modes, and the structured uncertainties from inaccuracies in the system modal frequencies and damping due to errors in the material properties, structural modeling, mass, and especially damping. This, of course, greatly influences the performance of attitude stabilization and vibration suppression schemes [2]. Complicating the control problem, in actual system operation due to word length limit, conversion accuracy, truncation error in numerical computations and other reasons, controller gain perturbations often appear [3]. Although any controller necessarily must tolerate small uncertainties in its parameters, actuator structural perturbations, which can also be regarded as controller perturbations, often occur during maneuvers, and may result in unsatisfactory performance or even instability [4,5]. Consequently, the development of nonfragile controllers with capacity for adjustment of their parameters is of current research interest [6,7]. As well, flexible spacecraft are often required to achieve high-precision pointing, rapid attitude maneuvers, and stabilization in the presence of external disturbances, performance constraints, and even input constraints [8,9]. Together, these challenges compound the control difficulties when one strives for high performance for attitude stabilization and vibration suppression of flexible spacecraft.

Spacecraft Attitude Control. DOI: https://doi.org/10.1016/B978-0-323-99005-9.00008-0

There have been many approaches for the attitude control of spacecraft over the past decades, including sliding mode control [10,11], adaptive control [12,13], H_∞ control [14], disturbance-observer based control [15,16], etc. Although these methods have demonstrated satisfactory results for some situations, there are few actual results which deal concurrently with the issues mentioned above. The majority of the existing literature is focused on rigid spacecraft assuming that the uncertainties are constant, not time-varying [17,18]. However, a growing number of flexible spacecraft are equipped with rapidly deployable appendages which, when coupled with rapid translational or rotational maneuvers, may cause appreciable time-varying changes in the model parameters. Thus time-varying model parameter uncertainty must be considered to achieve very precise and stable performance [19,20].

One approach for the suppression of flexible oscillations induced by spacecraft maneuvers is through the use of piezoelectric materials [21–24]. These actuators consist of films of piezoelectric material bonded to the surface of flexible appendages. With their inherently distributed nature, piezoelectric actuators seem, in some ways, more natural candidates for the elimination of undesirable vibrations than more localized actuators. In [21], a pulse width pulse frequency modulator and smart materials with positive position feedback were used for active vibration suppression for flexible spacecraft, where embedded piezoelectric ceramic patches were used as both sensors and actuators to detect and suppress vibrations. Ref. [22] presented an active vibration suppression-based dynamic controller with distributed piezoelectric actuators to attenuate the effects of disturbances and/or parameter variations, and damp out undesirable vibrations, but its main drawback was high energy consumption because the input control torque was more than 20 Nm with a saturation at 30 Nm. This will be compared with the active vibration suppression scheme for high natural frequencies in this chapter below. In [23], a nonlinear controller using a feedback linearization method and piezoelectric actuators was proposed to suppress panel flutter, but this work did not consider the uncertainties in an actual system. In addition to the above active vibration suppression schemes, a passive vibration suppression method using shunted piezoelectric transducers was presented in [24], which consisted of connecting piezoelectric transducers bonded to the flexible elements to electric circuits in such a way that the vibration energy, once converted into electrical energy, was transferred and partially dissipated into the electric circuit. However, some piezoelectric materials of piezoelectric transducers require moisture-proof measures, and the DC response of the output is poor, such that high input impedance circuits or charge amplifiers are required. Input-shaping techniques are also effective for reducing the residual vibrations of flexible systems [25]. Based on input-shaping technology, a control scheme for a slewing flexible spacecraft equipped with ON–OFF actuators was proposed in [26] to suppress deflections and eliminate residual oscillations. However, this technique cannot guarantee the convergence of modal variables, which may worsen the steady-state performance of an attitude control system. Another well-known method, H_∞ control, can guarantee robust stability in the presence of parameter uncertainties and external disturbances, and it has been exploited in much of the literature [27]. Refs. [28,29] made a detailed investigation of nonfragile controller design in terms of additive and multiplicative perturbations for rigid spacecraft attitude control systems, but did not consider flexible appendages and confined itself to restrictive cases compared with the controller presented below. Further complicating things, actuators for attitude control of flexible spacecraft often include mechanical and electrical components, thrusters, and environmental force actuators which must be accounted for, and input constraints are another critical issue that must be dealt with in controller design as pointed out in [30].

Despite the partial successes of the methods described above, high-precision control of flexible spacecraft remains challenging because the spacecraft dynamics and flexible dynamics are complicated and intrinsically nonlinear, and subject to various sources of uncertainties and disturbances. Thus it is clear that the problem of attitude stabilization and vibration suppression requires the development of correspondingly advanced control design strategies to deal with model parameter uncertainty, external disturbances, performance constraints, controller perturbations, and input constraints concurrently.

Motivated by the issues of the preceding discussion, this chapter introduces a novel nonfragile H_∞ control scheme which accounts for these problems simultaneously. In the chapter, two vibration suppression schemes, that is, passive vibration suppression and active vibration suppression, are employed to convert the dynamics model of flexible spacecraft into the same state-space form except for a few differences in two matrices of coefficients, which is convenient for controller design. Next, a Bernoulli random variable is introduced and its expectation is exploited for controller construction, so that this controller can tolerate uncertainties in the controller itself and is more general than traditional controllers.

Following this, a nonfragile H_∞ control scheme is developed that can simultaneously achieve attitude stabilization and vibration suppression with high-precision and rapid response in spite of the simultaneous presence of the issues mentioned previously. Simulations then provide some guidance for the choice of the passive or active suppression scheme for attitude stabilization and control in practice.

The chapter is organized as follows. Section 8.2 presents a general attitude model of flexible spacecraft and converts it into a state-space form in terms of passive and active vibration suppression schemes, and the control objective of this chapter is formulated taking system uncertainties and input constraint into account. Section 8.3 develops the design method of the nonfragile controller in terms of additive and multiplicative perturbations, and a stability analysis of the closed-loop system is performed via a Lyapunov approach. Numerical simulations of the above two cases are performed and comparisons between active and passive vibration suppression schemes are made for high and low natural frequencies in Section 8.4. Finally, conclusions are presented in Section 8.5.

8.2 Problem formulation

8.2.1 Attitude system description of flexible spacecraft

This section develops an attitude dynamic model of a flexible spacecraft which is a rigid body with two flexible appendages such as solar panels, antennas, or any other flexible structures. This model is nonlinear and includes rigid and flexible mode interactions. We define a local-vertical local-horizontal frame $F_o(X_o, Y_o, Z_o)$ as the frame of reference, with the origin as the mass center of the spacecraft. The axes of this reference frame are: roll axis X_o in the flight direction, pitch axis Y_o perpendicular to the orbital plane, and yaw axis Z_o pointing towards the center of the Earth. A body reference frame $F_B(X_B, Y_B, Z_B)$ is defined with origin at the mass center and axes fixed in the spacecraft body so as to coincide with the principal axes of inertia.

Then the spacecraft body and flexible dynamics can be described by the following differential equations [31,32]:

$$\begin{cases} \boldsymbol{J}\dot{\boldsymbol{\omega}} + \boldsymbol{\omega}^{\times}\left(\boldsymbol{J}\boldsymbol{\omega} + \tilde{\boldsymbol{\Delta}}\dot{\tilde{\eta}}\right) + \tilde{\boldsymbol{\Delta}}\ddot{\tilde{\eta}} = \boldsymbol{T}_c + \boldsymbol{T}_d \\ \ddot{\tilde{\eta}} + \tilde{\boldsymbol{C}}\dot{\tilde{\eta}} + \tilde{\boldsymbol{K}}\tilde{\eta} + \tilde{\boldsymbol{\Delta}}^T\dot{\boldsymbol{\omega}} = -\tilde{\boldsymbol{\Delta}}_p u_p \end{cases} \tag{8.1}$$

where $\boldsymbol{J} \in \mathbb{R}^{3\times3}$ is the inertia matrix; $\boldsymbol{\omega} = [\omega_x \quad \omega_y \quad \omega_z]^T$ is the angular velocity vector including roll, pitch and yaw angular rates of Eq. (8.2); $\tilde{\boldsymbol{\Delta}} \in \mathbb{R}^{3\times m}$ is the coupling coefficient matrix between the rigid and elastic structures; $\tilde{\boldsymbol{\eta}} \in \mathbb{R}^m$ is the modal coordinate vector relative to the main body; $\boldsymbol{T}_c \in \mathbb{R}^3$ and $\boldsymbol{T}_d \in \mathbb{R}^3$ denote the control input torque and external disturbance torque; $\tilde{\boldsymbol{C}} = \mathrm{diag}\{[2\tilde{\xi}_1\Omega_1 \quad 2\tilde{\xi}_2\Omega_2 \quad \cdots \quad 2\tilde{\xi}_m\Omega_m]\} \in \mathbb{R}^{m\times m}$ is the modal damping matrix, where $\tilde{\xi}_i, i = 1, 2, \ldots, m$ and $\Omega_i, i = 1, 2, \ldots, m$ are the damping ratios and the natural frequencies; $\tilde{\boldsymbol{K}} = \mathrm{diag}\{[\Omega_1^2 \quad \Omega_2^2 \quad \cdots \quad \Omega_m^2]\} \in \mathbb{R}^{m\times m}$ is the stiffness matrix, and m is the number of elastic modes considered. Here, $\boldsymbol{T}_d$ includes gravity-gradient torque, solar radiation pressure torque and aerodynamic torque; u_p is the piezoelectric input and $\tilde{\boldsymbol{\Delta}}_p$ is the corresponding coupling matrix.

The kinematics of such a flexible spacecraft can be described by:

$$\begin{bmatrix} \omega_x \\ \omega_y \\ \omega_z \end{bmatrix} = \begin{bmatrix} 1 & 0 & -\sin\theta \\ 0 & \cos\phi & \sin\phi\cos\theta \\ 0 & -\sin\phi & \cos\phi\cos\theta \end{bmatrix} \begin{bmatrix} \dot{\phi} \\ \dot{\theta} \\ \dot{\psi} \end{bmatrix} - \omega_0 \begin{bmatrix} \cos\theta\sin\psi \\ \sin\phi\sin\theta\sin\psi + \cos\phi\cos\psi \\ \cos\phi\sin\theta\sin\psi - \sin\phi\cos\psi \end{bmatrix} \tag{8.2}$$

where ω_x, ω_y, and ω_z denote the three components of angular velocity; ϕ, θ, and ψ denote the three components of attitude angle and ω_0 is the orbital frequency.

Defining an auxiliary variable $\boldsymbol{q} = \dot{\tilde{\boldsymbol{\eta}}} + \tilde{\boldsymbol{\Delta}}^T\boldsymbol{\omega}$, one has

$$\dot{\boldsymbol{q}} = \ddot{\tilde{\boldsymbol{\eta}}} + \tilde{\boldsymbol{\Delta}}^T\dot{\boldsymbol{\omega}} = -\tilde{\boldsymbol{C}}\boldsymbol{q} + \tilde{\boldsymbol{C}}\tilde{\boldsymbol{\Delta}}^T\boldsymbol{\omega} - \tilde{\boldsymbol{K}}\tilde{\boldsymbol{\eta}} - \tilde{\boldsymbol{\Delta}}_p u_p \tag{8.3}$$

So that the first term of Eq. (8.1) becomes

$$\left(\boldsymbol{J} - \tilde{\boldsymbol{\Delta}}\tilde{\boldsymbol{\Delta}}^T\right)\dot{\boldsymbol{\omega}} + \boldsymbol{\omega}^\times\left(\left(\boldsymbol{J} - \tilde{\boldsymbol{\Delta}}\tilde{\boldsymbol{\Delta}}^T\right)\boldsymbol{\omega} + \tilde{\boldsymbol{\Delta}}\boldsymbol{q}\right)$$

$$+\tilde{\boldsymbol{\Delta}}\left(-\tilde{\boldsymbol{C}}\boldsymbol{q} + \tilde{\boldsymbol{C}}\tilde{\boldsymbol{\Delta}}^T\boldsymbol{\omega} - \tilde{\boldsymbol{K}}\tilde{\boldsymbol{\eta}} - \tilde{\boldsymbol{\Delta}}_p u_p\right) = \boldsymbol{T}_c + \boldsymbol{T}_d \tag{8.4}$$

For small Euler angles, Eq. (8.2) is approximately

$$\dot{\boldsymbol{\Theta}} = \boldsymbol{D}_1\boldsymbol{\Theta} + \boldsymbol{\omega} + \boldsymbol{D}_2$$

where,

$$\boldsymbol{\Theta} = \begin{bmatrix} \phi \\ \theta \\ \psi \end{bmatrix}, \boldsymbol{D}_1 = \begin{bmatrix} 0 & 0 & \omega_0 \\ 0 & 0 & 0 \\ -\omega_0 & 0 & 0 \end{bmatrix}, \boldsymbol{D}_2 = \begin{bmatrix} 0 \\ \omega_0 \\ 0 \end{bmatrix}$$

At very high altitudes, the influence of orbital frequency can be neglected and Eq. (8.2) can be further simplified to Eq. (8.5) below

$$\dot{\boldsymbol{\Theta}} = \boldsymbol{\omega} \tag{8.5}$$

Denoting $\boldsymbol{J}_0 = \boldsymbol{J} - \tilde{\boldsymbol{\Delta}}\tilde{\boldsymbol{\Delta}}^T$, with $\tilde{\boldsymbol{\Delta}}\tilde{\boldsymbol{\Delta}}^T$ as the contribution of the flexible parts to the total inertia matrix, allows Eqs. (8.3)-(8.5) to be converted into the following state-space form:

$$\begin{cases} \dot{\boldsymbol{x}}(t) = \boldsymbol{A}\boldsymbol{x}(t) + \boldsymbol{B}_1\boldsymbol{u}(t) + \boldsymbol{B}_2\boldsymbol{w}(t) \\ \boldsymbol{y}(t) = \boldsymbol{C}_1\boldsymbol{x}(t) \end{cases} \tag{8.6}$$

8.2.2 Passive and active vibration suppression cases

The following two cases for active and passive vibration suppression (details below) are of interest in the sequel:

Case 1: When the passive vibration suppression scheme is used, $u_p = 0$, and the quantities in Eq. (8.6) are defined as follows:

$\boldsymbol{x}(t) = \left[\boldsymbol{\Theta}^T \quad \boldsymbol{\omega}^T \quad \boldsymbol{q}^T \quad \tilde{\boldsymbol{\eta}}^T\right]^T$ is the state vector, $\boldsymbol{y}(t) = \left[\boldsymbol{\Theta}^T \quad \boldsymbol{\omega}^T \quad \boldsymbol{q}^T \quad \tilde{\boldsymbol{\eta}}^T\right]^T$ is the output vector, $\boldsymbol{u}(t) = \boldsymbol{T}_c$ is the control input torque. Because $-\boldsymbol{\omega}^\times\left(\boldsymbol{J}_0\boldsymbol{\omega} + \tilde{\boldsymbol{\Delta}}\boldsymbol{q}\right)$ is bounded in a practical flexible spacecraft attitude system, it can be treated as a disturbance torque on the spacecraft so that the combined disturbance is

$$\boldsymbol{w}(t) = \boldsymbol{T}_d - \boldsymbol{\omega}^\times\left(\boldsymbol{J}_0\boldsymbol{\omega} + \tilde{\boldsymbol{\Delta}}\boldsymbol{q}\right) \tag{8.7}$$

The coefficient matrices are then

$$\boldsymbol{A} = \begin{bmatrix} 0 & \boldsymbol{I} & 0 & 0 \\ 0 & -\boldsymbol{J}_0^{-1}\tilde{\boldsymbol{\Delta}}\tilde{\boldsymbol{C}}\tilde{\boldsymbol{\Delta}}^T & \boldsymbol{J}_0^{-1}\tilde{\boldsymbol{\Delta}}\tilde{\boldsymbol{C}} & \boldsymbol{J}_0^{-1}\tilde{\boldsymbol{\Delta}}\tilde{\boldsymbol{K}} \\ 0 & \tilde{\boldsymbol{C}}\tilde{\boldsymbol{\Delta}}^T & -\tilde{\boldsymbol{C}} & -\tilde{\boldsymbol{K}} \\ 0 & -\tilde{\boldsymbol{\Delta}}^T & \boldsymbol{I} & 0 \end{bmatrix}, \boldsymbol{B}_1 = \boldsymbol{B}_2 = \begin{bmatrix} 0 \\ \boldsymbol{J}_0^{-1} \\ 0 \\ 0 \end{bmatrix}, \boldsymbol{C}_1 = \boldsymbol{I}_{6+2m}$$

Remark 8.1: The desired attitude motion represented by attitude angle and angular rate is bounded and differentiable in practice [33], such that the angular velocity $\boldsymbol{\omega}$ is bounded. For flexible spacecraft, the structural damping always exists in system, even if it may be small, such that the vibration energy can be dissipated with increasing time. Furthermore, some auxiliary materials or techniques, for example, piezoelectric actuators, are used to actively damp out the vibrations. Thus it is appropriate to assume that the magnitudes of the elastic displacement $\|\tilde{\eta}\|$ and its rate $\|\dot{\tilde{\eta}}\|$ are both bounded. For $q=\dot{\tilde{\eta}}+\tilde{\Delta}^T\omega$, it is easily concluded that q is bounded. Thus the boundedness of $-\omega^\times(J_0\omega+\tilde{\Delta}q)$ can be guaranteed. It should be noted that this assumption has been adopted in much literature [34,35].

Case 2: When the active vibration suppression scheme is used, $u_p \neq 0$. In this case, the piezoelectric input u_p is calculated by using a feedback with $u_p=F_1\tilde{\eta}+F_2q$; F_1 and F_2 are matrices introduced later, designed such that the eigenvalues of the controllable flexible dynamics are placed where desired. That this can be done is shown in the following brief development.

The second term of Eq. (8.1) is

$$
\begin{aligned}
\ddot{\tilde{\eta}} &= -\tilde{C}\dot{\tilde{\eta}}-\tilde{K}\tilde{\eta}-\tilde{\Delta}^T\dot{\omega}-\tilde{\Delta}_p u_p \\
&= -\tilde{C}\dot{\tilde{\eta}}-\tilde{K}\tilde{\eta}-\tilde{\Delta}^T\dot{\omega}-\tilde{\Delta}_p(F_1\tilde{\eta}+F_2(\dot{\tilde{\eta}}+\tilde{\Delta}^T\omega)) \\
&= -(\tilde{C}+\tilde{\Delta}_pF_2)\dot{\tilde{\eta}}-(\tilde{K}+\tilde{\Delta}_pF_1)\tilde{\eta}-\tilde{\Delta}^T\dot{\omega}-\tilde{\Delta}_p\tilde{\Delta}^T\omega
\end{aligned}
\tag{8.8}
$$

Once the attitude maneuver is completed, that is, $\omega\simeq 0$, $\dot{\omega}\simeq 0$, which can be achieved by Theorem 7.1 or Theorem 7.2 as shown below, Eq. (8.8) becomes

$$
\ddot{\tilde{\eta}} = -(\tilde{C}+\tilde{\Delta}_pF_2)\dot{\tilde{\eta}}-(\tilde{K}+\tilde{\Delta}_pF_1)\tilde{\eta} \tag{8.9}
$$

Defining an auxiliary variable $\tilde{x}=\begin{bmatrix}\tilde{\eta}^T & \dot{\tilde{\eta}}^T\end{bmatrix}^T$, one has

$$
\dot{\tilde{x}}=\tilde{A}\tilde{x} \tag{8.10}
$$

where,

$$
\tilde{A}=\begin{bmatrix} 0 & I \\ -(\tilde{K}+\tilde{\Delta}_pF_1) & -(\tilde{C}+\tilde{\Delta}_pF_2)\end{bmatrix}
$$

Choosing a Lyapunov function candidate as

$$
V(t)=\tilde{x}^T\tilde{P}\tilde{x} \tag{8.11}
$$

where $\tilde{P}$ is a symmetric positive definite matrix.

The time derivative of $V(t)$ along Eq. (8.10) yields

$$
\dot{V}(t)=\dot{\tilde{x}}^T\tilde{P}\tilde{x}+\tilde{x}^T\tilde{P}\dot{\tilde{x}}=\tilde{x}^T\left(\tilde{A}^T\tilde{P}+\tilde{P}\tilde{A}\right)\tilde{x}
$$

The system of Eq. (8.10) is Hurwitz stabilizable if

$$\tilde{A}^T\tilde{P}+\tilde{P}\tilde{A}<0 \tag{8.12}$$

By choosing an appropriate symmetric positive definite matrix $\tilde{\boldsymbol{P}}$, the matrices F_1 and F_2 can be obtained.

In this case, one has $u_p=\boldsymbol{F}_1\tilde{\eta}+\boldsymbol{F}_2\boldsymbol{q}$ so that Eq. (8.3) and Eq. (8.4) become the following two equations, respectively,

$$\dot{q}=\ddot{\tilde{\eta}}+\tilde{\boldsymbol{\Delta}}^T\dot{\omega}=\tilde{\boldsymbol{C}}\tilde{\boldsymbol{\Delta}}^T\omega-\left(\tilde{\boldsymbol{C}}+\tilde{\boldsymbol{\Delta}}_p\boldsymbol{F}_2\right)q-\left(\tilde{\boldsymbol{K}}+\tilde{\boldsymbol{\Delta}}_p\boldsymbol{F}_1\right)\tilde{\eta} \tag{8.13}$$

$$\left(\boldsymbol{J}-\tilde{\boldsymbol{\Delta}}\tilde{\boldsymbol{\Delta}}^T\right)\dot{\omega}+\omega^\times\left(\left(\boldsymbol{J}-\tilde{\boldsymbol{\Delta}}\tilde{\boldsymbol{\Delta}}^T\right)\omega+\tilde{\boldsymbol{\Delta}}q\right)$$

$$+\tilde{\boldsymbol{\Delta}}\left(-\tilde{\boldsymbol{C}}q+\tilde{\boldsymbol{C}}\tilde{\boldsymbol{\Delta}}^T\omega-\tilde{\boldsymbol{K}}\tilde{\eta}-\tilde{\boldsymbol{\Delta}}_p(\boldsymbol{F}_1\tilde{\eta}+\boldsymbol{F}_2\tilde{q})\right)=\boldsymbol{T}_c+\boldsymbol{T}_d \tag{8.14}$$

Denoting $\boldsymbol{J}_0=\boldsymbol{J}-\tilde{\boldsymbol{\Delta}}\tilde{\boldsymbol{\Delta}}^T$, with $\tilde{\boldsymbol{\Delta}}\tilde{\boldsymbol{\Delta}}^{\mathrm{T}}$ the contribution of the flexible portion to the total inertia matrix, allows Eqs. (8.5), (8.13), and (8.14) to be converted into the state-space form of Eq. (8.6).

where $\boldsymbol{x}(t)=\begin{bmatrix}\boldsymbol{\Theta}^{\mathrm{T}} & \omega^{\mathrm{T}} & q^{\mathrm{T}} & \tilde{\eta}^{\mathrm{T}}\end{bmatrix}^{\mathrm{T}}$ is the state vector, $\boldsymbol{y}(t)=\begin{bmatrix}\boldsymbol{\Theta}^T & \omega^T & q^T & \tilde{\eta}^T\end{bmatrix}$ is the output vector, $\boldsymbol{u}(t)=\boldsymbol{T}_c$ is the control input torque. Analogous to the analysis in Case 1, $\boldsymbol{w}(t)=\boldsymbol{T}_d-\omega^\times\left(\boldsymbol{J}_0\omega+\tilde{\boldsymbol{\Delta}}q\right)$ is the total combined disturbance. The coefficient matrices are

$$\boldsymbol{A}=\begin{bmatrix}0 & \boldsymbol{I} & 0 & 0\\ 0 & -\boldsymbol{J}_0^{-1}\tilde{\boldsymbol{\Delta}}\tilde{\boldsymbol{C}}\tilde{\boldsymbol{\Delta}}^T & \boldsymbol{J}_0^{-1}\tilde{\boldsymbol{\Delta}}\left(\tilde{\boldsymbol{C}}+\tilde{\boldsymbol{\Delta}}_p\boldsymbol{F}_2\right) & \boldsymbol{J}_0^{-1}\tilde{\boldsymbol{\Delta}}\left(\tilde{\boldsymbol{K}}+\tilde{\boldsymbol{\Delta}}_p\boldsymbol{F}_1\right)\\ 0 & \tilde{\boldsymbol{C}}\tilde{\boldsymbol{\Delta}}^T & -\left(\tilde{\boldsymbol{C}}+\tilde{\boldsymbol{\Delta}}_p\boldsymbol{F}_2\right) & -\left(\tilde{\boldsymbol{K}}+\tilde{\boldsymbol{\Delta}}_p\boldsymbol{F}_1\right)\\ 0 & -\tilde{\boldsymbol{\Delta}}^T & \boldsymbol{I} & 0\end{bmatrix},$$

$$\boldsymbol{B}_1=\boldsymbol{B}_2=\begin{bmatrix}0\\ \boldsymbol{J}_0^{-1}\\ 0\\ 0\end{bmatrix},\ \boldsymbol{C}_1=\boldsymbol{I}_{6+2m}$$

8.2.3 Brief introduction on piezoelectric actuators

If thin, homogeneous, and isotropic films of piezoelectric actuators are bonded to the flexible panel to provide input voltage u_p, the rigid ones represented by the first term of Eq. (8.1) will be influenced by means of $\tilde{\boldsymbol{\Delta}}\ddot{\eta}$ [22], more details about the physical properties of piezoelectric actuators are as follows.

Piezoelectric actuators are used to actively damp out the vibrations induced by attitude maneuvers. Fig. 8.1 shows a schematic drawing of a flexible panel, constituted by aluminum, whose deflection is to be controlled by the piezoelectric actuator bonded to the panel by bonding layer. The two interfacial nodes, that is, 1 and 2, can be subjected to external forces, moments, or both. Further, the degrees of freedom of each node can be limited to linear translations, angular rotations, or restrained completely depending on the nature of support at the node under consideration. In this chapter, the panel is assumed to have a rectangular cross-section of constant width. It is further assumed that its transverse deflection is due to the flexural action of the external forces and moments [36].

In Fig. 8.1, the piezoelectric actuator is shown bonded to the flexible panel to form a composite panel. When an electric field is applied across the film, it will expand if the field is aligned with the film polarization axis and it will contract if the two are opposed. The expansion or contraction of the film relative to the panel, by virtue of the piezoelectric effect, creates longitudinal bending stresses in the composite panel which tend to bend the panel in a manner very similar to a bimetallic thermostat.

With proper selection, placement, and control of the actuator, it would be possible to generate enough piezoelectric bending stresses to counterbalance the effect of the exciting forces and moments acting on the panel in a way that minimizes its structural vibrations.

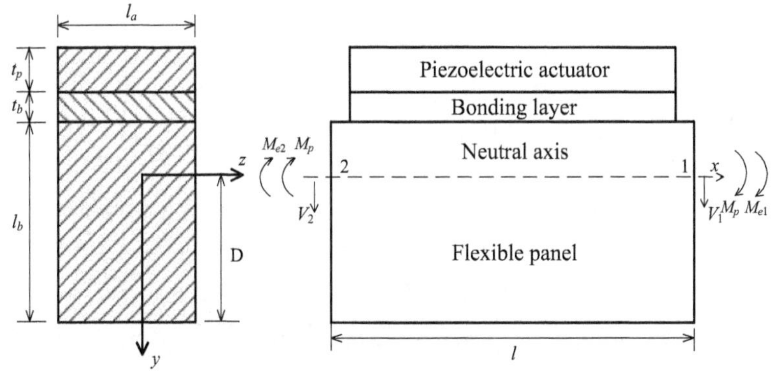

Figure 8.1 Schematic of an actuator bonded to a flexible panel.

If an input voltage u_p is applied across the film, a piezoelectric strain ε_p is introduced in the film, which is of the form below:

$$\varepsilon_p = \frac{d_p u_p}{t_p} \tag{8.15}$$

where d_p is the piezoelectric charge constant and t_p is the thickness of the piezoelectric actuator.

This strain results in a longitudinal stress σ_p given by

$$\sigma_p = E_p \frac{d_p u_p}{t_p} \tag{8.16}$$

where E_p is the Young's modulus of elasticity of the film.

Thus a bending moment M_p is generated around the neutral axis of the composite panel, that is,

$$M_p = \int_{-(t_p+t_b+l_b-D)}^{-(t_b+l_b-D)} \sigma_p(l_a y)dy \tag{8.17}$$

where, t_b is the thickness of the bonding layer, l_b is the thickness of the flexible panel, and l_a is the width of the panel, the bonding layer and the piezoelectric film. D is the distance of the neutral axis from the lower edge of the panel which can be determined by considering the force balance in the longitudinal direction x of the panel, that is,

$$\int_{film} \sigma_1 dS + \int_{bond} \sigma_2 dS + \int_{panel} \sigma_3 dS = 0 \tag{8.18}$$

As thin, homogeneous, and isotropic films of piezoelectric actuators are attached on the surface of the flexible panel, the longitudinal stresses $\sigma_{i=1,2,3}$ are very linearly with the distance y from the neutral axis, then Eq. (8.18) becomes

$$E_p l_a \int_{-(t_p+t_b+l_b-D)}^{-(t_b+l_b-D)} y dy + E_b l_a \int_{-(t_b+l_b-D)}^{-(l_b-D)} y dy + E l_a \int_{-(l_b-D)}^{D} y dy = 0 \tag{8.19}$$

where E_b is the Young's modulus of elasticity of the bonding layer and E is the Young's modulus of elasticity of the panel. With Eq. (8.19), one can obtain

$$D = \frac{E_p t_p^2 + E_b t_b^2 + E l_b^2 + 2E_p t_p (t_b + l_b) + 2E_b t_b l_b}{2(E_p t_p + E_b t_b + E l_b)} \tag{8.20}$$

Combining Eqs. (8.16), (8.17), and (8.20) yields the bending moment M_p generated by the piezoelectric actuator on the composite panel as follows:

$$M_p = d_p l_a E_p u_p \frac{E_b t_b (t_p + t_b) + E l_b (t_p + 2t_b + l_b)}{2(E_p t_p + E_b t_b + E l_b)} \tag{8.21}$$

Here, the characteristics are summarized as Table 8.1, which was reported in [22,37].

8.2.4 Improved model and control objective

In addition, the following important points should be noted as they figure into the development of the main results of this chapter.

1. Model parameter uncertainty

   Due to parameter perturbations, modeling errors, and environmental changes, model uncertainty is prevalent in physical systems. If model uncertainty is not considered during control system design, the controller obtained is highly likely to fail in real situations. Taking parameter uncertainty into account, $\boldsymbol{A}$ becomes $\boldsymbol{A} + \Delta \boldsymbol{A}$, where $\Delta \boldsymbol{A}$ denotes model parameter uncertainty, and satisfies a matching condition, that is, $\Delta \boldsymbol{A}(t) = \boldsymbol{M}_1 \boldsymbol{F}_1(t) \boldsymbol{N}_1$, $\left\| \boldsymbol{F}_1(t) \right\| \leq 1, \ \forall t$, where, $\boldsymbol{M}_1$ and $\boldsymbol{N}_1$ are known constant real matrices of appropriate dimension, $\boldsymbol{F}_1(\mathrm{t})$ is an unknown matrix function with Lebesgue measurable elements.

2. Measurement errors

   Due to the unexpected variations in the environment, limitations of measurement technique, and aging of measuring equipment, there may appear measurement errors $\boldsymbol{v}(t) \in L_2[0, \infty)$ between the system outputs and the measurement outputs, especially for electronic measurements [38].

Table 8.1 Characteristics of the flexible panel and the piezoelectric film and the bounding layer.

Flexible panel	Piezoelectric film	Bounding layer
l=5 m l_a=0.8 m l_b=0.1 m E=6.8 × 10^{10} N m^{-2}	dp=171 × 10^{-12} m V^{-1} E_p=139 × 10^9 N m^{-2} t_p=2.1 × 10^{-3} m	E_b=1.1 × 10^6 N m^{-2} t_b=2.1 × 10^{-3} m

3. Input constraints

In actual applications, actuator constraints are unavoidable due to the limits on power, mechanical constraints, saturation, and other physical limitations, which if unaccounted for, will usually negatively affect the stability and performance of an attitude control system designed for the "ideal" case. The optimization of control inputs is very important in practice, and as well as avoiding saturation, control inputs should be energy efficient and as small as possible to minimize energy consumption. Toward this end, a positive scalar λ_0 is introduced satisfying

$$\left\| \boldsymbol{u} \right\|_2^2 \leq \lambda_0 \tag{8.22}$$

where λ_0 denotes the norm squared of control input torque.

4. Controller gain perturbation

Certain controller errors and unknown actuator characteristics can lead to the problem of actuator perturbation or nominal controller perturbation, which is called controller gain perturbation. Taking the actuator perturbation to be represented by $\boldsymbol{\Xi}$ and including it in the control system produces the nominal controller gain perturbation $\Delta \boldsymbol{K}$ with norm $\left\| \Delta \boldsymbol{K} \right\| \leq \eta_0$. Two types are considered as follows:
When $\Delta \boldsymbol{K}$ is an additive perturbation, it satisfies

$$\Delta \boldsymbol{K} = \boldsymbol{M}_2 \boldsymbol{F}_2(t) \boldsymbol{N}_2, \quad \left\| \boldsymbol{F}_2(t) \right\| \leq 1 \tag{8.23}$$

When $\Delta \boldsymbol{K}$ is a multiplicative perturbation, it satisfies

$$\Delta \boldsymbol{K} = \boldsymbol{M}_3 \boldsymbol{F}_3(t) \boldsymbol{N}_3 \boldsymbol{K}, \quad \left\| \boldsymbol{F}_3(t) \right\| \leq 1 \tag{8.24}$$

The definitions of $\boldsymbol{M}_2$, $\boldsymbol{N}_2$, $\boldsymbol{M}_3$, $\boldsymbol{N}_3$, $\boldsymbol{F}_2(t)$ and $\boldsymbol{F}_3(t)$ are similar to the definition of $\Delta \boldsymbol{A}$ above.

The following assumptions and remarks provide theoretical support for further analysis.

Assumption 8.1: The external disturbance $\boldsymbol{T}_d$ is assumed to be bounded and belongs to $L_2[0, \infty)$.

Remark 8.2: Assumption 8.1 is reasonable. The main reason is that the external disturbances primarily include magnetic torque, solar radiation and aerodynamic torque, which are all bounded and continuous, which indicates that $\boldsymbol{T}_d$ belongs to $L_2[0, \infty)$.

Remark 8.3: To improve tolerance to uncertainties in the controller, the nonfragility of the controller has been taken into account as ΔK represents possible controller gain perturbation. A Bernoulli random variable $\overline{\rho}(t)$ is introduced here to express whether the controller perturbation occurs and $P(\overline{\rho}(t)=1)=\overline{\rho}$, so that increasing $\overline{\rho}$ will lead to system performance degradation, which may represent the maximum allowed perturbation parameter. In addition, the introduction of $\overline{\rho}$ models a randomly occurring gain perturbation. When $\overline{\rho}\rightarrow 0$, the nonfragile controller becomes a common fragile controller in [37]; When $\overline{\rho}\in(0,1)$, the controller perturbation randomly occurs; When $\overline{\rho}=1$, the controller becomes a common nonfragile controller in [39]. As these three cases are accounted for here, the controller in this chapter is more general than those in the references.

The controller is designed as

$$u(t) = -(K+\overline{\rho}\Delta K)x(t) \tag{8.25}$$

and is termed a *nonfragile* controller because it is able to tolerate uncertainties in the controller.

Considering the above factors, the closed-loop attitude control system of the flexible spacecraft can be expressed as:

$$\begin{cases} \dot{x}(t) = (A + \Delta A - B_1 K - B_1 \overline{\rho} \Delta K)x(t) + B_2 w(t) \\ y(t) = C_1 x(t) + v(t) \end{cases} \tag{8.26}$$

The objective of this chapter is to find a nonfragile controller such that the resulting closed-loop system (8.26) satisfies the following conditions:

1. The closed-loop system represented by Eq. (8.26) ($w(t)=0$, $v(t)=0$) is quadratically stable (see Definition 1 below).
2. For a given scalar $\gamma>0$, $\left\| G_{y\hat{w}}(s) \right\|_\infty < \gamma$ is guaranteed, where $G_{y\hat{w}}(s)$ denotes the closed-loop transfer function from $\hat{w}$ to y, and $\hat{w}$ is a constructed vector with norm $|\hat{w}| = \sqrt{w^2 + v^2}$. The control input is limited, that is, it cannot exceed its upper bound under the optimal conditions.

Definition 1: (quadratic stability) If there is a symmetric positive definite matrix $P>0$ and a positive constant α, for arbitrary uncertainty so that the time derivative of Lyapunov function $V(t)$ satisfies

$$\dot{V}(t) \leq -\alpha \left\| x(t) \right\|_2^2 \tag{8.27}$$

Then, the system represented by Eq. (8.26) ($\boldsymbol{w}(t) = 0$, $\boldsymbol{v}(t) = 0$) is said to be *quadratically stable.*

8.3 Nonfragile H_∞ control law

Before developing the main results, we introduce the following preliminary lemmas.

Lemma 8.1: [40] Let $\overline{\boldsymbol{H}}, \overline{\boldsymbol{E}}$, and $\overline{\boldsymbol{F}}(t)$ be real matrices of appropriate dimensions with $\|\overline{\boldsymbol{F}}(t)\| \leq 1$, then for any scalar $\overline{\xi} > 0$,

$$\overline{\boldsymbol{H}}\overline{\boldsymbol{F}}(t)\overline{\boldsymbol{E}} + \overline{\boldsymbol{E}}^T\overline{\boldsymbol{F}}(t)^T\overline{\boldsymbol{H}}^T \leq \overline{\xi}^{-1}\overline{\boldsymbol{H}}\overline{\boldsymbol{H}}^T + \overline{\xi}\overline{\boldsymbol{E}}^T\overline{\boldsymbol{E}}$$

Lemma 8.2: (Schur complement lemma) Let the partitioned matrix

$$\overline{\boldsymbol{A}} = \begin{bmatrix} \overline{\boldsymbol{A}}_{11} & \overline{\boldsymbol{A}}_{12} \\ * & \overline{\boldsymbol{A}}_{22} \end{bmatrix}$$

be symmetric. Then

$$\overline{\boldsymbol{A}} < 0 \Leftrightarrow \overline{\boldsymbol{A}}_{11} < 0, \overline{\boldsymbol{A}}_{22} - \overline{\boldsymbol{A}}_{12}^T\overline{\boldsymbol{A}}_{11}^{-1}\overline{\boldsymbol{A}}_{12} < 0 \Leftrightarrow \overline{\boldsymbol{A}}_{22} < 0, \overline{\boldsymbol{A}}_{11} - \overline{\boldsymbol{A}}_{12}\overline{\boldsymbol{A}}_{22}^{-1}\overline{\boldsymbol{A}}_{12}^T < 0$$

Lemma 8.3: [28] Assume that $\boldsymbol{M}$, $\boldsymbol{N}$ are real matrices of appropriate dimensions, then, for arbitrary $\varepsilon > 0$, one has

$$\begin{bmatrix} 0 & \boldsymbol{N}\boldsymbol{M}^T \\ * & 0 \end{bmatrix} \leq \begin{bmatrix} \varepsilon\boldsymbol{N}\boldsymbol{N}^T & 0 \\ * & \varepsilon^{-1}\boldsymbol{M}\boldsymbol{M}^T \end{bmatrix}$$

8.3.1 Sufficient conditions under additive perturbation

Theorem 8.1 now gives the existence conditions of the controller gain matrix $\boldsymbol{K}$ in terms of additive perturbation.

Theorem 8.1: For controller gain perturbation in the form of Eq. (8.23), given positive scalars $\xi_{i=1,2,3}, \overline{\rho}, \gamma_0, \varepsilon$, if there exists a symmetric positive definite matrix $\boldsymbol{X}$ and matrix $\boldsymbol{W}$, such that linear matrix inequalities (LMIs) (8.28)−(8.30) hold, then the closed-loop system represented by

Eq. (8.26) is quadratically stable under controller (8.25), $\boldsymbol{y}(t)$ satisfies an H_∞ performance constraint, and the control input constraint is satisfied.

$$\begin{bmatrix} \boldsymbol{\Theta}_{11} & \boldsymbol{\Theta}_{12} \\ * & \boldsymbol{\Theta}_{22} \end{bmatrix} < 0 \tag{8.28}$$

$$\begin{bmatrix} -\boldsymbol{X} & \boldsymbol{W}^T & \boldsymbol{X}^T \\ * & -\lambda_0\gamma_0^{-1}\boldsymbol{I} + \varepsilon^{-1}\overline{\rho}^2\eta_0^2\boldsymbol{I} & 0 \\ * & * & -\varepsilon^{-1}\boldsymbol{I} \end{bmatrix} < 0 \tag{8.29}$$

$$\begin{bmatrix} -\gamma_0 & \boldsymbol{x}(0)^T \\ * & -\boldsymbol{X} \end{bmatrix} < 0 \tag{8.30}$$

where,

$$\boldsymbol{\Theta}_{11} = \boldsymbol{AX} - \boldsymbol{B}_1\boldsymbol{W} + \boldsymbol{XA}^T - \boldsymbol{W}^T\boldsymbol{B}_1^T,$$
$$\boldsymbol{\Theta}_{12} = \begin{bmatrix} \boldsymbol{B}_2 & \boldsymbol{XC}_1^T & \boldsymbol{M}_1 & \boldsymbol{XN}_1^T & \boldsymbol{B}_1\boldsymbol{M}_2 & \boldsymbol{XN}_2^T & \boldsymbol{XC}_1^T \end{bmatrix},$$
$$\boldsymbol{\Theta}_{22} = \mathrm{diag}\left(\begin{bmatrix} -\overline{\gamma}\boldsymbol{I} & (1-\overline{\gamma})\boldsymbol{I} & -\xi_1^{-1}\boldsymbol{I} & -\xi_1\boldsymbol{I} & -\xi_2^{-1}\overline{\rho}^{-1}\boldsymbol{I} & -\xi_2\overline{\rho}^{-1}\boldsymbol{I} & -\boldsymbol{I} \end{bmatrix}\right)$$

By minimizing $\overline{\gamma}$, where $\overline{\gamma} = \gamma^2$, the optimal feasible solution of LMIs (8.28)–(8.30) are obtained, and one can obtain the controller gain matrix $\boldsymbol{K} = \boldsymbol{WX}^{-1}$.

Proof: First, one shows that the closed-loop system is quadratically stable.

Choose a Lyapunov function candidate

$$V(t) = \boldsymbol{x}(t)^T\boldsymbol{Px}(t)$$

where $\boldsymbol{P} = \boldsymbol{X}^{-1}$.

Then

$$\begin{aligned} \dot{V}(t) &= \dot{\boldsymbol{x}}(t)^T\boldsymbol{Px}(t) + \boldsymbol{x}(t)^T\boldsymbol{P}\dot{\boldsymbol{x}}(t) \\ &= \boldsymbol{x}(t)^T(\boldsymbol{A} + \Delta\boldsymbol{A} - \boldsymbol{B}_1\boldsymbol{K} - \boldsymbol{B}_1\overline{\rho}\Delta\boldsymbol{K})^T\boldsymbol{Px}(t) + \boldsymbol{w}(t)^T\boldsymbol{B}_2^T\boldsymbol{Px}(t) \\ &\quad + \boldsymbol{x}(t)^T\boldsymbol{P}(\boldsymbol{A} + \Delta\boldsymbol{A} - \boldsymbol{B}_1\boldsymbol{K} - \boldsymbol{B}_1\overline{\rho}\Delta\boldsymbol{K})\boldsymbol{x}(t) + \boldsymbol{x}(t)^T\boldsymbol{PB}_2\boldsymbol{w}(t) \\ &= \boldsymbol{x}(t)^T\left[(\boldsymbol{A} - \boldsymbol{B}_1\boldsymbol{K})^T\boldsymbol{P} + \boldsymbol{P}(\boldsymbol{A} - \boldsymbol{B}_1\boldsymbol{K})\right]\boldsymbol{x}(t) \\ &\quad + 2\boldsymbol{x}(t)^T\boldsymbol{PM}_1\boldsymbol{F}_1(t)\boldsymbol{N}_1\boldsymbol{x}(t) - 2\boldsymbol{x}(t)^T\boldsymbol{PB}_1\overline{\rho}\boldsymbol{M}_2\boldsymbol{F}_2(t)\boldsymbol{N}_2\boldsymbol{x}(t) \\ &\quad + 2\boldsymbol{x}(t)^T\boldsymbol{PB}_2\boldsymbol{w}(t) \end{aligned}$$

When $w(t) = 0$, according to Lemma 8.1 and the given constraint conditions, one has

$$\dot{V}(t) \le x(t)^T M_0 x(t)$$

where,

$$M_0 = (A - B_1K)^T P + P(A - B_1K) + \xi_1 PM_1M_1^T P + \xi_1^{-1} N_1^T N_1$$
$$+ \xi_2 \bar{\rho} PB_1M_2M_2^T B_1^T P + \xi_2^{-1} \bar{\rho} N_2^T N_2.$$

Considering (8.28), multiplied by $diag([P,\ I,\ I,\ I,\ I,\ I,\ I,\ I])$ and its transpose on both sides simultaneously, the time derivative of the Lyapunov function becomes

$$\begin{bmatrix} \tilde{\Theta}_{11} & \tilde{\Theta}_{12} \\ * & \tilde{\Theta}_{22} \end{bmatrix} < 0 \tag{8.31}$$

where,

$$\tilde{\Theta}_{11} = PA - PB_1K + A^T P - K^T B_1^T P,$$
$$\tilde{\Theta}_{12} = \begin{bmatrix} PB_2 & C_1^T & PM_1 & N_1^T & PB_1M_2 & N_2^T & C_1^T \end{bmatrix},$$
$$\tilde{\Theta}_{22} = diag\left(\begin{bmatrix} -\bar{\gamma} I & (1-\bar{\gamma}) I & -\xi_1^{-1} I & -\xi_1 I & -\xi_2^{-1} \bar{\rho}^{-1} I & -\xi_2 \bar{\rho}^{-1} I & -I \end{bmatrix}\right)$$

According to Lemma 8.2 and (8.31), one knows $M_0 < 0$, so that

$$\dot{V}(t) \le x(t)^T M_0 x(t) \le \lambda_{\max}(M_0) x(t)^T x(t)$$

Let $\alpha = -\lambda_{\max}(M_0) > 0$, then,

$$\dot{V}(t) \le -\alpha \left\| x(t) \right\|_2^2$$

that is, the closed-loop system is quadratically stable under controller (8.25).

Next one shows the output $y(t)$ satisfies an H_∞ performance constraint. To establish the $L_2\ [0,\ \infty)$ norm bound $\gamma^2 \left\| \hat{w}(t) \right\|_2^2$, consider the following functional:

$$J = \int_0^\infty [y(t)^T y(t) - \gamma^2 \hat{w}(t)^T \hat{w}(t)] dt$$

As the closed-loop system has quadratic stability, for arbitrary nonzero $\hat{w}(t) \in L_2[0,\ \infty)$, under zero initial conditions, one has $V(0) = 0$ and $V(\infty) \ge 0$, then,

$$J=\int_0^\infty [\boldsymbol{y}(t)^T\boldsymbol{y}(t)-\gamma^2\hat{\boldsymbol{w}}(t)^T\hat{\boldsymbol{w}}(t)+\dot{V}(t)]dt-V(\infty)+V(0)$$

$$\leq\int_0^\infty\left\{\begin{array}{l}\boldsymbol{x}(t)^T\left(\boldsymbol{M}_0+\boldsymbol{C}_1^T\boldsymbol{C}_1\right)\boldsymbol{x}(t)+2\boldsymbol{x}(t)^T\boldsymbol{C}_1^T\boldsymbol{v}(t)+\boldsymbol{v}(t)^T\boldsymbol{v}(t)\\-\gamma^2\left[\boldsymbol{w}(t)^T\boldsymbol{w}(t)+\boldsymbol{v}(t)^T\boldsymbol{v}(t)\right]+2\boldsymbol{x}(t)^T\boldsymbol{P}\boldsymbol{B}_2\boldsymbol{w}(t)\end{array}\right\}dt$$

$$\leq\int_0^\infty\left[\boldsymbol{x}(t)^T\quad\boldsymbol{w}(t)^T\quad\boldsymbol{v}(t)^T\right]\begin{bmatrix}\boldsymbol{M}_0+\boldsymbol{C}_1^T\boldsymbol{C}_1 & \boldsymbol{P}\boldsymbol{B}_2 & \boldsymbol{C}_1^T\\ * & -\gamma^2\boldsymbol{I} & 0\\ * & * & (1-\gamma^2)\boldsymbol{I}\end{bmatrix}\begin{bmatrix}\boldsymbol{x}(t)\\ \boldsymbol{w}(t)\\ \boldsymbol{v}(t)\end{bmatrix}dt$$

According to Lemma 8.2 and (8.31), $J<0$ holds, that is, $\boldsymbol{y}(t)$ satisfies an H_∞ performance constraint.

Finally, one shows the control input $\boldsymbol{u}(t)$ is constrained, that is, it satisfies the condition of (8.22).

The inequality (8.29) is equal to

$$\begin{bmatrix}-\boldsymbol{X}+\varepsilon\boldsymbol{X}^T\boldsymbol{X} & \boldsymbol{W}^T\\ * & -\lambda_0\gamma_0^{-1}\boldsymbol{I}+\varepsilon^{-1}\overline{\rho}^2\eta_0^2\boldsymbol{I}\end{bmatrix}<0 \tag{8.32}$$

For $\left\|\Delta\boldsymbol{K}\right\|\leq\eta_0$, and using $\boldsymbol{W}=\boldsymbol{K}\boldsymbol{X}$, one has

$$\begin{bmatrix}-\boldsymbol{X}+\varepsilon\boldsymbol{X}^T\boldsymbol{X} & \boldsymbol{X}^T\boldsymbol{K}^T\\ * & -\lambda_0\gamma_0^{-1}\boldsymbol{I}+\varepsilon^{-1}\overline{\rho}^2\Delta\boldsymbol{K}\Delta\boldsymbol{K}^T\end{bmatrix}<0 \tag{8.33}$$

Multiplied by diag$\{\boldsymbol{P},\boldsymbol{I}\}$ and its transpose on both sides of (8.33) simultaneously, (8.34) can be obtained.

$$\begin{bmatrix}-\boldsymbol{P}+\varepsilon\boldsymbol{I} & \boldsymbol{K}^T\\ * & -\lambda_0\gamma_0^{-1}\boldsymbol{I}+\varepsilon^{-1}\overline{\rho}^2\Delta\boldsymbol{K}\Delta\boldsymbol{K}^T\end{bmatrix}<0 \tag{8.34}$$

Using Lemma 8.3, one has

$$\begin{bmatrix}-\boldsymbol{P} & (\boldsymbol{K}+\overline{\rho}\Delta\boldsymbol{K})^T\\ \boldsymbol{K}+\overline{\rho}\Delta\boldsymbol{K} & -\lambda_0\gamma_0^{-1}\boldsymbol{I}\end{bmatrix}<0 \tag{8.35}$$

According to Lemma 8.2, one has

$$\gamma_0(\boldsymbol{K}+\overline{\rho}\Delta\boldsymbol{K})^T(\boldsymbol{K}+\overline{\rho}\Delta\boldsymbol{K})<\lambda_0\boldsymbol{P} \tag{8.36}$$

The condition $\dot{V}(t)\leq 0$ yields

$$V(t)\leq V(0)$$

Based on the definition of Lyapunov function, one has

$$\boldsymbol{x}(t)^T\boldsymbol{P}\boldsymbol{x}(t)\leq\boldsymbol{x}(0)^T\boldsymbol{P}\boldsymbol{x}(0)$$

According to (8.30),

$$\boldsymbol{x}(0)^T \boldsymbol{P}\boldsymbol{x}(0) < \gamma_0 \tag{8.37}$$

Therefore using (8.25), (8.36), and (8.37), we obtain

$$\begin{aligned}\|\boldsymbol{u}(t)\|_2^2 &= \boldsymbol{u}(t)^T\boldsymbol{u}(t)\\ &= \boldsymbol{x}(t)^T(\boldsymbol{K}+\overline{\rho}\Delta\boldsymbol{K})^T(K+\overline{\rho}\Delta\boldsymbol{K})\boldsymbol{x}(t)\\ &< \gamma_0^{-1}\lambda_0\boldsymbol{x}(t)^T\boldsymbol{P}\boldsymbol{x}(t)\\ &< \lambda_0\end{aligned} \tag{8.38}$$

This completes the proof.

8.3.2 Sufficient conditions under multiplicative perturbation

Theorem 8.2 now gives the existence conditions of the controller gain matrix $\boldsymbol{K}$ in terms of multiplicative perturbation.

Theorem 8.2: For controller gain perturbation in the form of Eq. (8.24), given positive scalars $\xi_{i=1,2,3}, \overline{\rho}, \gamma_0, \varepsilon$, if there exists a symmetric positive definite matrix $\boldsymbol{X}$ and matrix $\boldsymbol{W}$, such that the LMIs (8.39)-(8.41) hold, then the closed-loop system represented by Eq. (8.26) is quadratically stable under controller (8.25), $\boldsymbol{y}(t)$ satisfies an H_∞ performance constraint, and the control input constraint is satisfied.

$$\begin{bmatrix}\boldsymbol{\Theta}_{11} & \boldsymbol{\Theta}_{12}\\ * & \boldsymbol{\Theta}_{22}\end{bmatrix} < 0 \tag{8.39}$$

$$\begin{bmatrix}-\boldsymbol{X} & \boldsymbol{W}^T & \boldsymbol{X}^T\\ * & -\lambda_0\gamma_0^{-1}\boldsymbol{I}+\varepsilon^{-1}\overline{\rho}^2\eta_0^2\boldsymbol{I} & 0\\ * & * & -\varepsilon^{-1}\boldsymbol{I}\end{bmatrix} < 0 \tag{8.40}$$

$$\begin{bmatrix}-\gamma_0 & \boldsymbol{x}(0)^T\\ * & -\boldsymbol{X}\end{bmatrix} < 0 \tag{8.41}$$

where,

$$\begin{aligned}&\boldsymbol{\Theta}_{11} = \boldsymbol{A}\boldsymbol{X} - \boldsymbol{B}_1\boldsymbol{W} + \boldsymbol{X}\boldsymbol{A}^T - \boldsymbol{W}^T\boldsymbol{B}_1^T,\\ &\boldsymbol{\Theta}_{12} = \begin{bmatrix}\boldsymbol{B}_2 & \boldsymbol{X}\boldsymbol{C}_1^T & \boldsymbol{M}_1 & \boldsymbol{X}\boldsymbol{N}_1^T & \boldsymbol{B}_1\boldsymbol{M}_3 & \boldsymbol{W}^T\boldsymbol{N}_3^T & \boldsymbol{X}\boldsymbol{C}_1^T\end{bmatrix},\\ &\boldsymbol{\Theta}_{22} = \mathrm{diag}\left(\begin{bmatrix}-\overline{\gamma}\boldsymbol{I} & (1-\overline{\gamma})\boldsymbol{I} & -\xi_1^{-1}\boldsymbol{I} & -\xi_1\boldsymbol{I} & -\xi_2^{-1}\overline{\rho}^{-1}\boldsymbol{I} & -\xi_2\overline{\rho}^{-1}\boldsymbol{I} & -\boldsymbol{I}\end{bmatrix}\right)\end{aligned}$$

By minimizing $\overline{\gamma}$, where $\overline{\gamma}=\gamma^2$, the optimal feasible solution of LMIs (8.39)−(8.41) are feasible, one can obtain the controller gain matrix $\boldsymbol{K}=\boldsymbol{W}\boldsymbol{X}^{-1}$.

Proof: First, one shows that the closed-loop system is quadratically stable.

With the Lyapunov function candidate

$$V(t) = \boldsymbol{x}(t)^T \boldsymbol{P}\boldsymbol{x}(t)$$

where $\boldsymbol{P} = \boldsymbol{X}^{-1}$.

Then

$$\begin{aligned}\dot{V}(t) &= \dot{\boldsymbol{x}}(t)^T \boldsymbol{P}\boldsymbol{x}(t) + \boldsymbol{x}(t)^T \boldsymbol{P}\dot{\boldsymbol{x}}(t) \\ &= \boldsymbol{x}(t)^T (\boldsymbol{A} + \Delta\boldsymbol{A} - \boldsymbol{B}_1\boldsymbol{K} - \boldsymbol{B}_1\overline{\rho}\Delta\boldsymbol{K})^T \boldsymbol{P}\boldsymbol{x}(t) + \boldsymbol{w}(t)^T \boldsymbol{B}_2^T \boldsymbol{P}\boldsymbol{x}(t) \\ &\quad + \boldsymbol{x}(t)^T \boldsymbol{P}(\boldsymbol{A} + \Delta\boldsymbol{A} - \boldsymbol{B}_1\boldsymbol{K} - \boldsymbol{B}_1\overline{\rho}\Delta\boldsymbol{K})\boldsymbol{x}(t) + \boldsymbol{x}(t)^T \boldsymbol{P}\boldsymbol{B}_2\boldsymbol{w}(t) \\ &= \boldsymbol{x}(t)^T \left[(\boldsymbol{A} - \boldsymbol{B}_1\boldsymbol{K})^T \boldsymbol{P} + \boldsymbol{P}(\boldsymbol{A} - \boldsymbol{B}_1\boldsymbol{K})\right]\boldsymbol{x}(t) \\ &\quad + 2\boldsymbol{x}(t)^T \boldsymbol{P}\boldsymbol{M}_1\boldsymbol{F}_1(t)\boldsymbol{N}_1\boldsymbol{x}(t) - 2\boldsymbol{x}(t)^T \boldsymbol{P}\boldsymbol{B}_1\overline{\rho}\boldsymbol{M}_3\boldsymbol{F}_3(t)\boldsymbol{N}_3\boldsymbol{K}\boldsymbol{x}(t) \\ &\quad + 2\boldsymbol{x}(t)^T \boldsymbol{P}\boldsymbol{B}_2\boldsymbol{w}(t)\end{aligned}$$

When $\boldsymbol{w}(t)=0$, According to Lemma 8.1 and the given constraint conditions, one has

$$\dot{V}(t) \le \boldsymbol{x}(t)^T \boldsymbol{M}_0 \boldsymbol{x}(t)$$

where,

$$\boldsymbol{M}_0 = (\boldsymbol{A} - \boldsymbol{B}_1\boldsymbol{K})^T \boldsymbol{P} + \boldsymbol{P}(\boldsymbol{A} - \boldsymbol{B}_1\boldsymbol{K}) + \xi_1 \boldsymbol{P}\boldsymbol{M}_1\boldsymbol{M}_1^T \boldsymbol{P} + \xi_1^{-1}\boldsymbol{N}_1^T \boldsymbol{N}_1$$

$$+\xi_2\overline{\rho}\boldsymbol{P}\boldsymbol{B}_1\boldsymbol{M}_3\boldsymbol{M}_3^T \boldsymbol{B}_1^T \boldsymbol{P} + \xi_2^{-1}\overline{\rho}\boldsymbol{K}^T \boldsymbol{N}_3^T \boldsymbol{N}_3 \boldsymbol{K}.$$

Considering (8.39), multiplied by $diag([\boldsymbol{P}, \boldsymbol{I}, \boldsymbol{I}, \boldsymbol{I}, \boldsymbol{I}, \boldsymbol{I}, \boldsymbol{I}, \boldsymbol{I}])$ and its transpose on both sides simultaneously, the time derivative of the Lyapunov function becomes

$$\begin{bmatrix} \tilde{\boldsymbol{\Theta}}_{11} & \tilde{\boldsymbol{\Theta}}_{12} \\ * & \tilde{\boldsymbol{\Theta}}_{22} \end{bmatrix} < 0 \tag{8.42}$$

where,

$$\begin{aligned}\tilde{\boldsymbol{\Theta}}_{11} &= \boldsymbol{P}\boldsymbol{A} - \boldsymbol{P}\boldsymbol{B}_1\boldsymbol{K} + \boldsymbol{A}^T\boldsymbol{P} - \boldsymbol{K}^T\boldsymbol{B}_1^T\boldsymbol{P}, \\ \tilde{\boldsymbol{\Theta}}_{12} &= \begin{bmatrix} \boldsymbol{P}\boldsymbol{B}_2 & \boldsymbol{C}_1^T & \boldsymbol{P}\boldsymbol{M}_1 & \boldsymbol{N}_1^T & \boldsymbol{P}\boldsymbol{B}_1\boldsymbol{M}_3 & \boldsymbol{K}^T\boldsymbol{N}_3^T & \boldsymbol{C}_1^T \end{bmatrix}, \\ \tilde{\boldsymbol{\Theta}}_{22} &= \mathrm{diag}\left(\begin{bmatrix} -\overline{\gamma}\boldsymbol{I} & (1-\overline{\gamma})\boldsymbol{I} & -\xi_1^{-1}\boldsymbol{I} & -\xi_1\boldsymbol{I} & -\xi_2^{-1}\overline{\rho}^{-1}\boldsymbol{I} & -\xi_2\overline{\rho}^{-1}\boldsymbol{I} & -\boldsymbol{I} \end{bmatrix}\right)\end{aligned}$$

According to Lemma 8.2 and (8.42), one knows $\boldsymbol{M}_0 < 0$, so that

$$\dot{V}(t) \le \boldsymbol{x}(t)^T \boldsymbol{M}_0 \boldsymbol{x}(t) \le \lambda_{\max}(\boldsymbol{M}_0)\boldsymbol{x}(t)^T \boldsymbol{x}(t)$$

Let $\alpha = -\lambda_{\max}(\boldsymbol{M}_0) > 0$, then,

$$\dot{V}(t) \le -\alpha \left\| \boldsymbol{x}(t) \right\|_2^2$$

that is, the closed-loop system is quadratically stable.

Next, one shows the output $\boldsymbol{y}(t)$ satisfies an H_∞ performance constraint.

To establish the L_2 $[0, \infty)$ norm bound $\gamma^2\|\hat{\boldsymbol{w}}(t)\|_2^2$, consider the following functional:

$$J=\int_0^\infty [\boldsymbol{y}(t)^T\boldsymbol{y}(t)-\gamma^2\hat{\boldsymbol{w}}(t)^T\hat{\boldsymbol{w}}(t)]dt$$

As the closed-loop system has quadratic stability, for arbitrary nonzero $\hat{\boldsymbol{w}}(t)\in L_2[0,\infty)$, under zero initial conditions, one has $V(0)=0$ and $V(\infty)\geq 0$, then,

$$\begin{aligned}
J&=\int_0^\infty [\boldsymbol{y}(t)^T\boldsymbol{y}(t)-\gamma^2\hat{\boldsymbol{w}}(t)^T\hat{\boldsymbol{w}}(t)+\dot{V}(t)]dt-V(\infty)+V(0)\\
&\leq\int_0^\infty\left\{\begin{array}{l}\boldsymbol{x}(t)^T\left(\boldsymbol{M}_0+\boldsymbol{C}_1^T\boldsymbol{C}_1\right)\boldsymbol{x}(t)+2\boldsymbol{x}(t)^T\boldsymbol{C}_1^T\boldsymbol{v}(t)+\boldsymbol{v}(t)^T\boldsymbol{v}(t)\\ -\gamma^2\left[\boldsymbol{w}(t)^T\boldsymbol{w}(t)+\boldsymbol{v}(t)^T\boldsymbol{v}(t)\right]+2\boldsymbol{x}(t)^T\boldsymbol{P}\boldsymbol{B}_2\boldsymbol{w}(t)\end{array}\right\}dt\\
&\leq\int_0^\infty\left[\boldsymbol{x}(t)^T\quad\boldsymbol{w}(t)^T\quad\boldsymbol{v}(t)^T\right]\begin{bmatrix}\boldsymbol{M}_0+\boldsymbol{C}_1^T\boldsymbol{C}_1 & \boldsymbol{P}\boldsymbol{B}_2 & \boldsymbol{C}_1^T\\ * & -\gamma^2\boldsymbol{I} & 0\\ * & * & (1-\gamma^2)\boldsymbol{I}\end{bmatrix}\begin{bmatrix}\boldsymbol{x}(t)\\ \boldsymbol{w}(t)\\ \boldsymbol{v}(t)\end{bmatrix}dt
\end{aligned}$$

According to Lemma 8.2 and (8.42), $J<0$ holds, that is, $\boldsymbol{y}(t)$ satisfies an H_∞ performance constraint.

Finally, the control input $\boldsymbol{u}(t)$ is constrained, that is, it satisfies the condition of (8.22). The proof is the same as that under the condition of additive perturbation and is omitted here.

This completes the proof.

Remark 8.4: The scalar γ can be regarded as an optimization variable to obtain a reduction in the H_∞ performance constraint bound. Therefore the minimum of H_∞ performance constraint bound with admissible controllers can be readily found by solving the following convex optimization problem: Minimize γ subject to the LMIs in Theorem 8.1 or Theorem 8.2. In addition, $\varepsilon^{-1}\overline{\rho}^2\eta_0^2<\lambda_0\gamma_0^{-1}$ should be guaranteed while positive scalars are chosen.

8.4 Simulation test

In this section, the effectiveness of the developed controller is illustrated with simulations of a flexible spacecraft attitude control system, with the main parameters from [22,34]. It should be noted that the control task is to achieve attitude stabilization and vibration suppression simultaneously with control input constraints, and meanwhile analyze the characteristics of the two vibration suppression schemes. For this example the first four bending modes of flexible appendages have been considered for the implemented spacecraft model (see Table 8.2), since the low-order modes are generally dominant in a flexible spacecraft. The system states are assumed to be directly available, and two types of natural frequencies, that is, high natural frequency and low natural frequency, are exploited here to show which vibration suppression scheme is superior for different frequencies. The frequency uncertainty often appears in practice [41], and it can be included in model parameter uncertainty. Here, to illustrate the robustness of the proposed control scheme, the simulations with appropriate uncertainties (e.g., the maximum is one tenth of the nominal value for each) in natural frequency and damping values and the coupling matrices are performed in this section.

The inertia matrix of flexible spacecraft is taken to be

$$J = \begin{bmatrix} 350 & 3 & 4 \\ 3 & 280 & 10 \\ 4 & 10 & 190 \end{bmatrix} \text{kg} \cdot \text{m}^2$$

and the nominal coupling matrices are

$$\tilde{\boldsymbol{\Delta}} = \begin{bmatrix} 6.45637 & -1.25619 & 1.11687 & 1.23637 \\ 1.27814 & 0.91756 & 2.48901 & -0.83674 \\ 2.15629 & -1.67264 & -0.83674 & -1.12503 \end{bmatrix} \sqrt{\text{kg}} \cdot \text{m},$$

$$\tilde{\boldsymbol{\Delta}}_p = \begin{bmatrix} 2.342552 \\ -0.422537 \\ 3.912984 \\ 7.026176 \end{bmatrix} \times 10^{-2} \sqrt{\text{kg}} \cdot \text{m s}^{-2} V^{-1}$$

The initial attitude angle and angular velocity are arbitrarily chosen as $\Theta(0) = [0.08, 0.06, -0.06]^{\text{T}}$ rad and $\omega(0) = [-0.01, -0.01, 0.01]^{\text{T}}$ rad s^{-1},

Table 8.2 Parameters of flexible dynamics.

	High natural frequency (rad s^{-1})	**Low natural frequency (rad s^{-1})**	**Damping**
Mode 1	0.7681	0.7681 × 0.3	0.005607
Mode 2	1.1038	1.1038 × 0.2	0.008620
Mode 3	1.8733	1.8733 × 0.1	0.01283
Mode 4	2.5496	2.5496 × 0.1	0.02516

respectively. All the initial values of $\tilde{\eta}(0)$ and $\dot{\tilde{\eta}}(0)$ are assumed to be zero, that is, the flexible appendages are initially un-deformed. The maximum control torque is constrained by $\lambda_0 = 27$, and this constraint on control torque is considered in the simulation.

Ref. [42] has analyzed the environmental disturbances in detail, including gravity-gradient torque, aerodynamic torque, and Earth magnetic torque. As pointed out in this reference, the sinusoid is the prototype of some periodic disturbances, and the environmental disturbances such as magnetic torque and gravity-gradient torque are cyclic essentially and they can be represented by sinusoids. Combined with aerodynamic torque, the lumped external disturbances can contain constant terms [43]. Thus it is suitable to set the external disturbance $\boldsymbol{T}_d$ as

$$\boldsymbol{T}_d = 3 \times 10^{-4} \times \begin{bmatrix} 1 + \sin(0.11\pi t) \\ 1.5 + \cos(0.11\pi t) \\ 1 + \cos(0.11\pi t + \pi/3) \end{bmatrix} \mathrm{Nm}$$

Meanwhile, choose similar numbers to [29] and we take

$$\begin{aligned}
&\boldsymbol{M}_1 = 0.01 \times \begin{bmatrix} 8 & 11 & 13 & 15 & 16 & -18 & 8 & 11 & 13 & 15 & 16 & 18 & 8 & 11 \end{bmatrix}^T, \\
&\boldsymbol{N}_1 = 0.01 \times \begin{bmatrix} 1 & 2 & 3 & 4 & 2 & 10 & 1 & 2 & 3 & 4 & 2 & 10 & 1 & 2 \end{bmatrix}, \boldsymbol{F}_1(t) = \sin(0.11\pi t), \\
&\boldsymbol{M}_2 = 0.1 \times \boldsymbol{1}_{3\times 1}, \boldsymbol{N}_2 = 0.01 \times \boldsymbol{1}_{1\times 14}, \boldsymbol{F}_2(t) = \sin(0.11\pi t + \pi/4), \eta_{02} = 0.0083 \\
&\boldsymbol{M}_3 = 0.1 \times \boldsymbol{1}_{3\times 1}, \boldsymbol{N}_3 = 0.01 \times \boldsymbol{1}_{1\times 3}, \boldsymbol{F}_3(t) = \cos(0.11t), \eta_{03} = 0.8 \\
&\tilde{\boldsymbol{P}} = \boldsymbol{I}_8, \xi_1 = 0.002, \xi_2 = 0.22, \varepsilon = 0.05, \overline{p} = 0.6, \gamma_0 = 0.65
\end{aligned}$$

where η_{02} and η_{03} correspond to the bounded norm of the additive and multiplicative perturbations, respectively.

The simulation results of attitude stabilization based on passive and active vibration suppression under additive perturbation are first presented and compared. Following this, another simulation compares the performance of the control scheme with different initial conditions while the other parameters remain the same as previously defined. Finally, simulation results under multiplicative perturbation are shown.

8.4.1 Comparisons of control performance under additive perturbation

When the active vibration suppression scheme for high natural frequencies (AVSHNF) is used, solving inequality (8.12) gives

$$\boldsymbol{F}_1 = [1.3654 \ -0.1312 \ -13.9579 \ -54.9398],$$

$$\boldsymbol{F}_2 = [0.0938 \ -0.0044 \ -0.2822 \ -2.1095]$$

and solving LMIs(8.28)−(8.30) gives the controller gain matrix $\boldsymbol{K}_1$ with optimization variable γ=0.1361, and

$$\boldsymbol{K}_1=\begin{bmatrix} 9.7270 & -0.1144 & -2.4456 & 249.7563 & 18.6183 & -10.1614 & -14.7539 & 2.6689 & -7.3214 & -20.7563 & -7.1394 & 7.8063 & -24.6970 & -44.0799 \\ 0.8189 & 10.0592 & 0.0374 & 7.0715 & 118.9741 & -47.1614 & -0.8078 & -22.5528 & -6.5631 & -2.1134 & -3.5784 & -2.5434 & -1.0981 & 3.7911 \\ 2.3994 & -0.0007 & 9.2604 & -65.1947 & -25.2766 & 113.8060 & 4.4252 & -1.8283 & 22.6778 & 42.8815 & -15.8952 & 5.3167 & -6.1132 & -18.2546 \end{bmatrix}$$

When the active vibration suppression scheme for low natural frequencies (AVSLNF) is used, solving inequality (8.12) gives

$$\boldsymbol{F}_1=\begin{bmatrix} 3.1533 & -0.5714 & 5.3674 & 9.3389 \end{bmatrix},$$

$$\boldsymbol{F}_2=\begin{bmatrix} 1.0976 & -0.1965 & 1.8086 & 3.0873 \end{bmatrix}$$

and solving LMIs (8.28)−(8.30) gives the controller gain matrix $\boldsymbol{K}_2$ with optimization variable γ=0.1233, and

$$\boldsymbol{K}_2=\begin{bmatrix} 11.7797 & -0.2346 & -3.0321 & 193.5924 & 18.8632 & 21.8318 & -0.2900 & 8.2522 & -4.3351 & 3.7375 & -10.2710 & -5.1362 & 3.3059 & 0.1918 \\ 0.5153 & 12.7443 & -0.6729 & 1.7003 & 146.3028 & -12.0954 & 2.9843 & 5.2866 & 6.2470 & -2.8022 & -1.0830 & -2.2534 & -10.0298 & 6.0073 \\ 0.1751 & 0.3716 & 11.4783 & 44.9615 & -4.6367 & 112.9199 & -1.5559 & -11.7756 & 5.0992 & -1.2112 & -2.3072 & 11.1168 & -0.7927 & 3.5044 \end{bmatrix}$$

When the passive vibration suppression scheme for high natural frequencies (PVSHNF) is used, solving LMIs (8.28)−(8.30) gives the controller gain matrix $\boldsymbol{K}_3$ with optimization variable γ=0.1148, and

$$\boldsymbol{K}_3=\begin{bmatrix} 12.0490 & -0.1236 & -2.1818 & 165.9797 & 12.2905 & 11.7943 & -9.9638 & -6.5667 & 5.6766 & -1.1014 & -8.1934 & -3.3203 & 1.1853 & -1.2336 \\ -0.0141 & 12.2130 & 0.7691 & 23.8198 & 131.6451 & -14.5863 & -0.7863 & -9.9674 & -12.3984 & 2.3051 & -1.5059 & -3.0234 & -6.4138 & 1.7561 \\ 0.8638 & -0.3331 & 11.3353 & 83.1749 & 3.2513 & 112.0705 & -3.9197 & 12.3876 & -3.4850 & 1.5372 & -3.6394 & 7.8026 & 0.3579 & 1.3832 \end{bmatrix}$$

When the passive vibration suppression scheme for low natural frequencies (PVSLNF) is used, solving LMIs (8.28)−(8.30) gives the controller gain matrix $\boldsymbol{K}_4$ with optimization variable γ=0.1255, and

$$\boldsymbol{K}_4=\begin{bmatrix} 11.8334 & 0.4832 & -2.6300 & 207.7844 & 21.1272 & 5.8029 & -1.2426 & -0.4615 & 7.6149 & 3.1584 & -9.6181 & -2.0009 & -1.0978 & -5.9084 \\ 1.0468 & 13.4166 & -0.5795 & 3.2100 & 153.6436 & -7.1243 & 20.4526 & 58.9994 & -21.7799 & -12.6241 & 0.6444 & -0.5730 & -12.8359 & 6.3944 \\ -0.1798 & -0.0801 & 11.1485 & 57.1743 & 3.0130 & 125.4667 & -1.7676 & -4.6647 & -5.3570 & -3.5322 & -3.7493 & 7.9010 & 1.5494 & 6.9909 \end{bmatrix}$$

Applying these four controllers to the corresponding closed-loop attitude control systems of the flexible spacecraft produces the simulation results of Figs. 8.2-8.9. Figs. 8.2 and 8.3 display the attitude angle and angular velocity of the flexible spacecraft under the four conditions, (i.e., AVSHNF, AVSLNF, PVSHNF and PVSLNF), from which it can be seen that the attitude angle and angular velocity all converge to within a small bound. However, the convergence time and steady-state errors differ for the cases. The maximum convergence time and steady-state error in roll, pitch and yaw attitude angle are taken for the evaluation index, with details in Table 8.3. As can be seen, the PVSHNF requires the shortest convergence time but has the largest steady-state error, and next is the AVSHNF, followed by AVSLNF and PVSLNF. With respect to the

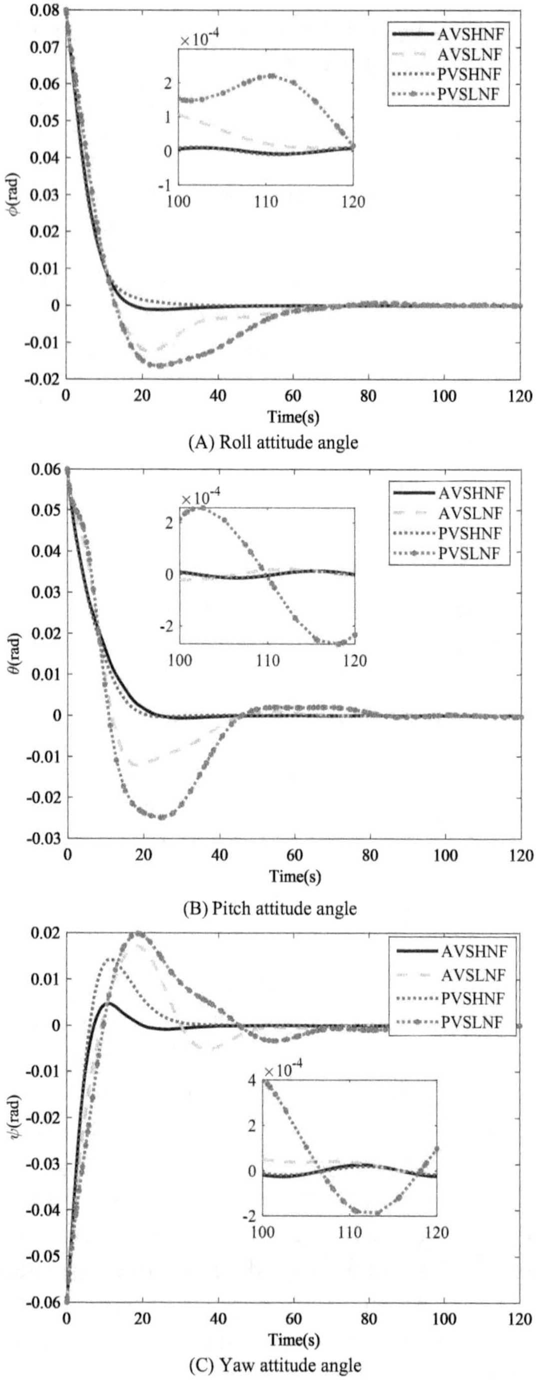

Figure 8.2 Time response of attitude angle under additive perturbation.

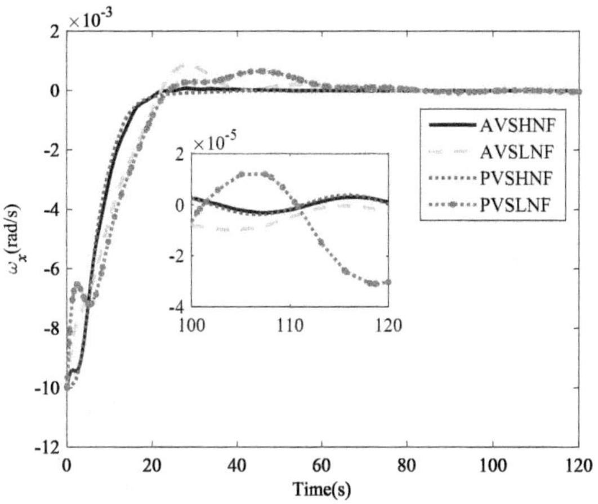

(A) Roll angular velocity

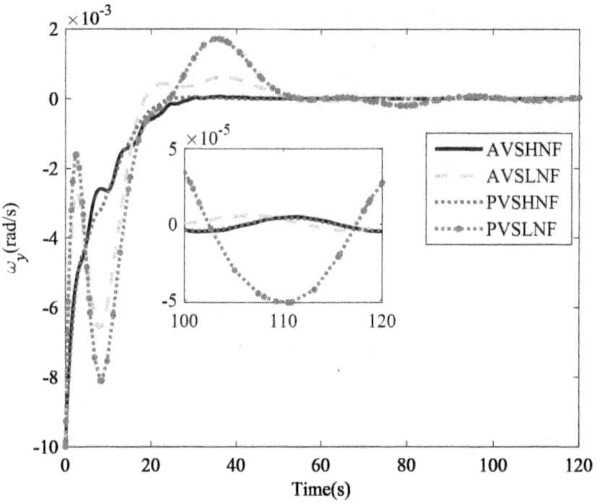

(B) Pitch angular velocity

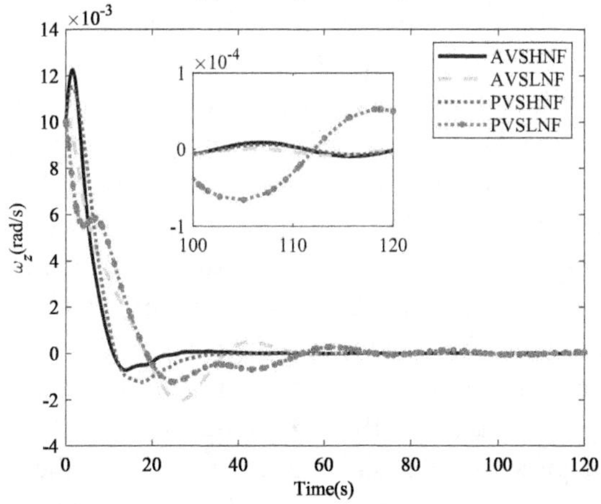

(C) Yaw angular velocity

Figure 8.3 Time response of angular velocity under additive perturbation.

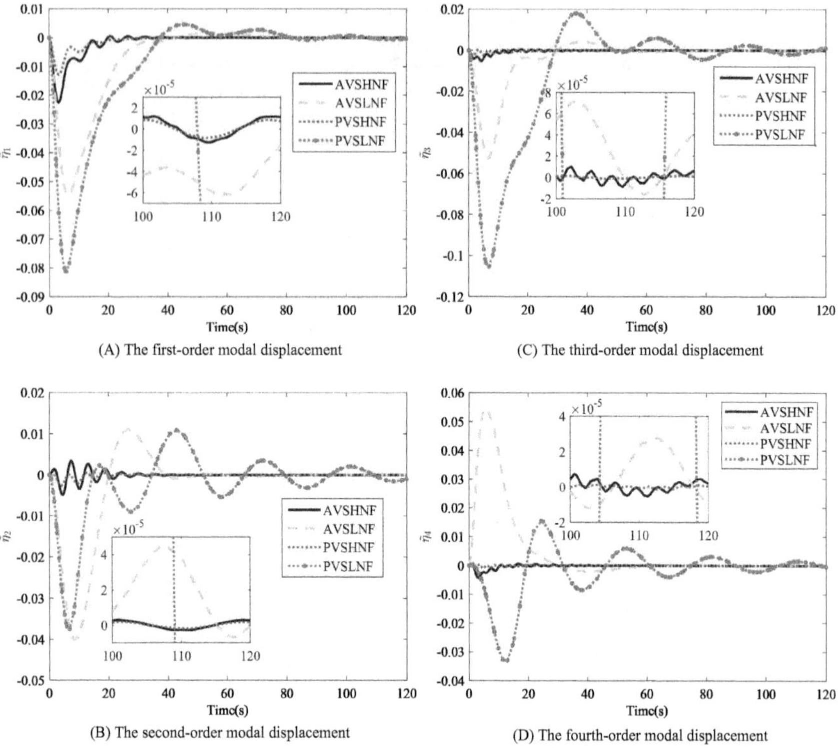

Figure 8.4 Time response of modal displacement under additive perturbation.

convergence speed and accuracy of attitude stabilization, it can be concluded that with the high natural frequencies, the passive vibration suppression scheme is superior to the active vibration suppression scheme; while for low natural frequencies, the active vibration suppression scheme is superior. In addition, for both the active and passive vibration suppression schemes, attitude stabilization requires much more time for low natural frequencies than for high natural frequencies. Fig. 8.4 shows the modal displacement of the flexible appendages. It is observed that the AVSHNF can effectively suppress the vibrations within 40 seconds while the PVSHNF damps the vibrations with better performance -within only 30 seconds, followed by AVSLNF with approximately 60 seconds and PVSLNF with more than 120 seconds. Thus the low natural frequencies have a greater influence on the performance of attitude stabilization and vibration suppression than do the high frequencies.

Fig. 8.5 depicts the combined disturbance from the external disturbance and flexible appendages, which indicates that the response tends to

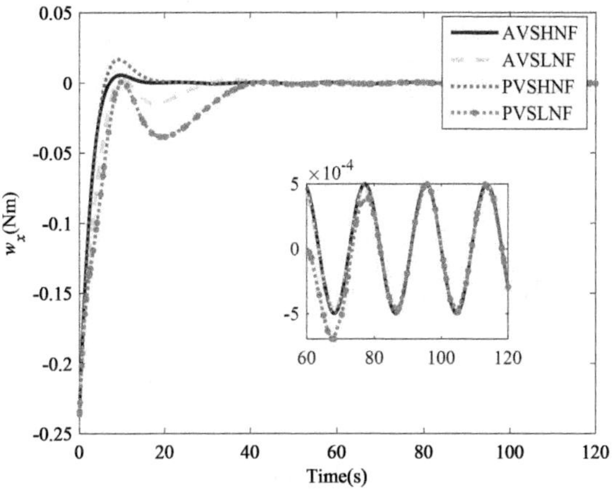

(A) Lumped disturbance along x axis

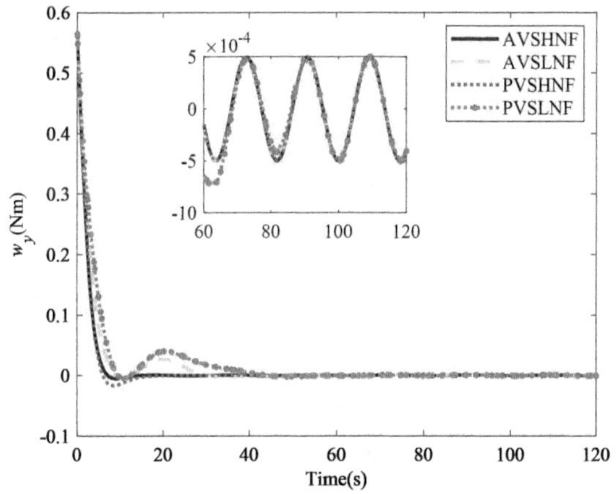

(B) Lumped disturbance along y axis

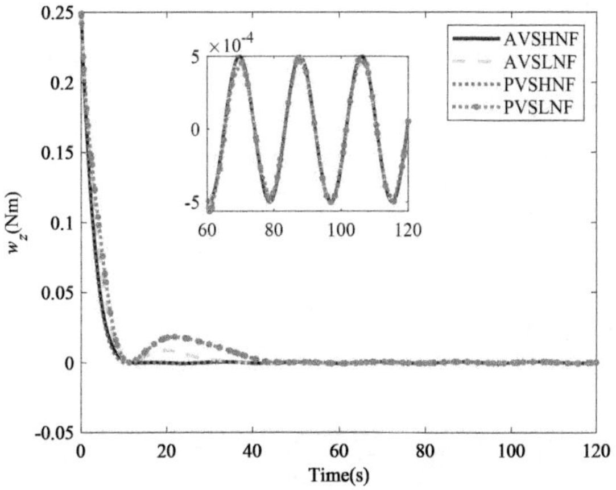

(C) Lumped disturbance along z axis

Figure 8.5 The lumped disturbance under additive perturbation.

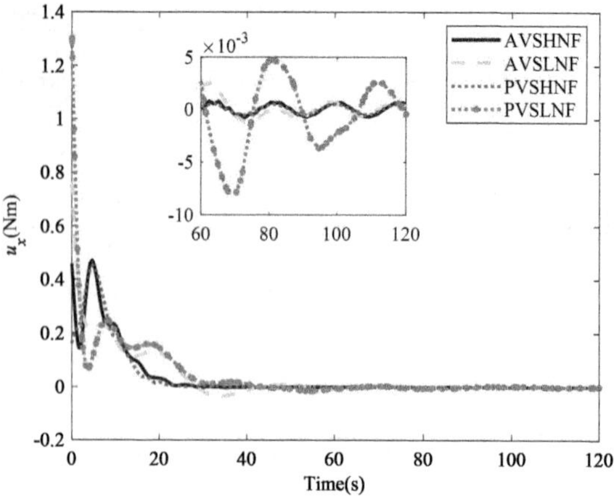

(A) Control torque along x axis

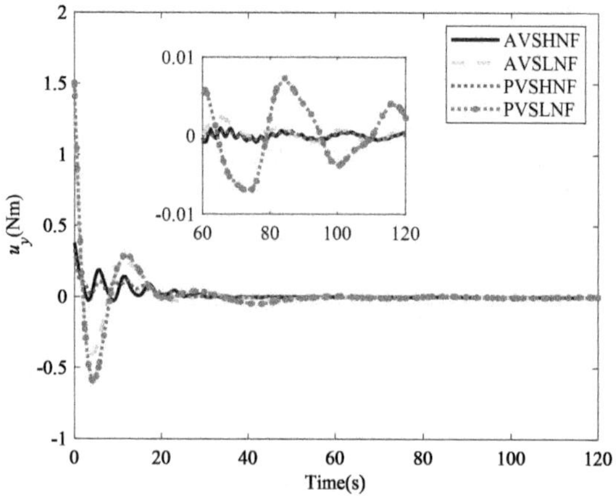

(B) Control torque along y axis

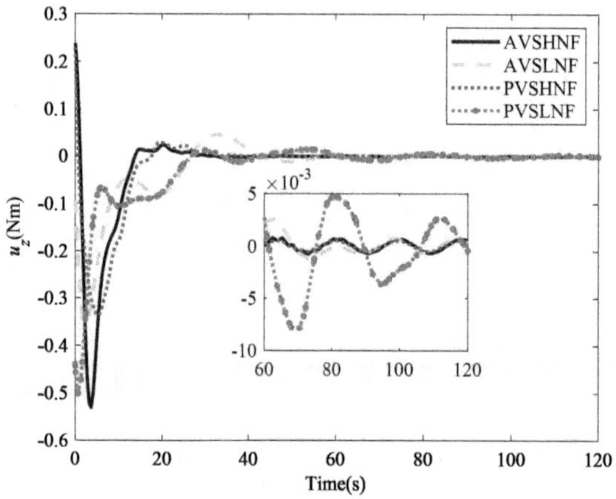

(C) Control torque along z axis

Figure 8.6 The control torque under additive perturbation.

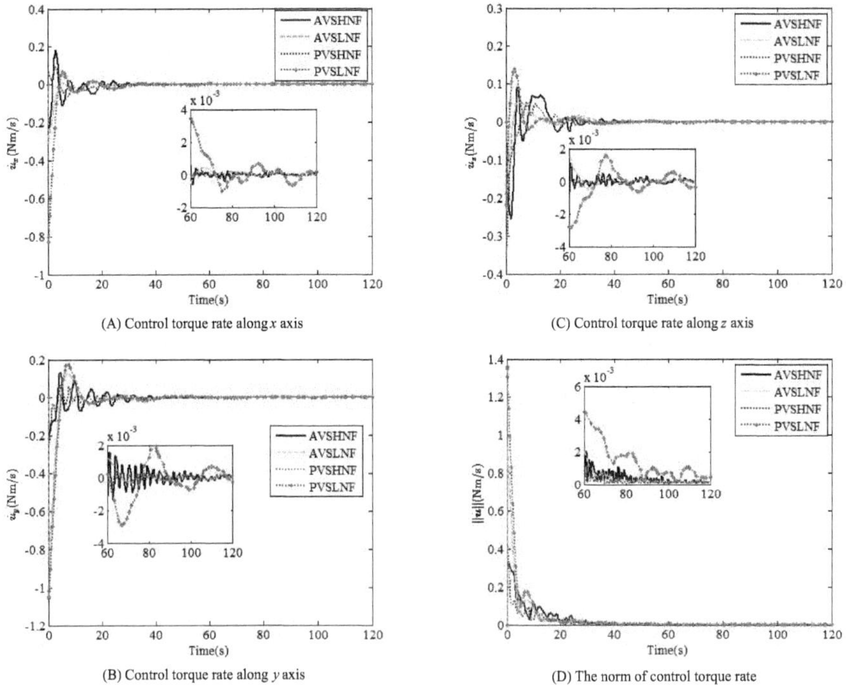

Figure 8.7 The control torque rate under additive perturbation.

track the external disturbance as time increases, and that the vibration suppression is effective. The time history of the control torque is plotted in Fig. 8.6, which reveals that the four schemes all require sustained control torque to compensate for the external disturbance and the sustained low amplitude vibrations of the flexible appendages. In particular, the influence of sustained vibration of the flexible appendage under PVSLNF is the greatest, which also corresponds to Fig. 8.4. The control torque under the PVSHNF converges faster than the other three schemes, that is, AVSHNF, AVSLNF and PVSLNF, because it improves the suppression effect of the vibration of the flexible appendage more effectively than the other three schemes. It is to be noted that the control torque is always less than its limit constraint. The rate of control torque and its norm are plotted in Fig. 8.7, from which one can see that it is less than 1.2 Nm s^{-1} with its norm less than 1.4 Nm s^{-1}.

Energy efficiency is an extremely important issue for achieving the target tasks and extending the mission operating life so that optimization of energy consumption must be taken into account for flexible spacecraft

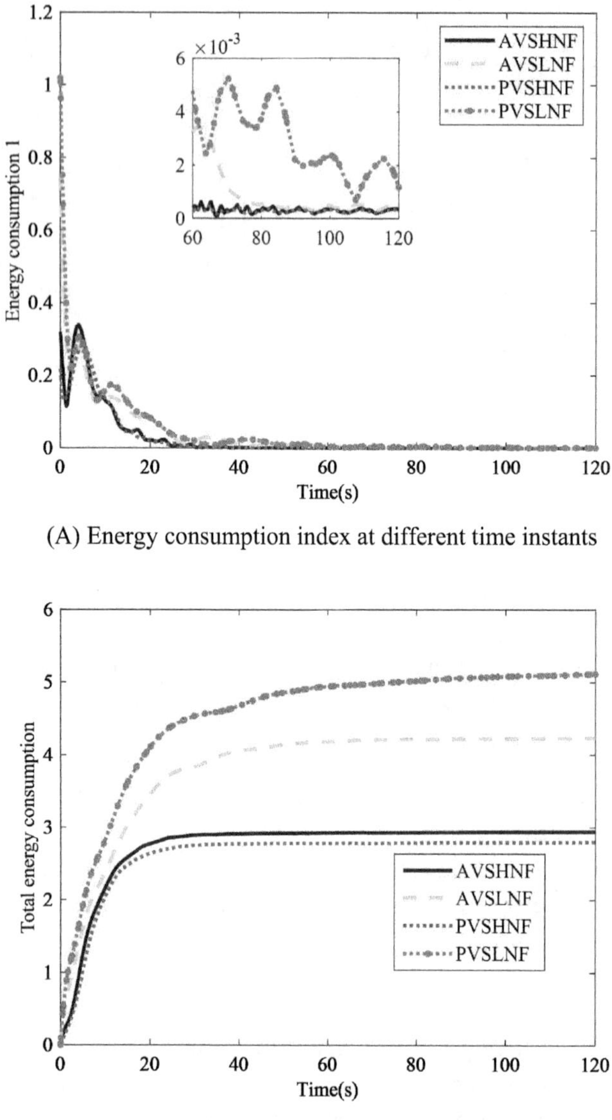

(A) Energy consumption index at different time instants

(B) Total Energy consumption index

Figure 8.8 The energy consumption index under additive perturbation.

attitude control system design. To analyze the total energy consumption, an energy index function defined by $E = \frac{1}{2}\int_0^T \|\boldsymbol{u}\| dt$ is used, where T denotes the simulation time, and T=120 seconds is chosen in the simulation. The time responses of the energy consumption index at different

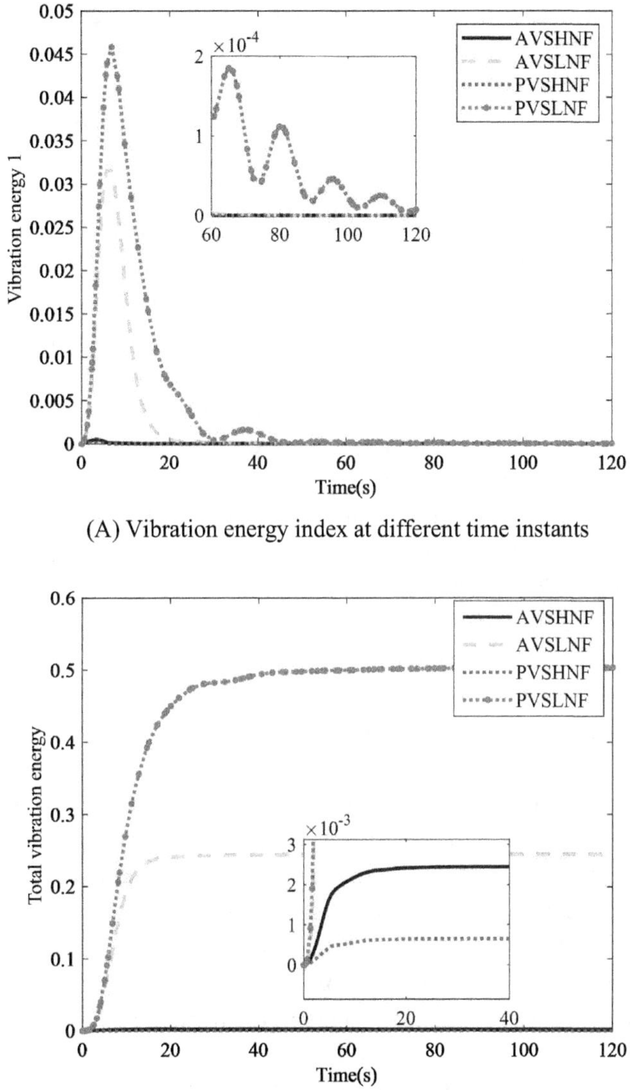

(A) Vibration energy index at different time instants

(B) Total vibration energy index

Figure 8.9 The vibration energy index under additive perturbation.

Table 8.3 The evaluation index under additive perturbation.

Type	Convergence time of Θ (s)	steady-state error of Θ (rad)
AVSHNF	50	5.0×10^{-5}
AVSLNF	98	1.5×10^{-4}
PVSHNF	38	3.8×10^{-5}
PVSLNF	99	2.6×10^{-4}

time instants and the total energy consumption index are shown in Fig. 8.8A and B, respectively. It is observed that the PVSLNF requires the greatest energy consumption while the PVSHNF requires the least energy consumption compared with the other two schemes. One can also see that the flexible spacecraft with low natural frequencies would evidently require much more energy compared to one with high natural frequencies. The energy consumptions of AVSHNF and PVSHNF are nearly the same. As the main parameters are from [22], the properties of flexible spacecraft and piezoelectric actuators are identical, thus the energy consumption in [22] can be compared with the case of AVSHNF in this chapter. According to the simulation result of control torque in [22], the input control torque was more than 20 Nm with saturation at 30 Nm, and for the initial action time of approximately 3 seconds, the energy consumption index would be more than 30, while it is about 3 for AVSHNF. However, of greater interest is comparison of vibration suppression capability. To further investigate the vibration suppression performance, we define a vibration energy index as $\tilde{E} = \int_0^T (\dot{\tilde{\eta}}^T \dot{\tilde{\eta}} + \tilde{\eta}\tilde{\boldsymbol{K}}\tilde{\eta})dt$, which physically means the vibration intensity during the whole simulation process, where T denotes the simulation time and T=120 seconds is chosen in the simulation. Fig. 8.9A and B present the vibration energy index at different time instants, and the total vibration energy, which verifies the effectiveness of the active vibration suppression technique for the low natural frequency circumstance and the effectiveness of the passive vibration suppression technique for the high natural frequencies in terms of vibration energy. It is clear that the vibration energy for low natural frequencies is much greater than that for high natural frequencies. It can be concluded that by the measures of comparison of this chapter, the proposed nonfragile controller here can achieve attitude stabilization and vibration suppression simultaneously for H_∞ performance and quadratic stability, and in the presence of model parameter uncertainty, additive perturbation, external disturbances, and input constraints.

8.4.2 Comparisons of control performance under multiplicative perturbation

When the active vibration suppression scheme for high natural frequencies (AVSHNF) is used, solving inequality (8.12) gives

$$\boldsymbol{F}_1 = \begin{bmatrix} 1.3654 & 0.1312 & -13.9579 & -54.9398 \end{bmatrix},$$

$$\boldsymbol{F}_2 = \begin{bmatrix} 0.0938 & -0.0044 & -0.2822 & -2.1095 \end{bmatrix}$$

and solving LMIs (8.39)–(8.41) gives the controller gain matrix $\boldsymbol{K}_1$ with optimization variable γ=0.1452, and

$$\boldsymbol{K}_1 = \begin{bmatrix} 9.3250 & -0.0298 & -2.2342 & 246.6528 & 20.0920 & -11.4940 & -14.7015 & 2.0512 & -7.6570 & -21.4337 & -7.3194 & 7.4838 & -25.2470 & -45.1424 \\ 0.7430 & 10.1249 & 0.5578 & 11.9592 & 117.3470 & -36.2758 & -1.1491 & -19.4155 & -6.7797 & 0.0126 & -1.5646 & 1.3465 & 1.5220 & 9.8035 \\ 2.2393 & 0.0783 & 8.8723 & -65.9720 & -24.3899 & 114.1348 & 4.2058 & -1.2805 & 22.4283 & 43.8070 & -15.4444 & 6.5622 & -5.9121 & -17.0490 \end{bmatrix}$$

When the active vibration suppression scheme for low natural frequencies (AVSLNF) is used, solving inequality (8.12) gives

$$\boldsymbol{F}_1 = \begin{bmatrix} 3.1533 & -0.5714 & 5.3674 & 9.3389 \end{bmatrix},$$

$$\boldsymbol{F}_2 = \begin{bmatrix} 1.0976 & -0.1965 & 1.8086 & 3.0873 \end{bmatrix}$$

and solving LMIs (8.39)–(8.41) gives the controller gain matrix $\boldsymbol{K}_2$ with optimization variable γ=0.1309, and

$$\boldsymbol{K}_2 = \begin{bmatrix} 11.2894 & 0.1670 & -2.8790 & 190.1758 & 22.5925 & 22.1348 & -0.9238 & 7.3166 & -3.3173 & 3.4663 & -9.7489 & -5.0207 & 2.8600 & 0.4187 \\ 0.3457 & 12.1009 & -0.6260 & -8.1620 & 142.3787 & -16.2217 & 3.8291 & 4.7807 & 4.8334 & -2.4468 & -0.6986 & -2.2368 & -9.7020 & 5.7266 \\ 0.2250 & 0.5041 & 10.9409 & 50.4445 & -1.4803 & 112.7944 & -2.4019 & -9.8588 & 4.6081 & -0.6047 & -2.3835 & 10.5690 & -0.9252 & 3.4333 \end{bmatrix}$$

When the passive vibration suppression scheme for high natural frequencies (PVSHNF) is used, solving LMIs (8.39)–(8.41) gives the controller gain matrix $\boldsymbol{K}_3$ with optimization variable γ=0.1214, and

$$\boldsymbol{K}_3 = \begin{bmatrix} 11.7161 & 0.0251 & -2.0226 & 158.6188 & 13.8183 & 10.4177 & -9.3122 & -6.2143 & 4.7532 & -0.9432 & -8.6544 & -2.7244 & 0.9076 & -1.1063 \\ -0.1399 & 11.8050 & 0.7605 & 27.3247 & 128.3546 & -11.8905 & -1.1343 & -10.0783 & -11.0794 & 2.1726 & 0.6723 & -3.8905 & -5.6829 & 1.5814 \\ 0.7141 & -0.0708 & 10.8835 & 88.1247 & 5.5671 & 111.6463 & -4.6762 & 11.5039 & -2.6807 & 1.4082 & -2.3086 & 6.5636 & 0.5255 & 1.2208 \end{bmatrix}$$

When the passive vibration suppression scheme for low natural frequencies (PVSLNF) is used, solving LMIs (8.39)-(8.41) gives the controller gain matrix $\boldsymbol{K}_4$ with optimization variable γ=0.1339, and

$$\boldsymbol{K}_4 = \begin{bmatrix} 11.3614 & 0.8188 & -2.4832 & 210.2706 & 26.8848 & 8.3336 & -1.3209 & 2.4342 & 5.6628 & 1.8805 & -9.0355 & -1.7698 & -1.6367 & -5.4556 \\ 0.6588 & 12.8976 & -0.6386 & -2.1824 & 147.5059 & -7.0008 & 17.5443 & 51.5136 & -17.9172 & -9.8365 & 0.5051 & -0.8552 & -11.7485 & 6.2051 \\ -0.0662 & -0.1232 & 10.7447 & 55.5605 & 5.3405 & 120.7995 & -1.2872 & -3.7063 & -6.0921 & -2.9245 & -3.8037 & 7.1606 & 1.6197 & 6.7981 \end{bmatrix}$$

Analogous to the sequence under additive perturbation, applying the four controllers to the corresponding closed-loop attitude control systems of the flexible spacecraft with multiplicative perturbations produces the simulation results of Figs. 8.10-8.17, which are nearly the same as those of Figs. 8.2 to 8.9. Figs. 8.10 and 8.11 display the attitude angle and angular velocity of flexible spacecraft under four different conditions, that is, AVSHNF, AVSLNF, PVSHNF, and PVSLNF, from which one can see that the attitude angle and angular velocity all converge to within a small bound. However, the convergence time and steady-state errors differ for these cases. For maximum convergence time and steady-state error in roll, pitch and yaw attitude angle, for example, the details are given in Table 8.4, which are similar to those in Table 8.3. As can be seen, the

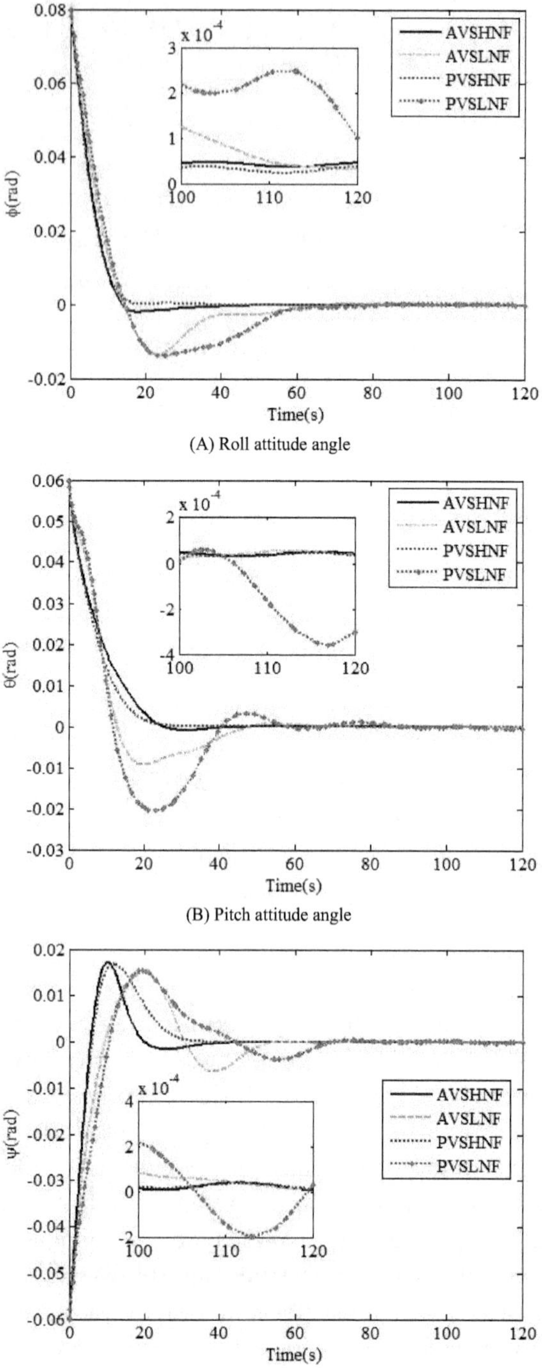

(A) Roll attitude angle
(B) Pitch attitude angle
(C) Yaw attitude angle

Figure 8.10 Time response of attitude angle under multiplicative perturbation.

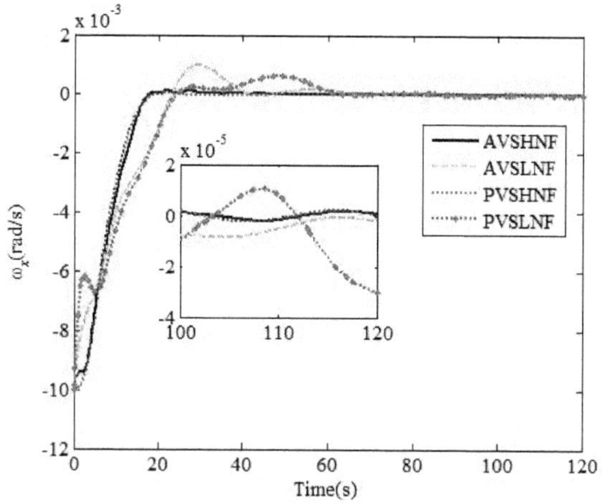

(A) Roll angular velocity

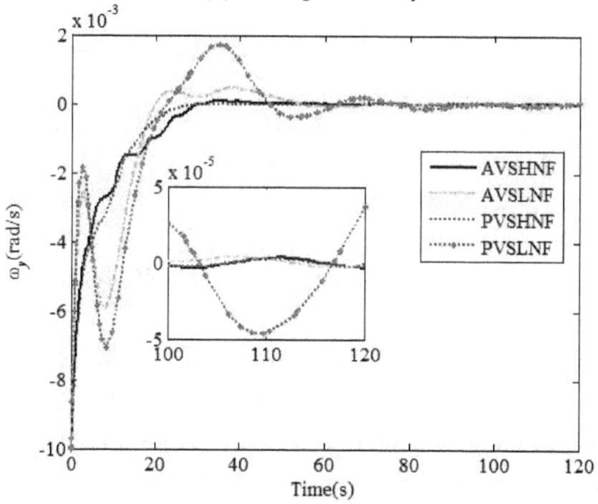

(B) Pitch angular velocity

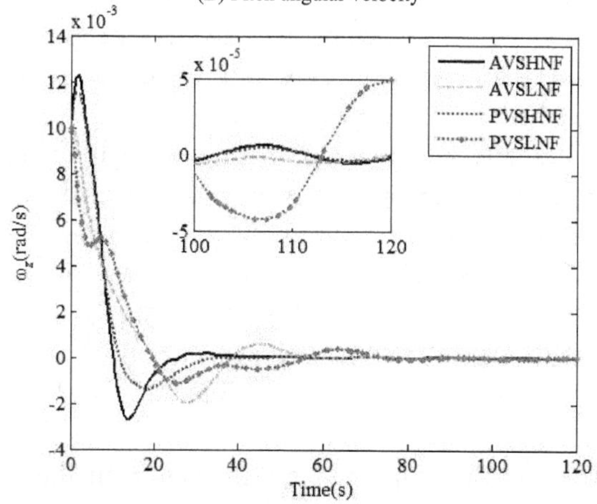

(C) Yaw angular velocity

Figure 8.11 Time response of angular velocity under multiplicative perturbation.

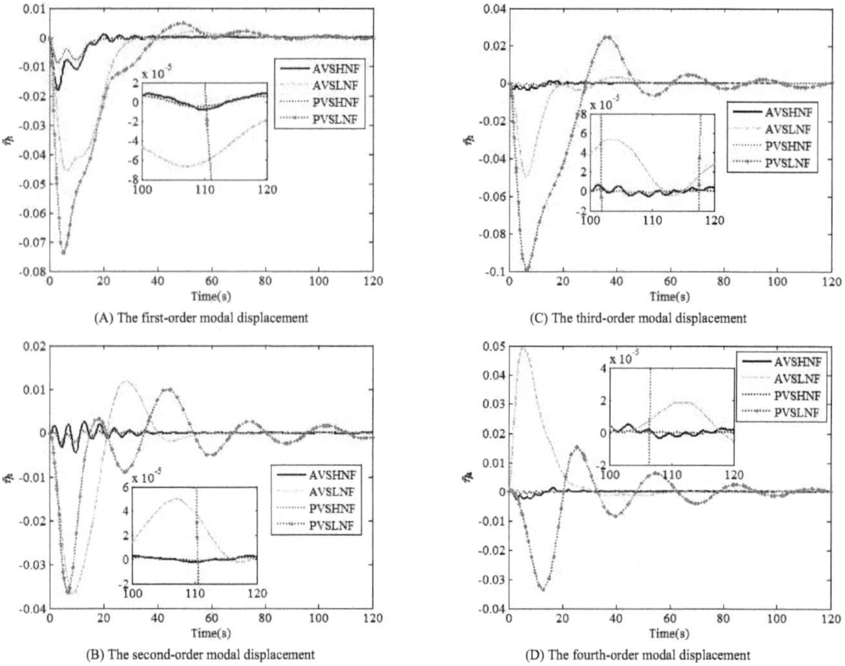

Figure 8.12 Time response of modal displacement under multiplicative perturbation.

PVSHNF requires the shortest convergence time but with the largest steady-state error, and then it is the AVSHNF, followed by AVSLNF and PVSLNF. With respect to the convergence speed and accuracy of attitude stabilization, it can be concluded that for high natural frequencies, the passive vibration suppression scheme is superior to the active vibration suppression scheme; while for low natural frequencies, the active vibration suppression scheme is superior. In addition, for active and passive vibration suppression schemes, the attitude stabilization requires much more time for low natural frequencies than for high natural frequencies. Fig. 8.12 shows the modal displacement of the flexible appendages. It is observed that the AVSHNF can effectively suppress the vibration within 40 seconds while the PVSHNF damps the vibration with better performance-within only 30 seconds, followed by AVSLNF with approximately 60 seconds and PVSLNF with more than 120 seconds. Thus the low natural frequencies have more influence on the performance of attitude stabilization and vibration suppression.

Fig. 8.13 depicts the combined disturbance from the external disturbance and flexible appendages, which indicates that the response tends to

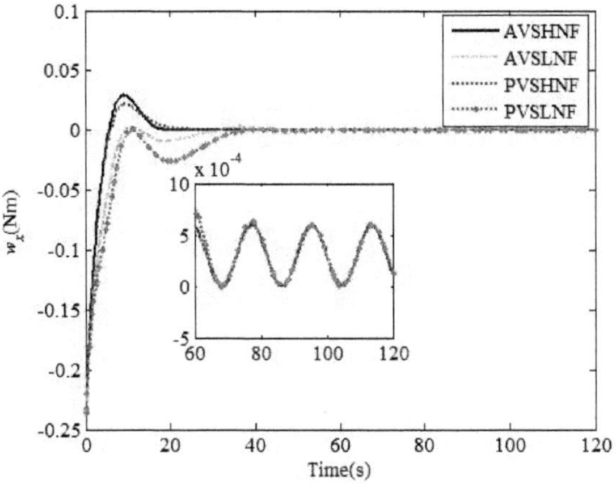

(A) Lumped disturbance along x axis

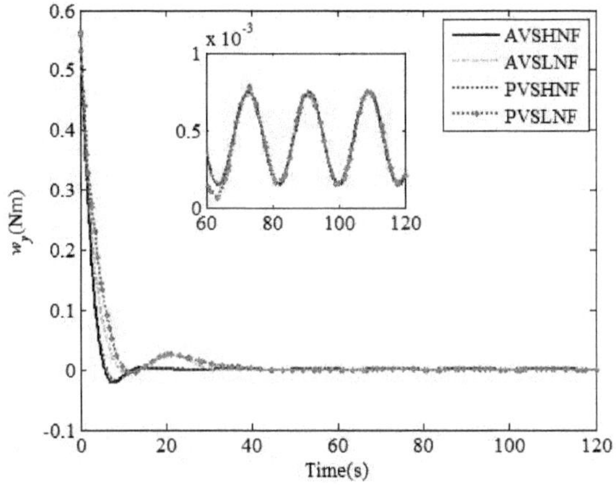

(B)Lumped disturbance along y axis

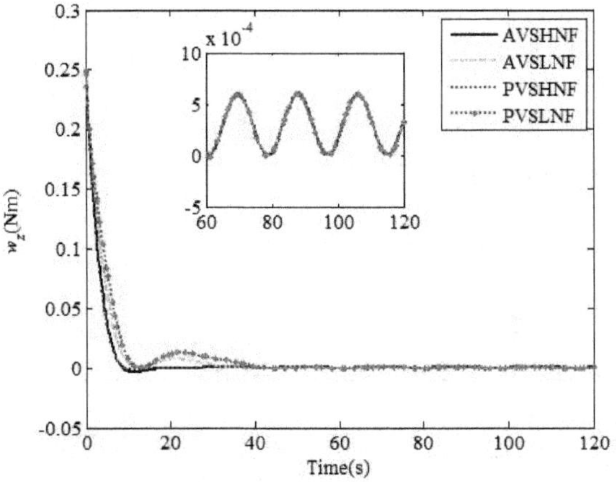

(C) Lumped disturbance along z axis

Figure 8.13 The lumped disturbance under multiplicative perturbation.

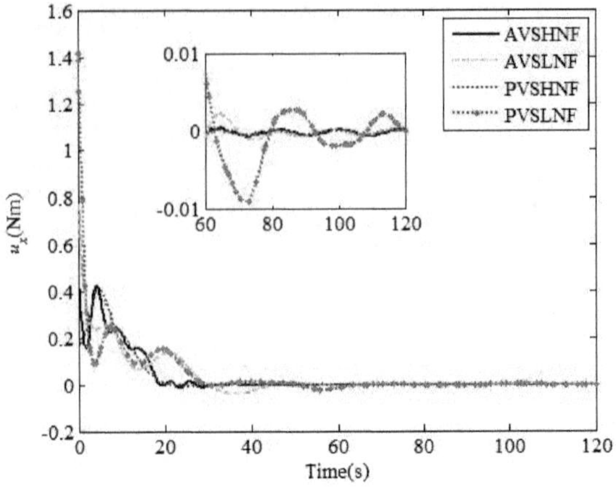

(A) Control torque along x axis

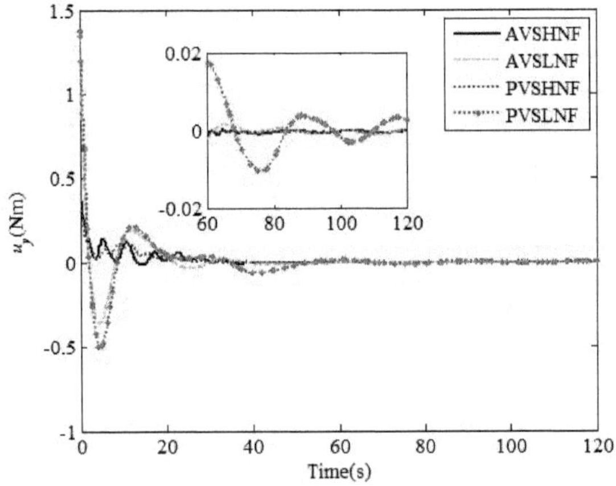

(B) Control torque along y axis

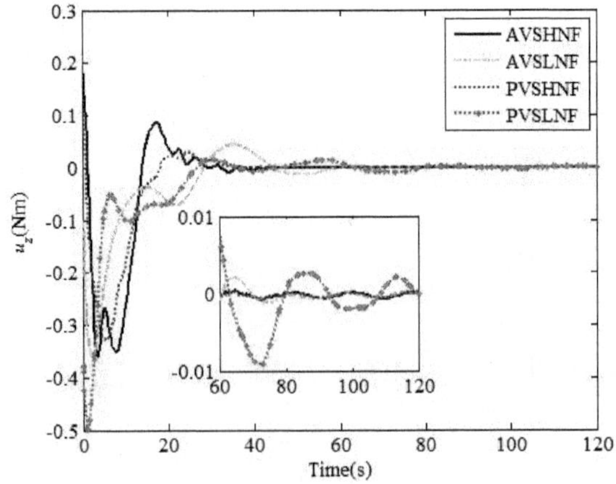

(C) Control torque along z axis

Figure 8.14 The control torque under multiplicative perturbation.

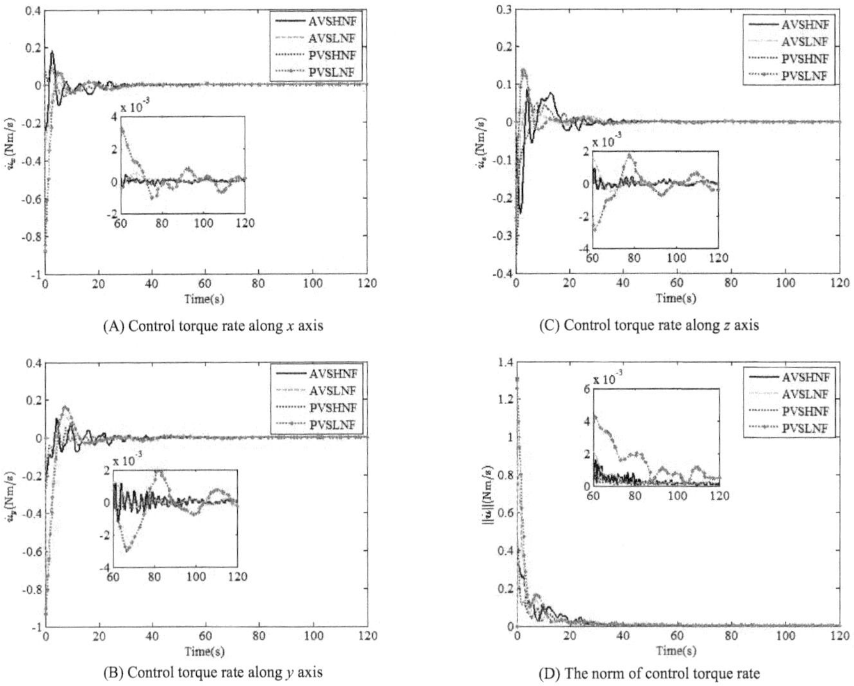

Figure 8.15 The control torque rate under multiplicative perturbation.

track the external disturbance as time increases, and that the vibration suppression is effective. The time history of the control torque is plotted in Fig. 8.14, which reveals that the four schemes all require sustained control torque, to compensate for the external disturbance and the sustained vibration of the flexible appendage. In particular, the influence of sustained vibrations of the flexible appendage under PVSLNF is the greatest, which also corresponds to Fig. 8.12. The control torque under the PVSHNF converges faster than for other three schemes, that is, AVSHNF, AVSLNF and PVSLNF, because it improves the suppression effect on the vibration of the flexible appendage more effectively than the other three schemes. It is to be noted that the control torque is always less than its constraint. The rate of control torque and its norm are plotted in Fig. 8.15, from which one can see that it is less than 1.0 Nm s^{-1} with its norm less than 1.4 Nm s^{-1}.

The time responses of the energy consumption index at different time instants and the total energy consumption index are shown in Fig. 8.16A and B, respectively. It is observed that the PVSLNF requires the greatest energy consumption while the PVSHNF requires the minimum energy

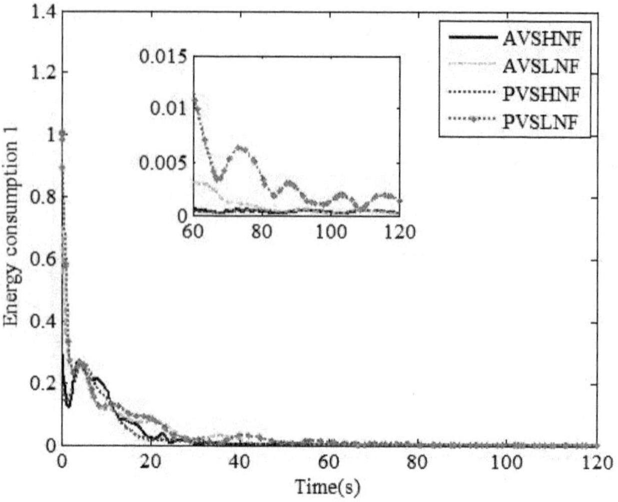

(A) Energy consumption index at different time instants

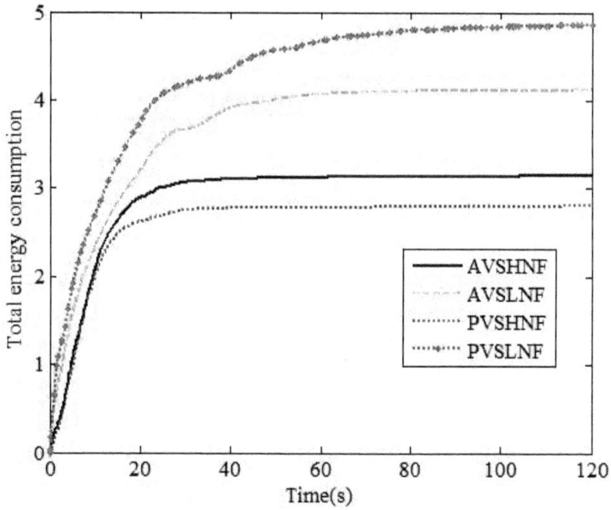

(B) Total Energy consumption index

Figure 8.16 The energy consumption index under multiplicative perturbation.

consumption compared with the other two schemes. One can also see that the flexible spacecraft for low natural frequencies would evidently require much more energy compared with the scheme for high natural frequencies. The energy consumptions of AVSHNF and PVSHNF are

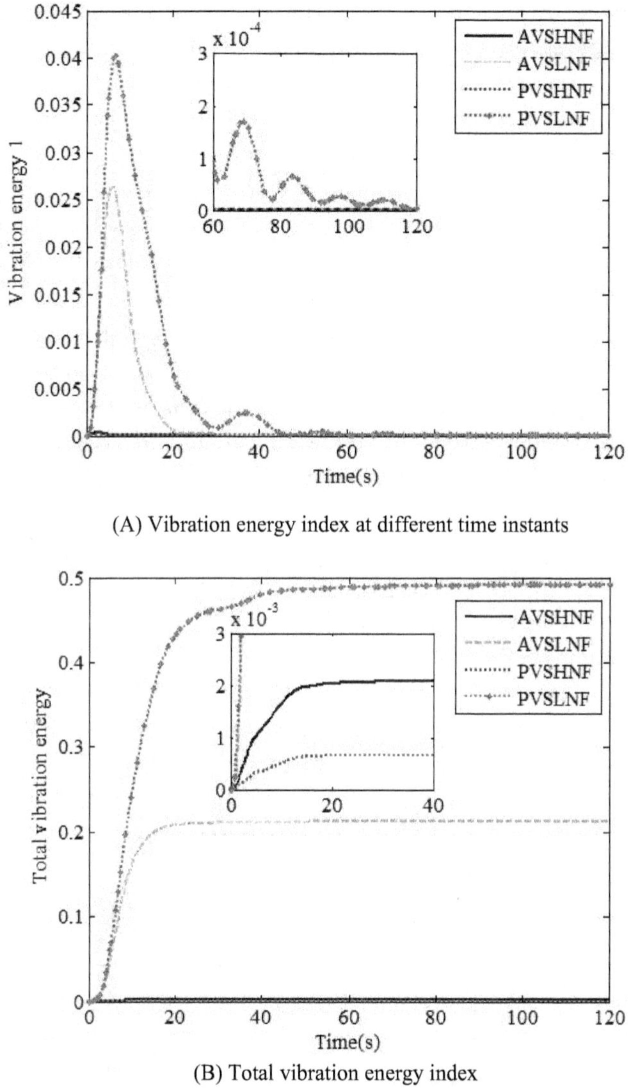

Figure 8.17 The vibration energy index under multiplicative perturbation.

Table 8.4 The evaluation index under multiplicative perturbation.

Type	Convergence time of Θ (s)	Steady-state error of Θ (rad)
AVSHNF	50	5.3×10^{-5}
AVSLNF	102	1.1×10^{-4}
PVSHNF	48	3.7×10^{-5}
PVSLNF	112	2.5×10^{-4}

nearly same. However, again of greater interest is the comparison of the vibration suppression effect. To further illustrate the vibration suppression performance, Fig. 8.17A and B present the vibration energy index at different time instants and the total vibration energy, which verifies the effectiveness of the active vibration suppression technique for the background of low natural frequencies and the effectiveness of the passive vibration suppression technique for high natural frequencies with respect to vibration energy. It is clear that the vibration energy for low natural frequencies is much greater than that for high natural frequencies. It can be concluded that by the measures of comparison of this chapter, the proposed nonfragile controller here can achieve attitude stabilization and vibration suppression simultaneously for H_∞ performance and quadratic stability, and in the presence of model parameter uncertainty, multiplicative perturbation, external disturbances, and input constraints.

8.4.3 Simulation comparison analysis

For comparative purposes, three different cases are employed for simulations, that is, the AVSHNF scheme, the ISAVS scheme presented in [32], and the case when the design scheme becomes ineffective. First, for the AVSHNF scheme and the ISAVS scheme presented in [32], the identical simulation parameters produce the following results, as shown in Figs. 8.18-8.23. Figs. 8.18 and 8.19 display the attitude angle and

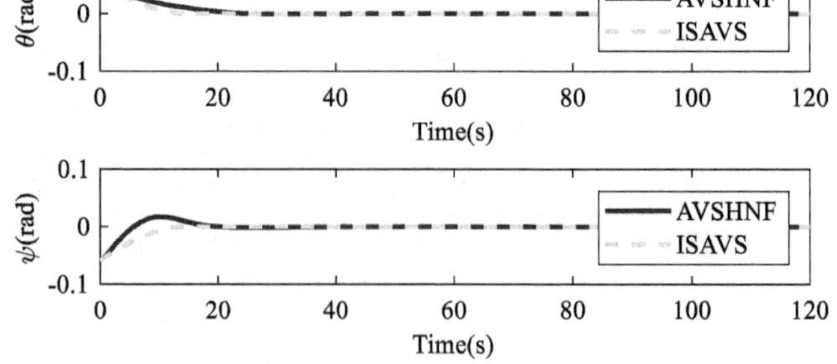

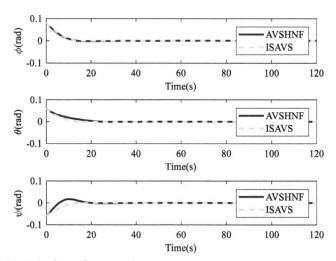

Figure 8.18 Attitude angle comparison.

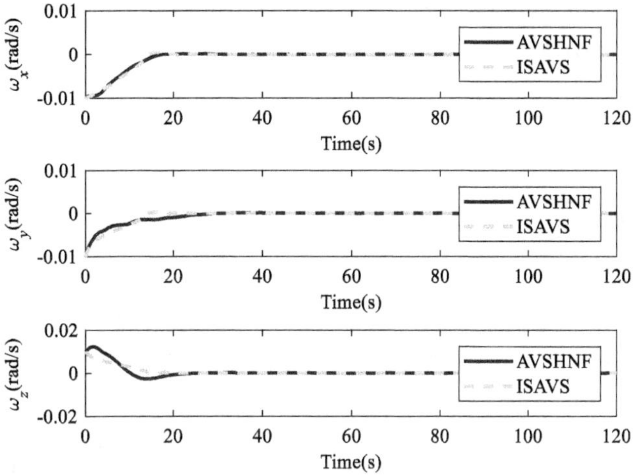

Figure 8.19 Angular velocity comparison.

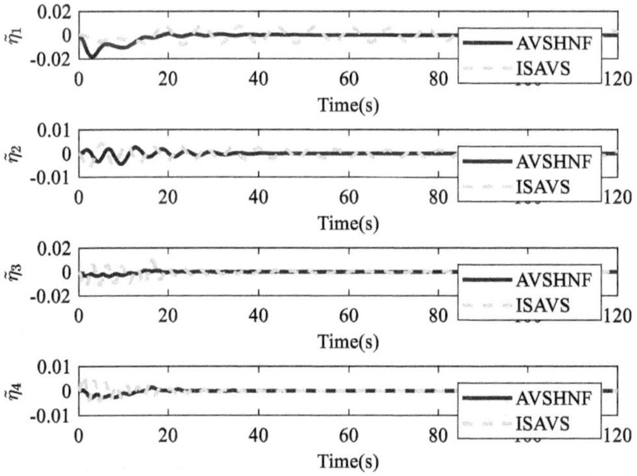

Figure 8.20 Modal displacement comparison.

angular velocity of flexible spacecraft under the AVSHNF scheme and the ISAVS scheme, from which one can see that the attitude angle and angular velocity all converge to within a small bound. It should be noted that they two have similar convergence time. Fig. 8.20 displays the modal displacement of the flexible appendages. It is observed that the response curves of modal displacement are quite different under different controllers, especially for the first and second order mode.

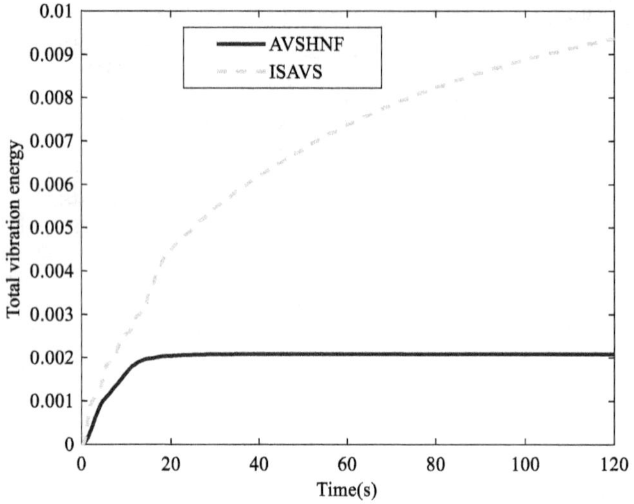

Figure 8.21 Total vibration energy index comparison.

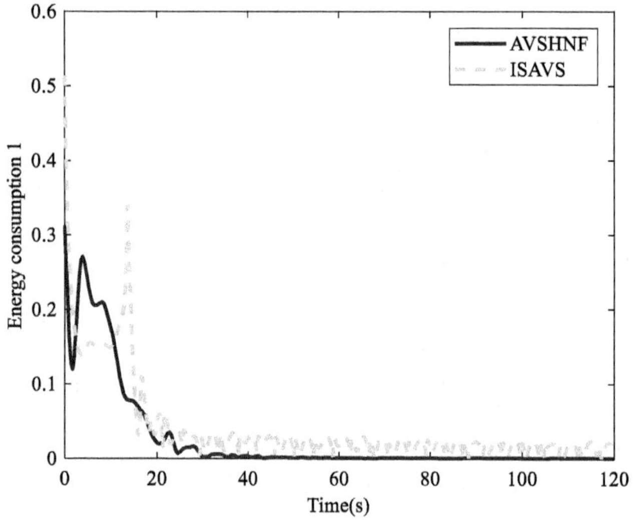

Figure 8.22 Energy consumption index comparison at different time instants.

Then, it is concluded that the AVSHNF scheme has a better performance in terms of vibration suppression. Furthermore, Fig. 8.21 also verifies the superiority of the proposed control scheme in performance of vibration suppression. The comparisons of the energy consumption index at different time instants and the total energy consumption index are shown in Figs. 8.22 and 8.23, respectively. It is observed ISAVS scheme would evidently require much more energy compared with the proposed control scheme. Thus the proposed control scheme is superior to the ISAVS scheme when it comes to performance of vibration suppression and total energy consumption.

Then, for the case when the design scheme becomes ineffective, the simulation results are shown in Figs. 8.24–8.27. It is observed that the

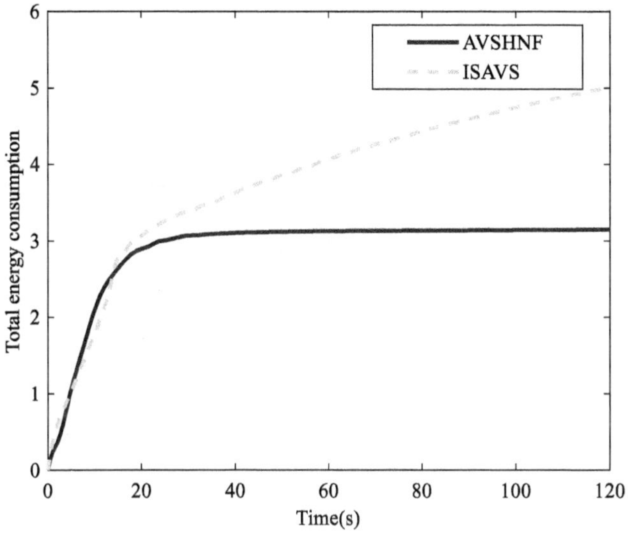

Figure 8.23 Total energy consumption index comparison.

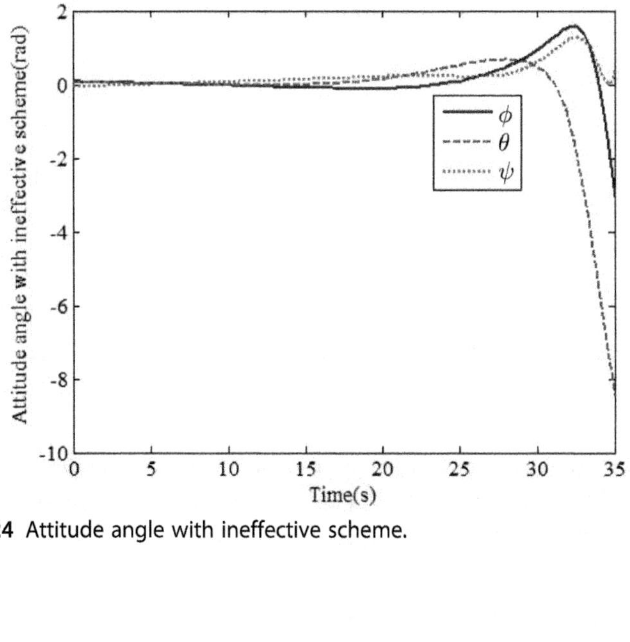

Figure 8.24 Attitude angle with ineffective scheme.

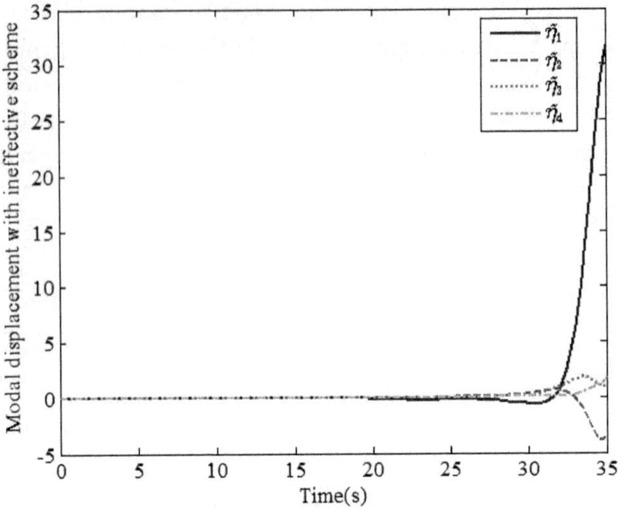

Figure 8.25 Modal displacement with ineffective scheme.

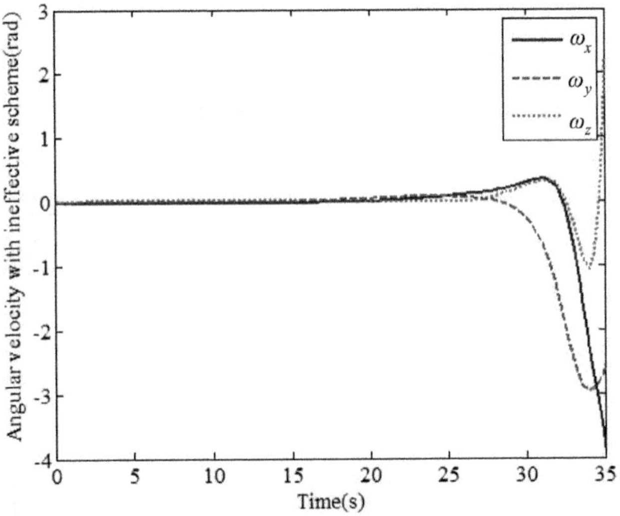

Figure 8.26 Angular velocity with ineffective scheme.

performance of attitude control system has a sharp decline after 25 seconds, and the energy consumption and total vibration energy increase rapidly with increasing time, leading to eventual failure of spacecraft.

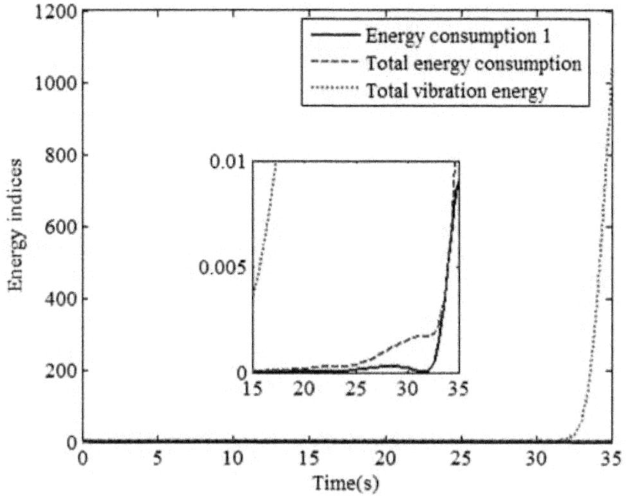

Figure 8.27 Energy indices with ineffective scheme.

8.5 Conclusions

In this work, the attitude stabilization and vibration suppression problem for flexible spacecraft in the presence of model parameter uncertainty, controller perturbations, external disturbances, and input constraint is investigated via a state feedback controller. The controller designed in this chapter is a nonfragile controller for improving tolerance to uncertainties in the controller, and it can control the attitude while eliminating the influence of unwanted flexible appendage vibrations and external disturbances. The controller gain is obtained by solving LMIs subject to input constraints, and the stability of flexible spacecraft closed-loop system is proved via Lyapunov methods. By using the nonfragile H_∞ controller developed here, attitude stabilization and vibration suppression is achieved with very good performance regardless of system uncertainties or input constraints. With respect to controller perturbations, two cases, that of additive and multiplicative perturbations, are taken into account in controller design and simulation analysis. The simulation results show that for both types of perturbation the passive vibration suppression scheme is superior for high natural frequencies while the active vibration suppression scheme is superior for low natural frequencies, and the low natural frequencies have a greater effect on the performance of attitude stabilization and vibration suppression than do the high frequencies. In addition, the superiority of the proposed control scheme is also verified by means of comparisons.

References

[1] Yuan Q, Liu Y, Qi N. Active vibration suppression for maneuvering spacecraft with high flexible appendages. Acta Astronautica 2017;139:512–20.

[2] Khot NS, Heise SA. Consideration of plant uncertainties in the optimum structural-control design. AIAA Journal 1994;32(3):610–15.

[3] Keel LH, Bhattacharyya SP. Robust, fragile, or optimal? IEEE Transactions on Automatic Control 1997;42(8):1098–105.

[4] Wu YH, Han F, Zheng M, He M, Chen ZM, Hua B, et al. Attitude control for on-orbit servicing spacecraft using hybrid actuator. Advances in Space Research 2018;61(6):1600–16.

[5] Wu YH, Han F, Hua B, Chen ZM. Null motion strategy for spacecraft large angle agile maneuvering using hybrid actuators. Acta Astronautica 2017;140:459–68.

[6] Sivaranjani K, Rakkiyappan R, Lakshmanan S, Lim CP. Robust non-fragile control for offshore steel jacket platform with nonlinear perturbations. Nonlinear Dynamics 2015;81(4):2043–57.

[7] Wu ZG, Dong S, Shi P, Su H, Huang T, Lu R. Fuzzy-model-based nonfragile guaranteed cost control of nonlinear Markov jump systems. IEEE Transactions on Systems, Man, and Cybernetics: Systems 2017;47(8):2388–97.

[8] Chen T, Chen G. Distributed adaptive tracking control of multiple flexible spacecraft under various actuator and measurement limitations. Nonlinear Dynamics 2018;91 (3):1571–86.

[9] Cao X, Yue C, Liu M, Wu B. Time efficient spacecraft maneuver using constrained torque distribution. Acta Astronautica 2016;123:320–9.

[10] Bang H, Ha CK, Kim JH. Flexible spacecraft attitude maneuver by application of sliding mode control. Acta Astronautica 2005;57(11):841–50.

[11] Cao L, Chen X, Misra AK. Minimum sliding mode error feedback control for fault tolerant reconfigurable satellite formations with J2 perturbations. Acta Astronautica 2014;96:201–16.

[12] Shahravi M, Kabganian M, Alasty A. Adaptive robust attitude control of a flexible spacecraft. International Journal of Robust and Nonlinear Control 2006;16 (6):287–302.

[13] Ma Y, Jiang B, Tao G, Cheng Y. Uncertainty decomposition-based fault-tolerant adaptive control of flexible spacecraft. IEEE Transactions on Aerospace and Electronic Systems 2015;51(2):1053–68.

[14] Charbonnel C. H_∞ and LMI attitude control design: towards performances and robustness enhancement. Acta Astronautica 2004;54(5):307–14.

[15] Yan R, Wu Z. Attitude stabilization of flexible spacecraft via extended disturbance observer based controller. Acta Astronautica 2017;133:73–80.

[16] Chak YC, Varatharajoo R, Razoumny Y. Disturbance observer-based fuzzy control for flexible spacecraft combined attitude & sun tracking system. Acta Astronautica 2017;133:302–10.

[17] Thakur D, Srikant S, Akella MR. Adaptive attitude-tracking control of spacecraft with uncertain time-varying inertia parameters. Journal of Guidance, Control, and Dynamics 2015;38(1):41–52.

[18] Lee D. Nonlinear disturbance observer-based robust control of attitude tracking of rigid spacecraft. Nonlinear Dynamics 2017;88(2):1317–28.

[19] Xiao B, Yin S, Wu L. A structure simple controller for satellite attitude tracking maneuver. IEEE Transactions on Industrial Electronics 2017;64(2):1436–46.

[20] Cao L, Xiao B, Golestani M. Robust fixed-time attitude stabilization control of flexible spacecraft with actuator uncertainty. Nonlinear Dynamics 2020;100 (3):2505–19.

[21] Song G, Agrawal BN. Vibration suppression of flexible spacecraft during attitude control. Acta Astronautica 2001;49(2):73–83.

[22] Di Gennaro S. Output stabilization of flexible spacecraft with active vibration suppression. IEEE Transactions on Aerospace and Electronic systems 2003;39 (3):747–59.

[23] Moon SH, Chwa D, Kim SJ. Feedback linearization control for panel flutter suppression with piezoelectric actuators. AIAA journal 2005;43(9):2069–73.

[24] Sales TP, Rade DA, De Souza LC. Passive vibration control of flexible spacecraft using shunted piezoelectric transducers. Aerospace Science and Technology 2013;29 (1):403–12.

[25] Sung YG, Singhose WE. Robustness analysis of input shaping commands for two-mode flexible systems. IET Control Theory & Applications 2009;3(6):722–30.

[26] Singhose WE, Banerjee AK, Seering WP. Slewing flexible spacecraft with deflection-limiting input shaping. Journal of Guidance, Control, and Dynamics 1997;20(2):291–8.

[27] Bai H, Zhou Y, Sun H, Zeng J. Observer-based non-linear H_∞ attitude control for a flexible satellite. IET Control Theory & Applications 2017;11(15):2403–11.

[28] Liu C, Sun Z, Shi K, Wang F. Robust non-fragile state feedback attitude control for uncertain spacecraft with input saturation. Proceedings of the Institution of Mechanical Engineers, Part G: Journal of Aerospace Engineering 2018;232 (2):246–59.

[29] Liu C, Sun Z, Shi K, Wang F. Robust dynamic output feedback control for attitude stabilization of spacecraft with nonlinear perturbations. Aerospace Science and Technology 2017;64:102–21.

[30] Wu YH, Han F, Zhang SJ, Hua B, Chen ZM. Attitude agile maneuvering control for spacecraft equipped with hybrid actuators. Journal of Guidance, Control, and Dynamics 2018;41(3):809–12.

[31] Zhong C, Chen Z, Guo Y. Attitude control for flexible spacecraft with disturbance rejection. IEEE Transactions on Aerospace and Electronic Systems 2017;53 (1):101–10.

[32] Cao X, Yue C, Liu M. Flexible satellite attitude maneuver via constrained torque distribution and active vibration suppression. Aerospace Science and Technology 2017;67:387–97.

[33] Boskovic JD, Li SM, Mehra RK. Robust tracking control design for spacecraft under control input saturation. Journal of Guidance, Control, and Dynamics 2004;27 (4):627–33.

[34] Xu S, Sun G, Sun W. Takagi–Sugeno fuzzy model based robust dissipative control for uncertain flexible spacecraft with saturated time-delay input. ISA Transactions 2017;66:105–21.

[35] Liu C, Ye D, Shi K, Sun Z. Robust high-precision attitude control for flexible spacecraft with improved mixed H_2/H_∞ control strategy under poles assignment constraint. Acta Astronautica 2017;136:166–75.

[36] Baz A, Poh S. Performance of an active control system with piezoelectric actuators. Journal of Sound and Vibration 1988;126(2):327–43.

[37] Gennaro SD. Active vibration suppression in flexible spacecraft attitude tracking. Journal of Guidance, Control, and Dynamics 1998;21(3):400–8.

[38] Liu C, Vukovich G, Shi K, Sun Z. Robust fault tolerant nonfragile H_∞ attitude control for spacecraft via stochastically intermediate observer. Advances in Space Research 2018;62(9):2631–48.

[39] Kumar SV, Raja R, Anthoni SM, Cao J, Tu Z. Robust finite-time non-fragile sampled-data control for TS fuzzy flexible spacecraft model with stochastic actuator faults. Applied Mathematics and Computation 2018;321:483–97.

[40] Zhou J, Li G, Wang H. Robust tracking controller design for non-Gaussian singular uncertainty stochastic distribution systems. Automatica 2014;50(4):1296–303.
[41] Hou Z, Geng Y, Huang S. Minimum residual vibrations for flexible satellites with frequency uncertainty. IEEE Transactions on Aerospace and Electronic Systems 2018;54(2):1029–38.
[42] Yang CD, Sun YP. Mixed H_2/H_∞ state-feedback design for microsatellite attitude control. Control Engineering Practice 2002;10(9):951–70.
[43] Liu C, Sun Z, Ye D, Shi K. Robust adaptive variable structure tracking control for spacecraft chaotic attitude motion. IEEE Access 2018;6:3851–7.

CHAPTER 9

Antidisturbance control with active vibration suppression

9.1 Introduction

Flexible appendages, for example, deployable antennas and solar panels, are widely used in spacecraft for their light weight, low energy consumption, and flexibility [1]. However, the disturbances caused by the elastic vibration of flexible appendages can considerably deteriorate the control performance and even lead to instability of the flexible spacecraft. The interactions of elastic motions with spacecraft attitude together with increasingly stringent requirements on attitude performance require that active vibration control be implemented for spacecraft maneuvers and stabilization [2]. As well, flexible spacecraft are often required to achieve high-precision pointing, rapid attitude maneuvers and stabilization in the presence of external disturbances, actuator faults, measurement errors, and even input constraints [3,4]. As stated in [5], parameter perturbations and external disturbances can be considered as uncertainties. Together, these challenges compound the control difficulties when one strives for high performance for attitude stabilization and vibration suppression of flexible spacecraft.

Recent years have witnessed many studies on attitude control of flexible spacecraft. For instance, an extended disturbance observer-based controller was proposed in [6], where the observer and backstepping feedback controller were combined. In [7], a fault-tolerant adaptive control scheme was developed for attitude tracking of flexible spacecraft with unknown inertia parameters, external disturbance, and actuator faults. The H_∞ control approach has been addressed in [8] and was proved to be an effective scheme to deal with external disturbances, parameter uncertainties, and imprecise collocation of sensors and actuators. In addition to the flexible spacecraft, the robust controller design problem for tail-sitters and quadrotors also considered external disturbances, parameter uncertainties, unmodeled uncertainties, or coupled dynamics [9,10]. To deal with these multiple uncertainties, the proposed controllers in [9] and [10] were both

Spacecraft Attitude Control. DOI: https://doi.org/10.1016/B978-0-323-99005-9.00009-2

partitioned into a position controller and an attitude controller based on the robust compensation theory and the backstepping approach, to achieve the desired translational and rotational motions, respectively. However, from the perspective of attitude controller design, the prior information on inertia matrix is needed. It should be noted that the measurement errors and input magnitude, and rate constraints (MRCs) are not taken into account in the aforementioned references. Despite the vast range of available techniques, however, the development of controllers for attitude stabilization of flexible spacecraft is a labor-intensive, time-consuming process. For applications in rapid response spacecraft, it is of great interest to employ control algorithms that are robust to inexact knowledge of the spacecraft inertia matrix and other uncertainties. In addition, it is often expensive to determine the mass properties with a high degree of accuracy [11]. To alleviate this requirement, the attitude controllers developed in [11] and [12] are inertia-free in the sense that no prior information on inertia matrix is required. However, only rigid spacecraft were discussed. The onboard actuators for attitude control of spacecraft, for example, thrusters, reaction wheels, and control moment gyros, are easy to suffer from input constraints [13]. Neglecting input constraints on both magnitude and dynamics can be source of undesirable or even catastrophic behavior for the closed-loop system. To cope with this critical issue, several methods have already been presented in the literature. For instance, a special Nussbaum-type function was introduced in [14] to deal with the time-varying nonlinear terms arising from input saturation. A standard hyperbolic tangent function was used in [15], and a saturation function was used in [16] to solve the input constraints problem. However, the aforementioned references only considered the input magnitude constraint. For safe use of actuators, the control constraint on input rate should also be considered [17]. Physical, safety, or technological constraints lead to the fact that actuators can neither provide unlimited magnitude signals nor unlimited speed of reaction.

Despite the partial successes of the methods described above, the problem of attitude stabilization and vibration suppression for flexible spacecraft remains challenging once the inertia matrix, and attitude and modal information are unknown. Furthermore, multiple uncertainties may appear in the attitude system of the flexible spacecraft, not to mention the high requirement resulting from input MRCs. It is thus clear that the problem of attitude stabilization and vibration suppression requires the development of effective control strategies to deal with the unknown and uncertain inertia matrix, external disturbances, actuator faults, measurement errors, and input MRCs concurrently.

Motivated by the preceding discussions, this chapter presents a novel intermediate observer-based control scheme without prior knowledge of inertia matrix, external disturbances and actuator faults to achieve attitude stabilization and vibration suppression of flexible spacecraft. The controller design is based on linear matrix inequalities (LMIs), which have become an effective technique in the field of stability and control [18]. Compared to previous studies on attitude control of flexible spacecraft, these new contributions are stated as follows. First, the nominal representative of a valid inertia matrix is introduced and the difference compared with the real inertia matrix is regarded as a source for lumped disturbances, such that the controller design is inertia-free. Second, a novel stochastic intermediate variable that contains the Bernoulli random variable is introduced, and its expectation is employed for observer construction. Third, the attitude, modal information, and the lumped disturbance can be estimated simultaneously with fast speed and high-precision. Fourth, by employing online estimation of the attitude and modal information and lumped disturbances, the objective of this chapter can be achieved without use of any prior fault information or measured attitude and modal information. Fifth, the active vibration suppression scheme and input MRCs are incorporated into controller design, such that the developed controller satisfies the given performance constraints.

The chapter is organized as follows. Section 9.2 presents a general attitude model of flexible spacecraft and converts it into a state-space form where an active vibration suppression scheme is incorporated, and the control objective of this chapter is formulated considering measurement errors, actuator faults and constraints. Section 9.3 presents the development of the H_∞ controller and the stability analysis of the closed-loop system via a Lyapunov approach. Numerical simulations are performed to illustrate the superior performance of the proposed controller with active vibration suppression scheme in Section 9.4. Finally, conclusions are presented in Section 9.5.

9.2 Problem formulation

9.2.1 Attitude dynamics modeling

This section describes the attitude model of a flexible spacecraft, which is a rigid body with two flexible appendages such as solar panels, antennas, or any other flexible structures. The attitude dynamics model is nonlinear and coupled with flexible motion. Here, a local-vertical local-horizontal frame $F_o(X_o, Y_o, Z_o)$ is defined as the reference frame with its origin located at the mass center of spacecraft. The axes of this reference frame are chosen such that the roll axis is in the flight direction, the pitch axis is perpendicular to the orbital plane, and the yaw axis points towards the center of the Earth. The body reference frame $F_B(X_B, Y_B, Z_B)$ is defined with its origin located at the mass center and its axes fixed in the spacecraft body to coincide with the principal axes of inertia. The kinematics of a flexible spacecraft can be described by [19]

$$\begin{bmatrix} \omega_x \\ \omega_y \\ \omega_z \end{bmatrix} = \begin{bmatrix} 1 & 0 & -\sin\theta \\ 0 & \cos\phi & \sin\phi\cos\theta \\ 0 & -\sin\phi & \cos\phi\cos\theta \end{bmatrix} \begin{bmatrix} \dot{\phi} \\ \dot{\theta} \\ \dot{\psi} \end{bmatrix} - \omega_0 \begin{bmatrix} \cos\theta\sin\psi \\ \sin\phi\sin\theta\sin\psi + \cos\phi\cos\psi \\ \cos\phi\sin\theta\sin\psi - \sin\phi\cos\psi \end{bmatrix} \tag{9.1}$$

where, ω_x, ω_y, and ω_z denote the three components of angular velocity; ϕ, θ, and ψ denote the three components of attitude angle and ω_0 is the orbital angular velocity.

If piezoelectric actuators are bonded to the surface of the flexible appendages to provide input voltage u_p that will generate deformation resulting in control torque, then the spacecraft body and flexible dynamics can be described by the following differential equations [20,21]:

$$\boldsymbol{J}\dot{\boldsymbol{\omega}} + \boldsymbol{\omega}^{\times}\left(\boldsymbol{J}\boldsymbol{\omega} + \tilde{\boldsymbol{\Delta}}\dot{\tilde{\eta}}\right) + \tilde{\boldsymbol{\Delta}}\ddot{\tilde{\eta}} = \boldsymbol{T}_c + \boldsymbol{T}_d \tag{9.2}$$

$$\ddot{\tilde{\eta}} + \tilde{\boldsymbol{C}}\dot{\tilde{\eta}} + \tilde{\boldsymbol{K}}\tilde{\eta} + \tilde{\boldsymbol{\Delta}}^T\dot{\boldsymbol{\omega}} = -\tilde{\boldsymbol{\Delta}}_p u_p \tag{9.3}$$

where $\boldsymbol{J}\in\mathbb{R}^{3\times 3}$ is the inertia matrix; $\boldsymbol{\omega}$ is the angular velocity vector including roll, pitch and yaw angular rates of Eq. (9.1); $\tilde{\boldsymbol{\Delta}}\in\mathbb{R}^{3\times m}$ is the coupling coefficient matrix between the rigid and elastic structures; $\tilde{\eta}\in\mathbb{R}^m$ is the modal coordinate vector relative to the main body; $\boldsymbol{T}_c\in\mathbb{R}^3$ and $\boldsymbol{T}_d\in\mathbb{R}^3$ denote the control input torque and external disturbance torque; $\tilde{\boldsymbol{C}} = diag\{[2\xi_1\Omega_1 \quad 2\xi_2\Omega_2 \quad \cdots \quad 2\xi_m\Omega_m]\}\in\mathbb{R}^{m\times m}$ is the modal damping matrix, where ξ_i and Ω_i $(i = 1, 2, \cdots, m)$ are the damping ratios and the natural frequencies; $\tilde{\boldsymbol{K}} = diag\{[\Omega_1^2 \quad \Omega_2^2 \quad \cdots \quad \Omega_m^2]\}\in\mathbb{R}^{m\times m}$ is the stiffness matrix, and m is the number of elastic modes considered. Here, $\boldsymbol{T}_d$ includes gravity-gradient torque, solar radiation pressure torque and aerodynamic torque; u_p is the piezoelectric input and $\tilde{\boldsymbol{\Delta}}_p$ is the corresponding coupling matrix.

Defining an auxiliary variable $\boldsymbol{q}=\dot{\tilde{\boldsymbol{\eta}}}+\tilde{\boldsymbol{\Delta}}^T\boldsymbol{\omega}$, one has

$$\dot{\boldsymbol{q}}=\ddot{\tilde{\boldsymbol{\eta}}}+\tilde{\boldsymbol{\Delta}}^T\dot{\boldsymbol{\omega}}=-\tilde{\boldsymbol{C}}\boldsymbol{q}+\tilde{\boldsymbol{C}}\tilde{\boldsymbol{\Delta}}^T\boldsymbol{\omega}-\tilde{\boldsymbol{K}}\tilde{\boldsymbol{\eta}}-\tilde{\boldsymbol{\Delta}}_p\boldsymbol{u}_p \tag{9.4}$$

Then, Eq. (9.2) becomes

$$\begin{aligned}&\left(\boldsymbol{J}-\tilde{\boldsymbol{\Delta}}\tilde{\boldsymbol{\Delta}}^T\right)\dot{\boldsymbol{\omega}}+\boldsymbol{\omega}^{\times}\left(\left(\boldsymbol{J}-\tilde{\boldsymbol{\Delta}}\tilde{\boldsymbol{\Delta}}^T\right)\boldsymbol{\omega}+\tilde{\boldsymbol{\Delta}}\boldsymbol{q}\right)\\&\qquad+\tilde{\boldsymbol{\Delta}}\left(-\tilde{\boldsymbol{C}}\boldsymbol{q}+\tilde{\boldsymbol{C}}\tilde{\boldsymbol{\Delta}}^T\boldsymbol{\omega}-\tilde{\boldsymbol{K}}\tilde{\boldsymbol{\eta}}-\tilde{\boldsymbol{\Delta}}_p\boldsymbol{u}_p\right)=\boldsymbol{T}_c+\boldsymbol{T}_d\end{aligned} \tag{9.5}$$

Denoting $\boldsymbol{\Theta}=\begin{bmatrix}\phi & \theta & \psi\end{bmatrix}^T$, for small Euler angles at very high altitudes, where the influence of ω_0 can be ignored, Eq. (9.1) is approximated as Eq. (9.6) below [22].

$$\dot{\boldsymbol{\Theta}}=\boldsymbol{\omega} \tag{9.6}$$

Denoting $\boldsymbol{J}_0=\boldsymbol{J}-\tilde{\boldsymbol{\Delta}}\tilde{\boldsymbol{\Delta}}^T$, with $\tilde{\boldsymbol{\Delta}}\tilde{\boldsymbol{\Delta}}^T$ as the contribution of the flexible parts to the total inertia matrix, this allows Eqs. (9.4)-(9.6) to be converted into the following state-space form:

$$\begin{cases}\dot{\boldsymbol{x}}(t)=(\boldsymbol{A}+\Delta\boldsymbol{A})\boldsymbol{x}(t)+(\boldsymbol{B}_1+\Delta\boldsymbol{B}_1)\boldsymbol{u}(t)+(\boldsymbol{B}_2+\Delta\boldsymbol{B}_2)\boldsymbol{w}_0(t)+(\boldsymbol{B}_3+\Delta\boldsymbol{B}_3)\boldsymbol{e}_x(t)\\ \boldsymbol{y}(t)=\boldsymbol{C}_1\boldsymbol{x}(t)\end{cases} \tag{9.7}$$

where $\boldsymbol{x}(t)=\begin{bmatrix}\boldsymbol{\Theta}^T & \boldsymbol{\omega}^T & \boldsymbol{q}^T & \tilde{\boldsymbol{\eta}}^T\end{bmatrix}^T$ is the state vector, $\boldsymbol{y}(t)=[\boldsymbol{\Theta}^T\ \boldsymbol{\omega}^T\boldsymbol{q}^T{}^T\tilde{\boldsymbol{\eta}}^T]^T$ is the output vector, $\boldsymbol{u}(t)=\boldsymbol{T}_c$ is the control input torque, $\boldsymbol{e}_x(t)$ is the estimation error of $\boldsymbol{x}(t)$ obtained via the proposed observer below. Because $-\boldsymbol{\omega}^{\times}\left(\boldsymbol{J}_0\boldsymbol{\omega}+\tilde{\boldsymbol{\Delta}}\boldsymbol{q}\right)$ is bounded in the practical flexible spacecraft attitude system, it can be treated as a disturbance on the spacecraft, then $\boldsymbol{w}_0(t)=\boldsymbol{T}_d-\boldsymbol{\omega}^{\times}\left(\boldsymbol{J}_0\boldsymbol{\omega}+\tilde{\boldsymbol{\Delta}}\boldsymbol{q}\right)$ is a lumped disturbance. The known coefficient matrices are denoted by

$$\boldsymbol{A}=\begin{bmatrix}0 & \boldsymbol{I} & 0 & 0\\ 0 & -\boldsymbol{J}_{0n}^{-1}\tilde{\boldsymbol{\Delta}}\tilde{\boldsymbol{C}}\tilde{\boldsymbol{\Delta}}^T & \boldsymbol{J}_{0n}^{-1}\tilde{\boldsymbol{\Delta}}\left(\tilde{\boldsymbol{C}}+\tilde{\boldsymbol{\Delta}}_p\boldsymbol{F}_2\right) & \boldsymbol{J}_{0n}^{-1}\tilde{\boldsymbol{\Delta}}\left(\tilde{\boldsymbol{K}}+\tilde{\boldsymbol{\Delta}}_p\boldsymbol{F}_1\right)\\ 0 & \tilde{\boldsymbol{C}}\tilde{\boldsymbol{\Delta}}^T & -\left(\tilde{\boldsymbol{C}}+\tilde{\boldsymbol{\Delta}}_p\boldsymbol{F}_2\right) & -\left(\tilde{\boldsymbol{K}}+\tilde{\boldsymbol{\Delta}}_p\boldsymbol{F}_1\right)\\ 0 & -\tilde{\boldsymbol{\Delta}}^T & \boldsymbol{I} & 0\end{bmatrix},$$

$$\boldsymbol{B}_1=\boldsymbol{B}_2=\begin{bmatrix}0\\ \boldsymbol{J}_{0n}^{-1}\\ 0\\ 0\end{bmatrix}$$

$$\boldsymbol{B}_3 = \begin{bmatrix} 0 & 0 & 0 & 0 \\ 0 & 0 & -\boldsymbol{J}_{0n}^{-1}\tilde{\boldsymbol{\Delta}}\tilde{\boldsymbol{\Delta}}_p\boldsymbol{F}_2 & -\boldsymbol{J}_{0n}^{-1}\tilde{\boldsymbol{\Delta}}\tilde{\boldsymbol{\Delta}}_p\boldsymbol{F}_1 \\ 0 & 0 & \tilde{\boldsymbol{\Delta}}_p\boldsymbol{F}_2 & \tilde{\boldsymbol{\Delta}}_p\boldsymbol{F}_1 \\ 0 & 0 & 0 & 0 \end{bmatrix}, \quad \boldsymbol{C}_1 = \boldsymbol{I}_{6+2m}$$

and the unknown coefficient matrices are denoted by

$$\Delta \boldsymbol{A} = \boldsymbol{A}(\boldsymbol{J}) - \boldsymbol{A}(\boldsymbol{J}_n)$$

$$= \begin{bmatrix} 0 & 0 & 0 & 0 \\ 0 & (\boldsymbol{J}_{0n}^{-1} - \boldsymbol{J}_0^{-1})\tilde{\boldsymbol{\Delta}}\tilde{\boldsymbol{C}}\tilde{\boldsymbol{\Delta}}^T & (\boldsymbol{J}_0^{-1} - \boldsymbol{J}_{0n}^{-1})\tilde{\boldsymbol{\Delta}}(\tilde{\boldsymbol{C}}+\tilde{\boldsymbol{\Delta}}_p\boldsymbol{F}_2) & (\boldsymbol{J}_0^{-1} - \boldsymbol{J}_{0n}^{-1})\tilde{\boldsymbol{\Delta}}(\tilde{\boldsymbol{K}}+\tilde{\boldsymbol{\Delta}}_p\boldsymbol{F}_1) \\ 0 & 0 & 0 & 0 \\ 0 & 0 & 0 & 0 \end{bmatrix}$$

$$\Delta \boldsymbol{B}_1 = \Delta \boldsymbol{B}_2 = \boldsymbol{B}_1(\boldsymbol{J}) - \boldsymbol{B}_1(\boldsymbol{J}_n)$$

$$= \begin{bmatrix} 0 \\ \boldsymbol{J}_0^{-1} - \boldsymbol{J}_{0n}^{-1} \\ 0 \\ 0 \end{bmatrix}$$

$$\Delta \boldsymbol{B}_3 = \boldsymbol{B}_3(\boldsymbol{J}) - \boldsymbol{B}_3(\boldsymbol{J}_n)$$

$$= \begin{bmatrix} 0 & 0 & 0 & 0 \\ 0 & 0 & (\boldsymbol{J}_{0n}^{-1} - \boldsymbol{J}_0^{-1})\tilde{\boldsymbol{\Delta}}\tilde{\boldsymbol{\Delta}}_p\boldsymbol{F}_2 & (\boldsymbol{J}_{0n}^{-1} - \boldsymbol{J}_0^{-1})\tilde{\boldsymbol{\Delta}}\tilde{\boldsymbol{\Delta}}_p\boldsymbol{F}_1 \\ 0 & 0 & 0 & 0 \\ 0 & 0 & 0 & 0 \end{bmatrix}$$

where $\boldsymbol{J}_{0n} = \boldsymbol{J}_n - \tilde{\boldsymbol{\Delta}}\tilde{\boldsymbol{\Delta}}^T$ and $\tilde{\boldsymbol{\Gamma}} = \tilde{\boldsymbol{\Gamma}}(\boldsymbol{J}_n)$ with $\tilde{\boldsymbol{\Gamma}} = \boldsymbol{A}, \boldsymbol{B}_1, \boldsymbol{B}_2, \boldsymbol{B}_3$.

The derivations are as follows:

In this case, the piezoelectric input u_p is calculated by using a feedback method with $u_p = \boldsymbol{F}_1\hat{\boldsymbol{\eta}} + \boldsymbol{F}_2\hat{\boldsymbol{q}}$, where $\hat{\boldsymbol{\eta}}$ is the estimation of $\tilde{\boldsymbol{\eta}}$ and $\hat{\boldsymbol{q}}$ is the estimation of $\boldsymbol{q}$; $\boldsymbol{F}_1$ and $\boldsymbol{F}_2$ are matrices introduced later, designed such that the eigenvalues of the controllable flexible dynamics are placed where desired. This is reasonable, with details as follows.

Note that the piezoelectric input u_p can be rewritten as

$$u_p = \boldsymbol{F}_1\hat{\boldsymbol{\eta}} + \boldsymbol{F}_2\hat{\boldsymbol{q}} = \boldsymbol{F}_1\tilde{\boldsymbol{\eta}} + \boldsymbol{F}_2\boldsymbol{q} - \boldsymbol{F}_1(\tilde{\boldsymbol{\eta}} - \hat{\boldsymbol{\eta}}) - \boldsymbol{F}_2(\boldsymbol{q} - \hat{\boldsymbol{q}})$$

If the controller designed can guarantee that the estimation errors $(\tilde{\boldsymbol{\eta}} - \hat{\boldsymbol{\eta}})$ and $(\boldsymbol{q} - \hat{\boldsymbol{q}})$ tend asymptotically to within a small neighborhood containing zero, the controllable eigenvalues of the flexible dynamics remain fixed. In fact, Eq. (9.3) is

$$\begin{aligned}\ddot{\tilde{\eta}} &= -\tilde{C}\dot{\tilde{\eta}} - \tilde{K}\tilde{\eta} - \tilde{\Delta}^T\dot{\omega} - \tilde{\Delta}_p u_p \\ &= -\tilde{C}\dot{\tilde{\eta}} - \tilde{K}\tilde{\eta} - \tilde{\Delta}^T\dot{\omega} - \tilde{\Delta}_p(F_1\tilde{\eta} + F_2(\dot{\tilde{\eta}} + \tilde{\Delta}^T\omega) \\ &\quad - F_1(\tilde{\eta} - \hat{\eta}) - F_2(q - \hat{q})) \\ &= -(\tilde{C}+\tilde{\Delta}_p F_2)\dot{\tilde{\eta}} - (\tilde{K}+\tilde{\Delta}_p F_1)\tilde{\eta} - \tilde{\Delta}^T\dot{\omega} - \tilde{\Delta}_p F_2\tilde{\Delta}^T\omega \\ &\quad + \tilde{\Delta}_p F_1(\tilde{\eta} - \hat{\eta}) + \tilde{\Delta}_p F_2(q - \hat{q})\end{aligned} \tag{9.8}$$

If the attitude maneuver is completed, that is, $\omega = 0$, $\dot{\omega} = 0$, and if the estimation errors $(\tilde{\eta} - \hat{\eta})$ and $(q - \hat{q})$ are zero, which can be achieved by Theorem 9.1 or Theorem 9.2, Eq. (9.8) becomes

$$\ddot{\tilde{\eta}} = -(\tilde{C}+\tilde{\Delta}_p F_2)\dot{\tilde{\eta}} - (\tilde{K}+\tilde{\Delta}_p F_1)\tilde{\eta} \tag{9.9}$$

Defining an auxiliary variable $\tilde{x} = [\,\tilde{\eta}^T \quad \dot{\tilde{\eta}}^T\,]^T$, one has

$$\dot{\tilde{x}} = \tilde{A}\tilde{x} \tag{9.10}$$

where,

$$\tilde{A} = \begin{bmatrix} 0 & I \\ -(\tilde{K}+\tilde{\Delta}_p F_1) & -(\tilde{C}+\tilde{\Delta}_p F_2) \end{bmatrix}$$

Choose a Lyapunov function candidate as

$$V(t) = \tilde{x}^T\tilde{P}\tilde{x}$$

The time derivative of $V(t)$ yields

$$\dot{V}(t) = \dot{\tilde{x}}^T\tilde{P}\tilde{x} + \tilde{x}^T\tilde{P}\dot{\tilde{x}} = \tilde{x}^T\left(\tilde{A}^T\tilde{P} + \tilde{P}\tilde{A}\right)\tilde{x}$$

Then, the system denoted by Eq. (9.10) is Hurwitz stabilizable if

$$\tilde{A}^T\tilde{P} + \tilde{P}\tilde{A} < 0 \tag{9.11}$$

By choosing an appropriate symmetric positive definite matrix $\tilde{P}$, the matrices F_1 and F_2 can be obtained.

In this case, one has $u_p = F_1\hat{\eta} + F_2\hat{q} = F_1\tilde{\eta} + F_2 q - F_1 e_{\tilde{\eta}} - F_2 e_q$, where $e_{\tilde{\eta}} = \tilde{\eta} - \hat{\eta}$ and $e_q = q - \hat{q}$, then Eqs. (9.4) and (9.5) become the following two equations, respectively, that is,

$$\begin{aligned}\dot{q} = \ddot{\tilde{\eta}} + \tilde{\Delta}^T\dot{\omega} &= \tilde{C}\tilde{\Delta}^T\omega - \left(\tilde{C}+\tilde{\Delta}_p F_2\right)q \\ &\quad - \left(\tilde{K}+\tilde{\Delta}_p F_1\right)\tilde{\eta} + \tilde{\Delta}_p\left(F_1 e_{\tilde{\eta}} + F_2 e_q\right)\end{aligned} \tag{9.12}$$

$$\left(J-\tilde{\Delta}\tilde{\Delta}^{T}\right)\dot{\omega}+\omega^{\times}\left(\left(J-\tilde{\Delta}\tilde{\Delta}^{T}\right)\omega+\tilde{\Delta}q\right)+\tilde{\Delta}\tilde{\Delta}_{p}\left(F_{1}e_{\tilde{\eta}}+F_{2}e_{q}\right)$$
$$+\tilde{\Delta}\left(-\tilde{C}q+\tilde{C}\tilde{\Delta}^{T}\omega-\tilde{K}\tilde{\eta}-\tilde{\Delta}_{p}(F_{1}\tilde{\eta}+F_{2}q)\right)=T_{c}+T_{d} \tag{9.13}$$

Denoting $J_0 = J - \tilde{\Delta}\tilde{\Delta}^T$, with $\tilde{\Delta}\tilde{\Delta}^T$ as the contribution of the flexible parts to the total inertia matrix, and $w_0(t) = T_d - \omega^\times\left(J_0\omega + \tilde{\Delta}q\right)$ as the lumped disturbance, this allows Eqs. (9.6), (9.12), and (9.13) to be converted into the following state-space form:

$$\begin{cases} \dot{x}(t) = A(J)x(t) + B_1(J)u(t) + B_2(J)w_0(t) + B_3(J)e_x(t) \\ y(t) = C_1 x(t) \end{cases} \tag{9.14}$$

where $x(t) = [\Theta^T \quad \omega^T \quad q^T \quad \tilde{\eta}^T]^T$ is the state vector, $y(t) = [\Theta^T \ \omega^T q^T \tilde{\eta}^T]^T$ is the output vector, $u(t) = T_c$ is the control input torque, and $e_x(t) = [e_\Theta{}^T \quad e_\omega{}^T \quad e_q{}^T \quad e_{\tilde{\eta}}{}^T]^T$ is the estimation error of state vector. The coefficient matrices are denoted by

$$A(J)=\begin{bmatrix} 0 & I & 0 & 0 \\ 0 & -J_0^{-1}\tilde{\Delta}\tilde{C}\tilde{\Delta}^T & J_0^{-1}\tilde{\Delta}(\tilde{C}+\tilde{\Delta}_pF_2) & J_0^{-1}\tilde{\Delta}(\tilde{K}+\tilde{\Delta}_pF_1) \\ 0 & \tilde{C}\tilde{\Delta}^T & -(\tilde{C}+\tilde{\Delta}_pF_2) & -(\tilde{K}+\tilde{\Delta}_pF_1) \\ 0 & -\tilde{\Delta}^T & I & 0 \end{bmatrix}, B_1(J)=\begin{bmatrix} 0 \\ J_0^{-1} \\ 0 \\ 0 \end{bmatrix},$$

$$B_2(J)=B_1(J), B_3(J)=\begin{bmatrix} 0 & 0 & 0 & 0 \\ 0 & 0 & -J_0^{-1}\tilde{\Delta}\tilde{\Delta}_pF_2 & -J_0^{-1}\tilde{\Delta}\tilde{\Delta}_pF_1 \\ 0 & 0 & \tilde{\Delta}_pF_2 & \tilde{\Delta}_pF_1 \\ 0 & 0 & 0 & 0 \end{bmatrix}, C_1 = I_{6+2m}$$

Remark 9.1: Due to the parameter perturbations, modeling errors and environmental changes, inertia uncertainties widely exist in physical systems. If inertial uncertainties are not considered in the control system design, the controller obtained is highly likely to fail in practical situations. In particular, the prior information on inertia matrix may not be exactly known, hence the nominal representative of a valid inertia matrix J_n can be used for the controller design, then the inertia uncertainty resulting from the difference between J_n and J will appear. Taking into account the uncertain inertia matrix which can be also unknown, system matrices related to inertia matrix will have uncertainties.

In addition, the following important points should be noted.

1. Measurement errors

   Due to the unexpected variations in external surroundings, limitations of measurement technique, and the obsolescence problem for most of measuring equipment, there may appear measurement errors $v(t) \in L_2[0, +\infty)$ between the system outputs and the measurement outputs, especially for electronic measurements [23].

2. Actuator faults

   Since the performance of attitude control system can be severely deteriorated by improper actuator actions, actuator fault has been considered to be one of the most critical challenges to be solved. Taking this issue into account, let $\boldsymbol{E}$ with full-column rank, which has a similar structure with $\boldsymbol{B}_1$, denote the distribution matrix of fault signal $\boldsymbol{f}(t)$ appearing in the input. Specifically, it represents the process faults if $\boldsymbol{E} \neq \boldsymbol{B}_1$, and it also represents actuator faults if $\boldsymbol{E} = \boldsymbol{B}_1$.

3. Actuator constraints

The magnitude and rate constraints on control input that the actuators can provide are ubiquitous phenomena due to the limits on power capacity of actuators in real applications and other physical limitations, if unaccounted for, will often have a negative impact on the stability and performance of an attitude control system designed for the "ideal" case. To not violate the requirements necessary for input MRCs, two positive scalars λ_1 and λ_2 are introduced satisfying

$$\|\boldsymbol{u}\| < \lambda_1, \ \|\dot{\boldsymbol{u}}\| < \lambda_2 \tag{9.15}$$

where λ_1 and λ_2 denote the 2-norms of control input magnitude and rate, respectively.

Considering the above factors, the new form of spacecraft attitude control system can be expressed as:

$$\dot{\boldsymbol{x}}(t) = \boldsymbol{A}\boldsymbol{x}(t) + \boldsymbol{B}_1\boldsymbol{u}(t) + \boldsymbol{B}_2\boldsymbol{w}(t) + \boldsymbol{B}_3\boldsymbol{e}_x(t) \tag{9.16}$$

$$\boldsymbol{y}(t) = \boldsymbol{C}_1\boldsymbol{x}(t) + \boldsymbol{v}(t) \tag{9.17}$$

where,
$\boldsymbol{w}(t) = \boldsymbol{w}_0(t) + \boldsymbol{B}_2^*(\boldsymbol{\Delta A}\boldsymbol{x}(t) + \boldsymbol{\Delta B}_1\boldsymbol{u}(t) + \boldsymbol{\Delta B}_2\boldsymbol{w}_0(t) + \boldsymbol{\Delta B}_3\boldsymbol{e}_x(t) + \boldsymbol{E}\boldsymbol{f}(t))$ is a new lumped disturbance considering inertia uncertainties and fault signals, which is to be considered in controller design, and where the matrix $\boldsymbol{B}_2^*$ is the pseudoinverse of $\boldsymbol{B}_2$.

Remark 9.2: It is obvious that the uncertain inertia matrix $\boldsymbol{J}$, which can be also unknown, only affects the second channel of $\boldsymbol{x}(t)$, that is, $\boldsymbol{\omega}$. Thus the related combined uncertain term, that is, $\boldsymbol{\Delta A}\boldsymbol{x}(t) + \boldsymbol{\Delta B}_1\boldsymbol{u}(t) +$

$\Delta B_2 w_0(t) + \Delta B_3 e_x(t)$, will contribute to the disturbance, and the same applies to $Ef(t)$. Together they form the lumped disturbance $w(t)$.

9.2.2 Preliminaries

The subsection presents the following necessary assumptions, remarks, and lemmas, which are needed to provide theoretical support before further analysis.

Assumption 9.1: The disturbance torque $w_0(t)$ and its first time derivative are assumed to be bounded.

Remark 9.3: Assumption 9.1 is reasonable. The main reason is that the external disturbances primarily include magnetic torque, solar radiation, and aerodynamic torque, which are all continuous and bounded [24], [25]. In addition, because the total control authority is limited, all three available control input torques should be continuous and bounded, which is reasonable in practice [26]. It is worth noting that the desired attitude motion represented by attitude angle and angular rate is also bounded and differentiable in practice [27]. In particular, the derivative of $f(t)$ with respect to time is often assumed to be norm-bounded [28,29]. They together result in the fact that the lumped disturbance torque is bounded and differentiable, and the first derivative of $w(t)$ satisfies $\|\dot{w}(t)\| \le \kappa_1$, where $\kappa_1 \ge 0$. It should be noted that the norm of lumped disturbance $w(t)$ and its components could be unknown. Thus it is more general than the fault-tolerant sliding-mode-observer method in [30], which requires the preliminary knowledge of the bound of fault signals.

Assumption 9.2: The measurement error signal $v(t)$ satisfies $\|v(t)\| \le \kappa_2$ with $\kappa_2 \ge 0$, which describes the boundary of measurement errors.

Remark 9.4: Assumption 9.2 is common in literature [23,31]. For κ_2 could be unknown when H_∞ control method is applied, it is more general compared with [23] where κ_2 should be known.

Assumption 9.3: $\mathrm{rank}(B_1, E) = \mathrm{rank}(B_1)$, and there exists a matrix B_1^* such that $(I - B_1 B_1{}^*)E = 0$.

Remark 9.5: Assumption 9.3 is general in [32], which means that the faults appear in the actuator and can be compensated for by the control input.

Lemma 9.1: [8] Let $\overline{H}, \overline{E}$ and $\overline{F}(t)$ be real matrices of appropriate dimensions with $\|\overline{F}(t)\| \le 1$, then for any scalar $\tilde{\xi} > 0$, one has

$$\overline{H}\overline{F}(t)\overline{E} + \overline{E}^T\overline{F}(t)^T\overline{H}^T \le \tilde{\xi}^{-1}\overline{H}\overline{H}^T + \tilde{\xi}\overline{E}^T\overline{E}$$

Lemma 9.2: (Schur complement lemma) Let the partitioned matrix

$$A = \begin{bmatrix} A_{11} & A_{12} \\ * & A_{22} \end{bmatrix}$$

be symmetric. Then

$$A < 0 \Leftrightarrow A_{11} < 0, A_{22} - A_{12}^T A_{11}^{-1} A_{12} < 0 \Leftrightarrow A_{22} < 0, A_{11} - A_{12} A_{22}^{-1} A_{12}^T < 0$$

Lemma 9.3: [33] Let function $V(t) \ge 0$ be a continuous function defined $\forall t \ge 0$ and $V(0)$ bounded, if the following inequality holds:

$$\dot{V}(t) \le -c_1 V(t) + c_2$$

where $c_1 > 0, c_2$ are constants, it can be concluded that $V(t)$ is bounded.

9.2.3 Control objective

Based on the state-space Eq. (9.16), one should design an effective control scheme to achieve the following objectives: (1) the states of the closed-loop system are stabilized in the presence of inertia uncertainties, external disturbances, actuator faults, measurement errors and input MRCs; (2) the input MRCs, that is, $\|u\| < \lambda_1$ and $\|\dot{u}\| < \lambda_2$, are satisfied, where λ_1 and λ_2 are known positive constants; (3) the attitude and modal information as well as the lumped disturbance are estimated with high-precision.

9.3 Antidisturbance control law with input magnitude, and rate constraints

9.3.1 Stochastically intermediate observer design

In this section, an intermediate observer with stochastic failure is constructed for system (9.16). To introduce the observer, a stochastic intermediate variable is introduced first:

$$\overline{\xi}(t) = w(t) - \hat{\omega}(t)\Omega x(t) \tag{9.18}$$

and $\boldsymbol{\Omega}$ is designed as follows:

$$\boldsymbol{\Omega} = \eta \boldsymbol{B}_2^T \tag{9.19}$$

where η is a scalar.

The variable $\hat{\omega}(t)$ is a Bernoulli random variable and described as

$$\hat{\omega}(t) = \begin{cases} 1, & \textit{If success of } \boldsymbol{\Omega} x(t) \textit{ occurs} \\ 0, & \textit{If failure of } \boldsymbol{\Omega} x(t) \textit{ occurs} \end{cases} \tag{9.20}$$

One can write $\hat{\omega}(t) \sim Ber(\varpi)$, where ϖ is the probability of success, and the random variable $\hat{\omega}(t)$ has the following distribution

$$P(\hat{\omega}(t) = 1) = \varpi, P(\hat{\omega}(t) = 0) = 1 - \varpi \tag{9.21}$$

Let $\boldsymbol{\xi}(t)$ be the expectation of $\overline{\boldsymbol{\xi}}(t)$, one can conclude

$$\boldsymbol{\xi}(t) = \boldsymbol{w}(t) - \varpi \eta \boldsymbol{B}_2^T \boldsymbol{x}(t) \tag{9.22}$$

$$\dot{\boldsymbol{\xi}}(t) = \dot{\boldsymbol{w}}(t) - \varpi \eta \boldsymbol{B}_2^T \left[\boldsymbol{A}\boldsymbol{x}(t) + \boldsymbol{B}_1 \boldsymbol{u}(t) + \boldsymbol{B}_2 \boldsymbol{\xi}(t) + \varpi \eta \boldsymbol{B}_2 \boldsymbol{B}_2^T \boldsymbol{x}(t) + \boldsymbol{B}_3 \boldsymbol{e}_x(t) \right] \tag{9.23}$$

For $\boldsymbol{w}(t)$, $\boldsymbol{x}(t)$ and their first derivative are bounded, the intermediate variable $\boldsymbol{\xi}(t)$ and its first derivative are also bounded, which satisfy $\|\boldsymbol{\xi}(t)\| \le \kappa_3$, $\|\dot{\boldsymbol{\xi}}(t)\| \le \kappa_4$ with $\kappa_3 > 0$, $\kappa_4 \ge 0$. Then, the intermediate observer is designed based on

$$\dot{\hat{\boldsymbol{x}}}(t) = \boldsymbol{A}\hat{\boldsymbol{x}}(t) + \boldsymbol{B}_1 \boldsymbol{u}(t) + \boldsymbol{B}_2 \hat{\boldsymbol{w}}(t) + \boldsymbol{L}(\boldsymbol{y}(t) - \hat{\boldsymbol{y}}(t)) \tag{9.24}$$

$$\dot{\hat{\boldsymbol{\xi}}}(t) = -\varpi \eta \boldsymbol{B}_2^T \left[\boldsymbol{A}\hat{\boldsymbol{x}}(t) + \boldsymbol{B}_1 \boldsymbol{u}(t) + \boldsymbol{B}_2 \hat{\boldsymbol{\xi}}(t) + \varpi \eta \boldsymbol{B}_2 \boldsymbol{B}_2^T \hat{\boldsymbol{x}}(t) \right] \tag{9.25}$$

$$\hat{\boldsymbol{y}}(t) = \boldsymbol{C}_1 \hat{\boldsymbol{x}}(t) \tag{9.26}$$

$$\hat{\boldsymbol{w}}(t) = \hat{\boldsymbol{\xi}}(t) + \varpi \eta \boldsymbol{B}_2^T \hat{\boldsymbol{x}}(t) \tag{9.27}$$

where $\hat{\boldsymbol{x}}(t)$, $\hat{\boldsymbol{\xi}}(t)$, $\hat{\boldsymbol{y}}(t)$, and $\hat{\boldsymbol{w}}(t)$ are the estimates of $\boldsymbol{x}(t)$, $\boldsymbol{\xi}(t)$, $\boldsymbol{y}(t)$, and $\boldsymbol{w}(t)$, respectively, and $\boldsymbol{L}$ is chosen such that $\boldsymbol{A} - \boldsymbol{L}\boldsymbol{C}_1$ is Hurwitz.

Remark 9.6: Compared with the traditional methods, the approach in this chapter is less conservative. For example, the methods in [34] and [35] have equation constraints, while the methods in [28] and [29] require observer matching condition. In contrast, the scheme in this chapter requires neither equation constraints nor observer matching condition. The intermediate variable introduced is stochastic, that is, it can tolerate stochastic failures. For the intermediate variable contains random variable $\varpi(t)$, its traditional derivative is meaningless, where the method in [31] cannot be applied directly. To tackle this difficulty, another intermediate variable based on the expectation of such a stochastic intermediate variable is defined, which is of great importance for the observer construction.

9.3.2 Antidisturbance controller design

Based on the above estimation, an antidisturbance controller is designed as

$$\boldsymbol{u}(t) = -\boldsymbol{K}\hat{\boldsymbol{x}}(t) - \boldsymbol{B}_1^*\boldsymbol{B}_2\hat{\boldsymbol{w}}(t) \tag{9.28}$$

where $\boldsymbol{K}$ is the controller gain matrix and $\boldsymbol{B}_1^*\boldsymbol{B}_2\hat{\boldsymbol{w}}(t)$ is added to the input to compensate for the effect of the lumped disturbance.

Denoting $\boldsymbol{e}_x(t) = \boldsymbol{x}(t) - \hat{\boldsymbol{x}}(t)$ and $\boldsymbol{e}_\xi(t) = \boldsymbol{\xi}(t) - \hat{\boldsymbol{\xi}}(t)$, substituting Eq. (9.28) into Eq. (9.16) yields

$$\dot{\boldsymbol{x}}(t) = (\boldsymbol{A} - \boldsymbol{B}_1\boldsymbol{K})\boldsymbol{x}(t) + \boldsymbol{B}_1\boldsymbol{K}\boldsymbol{e}_x(t) + \boldsymbol{B}_2\boldsymbol{e}_\xi(t) + \varpi\eta\boldsymbol{B}_2\boldsymbol{B}_2^T\boldsymbol{e}_x(t) + \boldsymbol{B}_3\boldsymbol{e}_x(t) \tag{9.29}$$

Then, it can be concluded that

$$\dot{\boldsymbol{e}}_x(t) = (\boldsymbol{A} - \boldsymbol{L}\boldsymbol{C}_1)\boldsymbol{e}_x(t) + \boldsymbol{B}_2\boldsymbol{e}_\xi(t) + \varpi\eta\boldsymbol{B}_2\boldsymbol{B}_2^T\boldsymbol{e}_x(t) + \boldsymbol{B}_3\boldsymbol{e}_x(t) - \boldsymbol{L}\boldsymbol{v}(t) \tag{9.30}$$

$$\dot{\boldsymbol{e}}_\xi(t) = \dot{\boldsymbol{w}}(t) - \varpi\eta\boldsymbol{B}_2^T\left[\boldsymbol{A}\boldsymbol{e}_x(t) + \boldsymbol{B}_2\boldsymbol{e}_\xi(t) + \varpi\eta\boldsymbol{B}_2\boldsymbol{B}_2^T\boldsymbol{e}_x(t) + \boldsymbol{B}_3\boldsymbol{e}_x(t)\right] \tag{9.31}$$

Denoting the aggregate state and disturbance as $\chi(t) = [\,\boldsymbol{x}(t)^T \; \boldsymbol{e}_x(t)^T \, \boldsymbol{e}_\xi(t)^T]^T$ and $\boldsymbol{w}_v(t) = [\,\boldsymbol{v}(t)^T \quad \dot{\boldsymbol{w}}(t)^T\,]^T$, one has

$$\dot{\chi}(t) = \boldsymbol{A}_\chi\chi(t) + \boldsymbol{B}_\chi\boldsymbol{w}_v(t) \tag{9.32}$$

where,

$$\boldsymbol{A}_\chi = \begin{bmatrix} \boldsymbol{A} - \boldsymbol{B}_1\boldsymbol{K} & \boldsymbol{B}_1\boldsymbol{K} + \varpi\eta\boldsymbol{B}_2\boldsymbol{B}_2^T + \boldsymbol{B}_3 & \boldsymbol{B}_2 \\ 0 & \boldsymbol{A} - \boldsymbol{L}\boldsymbol{C}_1 + \varpi\eta\boldsymbol{B}_2\boldsymbol{B}_2^T + \boldsymbol{B}_3 & \boldsymbol{B}_2 \\ 0 & -\varpi\eta\boldsymbol{B}_2^T\boldsymbol{A} - \varpi^2\eta^2\boldsymbol{B}_2^T\boldsymbol{B}_2\boldsymbol{B}_2^T - \varpi\eta\boldsymbol{B}_2^T\boldsymbol{B}_3 & -\varpi\eta\boldsymbol{B}_2^T\boldsymbol{B}_2 \end{bmatrix},$$

$$\boldsymbol{B}_\chi = \begin{bmatrix} 0 & 0 \\ -\boldsymbol{L} & 0 \\ 0 & \boldsymbol{I} \end{bmatrix}$$

Remark 9.7: As a consequence of the fact that $\boldsymbol{w}(t)$ includes uncertain magnitudes, the initial estimation errors $\boldsymbol{e}_x(0)$ and $\boldsymbol{e}_\xi(0)$ are not easy to be exactly known, and the initial state of flexible spacecraft may be unavailable, but they should be norm-bounded in practice, that is, there exist three positive scalars $\overline{\kappa}_k, k=1,2,3$ such that the infinity norms $\left\|\chi_k(0)\right\|_\infty \leq \overline{\kappa}_k$, where $\chi_1(0)=\boldsymbol{x}(0)$, $\chi_2(0)=\boldsymbol{e}_x(0)$, and $\chi_3(0)=\boldsymbol{e}_\xi(0)$. Then, for any symmetric positive definite matrix $\tilde{\boldsymbol{P}}_0$ of appropriate dimensions, it can be concluded that $\chi_k(0)^T\tilde{\boldsymbol{P}}_0\chi_k(0)\leq\overline{\kappa}_k^2\boldsymbol{1}_{1\times n}\tilde{\boldsymbol{P}}_0\boldsymbol{1}_{n\times 1}$, where $\boldsymbol{1}_{i\times j}$ denotes an i by j matrix (i rows, j columns) whose elements are all one, and n is the number of elements for $\chi_k(0), k=1,2,3$.

9.3.3 Sufficient conditions for uniform ultimate boundedness

Theorem 9.1: The closed-loop system represented by Eq. (9.32) is uniformly ultimately bounded, if for given scalars $\varpi\in[0,1]$, $\eta>0$, $\xi_i>0$, $i=1, 2, 3$, and given symmetric positive definite matrix $\boldsymbol{P}_1$ and matrix $\boldsymbol{L}$ satisfying Hurwitz condition, then there exist symmetric positive definite matrices $\boldsymbol{P}_2, \boldsymbol{P}_3$ and matrix $\boldsymbol{K}$, such that the following LMI holds:

$$\begin{bmatrix} \boldsymbol{\Pi}_{11}+\xi_3\boldsymbol{P}_1 & \boldsymbol{\Pi}_{12} & \boldsymbol{P}_1\boldsymbol{B}_2 & 0 & 0 \\ * & \boldsymbol{\Pi}_{22}+\xi_3\boldsymbol{P}_2 & \boldsymbol{\Pi}_{23} & \boldsymbol{P}_2\boldsymbol{L} & 0 \\ * & * & \boldsymbol{\Pi}_{33}+\xi_3\boldsymbol{P}_3 & 0 & \boldsymbol{P}_3 \\ * & * & * & -\xi_1\boldsymbol{I} & 0 \\ * & * & * & * & -\xi_2\boldsymbol{I} \end{bmatrix}<0 \tag{9.33}$$

where,

$$\begin{aligned}
\boldsymbol{\Pi}_{11} &= \boldsymbol{A}^T\boldsymbol{P}_1-\boldsymbol{K}^T\boldsymbol{B}_1^T\boldsymbol{P}_1+\boldsymbol{P}_1\boldsymbol{A}-\boldsymbol{P}_1\boldsymbol{B}_1\boldsymbol{K}\\
\boldsymbol{\Pi}_{12} &= \boldsymbol{P}_1\boldsymbol{B}_1K+\varpi\eta\boldsymbol{P}_1\boldsymbol{B}_2\boldsymbol{B}_2^T+\boldsymbol{P}_1\boldsymbol{B}_3\\
\boldsymbol{\Pi}_{22} &= \boldsymbol{A}^T\boldsymbol{P}_2-\boldsymbol{C}_1^T\boldsymbol{L}^T\boldsymbol{P}_2+\boldsymbol{P}_2\boldsymbol{A}-\boldsymbol{P}_2\boldsymbol{L}\boldsymbol{C}_1+\varpi\eta\boldsymbol{P}_2\boldsymbol{B}_2\boldsymbol{B}_2^T\\
&\quad +\varpi\eta\boldsymbol{B}_2\boldsymbol{B}_2^T\boldsymbol{P}_2+\boldsymbol{P}_2\boldsymbol{B}_3+\boldsymbol{B}_3^T\boldsymbol{P}_2\\
\boldsymbol{\Pi}_{23} &= \boldsymbol{P}_2\boldsymbol{B}_2-\varpi\eta\boldsymbol{A}^T\boldsymbol{B}_2\boldsymbol{P}_3-\varpi^2\eta^2\boldsymbol{B}_2\boldsymbol{B}_2^T\boldsymbol{B}_2\boldsymbol{P}_3-\varpi\eta\boldsymbol{B}_3^T\boldsymbol{B}_2\boldsymbol{P}_3\\
\boldsymbol{\Pi}_{33} &= -\varpi\eta(\boldsymbol{P}_3\boldsymbol{B}_2^T\boldsymbol{B}_2+\boldsymbol{B}_2^T\boldsymbol{B}_2\boldsymbol{P}_3)
\end{aligned}$$

When inequality (9.33) is feasible, the controller gain matrix $\boldsymbol{K}$ is obtained.

Proof: Choose a Lyapunov function candidate as

$$V(t) = x(t)^T P_1 x(t) + e_x(t)^T P_2 e_x(t) + e_\xi(t)^T P_3 e_\xi(t) \tag{9.34}$$

Then

$$\begin{aligned}
\dot{V}(t) &= \dot{x}(t)^T P_1 x(t) + x(t)^T P_1 \dot{x}(t) + \dot{e}_x(t)^T P_2 e_x(t) \\
&\quad + e_x(t)^T P_2 \dot{e}_x(t) + \dot{e}_\xi(t)^T P_3 e_\xi(t) + e_\xi(t)^T P_3 \dot{e}_\xi(t) \\
&= x(t)^T \left[(A - B_1 K)^T P_1 + P_1 (A - B_1 K)\right] x(t) \\
&\quad + 2x(t)^T P_1 B_1 K e_x(t) + 2x(t)^T P_1 B_2 e_\xi(t) \\
&\quad + 2\varpi\eta x(t)^T P_1 B_2 B_2^T e_x(t) + 2x(t)^T P_1 B_3 e_x(t) \\
&\quad + e_x(t)^T \left[(A - LC_1)^T P_2 + P_2 (A - LC_1)\right] e_x(t) \\
&\quad + 2e_x(t)^T P_2 B_2 e_\xi(t) + e_x(t)^T (\varpi\eta P_2 B_2 B_2^T + \varpi\eta B_2 B_2^T P_2) e_x(t) \\
&\quad + e_x(t)^T (P_2 B_3 + B_3^T P_2) e_x(t) - 2e_x(t)^T P_2 L v(t) \\
&\quad + 2e_\xi(t)^T P_3 \dot{w}(t) - \varpi\eta e_\xi(t)^T (P_3 B_2^T B_2 + B_2^T B_2 P_3) e_\xi(t) \\
&\quad - 2\varpi\eta e_\xi(t)^T P_3 B_2^T \left[A e_x(t) + \varpi\eta B_2 B_2^T e_x(t) + B_3 e_x(t)\right]
\end{aligned} \tag{9.35}$$

According to Lemma 9.1 and the given constraint conditions, one has

$$\begin{aligned}
\dot{V}(t) &\le x(t)^T \left[(A - B_1 K)^T P_1 + P_1 (A - B_1 K)\right] x(t) \\
&\quad + 2x(t)^T P_1 B_1 K e_x(t) + 2x(t)^T P_1 B_2 e_\xi(t) \\
&\quad + 2\varpi\eta x(t)^T P_1 B_2 B_2^T e_x(t) + 2x(t)^T P_1 B_3 e_x(t) \\
&\quad + e_x(t)^T \left[(A - LC_1)^T P_2 + P_2 (A - LC_1)\right] e_x(t) \\
&\quad + 2e_x(t)^T P_2 B_2 e_\xi(t) + e_x(t)^T (\varpi\eta P_2 B_2 B_2^T + \varpi\eta B_2 B_2^T P_2) e_x(t) \\
&\quad + e_x(t)^T (P_2 B_3 + B_3^T P_2) e_x(t) + \xi_1^{-1} e_x(t)^T P_2 L L^T P_2 e_x(t) \\
&\quad + \xi_1 v(t)^T v(t) + \xi_2^{-1} e_\xi(t)^T P_3 P_3 e_\xi(t) + \xi_2 \dot{w}(t)^T \dot{w}(t) \\
&\quad - \varpi\eta e_\xi(t)^T (P_3 B_2^T B_2 + B_2^T B_2 P_3) e_\xi(t) \\
&\quad - 2\varpi\eta e_\xi(t)^T P_3 B_2^T A e_x(t) - 2\varpi^2\eta^2 e_\xi(t)^T P_3 B_2^T B_2 B_2^T e_x(t) \\
&\quad - 2\varpi\eta e_\xi(t)^T P_3 B_2^T B_3 e_x(t)
\end{aligned}$$

Therefore, the first derivative of the Lyapunov function becomes

$$\dot{V}(t) \le \chi(t)^T \begin{bmatrix} \Pi_{11} & \Pi_{12} & P_1 B_2 \\ * & \Pi_{22} + \xi_1^{-1} P_2 L L^T P_2 & \Pi_{23} \\ * & * & \Pi_{33} + \xi_2^{-1} P_3 P_3 \end{bmatrix} \chi(t) + \beta_t$$

where,

$$\begin{aligned}
\boldsymbol{\Pi}_{11} &= (\boldsymbol{A}-\boldsymbol{B}_1\boldsymbol{K})^T\boldsymbol{P}_1 + \boldsymbol{P}_1(\boldsymbol{A}-\boldsymbol{B}_1\boldsymbol{K})\\
\boldsymbol{\Pi}_{12} &= \boldsymbol{P}_1\boldsymbol{B}_1\boldsymbol{K}+\varpi\eta\boldsymbol{P}_1\boldsymbol{B}_2\boldsymbol{B}_2^T + \boldsymbol{P}_1\boldsymbol{B}_3\\
\boldsymbol{\Pi}_{22} &= (\boldsymbol{A}-\boldsymbol{L}\boldsymbol{C}_1)^T\boldsymbol{P}_2 + \boldsymbol{P}_2(\boldsymbol{A}-\boldsymbol{L}\boldsymbol{C}_1) + \varpi\eta\boldsymbol{P}_2\boldsymbol{B}_2\boldsymbol{B}_2^T\\
&\quad + \varpi\eta\boldsymbol{B}_2\boldsymbol{B}_2^T\boldsymbol{P}_2 + \boldsymbol{P}_2\boldsymbol{B}_3 + \boldsymbol{B}_3^T\boldsymbol{P}_2\\
\boldsymbol{\Pi}_{23} &= \boldsymbol{P}_2\boldsymbol{B}_2 - \varpi\eta\boldsymbol{A}^T\boldsymbol{B}_2\boldsymbol{P}_3 - \varpi^2\eta^2\boldsymbol{B}_2\boldsymbol{B}_2^T\boldsymbol{B}_2\boldsymbol{P}_3 - \varpi\eta\boldsymbol{B}_3^T\boldsymbol{B}_2\boldsymbol{P}_3\\
\boldsymbol{\Pi}_{33} &= -\varpi\eta(\boldsymbol{P}_3\boldsymbol{B}_2^T\boldsymbol{B}_2+\boldsymbol{B}_2^T\boldsymbol{B}_2\boldsymbol{P}_3)\\
\beta_t &= \xi_1\boldsymbol{v}(t)^T\boldsymbol{v}(t) + \xi_2\dot{\boldsymbol{w}}(t)^T\dot{\boldsymbol{w}}(t)
\end{aligned}$$

Denoting

$$\boldsymbol{\Theta}_N = \begin{bmatrix} \boldsymbol{\Pi}_{11} & \boldsymbol{\Pi}_{12} & \boldsymbol{P}_1\boldsymbol{B}_2 \\ * & \boldsymbol{\Pi}_{22} + \xi_1^{-1}\boldsymbol{P}_2\boldsymbol{L}\boldsymbol{L}^T\boldsymbol{P}_2 & \boldsymbol{\Pi}_{23} \\ * & * & \boldsymbol{\Pi}_{33} + \xi_2^{-1}\boldsymbol{P}_3\boldsymbol{P}_3 \end{bmatrix}, \ \beta = \xi_1\kappa_2^2 + \xi_2\kappa_1^2$$

One has

$$\dot{V}(t) \le \boldsymbol{\chi}(t)^T\boldsymbol{\Theta}_N\boldsymbol{\chi}(t) + \beta$$

Defining $\boldsymbol{P} = diag\{\boldsymbol{P}_1 \quad \boldsymbol{P}_2 \quad \boldsymbol{P}_3\}$, according to inequality (9.33) and Lemma 9.2, one has

$$\boldsymbol{\Theta}_N + \xi_3\boldsymbol{P} < 0 \tag{9.36}$$

Furthermore, one has

$$\dot{V}(t) < -\xi_3\boldsymbol{\chi}(t)^T\boldsymbol{P}\boldsymbol{\chi}(t) + \beta \tag{9.37}$$

that is,

$$\dot{V}(t) < -\xi_3 V(t) + \beta \tag{9.38}$$

According to Lemma 9.3, it is obtained that the closed-loop system (9.32) is uniformly ultimately bounded.

This completes the proof.

Multiplying inequality (9.38) by $e^{\xi_3 t}$ yields

$$\frac{d}{dt}\left(V(t)e^{\xi_3 t}\right) < \beta e^{\xi_3 t} \tag{9.39}$$

Integrating inequality (9.39) over [0, t] leads to

$$0 \le V(t) < V(0)e^{-\xi_3 t} - \xi_3^{-1}\beta e^{-\xi_3 t} + \xi_3^{-1}\beta < V(0) + \xi_3^{-1}\beta \tag{9.40}$$

9.3.4 Sufficient conditions for H_∞ control strategy

It should be noted that the feasible solutions for inequality (9.33) are difficult to find due to the complex structure of the lumped disturbance $\boldsymbol{w}(t)$ and the existence of ξ_3. To improve the feasibility and reduce the difficulty in obtaining the controller gain matrix $\boldsymbol{K}$, a robust H_∞ control strategy based on Theorem 9.1 can be further concluded.

Theorem 9.2: For given scalars $\varpi \in [0,1]$, $\eta > 0$, symmetric positive definite matrix $\boldsymbol{P}_1$ and matrix $\boldsymbol{L}$ satisfying Hurwitz condition, the closed-loop system represented by Eq. (9.32) is quadratically stable under controller (9.28), and $\boldsymbol{y}(t)$ satisfies an H_∞ performance constraint if there exist a positive scalar $\gamma_2 = \gamma^2$, symmetric positive definite matrices $\boldsymbol{P}_2, \boldsymbol{P}_3$ and matrix $\boldsymbol{K}$, such that the following LMI holds:

$$\begin{bmatrix} \boldsymbol{\Pi}_{11} & \boldsymbol{\Pi}_{12} & \boldsymbol{P}_1\boldsymbol{B}_2 & 0 & \boldsymbol{C}_1^T & \boldsymbol{C}_1^T \\ * & \boldsymbol{\Pi}_{22} & \boldsymbol{\Pi}_{23} & 0 & -\boldsymbol{P}_2\boldsymbol{L} & 0 \\ * & * & \boldsymbol{\Pi}_{33} & \boldsymbol{P}_3 & 0 & 0 \\ * & * & * & -\gamma_2\boldsymbol{I} & 0 & 0 \\ * & * & * & * & (1-\gamma_2)\boldsymbol{I} & 0 \\ * & * & * & * & * & -\boldsymbol{I} \end{bmatrix} < 0 \tag{9.41}$$

When inequality (9.41) is feasible, the controller gain matrix $\boldsymbol{K}$ is obtained.

Proof: First, it is shown that the closed-loop system (9.32) is quadratically stable while $\boldsymbol{v}(t) = 0$, $\dot{\boldsymbol{w}}(t) = 0$, and the definition of quadratic stability has been stated in [36].

Choose the same Lyapunov function candidate as Eq. (9.34), according to Eq. (9.35), one has

$$\begin{aligned} \dot{V}(t) = {} & \boldsymbol{x}(t)^T\left[(\boldsymbol{A}-\boldsymbol{B}_1\boldsymbol{K})^T\boldsymbol{P}_1 + \boldsymbol{P}_1(\boldsymbol{A}-\boldsymbol{B}_1\boldsymbol{K})\right]\boldsymbol{x}(t) \\ & + 2\boldsymbol{x}(t)^T\boldsymbol{P}_1\boldsymbol{B}_1\boldsymbol{K}\boldsymbol{e}_x(t) + 2\boldsymbol{x}(t)^T\boldsymbol{P}_1\boldsymbol{B}_2\boldsymbol{e}_\xi(t) \\ & + 2\varpi\eta\boldsymbol{x}(t)^T\boldsymbol{P}_1\boldsymbol{B}_2\boldsymbol{B}_2^T\boldsymbol{e}_x(t) + 2\boldsymbol{x}(t)^T\boldsymbol{P}_1\boldsymbol{B}_3\boldsymbol{e}_x(t) \\ & + \boldsymbol{e}_x(t)^T\left[(\boldsymbol{A}-\boldsymbol{L}\boldsymbol{C}_1)^T\boldsymbol{P}_2 + \boldsymbol{P}_2(\boldsymbol{A}-\boldsymbol{L}\boldsymbol{C}_1)\right]\boldsymbol{e}_x(t) \\ & + 2\boldsymbol{e}_x(t)^T\boldsymbol{P}_2\boldsymbol{B}_2\boldsymbol{e}_\xi(t) + \boldsymbol{e}_x(t)^T(\varpi\eta\boldsymbol{P}_2\boldsymbol{B}_2\boldsymbol{B}_2^T + \varpi\eta\boldsymbol{B}_2\boldsymbol{B}_2^T\boldsymbol{P}_2)\boldsymbol{e}_x(t) \\ & + \boldsymbol{e}_x(t)^T(\boldsymbol{P}_2\boldsymbol{B}_3 + \boldsymbol{B}_3^T\boldsymbol{P}_2)\boldsymbol{e}_x(t) \\ & - \varpi\eta\boldsymbol{e}_\xi(t)^T(\boldsymbol{P}_3\boldsymbol{B}_2^T\boldsymbol{B}_2 + \boldsymbol{B}_2^T\boldsymbol{B}_2\boldsymbol{P}_3)\boldsymbol{e}_\xi(t) - 2\varpi\eta\boldsymbol{e}_\xi(t)^T\boldsymbol{P}_3\boldsymbol{B}_2^T\boldsymbol{A}\boldsymbol{e}_x(t) \\ & - 2\varpi^2\eta^2\boldsymbol{e}_\xi(t)^T\boldsymbol{P}_3\boldsymbol{B}_2^T\boldsymbol{B}_2\boldsymbol{B}_2^T\boldsymbol{e}_x(t) - 2\varpi\eta\boldsymbol{e}_\xi(t)^T\boldsymbol{P}_3\boldsymbol{B}_2^T\boldsymbol{B}_3\boldsymbol{e}_x(t) \end{aligned}$$

Denote

$$\overline{\boldsymbol{\Theta}}_N = \begin{bmatrix} \boldsymbol{\Pi}_{11} & \boldsymbol{\Pi}_{12} & \boldsymbol{P}_1\boldsymbol{B}_2 \\ * & \boldsymbol{\Pi}_{22} & \boldsymbol{\Pi}_{23} \\ * & * & \boldsymbol{\Pi}_{33} \end{bmatrix}$$

where,

$$\begin{aligned}
\boldsymbol{\Pi}_{11} &= (\boldsymbol{A}-\boldsymbol{B}_1\boldsymbol{K})^T\boldsymbol{P}_1 + \boldsymbol{P}_1(\boldsymbol{A}-\boldsymbol{B}_1\boldsymbol{K}) \\
\boldsymbol{\Pi}_{12} &= \boldsymbol{P}_1\boldsymbol{B}_1 K + \varpi\eta\boldsymbol{P}_1\boldsymbol{B}_2\boldsymbol{B}_2^T + \boldsymbol{P}_1\boldsymbol{B}_3 \\
\boldsymbol{\Pi}_{22} &= (\boldsymbol{A}-\boldsymbol{L}\boldsymbol{C}_1)^T\boldsymbol{P}_2 + \boldsymbol{P}_2(\boldsymbol{A}-\boldsymbol{L}\boldsymbol{C}_1) \\
&\quad + \varpi\eta\boldsymbol{P}_2\boldsymbol{B}_2\boldsymbol{B}_2^T + \varpi\eta\boldsymbol{B}_2\boldsymbol{B}_2^T\boldsymbol{P}_2 + \boldsymbol{P}_2\boldsymbol{B}_3 + \boldsymbol{B}_3^T\boldsymbol{P}_2 \\
\boldsymbol{\Pi}_{23} &= \boldsymbol{P}_2\boldsymbol{B}_2 - \varpi\eta\boldsymbol{A}^T\boldsymbol{B}_2\boldsymbol{P}_3 - \varpi^2\eta^2\boldsymbol{B}_2\boldsymbol{B}_2^T\boldsymbol{B}_2\boldsymbol{P}_3 - \varpi\eta\boldsymbol{B}_3^T\boldsymbol{B}_2\boldsymbol{P}_3 \\
\boldsymbol{\Pi}_{33} &= -\varpi\eta(\boldsymbol{P}_3\boldsymbol{B}_2^T\boldsymbol{B}_2 + \boldsymbol{B}_2^T\boldsymbol{B}_2\boldsymbol{P}_3)
\end{aligned}$$

The first derivative of the Lyapunov function becomes

$$\dot{V}(t) = \boldsymbol{\chi}(t)^T\overline{\boldsymbol{\Theta}}_N\boldsymbol{\chi}(t)$$

According to inequality (9.41) and Lemma 9.2, one has $\overline{\boldsymbol{\Theta}}_N < 0$, so that

$$\dot{V}(t) = \boldsymbol{\chi}(t)^T\overline{\boldsymbol{\Theta}}_N\boldsymbol{\chi}(t) \le \lambda_{\max}(\overline{\boldsymbol{\Theta}}_N)\left\|\boldsymbol{\chi}\right\|^2$$

Furthermore, let $\overline{\alpha} = -\lambda_{\max}(\overline{\boldsymbol{\Theta}}_N) > 0$, then,

$$\dot{V}(t) \le -\overline{\alpha}\left\|\boldsymbol{\chi}\right\|^2$$

that is, the closed-loop system (9.32) is quadratically stable under the condition of controller (9.28).

Next, it is shown that the output $\boldsymbol{y}(t)$ satisfies H_∞ performance constraint.

To establish the L_2 $[0, \infty)$ norm bound $\gamma^2\left\|\tilde{\boldsymbol{w}}(t)\right\|_2^2$, where $\tilde{\boldsymbol{w}}(t)$ is a constructed vector with norm $\left\|\tilde{\boldsymbol{w}}\right\| = \sqrt{\boldsymbol{w}^2 + \boldsymbol{v}^2}$, consider the following functional:

$$J = \int_0^\infty [\boldsymbol{y}(t)^T\boldsymbol{y}(t) - \gamma^2\tilde{\boldsymbol{w}}(t)^T\tilde{\boldsymbol{w}}(t)]dt$$

As the closed-loop system has quadratic stability, for arbitrary nonzero $\tilde{w}(t) \in L_2[0, \infty)$, under zero initial conditions,

$$J = \int_0^\infty [y(t)^T y(t) - \gamma^2 \tilde{w}(t)^T \tilde{w}(t) + \dot{V}(t)]dt - V(\infty) + V(0)$$
$$\leq \int_0^\infty \left\{ \begin{array}{c} x(t)^T C_1^T C_1 x(t) + 2x(t)^T C_1^T v(t) + v(t)^T v(t) \\ -\gamma^2 \left[\dot{w}(t)^T \dot{w}(t) + v(t)^T v(t) \right] + \chi(t)^T \overline{\Theta}_N \chi(t) \\ -2e_x(t)^T P_2 L v(t) + 2e_\xi(t)^T P_3 \dot{w}(t) \end{array} \right\} dt$$
$$\leq \int_0^\infty \left[x(t)^T \quad e_x(t)^T \quad e_\xi(t)^T \quad \dot{w}(t)^T \quad v(t)^T \right]$$

$$\begin{bmatrix} \Pi_{11} + C_1^T C_1 & \Pi_{12} & P_1 B_2 & 0 & C_1^T \\ * & \Pi_{22} & \Pi_{23} & 0 & -P_2 L \\ * & * & \Pi_{33} & P_3 & 0 \\ * & * & * & -\gamma^2 I & 0 \\ * & * & * & * & (1-\gamma^2) I \end{bmatrix} \begin{bmatrix} x(t) \\ e_x(t) \\ e_\xi(t) \\ \dot{w}(t) \\ v(t) \end{bmatrix} dt$$

According to Lemma 9.2 and inequality (9.41), $J < 0$ holds, that is, $y(t)$ satisfies H_∞ performance constraint.

This completes the proof.

Remark 9.8: As stated in [37], an important issue related to the vibration suppression of flexible structure is spillover effects. The finite number of modes, that is, modeled modes, should be referred to for there is always an infinity of residual modes. According to Theorem 9.2, the proposed controller can achieve the control task for the flexible spacecraft with the controlled modes of number m, where m can be chosen large enough such that the effect of residual modes on system performance becomes ignorable [38]. Hence, the proposed controller in this chapter can ignore the spillover effects.

9.3.5 Sufficient conditions for input magnitude, and rate constraints

Then, considering the input MRCs, one has the following theorem.

Theorem 9.3: Supposing that there exist symmetric positive definite matrices P_2, P_3 and matrix K satisfying LMI (9.41), then the control input

MRCs expressed by Eq. (9.15) are satisfied, if for given positive scalars $\varpi \in [0,1]$, $\eta > 0$, $\xi_i > 0, i = 3, \ldots, 21$, $\kappa_j > 0, j = 1,2,3,4$, $\overline{\kappa}_k > 0$, $k = 1,2,3$, $\varsigma > \lambda_{\max}(\boldsymbol{P}_2)$, λ_1, λ_2, and given symmetric positive definite matrix $\boldsymbol{P}_1$ and matrix $\boldsymbol{L}$, the following LMIs hold:

$$\begin{bmatrix} \xi_3^{-1}\beta - \alpha & \overline{\kappa}_1 \boldsymbol{1}_{1\times 14}\boldsymbol{P}_1 & \overline{\kappa}_2 \boldsymbol{1}_{1\times 14}\boldsymbol{P}_2 & \overline{\kappa}_3 \boldsymbol{1}_{1\times 3}\boldsymbol{P}_3 \\ * & -\boldsymbol{P}_1 & 0 & 0 \\ * & * & -\boldsymbol{P}_2 & 0 \\ * & * & * & -\boldsymbol{P}_3 \end{bmatrix} < 0 \tag{9.42}$$

$$\begin{bmatrix} -k_0\boldsymbol{P}_1 & 0 & 0 & 0 & \boldsymbol{K}^T + \varpi\eta\boldsymbol{B}_2 \\ * & -k_0\boldsymbol{P}_2 & 0 & 0 & -\boldsymbol{K}^T - \varpi\eta\boldsymbol{B}_2 \\ * & * & -k_0\boldsymbol{P}_3 & 0 & -\boldsymbol{I} \\ * & * & * & -k_0\kappa_3^{-2}\boldsymbol{I} & \boldsymbol{I} \\ * & * & * & * & -\boldsymbol{I} \end{bmatrix} < 0 \tag{9.43}$$

$$\begin{bmatrix} -k_1\boldsymbol{P}_1 & \boldsymbol{A}^T\boldsymbol{P}_1 - \boldsymbol{K}^T\boldsymbol{B}_1^T\boldsymbol{P}_1 \\ * & -\boldsymbol{P}_1 \end{bmatrix} < 0 \tag{9.44}$$

$$\begin{bmatrix} -k_1\boldsymbol{P}_2 & \boldsymbol{C}_1^T\boldsymbol{L}^T\boldsymbol{P}_2 \\ * & -\boldsymbol{P}_2 \end{bmatrix} < 0 \tag{9.45}$$

$$\begin{bmatrix} \boldsymbol{\Gamma}_1 - \boldsymbol{\Gamma}_2 & \boldsymbol{\Gamma}_3^T \\ * & -(1+\xi_{20}+\xi_{21})^{-1}\boldsymbol{I} \end{bmatrix} < 0 \tag{9.46}$$

$$\begin{bmatrix} \boldsymbol{\Gamma}_4 & \boldsymbol{B}_2^T\boldsymbol{B}_2 \\ * & -\boldsymbol{I} \end{bmatrix} < 0 \tag{9.47}$$

where $k_0 = (\alpha+1)^{-1}\lambda_1^2$, $k_1 = (k_{11}+k_{14})^{-1}(\lambda_2^2 - k_{12} - k_{13})$, and where

$$\begin{aligned} k_{11} &= (1 + \xi_6 + \xi_{10} + \xi_{13} + \xi_{16}^{-1} + \xi_{17}^{-1})k_0\varsigma\kappa_2^2 \\ k_{12} &= (1 + \xi_7 + \xi_{11} + \xi_{14} + \xi_{16} + \xi_{18}^{-1})\kappa_1^2 \\ k_{13} &= (1 + \xi_8 + \xi_{12} + \xi_{15} + \xi_{17} + \xi_{18})\kappa_4^2 \\ k_{14} &= (1 + \xi_4^{-1} + \xi_5^{-1} + \xi_6^{-1} + \xi_7^{-1} + \xi_8^{-1})k_0\alpha \end{aligned}$$

and the corresponding matrices are

$$\Gamma_1 = (1 + \xi_{19}^{-1} + \xi_{20}^{-1})k_0k_1\boldsymbol{P}_1 + (1 + \xi_{19} + \xi_{21}^{-1})k_0k_1\boldsymbol{P}_2$$

$$\Gamma_2 = \frac{1 + \xi_4^{-1} + \xi_5^{-1} + \xi_6^{-1} + \xi_7^{-1} + \xi_8^{-1}}{1 + \xi_4 + \xi_9^{-1} + \xi_{10}^{-1} + \xi_{11}^{-1} + \xi_{12}^{-1}} k_0k_1\boldsymbol{P}_2$$

$$\Gamma_3 = \varpi\eta\boldsymbol{B}_2^T\boldsymbol{A} + \varpi^2\eta^2\boldsymbol{B}_2^T\boldsymbol{B}_2\boldsymbol{B}_2^T + \varpi\eta\boldsymbol{B}_2^T\boldsymbol{B}_3$$

$$\Gamma_4 = -\frac{1 + \xi_4^{-1} + \xi_5^{-1} + \xi_6^{-1} + \xi_7^{-1} + \xi_8^{-1}}{(1 + \xi_5 + \xi_9 + \xi_{13}^{-1} + \xi_{14}^{-1} + \xi_{15}^{-1})\varpi^2\eta^2} k_0k_1\boldsymbol{P}_3$$

Proof: First it is shown that the input magnitude constraint, that is, $\|\boldsymbol{u}\| < \lambda_1$, is satisfied.

According to Lemma 9.2, the inequality (9.42) is equivalent to

$$\overline{\kappa}_1^2\boldsymbol{1}_{1\times14}\boldsymbol{P}_1\boldsymbol{1}_{14\times1} + \overline{\kappa}_2^2\boldsymbol{1}_{1\times14}\boldsymbol{P}_2\boldsymbol{1}_{14\times1} + \overline{\kappa}_3^2\boldsymbol{1}_{1\times3}\boldsymbol{P}_3\boldsymbol{1}_{3\times1} + \xi_3^{-1}\beta < \alpha \tag{9.48}$$

Due to the fact pointed out in Remark 9.7, it is easily obtained by

$$\boldsymbol{x}(0)^T\boldsymbol{P}_1\boldsymbol{x}(0) + \boldsymbol{e}_x(0)^T\boldsymbol{P}_2\boldsymbol{e}_x(0) + \boldsymbol{e}_\xi(0)^T\boldsymbol{P}_3\boldsymbol{e}_\xi(0) + \xi_3^{-1}\beta < \alpha \tag{9.49}$$

It is obvious that inequality (9.49) is equivalent to the inequality $V(0) + \xi_3^{-1}\beta < \alpha$, which implies that the state trajectory of the closed-loop system represented by Eq. (9.32) is uniformly and ultimately bounded.

Using Lemma 9.2, the inequality (9.43) is equivalent to

$$\begin{bmatrix} (\boldsymbol{K}+\varpi\eta\boldsymbol{B}_2^T)^T \\ -(\boldsymbol{K}+\varpi\eta\boldsymbol{B}_2^T)^T \\ -\boldsymbol{I} \\ \boldsymbol{I} \end{bmatrix} \begin{bmatrix} \boldsymbol{K}+\varpi\eta\boldsymbol{B}_2^T & -(\boldsymbol{K}+\varpi\eta\boldsymbol{B}_2^T) & -\boldsymbol{I} & \boldsymbol{I} \end{bmatrix}$$

$$< \begin{bmatrix} k_0\boldsymbol{P}_1 & 0 & 0 & 0 \\ * & k_0\boldsymbol{P}_2 & 0 & 0 \\ * & * & k_0\boldsymbol{P}_3 & 0 \\ * & * & * & k_0\kappa_3^{-2}\boldsymbol{I} \end{bmatrix} \tag{9.50}$$

For $\boldsymbol{B}_1 = \boldsymbol{B}_2$, the control input (9.28) becomes

$$\boldsymbol{u}(t) = -\boldsymbol{K}\hat{\boldsymbol{x}}(t) - \hat{\boldsymbol{w}}(t) \tag{9.51}$$

Then, one has

$$
\begin{aligned}
\left\|\boldsymbol{u}(t)\right\|_2^2 &= \boldsymbol{u}(t)^T\boldsymbol{u}(t)\\
&= [-\boldsymbol{K}\hat{\boldsymbol{x}}(t)-\hat{\boldsymbol{w}}(t)]^T[-\boldsymbol{K}\hat{\boldsymbol{x}}(t)-\hat{\boldsymbol{w}}(t)]\\
&= \left[\boldsymbol{K}(\boldsymbol{x}(t)-\boldsymbol{e}_x(t))+\boldsymbol{\xi}(t)-\boldsymbol{e}_\xi(t)+\varpi\eta\boldsymbol{B}_2^T(\boldsymbol{x}(t)-\boldsymbol{e}_x(t))\right]^T\\
&\quad \left[\boldsymbol{K}(\boldsymbol{x}(t)-\boldsymbol{e}_x(t))+\boldsymbol{\xi}(t)-\boldsymbol{e}_\xi(t)+\varpi\eta\boldsymbol{B}_2^T(\boldsymbol{x}(t)-\boldsymbol{e}_x(t))\right]\\
&= \begin{bmatrix}\boldsymbol{x}(t)^T & \boldsymbol{e}_x(t)^T & \boldsymbol{e}_\xi(t)^T & \boldsymbol{\xi}(t)^T\end{bmatrix}\begin{bmatrix}(\boldsymbol{K}+\varpi\eta\boldsymbol{B}_2^T)^T\\ -(\boldsymbol{K}+\varpi\eta\boldsymbol{B}_2^T)^T\\ -\boldsymbol{I}\\ \boldsymbol{I}\end{bmatrix}\\
&\quad \begin{bmatrix}\boldsymbol{K}+\varpi\eta\boldsymbol{B}_2^T & -(\boldsymbol{K}+\varpi\eta\boldsymbol{B}_2^T) & -\boldsymbol{I} & \boldsymbol{I}\end{bmatrix}\begin{bmatrix}\boldsymbol{x}(t)\\ \boldsymbol{e}_x(t)\\ \boldsymbol{e}_\xi(t)\\ \boldsymbol{\xi}(t)\end{bmatrix}\\
&< \begin{bmatrix}\boldsymbol{x}(t)^T & \boldsymbol{e}_x(t)^T & \boldsymbol{e}_\xi(t)^T & \boldsymbol{\xi}(t)^T\end{bmatrix}\begin{bmatrix}k_0\boldsymbol{P}_1 & 0 & 0 & 0\\ * & k_0\boldsymbol{P}_2 & 0 & 0\\ * & * & k_0\boldsymbol{P}_3 & 0\\ * & * & * & k_0\kappa_3^{-2}\boldsymbol{I}\end{bmatrix}\begin{bmatrix}\boldsymbol{x}(t)\\ \boldsymbol{e}_x(t)\\ \boldsymbol{e}_\xi(t)\\ \boldsymbol{\xi}(t)\end{bmatrix}\\
&= k_0\left(\boldsymbol{x}(t)^T\boldsymbol{P}_1\boldsymbol{x}(t)+\boldsymbol{e}_x(t)^T\boldsymbol{P}_2\boldsymbol{e}_x(t)+\boldsymbol{e}_\xi(t)^T\boldsymbol{P}_3\boldsymbol{e}_\xi(t)+\kappa_3^{-2}\boldsymbol{\xi}(t)^T\boldsymbol{\xi}(t)\right)\\
&\le k_0(V(t)+1)
\end{aligned}
$$

According to inequalities (9.40) and (9.49), one knows

$$\left\|\boldsymbol{u}(t)\right\|_2^2 < k_0(\alpha+1)=\lambda_1^2 \tag{9.52}$$

Next it is shown that the input rate constraint, that is, $\left\|\dot{\boldsymbol{u}}\right\| < \lambda_2$, is satisfied.

According to Lemma 9.2, the inequality (9.44) is equivalent to

$$(\boldsymbol{A}-\boldsymbol{B}_1\boldsymbol{K})^T\boldsymbol{P}_1(\boldsymbol{A}-\boldsymbol{B}_1\boldsymbol{K}) < k_1\boldsymbol{P}_1 \tag{9.53}$$

Inequality (9.50) leads to

$$(\boldsymbol{K}+\varpi\eta\boldsymbol{B}_2^T)^T(\boldsymbol{K}+\varpi\eta\boldsymbol{B}_2^T) < k_0\boldsymbol{P}_1 \tag{9.54}$$

Combining inequalities (9.53) and (9.54) yields

$$(\boldsymbol{A}-\boldsymbol{B}_1\boldsymbol{K})^T(\boldsymbol{K}+\varpi\eta\boldsymbol{B}_2^T)^T(\boldsymbol{K}+\varpi\eta\boldsymbol{B}_2^T)(\boldsymbol{A}-\boldsymbol{B}_1\boldsymbol{K}) < k_0k_1\boldsymbol{P}_1 \tag{9.55}$$

In addition, inequality (9.50) leads to

$$(\boldsymbol{K}+\varpi\eta\boldsymbol{B}_2^T)^T(\boldsymbol{K}+\varpi\eta\boldsymbol{B}_2^T) < k_0\boldsymbol{P}_2 \tag{9.56}$$

According to Lemma 9.2, the inequality (9.45) is equivalent to

$$C_1^T L^T P_2 L C_1 < k_1 P_2 \tag{9.57}$$

Combining inequalities (9.56) and (9.57) yields

$$C_1^T L^T (K+\varpi\eta B_2^T)^T (K+\varpi\eta B_2^T) L C_1 < k_0 k_1 P_2 \tag{9.58}$$

Hence, one has

$$\begin{aligned}
&\begin{bmatrix} (K+\varpi\eta B_2^T)(A-B_1K)-(K+\varpi\eta B_2^T)LC_1 \\ -(\varpi\eta B_2^T A+\varpi^2\eta^2 B_2^T B_2 B_2^T+\varpi\eta B_2^T B_3) \end{bmatrix}^T \\
&\begin{bmatrix} (K+\varpi\eta B_2^T)(A-B_1K)-(K+\varpi\eta B_2^T)LC_1 \\ -(\varpi\eta B_2^T A+\varpi^2\eta^2 B_2^T B_2 B_2^T+\varpi\eta B_2^T B_3) \end{bmatrix} \\
&= (A-B_1K)^T (K+\varpi\eta B_2^T)^T (K+\varpi\eta B_2^T)(A-B_1K) \\
&\quad + C_1^T L^T (K+\varpi\eta B_2^T)^T (K+\varpi\eta B_2^T) L C_1 \\
&\quad + (\varpi\eta B_2^T A+\varpi^2\eta^2 B_2^T B_2 B_2^T+\varpi\eta B_2^T B_3)^T (\varpi\eta B_2^T A+\varpi^2\eta^2 B_2^T B_2 B_2^T+\varpi\eta B_2^T B_3) \\
&\quad - (A-B_1K)^T (K+\varpi\eta B_2^T)^T (K+\varpi\eta B_2^T) L C_1 \\
&\quad - C_1^T L^T (K+\varpi\eta B_2^T)^T (K+\varpi\eta B_2^T)(A-B_1K) \\
&\quad - (A-B_1K)^T (K+\varpi\eta B_2^T)^T (\varpi\eta B_2^T A+\varpi^2\eta^2 B_2^T B_2 B_2^T+\varpi\eta B_2^T B_3) \\
&\quad - (\varpi\eta B_2^T A+\varpi^2\eta^2 B_2^T B_2 B_2^T+\varpi\eta B_2^T B_3)^T (K+\varpi\eta B_2^T)(A-B_1K) \\
&\quad + C_1^T L^T (K+\varpi\eta B_2^T)^T (\varpi\eta B_2^T A+\varpi^2\eta^2 B_2^T B_2 B_2^T+\varpi\eta B_2^T B_3) \\
&\quad + (\varpi\eta B_2^T A+\varpi^2\eta^2 B_2^T B_2 B_2^T+\varpi\eta B_2^T B_3)^T (K+\varpi\eta B_2^T) L C_1 \\
&\le (1+\xi_{19}^{-1}+\xi_{20}^{-1})(A-B_1K)^T (K+\varpi\eta B_2^T)^T (K+\varpi\eta B_2^T)(A-B_1K) \\
&\quad + (1+\xi_{19}+\xi_{21}^{-1}) C_1^T L^T (K+\varpi\eta B_2^T)^T (K+\varpi\eta B_2^T) L C_1 \\
&\quad + (1+\xi_{20}+\xi_{21})(\varpi\eta B_2^T A+\varpi^2\eta^2 B_2^T B_2 B_2^T+\varpi\eta B_2^T B_3)^T \\
&\quad (\varpi\eta B_2^T A+\varpi^2\eta^2 B_2^T B_2 B_2^T+\varpi\eta B_2^T B_3) \\
&< (1+\xi_{19}^{-1}+\xi_{20}^{-1}) k_0 k_1 P_1 \\
&\quad + (1+\xi_{19}+\xi_{21}^{-1}) k_0 k_1 P_2 \\
&\quad + (1+\xi_{20}+\xi_{21})(\varpi\eta B_2^T A+\varpi^2\eta^2 B_2^T B_2 B_2^T+\varpi\eta B_2^T B_3)^T \\
&\quad (\varpi\eta B_2^T A+\varpi^2\eta^2 B_2^T B_2 B_2^T+\varpi\eta B_2^T B_3)
\end{aligned} \tag{9.59}$$

According to inequality (9.46), one has

$$\begin{aligned}
&\begin{bmatrix} (K+\varpi\eta B_2^T)(A-B_1K)-(K+\varpi\eta B_2^T)LC_1 \\ -(\varpi\eta B_2^T A+\varpi^2\eta^2 B_2^T B_2 B_2^T+\varpi\eta B_2^T B_3) \end{bmatrix}^T \\
&\begin{bmatrix} (K+\varpi\eta B_2^T)(A-B_1K)-(K+\varpi\eta B_2^T)LC_1 \\ -(\varpi\eta B_2^T A+\varpi^2\eta^2 B_2^T B_2 B_2^T+\varpi\eta B_2^T B_3) \end{bmatrix} \\
&< \frac{1+\xi_4^{-1}+\xi_5^{-1}+\xi_6^{-1}+\xi_7^{-1}+\xi_8^{-1}}{1+\xi_4+\xi_9^{-1}+\xi_{10}^{-1}+\xi_{11}^{-1}+\xi_{12}^{-1}} k_0 k_1 P_2
\end{aligned} \tag{9.60}$$

According to Lemma 9.2, the inequality (9.47) is equivalent to

$$\boldsymbol{B}_2^T\boldsymbol{B}_2\boldsymbol{B}_2^T\boldsymbol{B}_2 < \frac{1+\xi_4^{-1}+\xi_5^{-1}+\xi_6^{-1}+\xi_7^{-1}+\xi_8^{-1}}{(1+\xi_5+\xi_9+\xi_{13}^{-1}+\xi_{14}^{-1}+\xi_{15}^{-1})\varpi^2\eta^2}k_0k_1\boldsymbol{P}_3 \tag{9.61}$$

For $\boldsymbol{C}_1 = \boldsymbol{I}_{6+2m}$, inequality (9.58) is equivalent to

$$\boldsymbol{L}^T(\boldsymbol{K}+\varpi\eta\boldsymbol{B}_2^T)^T(\boldsymbol{K}+\varpi\eta\boldsymbol{B}_2^T)\boldsymbol{L} < k_0k_1\boldsymbol{P}_2 \tag{9.62}$$

Considering Eqs. (9.32) and (9.51), one has

$$\begin{aligned}
\dot{\boldsymbol{u}}(t) =& -\boldsymbol{K}\dot{\hat{\boldsymbol{x}}}(t) - \dot{\hat{\boldsymbol{w}}}(t) \\
=& -\boldsymbol{K}(\dot{\boldsymbol{x}}(t)-\dot{\boldsymbol{e}}_x(t)) - \dot{\boldsymbol{\xi}}(t) - \varpi\eta\boldsymbol{B}_2^T\dot{\boldsymbol{x}}(t) + \dot{\boldsymbol{e}}_\xi(t) + \varpi\eta\boldsymbol{B}_2^T\dot{\boldsymbol{e}}_x(t) \\
=& -(\boldsymbol{K}+\varpi\eta\boldsymbol{B}_2^T)\dot{\boldsymbol{x}}(t) + (\boldsymbol{K}+\varpi\eta\boldsymbol{B}_2^T)\dot{\boldsymbol{e}}_x(t) + \dot{\boldsymbol{e}}_\xi(t) - \dot{\boldsymbol{\xi}}(t) \\
=& -(\boldsymbol{K}+\varpi\eta\boldsymbol{B}_2^T)((\boldsymbol{A}-\boldsymbol{B}_1\boldsymbol{K})\boldsymbol{x}(t) \\
&+ (\boldsymbol{B}_1\boldsymbol{K}+\varpi\eta\boldsymbol{B}_2\boldsymbol{B}_2^T+\boldsymbol{B}_3)\boldsymbol{e}_x(t) + \boldsymbol{B}_2\boldsymbol{e}_\xi(t)) \\
&+ (\boldsymbol{K}+\varpi\eta\boldsymbol{B}_2^T)((\boldsymbol{A}-\boldsymbol{L}\boldsymbol{C}_1+\varpi\eta\boldsymbol{B}_2\boldsymbol{B}_2^T+\boldsymbol{B}_3)\boldsymbol{e}_x(t) + \boldsymbol{B}_2\boldsymbol{e}_\xi(t) - \boldsymbol{L}\boldsymbol{v}(t)) \\
&- ((\varpi\eta\boldsymbol{B}_2^T\boldsymbol{A}+\varpi^2\eta^2\boldsymbol{B}_2^T\boldsymbol{B}_2\boldsymbol{B}_2^T+\varpi\eta\boldsymbol{B}_2^T\boldsymbol{B}_3)\boldsymbol{e}_x(t) \\
&+ \varpi\eta\boldsymbol{B}_2^T\boldsymbol{B}_2\boldsymbol{e}_\xi(t)) + \dot{\boldsymbol{w}}(t) - \dot{\boldsymbol{\xi}}(t) \\
=& -(\boldsymbol{K}+\varpi\eta\boldsymbol{B}_2^T)(\boldsymbol{A}-\boldsymbol{B}_1\boldsymbol{K})\boldsymbol{x}(t) \\
&+ \begin{bmatrix} -(\boldsymbol{K}+\varpi\eta\boldsymbol{B}_2^T)(\boldsymbol{B}_1\boldsymbol{K}+\varpi\eta\boldsymbol{B}_2\boldsymbol{B}_2^T+\boldsymbol{B}_3) \\ +(\boldsymbol{K}+\varpi\eta\boldsymbol{B}_2^T)(\boldsymbol{A}-\boldsymbol{L}\boldsymbol{C}_1+\varpi\eta\boldsymbol{B}_2\boldsymbol{B}_2^T+\boldsymbol{B}_3) \\ -(\varpi\eta\boldsymbol{B}_2^T\boldsymbol{A}+\varpi^2\eta^2\boldsymbol{B}_2^T\boldsymbol{B}_2\boldsymbol{B}_2^T+\varpi\eta\boldsymbol{B}_2^T\boldsymbol{B}_3) \end{bmatrix}\boldsymbol{e}_x(t) \\
&+ [-(\boldsymbol{K}+\varpi\eta\boldsymbol{B}_2^T)\boldsymbol{B}_2 + (\boldsymbol{K}+\varpi\eta\boldsymbol{B}_2^T)\boldsymbol{B}_2 - \varpi\eta\boldsymbol{B}_2^T\boldsymbol{B}_2]\boldsymbol{e}_\xi(t) \\
&- (\boldsymbol{K}+\varpi\eta\boldsymbol{B}_2^T)\boldsymbol{L}\boldsymbol{v}(t) + \dot{\boldsymbol{w}}(t) - \dot{\boldsymbol{\xi}}(t) \\
=& -(\boldsymbol{K}+\varpi\eta\boldsymbol{B}_2^T)(\boldsymbol{A}-\boldsymbol{B}_1\boldsymbol{K})\boldsymbol{x}(t) - [(\boldsymbol{K}+\varpi\eta\boldsymbol{B}_2^T)(\boldsymbol{B}_1\boldsymbol{K}-\boldsymbol{A}+\boldsymbol{L}\boldsymbol{C}_1) \\
&+ (\varpi\eta\boldsymbol{B}_2^T\boldsymbol{A}+\varpi^2\eta^2\boldsymbol{B}_2^T\boldsymbol{B}_2\boldsymbol{B}_2^T+\varpi\eta\boldsymbol{B}_2^T\boldsymbol{B}_3)]\boldsymbol{e}_x(t) \\
&- \varpi\eta\boldsymbol{B}_2^T\boldsymbol{B}_2\boldsymbol{e}_\xi(t) - (\boldsymbol{K}+\varpi\eta\boldsymbol{B}_2^T)\boldsymbol{L}\boldsymbol{v}(t) + \dot{\boldsymbol{w}}(t) - \dot{\boldsymbol{\xi}}(t)
\end{aligned} \tag{9.63}$$

Then,

$$
\begin{aligned}
\left\|\dot{\boldsymbol{u}}(t)\right\|_2^2 &= \dot{\boldsymbol{u}}(t)^T\dot{\boldsymbol{u}}(t)\\
&= \boldsymbol{x}(t)^T(\boldsymbol{A}-\boldsymbol{B}_1\boldsymbol{K})^T(\boldsymbol{K}+\varpi\eta\boldsymbol{B}_2^T)^T(\boldsymbol{K}+\varpi\eta\boldsymbol{B}_2^T)(\boldsymbol{A}-\boldsymbol{B}_1\boldsymbol{K})\boldsymbol{x}(t)\\
&\quad + 2\boldsymbol{x}(t)^T(\boldsymbol{A}-\boldsymbol{B}_1\boldsymbol{K})^T(\boldsymbol{K}+\varpi\eta\boldsymbol{B}_2^T)^T[(\boldsymbol{K}+\varpi\eta\boldsymbol{B}_2^T)(\boldsymbol{B}_1\boldsymbol{K}-\boldsymbol{A}+\boldsymbol{L}\boldsymbol{C}_1)\\
&\quad + (\varpi\eta\boldsymbol{B}_2^T\boldsymbol{A}+\varpi^2\eta^2\boldsymbol{B}_2^T\boldsymbol{B}_2\boldsymbol{B}_2^T+\varpi\eta\boldsymbol{B}_2^T\boldsymbol{B}_3)]\boldsymbol{e}_x(t)\\
&\quad + 2\boldsymbol{x}(t)^T(\boldsymbol{A}-\boldsymbol{B}_1\boldsymbol{K})^T(\boldsymbol{K}+\varpi\eta\boldsymbol{B}_2^T)^T\varpi\eta\boldsymbol{B}_2^T\boldsymbol{B}_2\boldsymbol{e}_\xi(t)\\
&\quad + 2\boldsymbol{x}(t)^T(\boldsymbol{A}-\boldsymbol{B}_1\boldsymbol{K})^T(\boldsymbol{K}+\varpi\eta\boldsymbol{B}_2^T)^T(\boldsymbol{K}+\varpi\eta\boldsymbol{B}_2^T)\boldsymbol{L}\boldsymbol{v}(t)\\
&\quad - 2\boldsymbol{x}(t)^T(\boldsymbol{A}-\boldsymbol{B}_1\boldsymbol{K})^T(\boldsymbol{K}+\varpi\eta\boldsymbol{B}_2^T)^T\dot{\boldsymbol{w}}(t)\\
&\quad + 2\boldsymbol{x}(t)^T(\boldsymbol{A}-\boldsymbol{B}_1\boldsymbol{K})^T(\boldsymbol{K}+\varpi\eta\boldsymbol{B}_2^T)^T\dot{\boldsymbol{\xi}}(t)\\
&\quad + \boldsymbol{e}_x(t)^T\begin{bmatrix}(\boldsymbol{K}+\varpi\eta\boldsymbol{B}_2^T)(\boldsymbol{B}_1\boldsymbol{K}-\boldsymbol{A}+\boldsymbol{L}\boldsymbol{C}_1)\\ +(\varpi\eta\boldsymbol{B}_2^T\boldsymbol{A}+\varpi^2\eta^2\boldsymbol{B}_2^T\boldsymbol{B}_2\boldsymbol{B}_2^T+\varpi\eta\boldsymbol{B}_2^T\boldsymbol{B}_3)\end{bmatrix}^T\\
&\quad \begin{bmatrix}(\boldsymbol{K}+\varpi\eta\boldsymbol{B}_2^T)(\boldsymbol{B}_1\boldsymbol{K}-\boldsymbol{A}+\boldsymbol{L}\boldsymbol{C}_1)\\ +(\varpi\eta\boldsymbol{B}_2^T\boldsymbol{A}+\varpi^2\eta^2\boldsymbol{B}_2^T\boldsymbol{B}_2\boldsymbol{B}_2^T+\varpi\eta\boldsymbol{B}_2^T\boldsymbol{B}_3)\end{bmatrix}\boldsymbol{e}_x(t)\\
&\quad + 2\boldsymbol{e}_x(t)^T[(\boldsymbol{K}+\varpi\eta\boldsymbol{B}_2^T)(\boldsymbol{B}_1\boldsymbol{K}-\boldsymbol{A}+\boldsymbol{L}\boldsymbol{C}_1)\\
&\quad +(\varpi\eta\boldsymbol{B}_2^T\boldsymbol{A}+\varpi^2\eta^2\boldsymbol{B}_2^T\boldsymbol{B}_2\boldsymbol{B}_2^T+\varpi\eta\boldsymbol{B}_2^T\boldsymbol{B}_3)]^T\varpi\eta\boldsymbol{B}_2^T\boldsymbol{B}_2\boldsymbol{e}_\xi(t)\\
&\quad + 2\boldsymbol{e}_x(t)^T[(\boldsymbol{K}+\varpi\eta\boldsymbol{B}_2^T)(\boldsymbol{B}_1\boldsymbol{K}-\boldsymbol{A}+\boldsymbol{L}\boldsymbol{C}_1)\\
&\quad +(\varpi\eta\boldsymbol{B}_2^T\boldsymbol{A}+\varpi^2\eta^2\boldsymbol{B}_2^T\boldsymbol{B}_2\boldsymbol{B}_2^T+\varpi\eta\boldsymbol{B}_2^T\boldsymbol{B}_3)]^T(\boldsymbol{K}+\varpi\eta\boldsymbol{B}_2^T)\boldsymbol{L}\boldsymbol{v}(t)\\
&\quad - 2\boldsymbol{e}_x(t)^T[(\boldsymbol{K}+\varpi\eta\boldsymbol{B}_2^T)(\boldsymbol{B}_1\boldsymbol{K}-\boldsymbol{A}+\boldsymbol{L}\boldsymbol{C}_1)\\
&\quad +(\varpi\eta\boldsymbol{B}_2^T\boldsymbol{A}+\varpi^2\eta^2\boldsymbol{B}_2^T\boldsymbol{B}_2\boldsymbol{B}_2^T+\varpi\eta\boldsymbol{B}_2^T\boldsymbol{B}_3)]^T\dot{\boldsymbol{w}}(t)\\
&\quad + 2\boldsymbol{e}_x(t)^T[(\boldsymbol{K}+\varpi\eta\boldsymbol{B}_2^T)(\boldsymbol{B}_1\boldsymbol{K}-\boldsymbol{A}+\boldsymbol{L}\boldsymbol{C}_1)\\
&\quad +(\varpi\eta\boldsymbol{B}_2^T\boldsymbol{A}+\varpi^2\eta^2\boldsymbol{B}_2^T\boldsymbol{B}_2\boldsymbol{B}_2^T+\varpi\eta\boldsymbol{B}_2^T\boldsymbol{B}_3)]^T\dot{\boldsymbol{\xi}}(t)\\
&\quad + \varpi^2\eta^2\boldsymbol{e}_\xi(t)^T\boldsymbol{B}_2^T\boldsymbol{B}_2\boldsymbol{B}_2^T\boldsymbol{B}_2\boldsymbol{e}_\xi(t) + 2\varpi\eta\boldsymbol{e}_\xi(t)^T\boldsymbol{B}_2^T\boldsymbol{B}_2(\boldsymbol{K}+\varpi\eta\boldsymbol{B}_2^T)\boldsymbol{L}\boldsymbol{v}(t)\\
&\quad - 2\varpi\eta\boldsymbol{e}_\xi(t)^T\boldsymbol{B}_2^T\boldsymbol{B}_2\dot{\boldsymbol{w}}(t) + 2\varpi\eta\boldsymbol{e}_\xi(t)^T\boldsymbol{B}_2^T\boldsymbol{B}_2\dot{\boldsymbol{\xi}}(t)\\
&\quad + \boldsymbol{v}(t)^T\boldsymbol{L}^T(\boldsymbol{K}+\varpi\eta\boldsymbol{B}_2^T)^T(\boldsymbol{K}+\varpi\eta\boldsymbol{B}_2^T)\boldsymbol{L}\boldsymbol{v}(t)\\
&\quad - 2\boldsymbol{v}(t)^T\boldsymbol{L}^T(\boldsymbol{K}+\varpi\eta\boldsymbol{B}_2^T)^T\dot{\boldsymbol{w}}(t) + 2\boldsymbol{v}(t)^T\boldsymbol{L}^T(\boldsymbol{K}+\varpi\eta\boldsymbol{B}_2^T)^T\dot{\boldsymbol{\xi}}(t)\\
&\quad + \dot{\boldsymbol{w}}(t)^T\dot{\boldsymbol{w}}(t) - 2\dot{\boldsymbol{w}}(t)^T\dot{\boldsymbol{\xi}}(t) + \dot{\boldsymbol{\xi}}(t)^T\dot{\boldsymbol{\xi}}(t)
\end{aligned}
$$

According to Lemma 9.1 and the given constraint conditions, one has

$$\begin{aligned}
\|\dot{u}(t)\|_2^2 \leq & (1+\xi_4^{-1}+\xi_5^{-1}+\xi_6^{-1}+\xi_7^{-1}+\xi_8^{-1})k_0k_1\boldsymbol{x}(t)^T\boldsymbol{P}_1\boldsymbol{x}(t) \\
& +(1+\xi_4+\xi_9^{-1}+\xi_{10}^{-1}+\xi_{11}^{-1}+\xi_{12}^{-1})\boldsymbol{e}_x(t)^{\mathrm{T}} \\
& \begin{bmatrix}(\boldsymbol{K}+\varpi\eta\boldsymbol{B}_2^T)(\boldsymbol{B}_1\boldsymbol{K}-\boldsymbol{A}+\boldsymbol{L}\boldsymbol{C}_1) \\ +(\varpi\eta\boldsymbol{B}_2^T\boldsymbol{A}+\varpi^2\eta^2\boldsymbol{B}_2^T\boldsymbol{B}_2\boldsymbol{B}_2^T+\varpi\eta\boldsymbol{B}_2^T\boldsymbol{B}_3)\end{bmatrix}^T \\
& \begin{bmatrix}(\boldsymbol{K}+\varpi\eta\boldsymbol{B}_2^T)(\boldsymbol{B}_1\boldsymbol{K}-\boldsymbol{A}+\boldsymbol{L}\boldsymbol{C}_1) \\ +(\varpi\eta\boldsymbol{B}_2^T\boldsymbol{A}+\varpi^2\eta^2\boldsymbol{B}_2^T\boldsymbol{B}_2\boldsymbol{B}_2^T+\varpi\eta\boldsymbol{B}_2^T\boldsymbol{B}_3)\end{bmatrix}\boldsymbol{e}_x(t) \\
& +(1+\xi_5+\xi_9+\xi_{13}^{-1}+\xi_{14}^{-1}+\xi_{15}^{-1})\varpi^2\eta^2\boldsymbol{e}_\xi(t)^T\boldsymbol{B}_2^T\boldsymbol{B}_2\boldsymbol{B}_2^T\boldsymbol{B}_2\boldsymbol{e}_\xi(t) \\
& +(1+\xi_6+\xi_{10}+\xi_{13}+\xi_{16}^{-1}+\xi_{17}^{-1})\boldsymbol{v}(t)^T\boldsymbol{L}^T(\boldsymbol{K}+\varpi\eta\boldsymbol{B}_2^T)^T(\boldsymbol{K}+\varpi\eta\boldsymbol{B}_2^T)\boldsymbol{L}\boldsymbol{v}(t) \\
& +(1+\xi_7+\xi_{11}+\xi_{14}+\xi_{16}+\xi_{18}^{-1})\dot{\boldsymbol{w}}(t)^T\dot{\boldsymbol{w}}(t) \\
& +(1+\xi_8+\xi_{12}+\xi_{15}+\xi_{17}+\xi_{18})\dot{\boldsymbol{\xi}}(t)^T\dot{\boldsymbol{\xi}}(t) \\
< & (1+\xi_4^{-1}+\xi_5^{-1}+\xi_6^{-1}+\xi_7^{-1}+\xi_8^{-1})k_0k_1\boldsymbol{x}(t)^T\boldsymbol{P}_1\boldsymbol{x}(t) \\
& +(1+\xi_4^{-1}+\xi_5^{-1}+\xi_6^{-1}+\xi_7^{-1}+\xi_8^{-1})k_0k_1\boldsymbol{e}_x(t)^T\boldsymbol{P}_2\boldsymbol{e}_x(t) \\
& +(1+\xi_4^{-1}+\xi_5^{-1}+\xi_6^{-1}+\xi_7^{-1}+\xi_8^{-1})k_0k_1\boldsymbol{e}_\xi(t)^T\boldsymbol{P}_3\boldsymbol{e}_\xi(t) \\
& +(1+\xi_6+\xi_{10}+\xi_{13}+\xi_{16}^{-1}+\xi_{17}^{-1})k_0k_1\boldsymbol{v}(t)^T\boldsymbol{P}_2\boldsymbol{v}(t) \\
& +(1+\xi_7+\xi_{11}+\xi_{14}+\xi_{16}+\xi_{18}^{-1})\dot{\boldsymbol{w}}(t)^T\dot{\boldsymbol{w}}(t) \\
& +(1+\xi_8+\xi_{12}+\xi_{15}+\xi_{17}+\xi_{18})\dot{\boldsymbol{\xi}}(t)^T\dot{\boldsymbol{\xi}}(t) \\
< & (1+\xi_4^{-1}+\xi_5^{-1}+\xi_6^{-1}+\xi_7^{-1}+\xi_8^{-1})k_0k_1V(t) \\
& +(1+\xi_6+\xi_{10}+\xi_{13}+\xi_{16}^{-1}+\xi_{17}^{-1})k_0k_1\boldsymbol{v}(t)^T\boldsymbol{P}_2\boldsymbol{v}(t) \\
& +(1+\xi_7+\xi_{11}+\xi_{14}+\xi_{16}+\xi_{18}^{-1})\kappa_1^2+(1+\xi_8+\xi_{12}+\xi_{15}+\xi_{17}+\xi_{18})\kappa_4^2
\end{aligned}$$

According to inequalities (9.40) and (9.49), one knows

$$\begin{aligned}
\|\dot{u}(t)\|_2^2 < & (1+\xi_4^{-1}+\xi_5^{-1}+\xi_6^{-1}+\xi_7^{-1}+\xi_8^{-1})k_0k_1\alpha \\
& +(1+\xi_6+\xi_{10}+\xi_{13}+\xi_{16}^{-1}+\xi_{17}^{-1})k_0k_1\lambda_{\max}(\boldsymbol{P}_2)\kappa_2^2 \\
& +(1+\xi_7+\xi_{11}+\xi_{14}+\xi_{16}+\xi_{18}^{-1})\kappa_1^2 \\
& +(1+\xi_8+\xi_{12}+\xi_{15}+\xi_{17}+\xi_{18})\kappa_4^2 \\
< & (1+\xi_4^{-1}+\xi_5^{-1}+\xi_6^{-1}+\xi_7^{-1}+\xi_8^{-1})k_0k_1\alpha \\
& +(1+\xi_6+\xi_{10}+\xi_{13}+\xi_{16}^{-1}+\xi_{17}^{-1})k_0k_1\varsigma\kappa_2^2 \\
& +(1+\xi_7+\xi_{11}+\xi_{14}+\xi_{16}+\xi_{18}^{-1})\kappa_1^2 \\
& +(1+\xi_8+\xi_{12}+\xi_{15}+\xi_{17}+\xi_{18})\kappa_4^2 \\
= & \lambda_2^2
\end{aligned} \tag{9.64}$$

This completes the proof.

Remark 9.9: It should be noted that the feasibility of the corresponding LMIs can be improved by adjusting the aforementioned parameters, that is, $\varpi\in[0,1]$, $\eta>0$, $\xi_i>0(i=3,\ldots,21)$, $\kappa_j>0(j=1,2,3,4)$, $\overline{\kappa}_k>0(k=1,2,3)$, ς, λ_1, λ_2, and the given symmetric positive definite matrix $\boldsymbol{P}_1$ and matrix $\boldsymbol{L}$. It is worth noting that ς is set in advance and evaluated after the solutions of LMIs (9.41)–(9.47) to see whether $\varsigma>\lambda_{\max}(\boldsymbol{P}_2)$ is satisfied. The LMIs above just give sufficient condition for the existence of controller satisfying given constraints.

9.4 Simulation test

In this section, the effectiveness of the proposed controller is illustrated through simulations. It should be noted that the control task is to achieve attitude stabilization and vibration suppression simultaneously with control input constraints. In these simulations, four bending modes of the flexible appendages have been considered, see Table 9.1, since the low-order modes are generally dominant in a flexible spacecraft. The main parameters of the flexible spacecraft are chosen from [39]. For comparison, two cases are studied in this section. In the first case, the proposed observer-based controller is applied for attitude stabilization. In the second case, the mixed H_2/H_∞ controller introduced in [8] is employed. All simulation parameters are identical in these two cases.

The nominal inertia matrix of the flexible spacecraft is taken to be

$$J_n = \begin{bmatrix} 350 & 3 & 4 \\ 3 & 280 & 10 \\ 4 & 10 & 190 \end{bmatrix} \text{kg} \cdot \text{m}^2$$

and the coupling matrices are

$$\tilde{\boldsymbol{\Delta}} = \begin{bmatrix} 6.45637 & -1.25619 & 1.11687 & 1.23637 \\ 1.27814 & 0.91756 & 2.48901 & -0.83674 \\ 2.15629 & -1.67264 & -0.83674 & -1.12503 \end{bmatrix} \sqrt{\text{kg}} \cdot \text{m}$$

$$\tilde{\boldsymbol{\Delta}}_p = \begin{bmatrix} 2.342552 \\ -0.422537 \\ 3.912984 \\ 7.026176 \end{bmatrix} \times 10^{-2} \sqrt{\text{kg}} \cdot \text{m s}^{-2} \text{V}^{-1}$$

Table 9.1 Parameters of the flexible dynamics.

	Natural frequency (rad s^{-1})	Damping
Mode 1	0.7681	0.005607
Mode 2	1.1038	0.008620
Mode 3	1.8733	0.01283
Mode 4	2.5496	0.02516

The unknown and uncertain part of the inertia matrix is assumed to be $\boldsymbol{\Delta J} = (0.1 + 0.02\sin(0.11\pi t))\boldsymbol{J}_n$. The initial attitude angle and angular velocity are chosen as $\boldsymbol{\Theta}(0) = [0.18,\ 0.15,\ -0.15]^{\mathrm{T}}$ rad and $\boldsymbol{\omega}(0)=[-0.02,\ -0.02,\ 0.02]^{\mathrm{T}}$ rad s^{-1}, respectively. All the initial values of $\tilde{\eta}(0)$ and $\dot{\tilde{\eta}}(0)$ are assumed to be zero, that is, the flexible appendages are initially undeformed. The estimation errors are 2.5 times of the initial states and the control torque is constrained by $\lambda_1 = 2$ Nm, $\lambda_2 = 10$ Nm second^{-1}. This constraint on control torque is considered in the simulation. Considering that the observer gain matrix is calculated before the simulation results, and it is a constant matrix, where only the symmetric positive definite matrices $\boldsymbol{P}_2$, $\boldsymbol{P}_3$ and controller gain matrix $\boldsymbol{K}$ are to be solved.

Ref. [40] has analyzed the environmental disturbances in detail, including gravity-gradient torque, aerodynamic torque, and Earth magnetic torque. As pointed out in this work, the sinusoid is the prototype of some periodic disturbances, and the environmental disturbances such as magnetic torque and gravity-gradient torque are cyclic essentially and they can be represented by sinusoids. Thus it is suitable to set the external disturbance $\boldsymbol{T}_d$ as

$$\boldsymbol{T}_d = 5\times10^{-4}\times\begin{bmatrix}\sin(0.11\pi t)\\ \cos(0.11\pi t)\\ \cos(0.11\pi t+\pi/3)\end{bmatrix}\ \mathrm{Nm}$$

The following parameters are also chosen:

$$\begin{aligned}&\varpi = 0.5, \eta = 10^5, \xi_3 = 10^{-5}, \xi_{4,\ldots,21} = 0.01, \kappa_1 = 4, \kappa_2 = 0.001, \kappa_3 = 0.8,\\ &\kappa_4 = 0.1, \overline{\kappa}_1 = 0.09, \overline{\kappa}_2 = 0.104, \overline{\kappa}_3 = 0.04, \lambda_1 = 2, \lambda_2 = 10, \varsigma = 7,\\ &\boldsymbol{P}_1 = diag([\,1000\quad 800\quad 850\quad 1\quad 1\quad 1\quad 1\quad 1\quad 1\quad 1\quad 1\quad 1\quad 1\quad 1])\end{aligned}$$

Here, the fault signal is assumed to be $\boldsymbol{f}(t) = [0.005\ 0.006\ 0.007]^{\mathrm{T}}\overline{f}(t)$, and the form of $\overline{f}(t)$ is shown in Table 9.2.

The measurement errors are assumed to be

$$\begin{aligned}&\boldsymbol{v}(t) = 10^{-4}\times\\ &\begin{bmatrix}4 & 5 & 6 & 0.2 & 0.2 & 0.2 & 0.2 & 0.1 & 0.2 & 0.2 & 0.2 & 0.2 & 0.1 & 0.2\end{bmatrix}^{\mathrm{T}}\sin(0.01\pi t)\end{aligned}$$

Table 9.2 The form of $\overline{f}(t)$.

t/s	[0,10]	(10,45]	(45,60]	(60,75]	(75,100]
$\overline{f}(t)$/Nm	0	$\sin(0.1\pi t)$	1	$\cos(0.1\pi t)$	0

An actuator fault is assumed to occur, and the probability of success is selected to be $\overline{\omega} = 0.5$.

9.4.1 Simulation results using an antidisturbance controller

Choose

$$\begin{aligned} \boldsymbol{F}_1 &= \begin{bmatrix} 1.1443 \times 10^3 & -6.7261 \times 10^2 & 1.1088 \times 10^2 & -0.0061 \times 10^2 \end{bmatrix} \\ \boldsymbol{F}_2 &= \begin{bmatrix} 6.5280 \times 10^2 & 8.3129 \times 10^2 & -1.6665 \times 10^2 & 0.0265 \times 10^2 \end{bmatrix} \end{aligned}$$

By choosing appropriate symmetric positive definite matrix, the observer gain matrix used for controller design is designed as $\boldsymbol{L} = 10^3 \times \begin{bmatrix} \boldsymbol{L}_1 & \boldsymbol{L}_2 \end{bmatrix}$, where,

$$\boldsymbol{L}_1 = \begin{bmatrix} 6.0121 & 0.0007 & 0.0005 & -0.0006 & 0.0019 & 0.0035 & 0.0065 \\ 0.0007 & 6.0084 & -0.0001 & 0.0006 & 0.0018 & -0.0005 & 0.0005 \\ 0.0005 & -0.0001 & 6.0083 & -0.0010 & -0.0007 & 0.0010 & 0.0015 \\ -0.0006 & 0.0006 & -0.0010 & 6.0212 & 0.0030 & 0.0044 & -0.0076 \\ 0.0019 & 0.0018 & -0.0007 & 0.0030 & 6.0181 & -0.0053 & -0.0006 \\ 0.0035 & -0.0005 & 0.0010 & 0.0044 & -0.0053 & 6.0246 & 0.0044 \\ 0.0065 & 0.0005 & 0.0015 & -0.0076 & -0.0006 & 0.0044 & 6.0350 \\ -0.0038 & -0.0018 & 0.0003 & -0.0021 & -0.0079 & 0.0020 & 0.0126 \\ 0.0134 & 0.0032 & 0.0014 & -0.0077 & 0.0114 & 0.0002 & 0.0198 \\ 0.0177 & 0.0039 & 0.0020 & -0.0088 & 0.0128 & 0.0041 & 0.0301 \\ -0.0055 & -0.0030 & 0.0011 & -0.0129 & -0.0134 & -0.0027 & 0.0224 \\ 0.0062 & 0.0017 & 0.0003 & 0.0029 & 0.0056 & 0.005 & -0.0057 \\ 0.0002 & -0.0007 & 0.0005 & -0.0007 & -0.0046 & 0.0052 & 0.0037 \\ 0.0023 & 0 & 0.0007 & -0.0049 & -0.0007 & 0.0005 & 0.0064 \end{bmatrix}$$

and

$$\boldsymbol{L}_2 = \begin{bmatrix} -0.0039 & 0.0135 & 0.0177 & -0.0056 & 0.0062 & 0.0002 & 0.0023 \\ -0.0018 & 0.0032 & 0.0039 & -0.0030 & 0.0017 & -0.0007 & 0 \\ 0.0003 & 0.0014 & 0.0020 & 0.0010 & 0.0003 & 0.0005 & 0.0007 \\ -0.0021 & -0.0077 & -0.0089 & -0.0129 & 0.0029 & -0.0007 & -0.0049 \\ -0.0080 & 0.0115 & 0.0129 & -0.0135 & 0.0057 & -0.0046 & -0.0007 \\ 0.0020 & 0.0002 & 0.0042 & -0.0028 & 0.0051 & 0.0052 & 0.0005 \\ 0.0124 & 0.0199 & 0.0303 & 0.0222 & -0.0055 & 0.0037 & 0.0065 \\ 6.0367 & -0.0207 & -0.0203 & 0.0415 & -0.0245 & -0.0059 & -0.0008 \\ -0.0209 & 6.0629 & 0.0679 & -0.0243 & 0.0236 & -0.0042 & 0.0095 \\ -0.0205 & 0.0680 & 6.0944 & -0.0237 & 0.0266 & -0.0030 & 0.0115 \\ 0.0416 & -0.0241 & -0.0234 & 6.0761 & -0.0371 & 0.0083 & 0.0026 \\ -0.0245 & 0.0235 & 0.0264 & -0.0372 & 6.0324 & -0.0032 & 0.0019 \\ 0.0059 & -0.0042 & -0.0029 & 0.0083 & -0.0032 & 6.0111 & 0.0006 \\ -0.0008 & 0.0095 & 0.0115 & 0.0025 & 0.0019 & 0.0006 & 6.0113 \end{bmatrix}$$

Solving LMIs (9.41)−(9.47) gives the following controller gain matrix $\boldsymbol{K}$:

$$\boldsymbol{K}=\begin{bmatrix} 1.2131 & 0.0204 & 0.0569 & 26.6479 & 0.5282 & 1.4103 & 0.0029 \\ 0.0238 & 1.1660 & -0.0656 & 0.5241 & 30.1739 & -1.6257 & 0.0011 \\ 0.0637 & -0.0628 & 1.8271 & 1.4018 & -1.6286 & 45.1921 & -0.0013 \end{bmatrix.$$
$$\begin{matrix} 0.0037 & -0.0007 & 0.0001 & -0.0057 & -0.0008 & -0.0013 & -0.0018 \\ 0.0014 & -0.0000 & -0.0002 & -0.0005 & -0.0029 & -0.0044 & 0.0013 \\ -0.0017 & 0.0000 & -0.0007 & -0.0086 & 0.0061 & 0.0021 & 0.0027 \end{matrix}\right]$$

Applying the controller gain $\boldsymbol{K}$ and observer gain $\boldsymbol{L}$ to the attitude control system represented by Eq. (9.32) produces the simulation results of Figs. 9.1-9.10. Figs. 9.1 and 9.2 display the attitude angle and angular velocity of the flexible spacecraft, from which it can be seen that the attitude angle and angular velocity both converge to within a small bound under controller (9.28). Fig. 9.3 shows the modal displacement of the flexible appendages, from which it is observed that the active vibration suppression method can effectively suppress the vibrations within 100 seconds. Figs. 9.4 to 9.6 plot the estimation errors of attitude angle, angular velocity and modal displacement, which suggest that they are estimated within 2×10^{-3} seconds. It can be concluded that the intermediate observer proposed in this chapter can effectively estimate attitude angle,

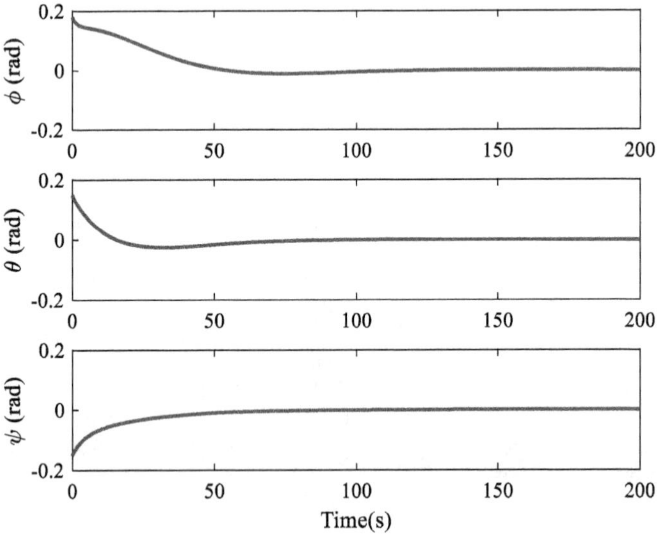

Figure 9.1 Time response of attitude angle.

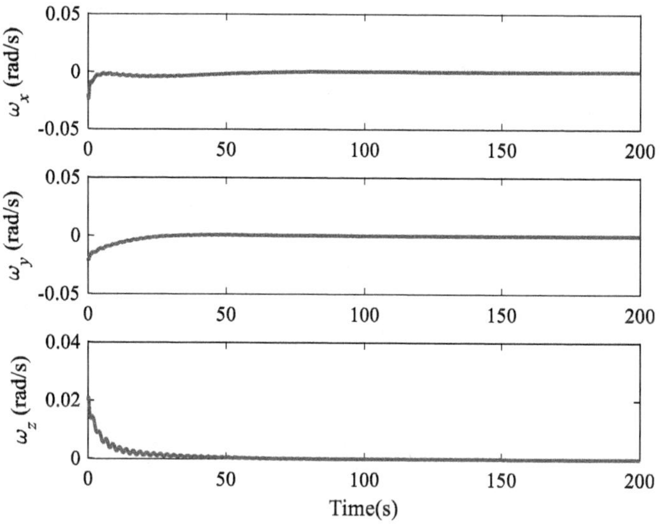

Figure 9.2 Time response of angular velocity.

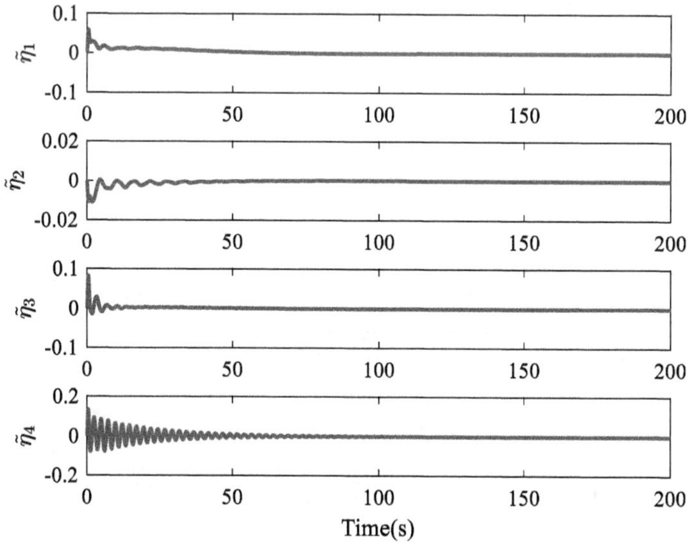

Figure 9.3 Time response of modal displacement.

angular velocity and modal displacement, with a very fast convergence speed.

Fig. 9.7 displays the lumped disturbance and its estimation, which are less than 10 Nm. The figure shows that the proposed observer can effectively estimate the lumped disturbances. Furthermore, Fig. 9.8 implies

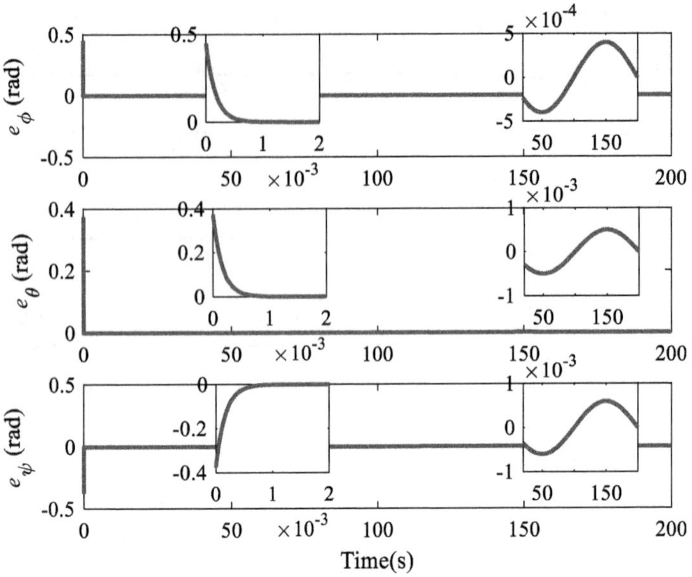

Figure 9.4 Estimation errors of attitude angle.

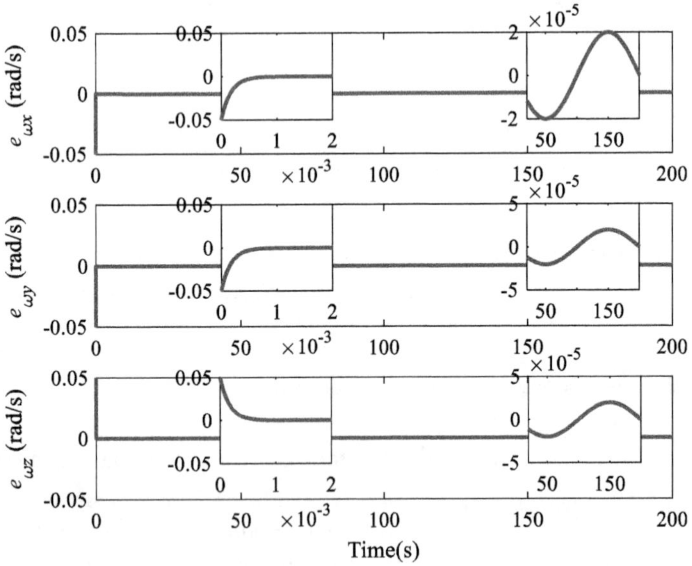

Figure 9.5 Estimation errors of angular velocity.

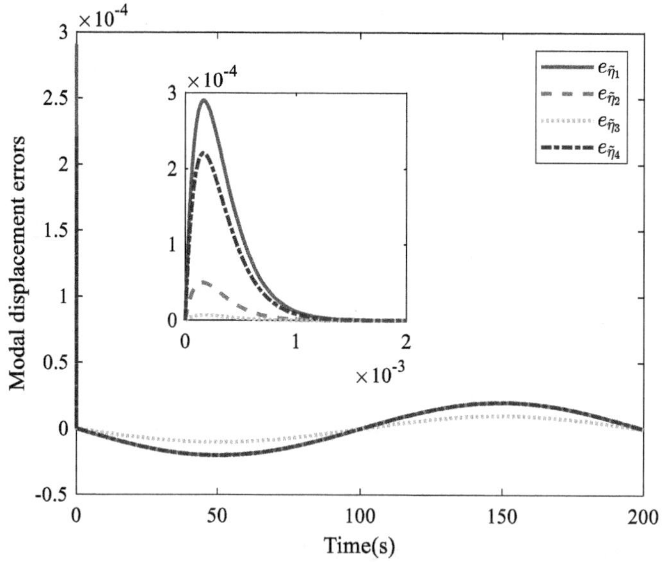

Figure 9.6 Estimation errors of modal displacement.

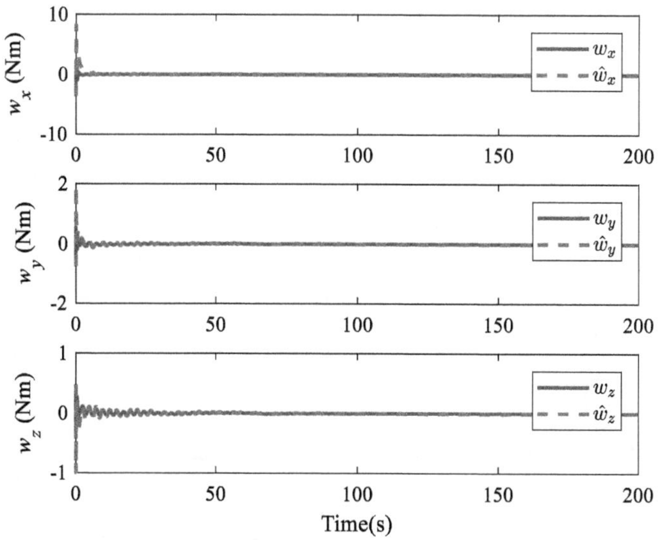

Figure 9.7 Lumped disturbance and its estimation.

that the estimation error of the lumped disturbance will converge to a small neighborhood containing equilibrium, which also validates the effectiveness of the observer. The magnitude and rate of the control torque are plotted in Figs. 9.9 and 9.10, from which it is observed that the

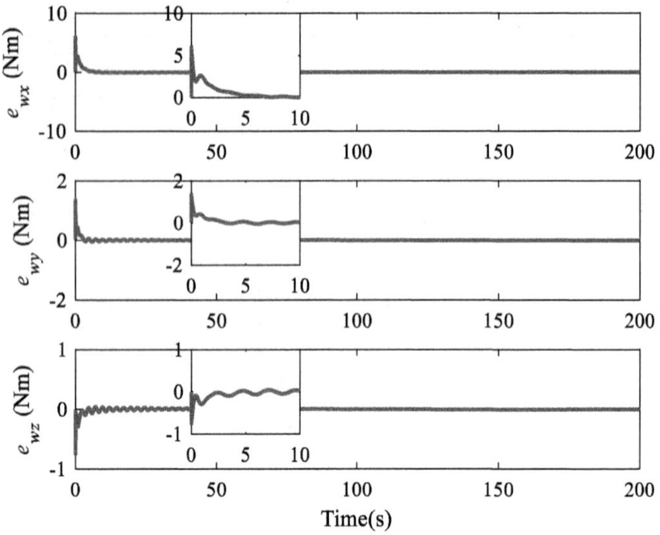

Figure 9.8 Estimation errors of lumped disturbance.

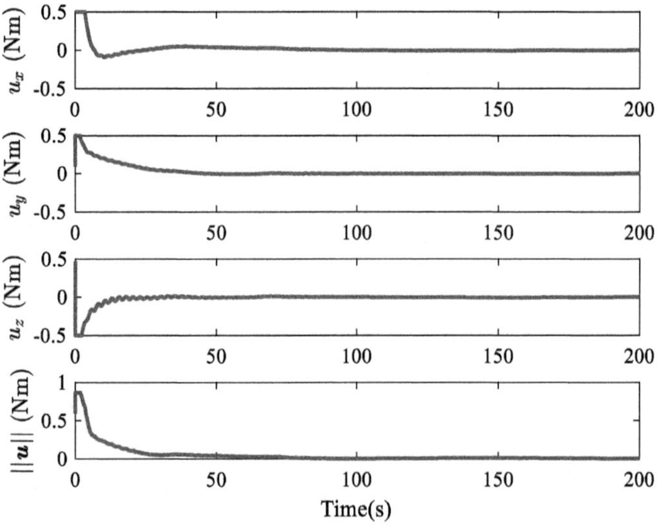

Figure 9.9 Time response of control torque magnitude.

magnitude and rate of the control torque along each axis are less than 0.5 Nm and 0.5 Nm second^{-1} respectively. It is obvious that the norms of magnitude and rate of control torque are less than 1 Nm and 0.6 Nm second^{-1}, respectively, which satisfies their limit constraints.

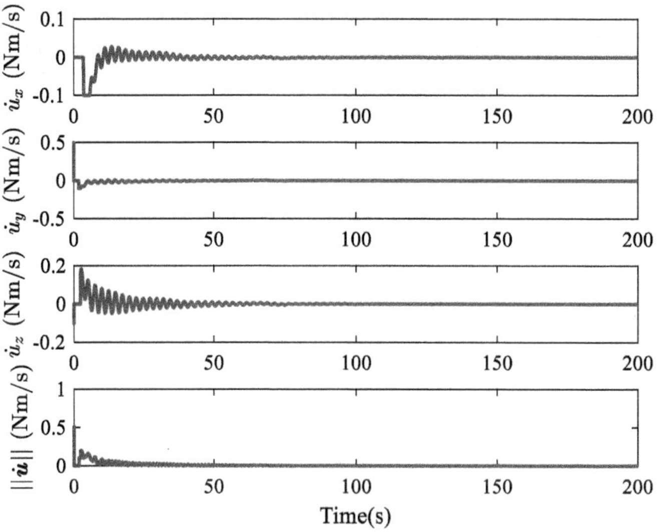

Figure 9.10 Time response of control torque rate.

9.4.2 Simulation results using a mixed H_2/H_∞ controller

In this case, the controller gain matrix $\boldsymbol{K}$ is

$$\boldsymbol{K} = \begin{matrix} \left[\begin{matrix} -0.1159 & -0.0012 & -0.0017 & -44.8950 & -6.4382 & -14.3781 & 5.9867 \\ -0.0009 & -0.1190 & 0.0019 & -7.9045 & -4.4401 & -4.6577 & 1.3985 \\ -0.0039 & 0.0012 & -0.1371 & 35.8916 & -7.6579 & -17.5341 & 5.2514 \end{matrix}\right. \\ \left.\begin{matrix} 0.3897 & -1.0375 & -1.3755 & -3.4081 & 2.0844 & -4.3014 & -8.1894 \\ -0.3938 & -2.3309 & 0.9342 & -0.6989 & -1.0046 & -8.9040 & 5.4387 \\ 0.0383 & 0.7514 & 1.2494 & -0.8448 & 2.3513 & 2.7848 & 7.2990 \end{matrix}\right] \end{matrix}$$

Applying this controller to the corresponding attitude control system represented by Eq. (9.32) produces the simulation results of Figs. 9.11-9.14. Figs. 9.11 and 9.12 plot the response curves of attitude angle and angular velocity of the flexible spacecraft, from which it can be seen that both of the attitude angle and angular velocity also converge to the vicinity of the expected states with smaller steady-state errors. However, compared with Figs. 9.1 and 9.2, the convergence speed is much slower with larger overshoot. Fig. 9.13 shows the modal displacement in this case, from which it is obviously seen that the vibration suppression is achieved at approximately 500 seconds. Thus the mixed H_2/H_∞ controller requires much longer time than the observer-based controller, although the maximum modal displacement during attitude maneuver is smaller. Fig. 9.14 depicts the response curve of control torque using mixed H_2/H_∞

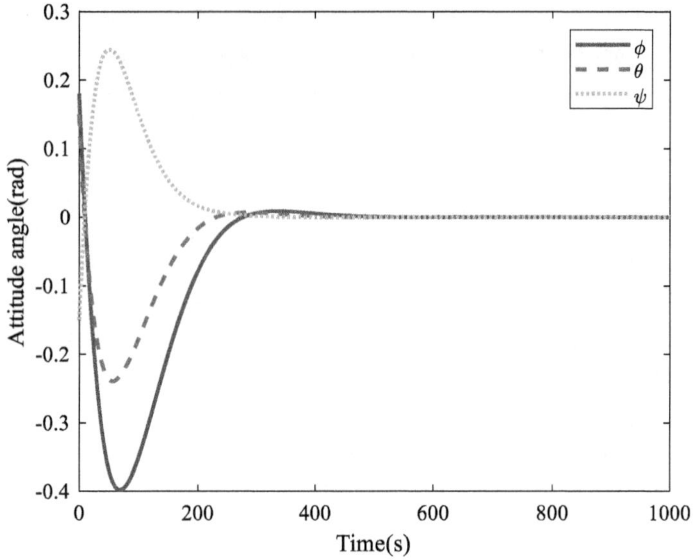

Figure 9.11 Time response of attitude angle using mixed H_2/H_∞ controller.

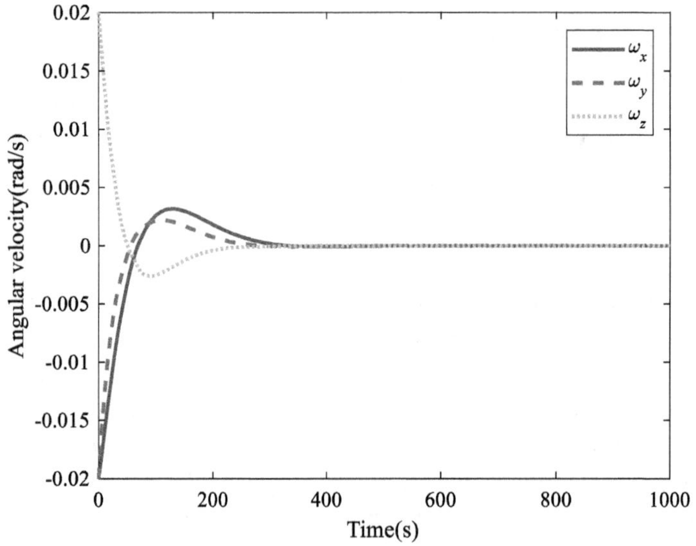

Figure 9.12 Time response of angular velocity using mixed H_2/H_∞ controller.

controller, which suggests that the maximum control torques for three directions are approximately 0.2 Nm, much smaller than the input limited magnitude. It can be concluded that the mixed H_2/H_∞ controller does not make effective use of the control torque that the actuator can provide,

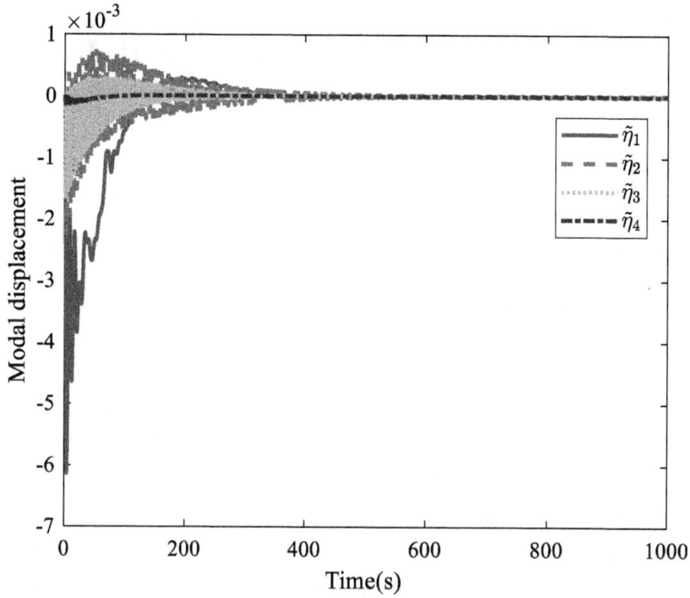

Figure 9.13 Time response of modal displacement using mixed H_2/H_∞ controller.

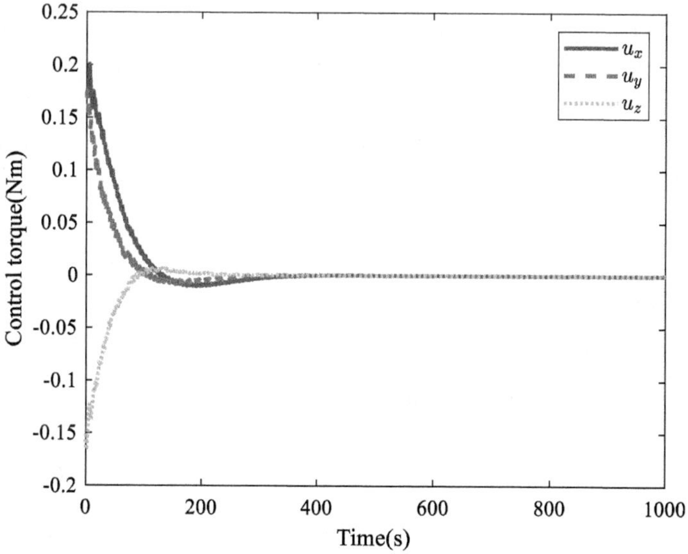

Figure 9.14 Time response of control torque using mixed H_2/H_∞ controller.

leading to much slower speed of convergence and larger overshoot, thus the observer-based controller is superior to the mixed H_2/H_∞ controller from this point of view. More importantly, the observer-based controller doesn't require the prior information on inertia matrix while the mixed H_2/H_∞ controller does.

9.5 Conclusions

In this work, an observer-based inertia-free attitude controller is developed for flexible spacecraft in the presence of inertia uncertainties, external disturbances, actuator faults, measurement errors, and input magnitude and rate constraints. By analyzing the effects of the aforementioned multiple uncertain factors, a lumped disturbance is constructed to facilitate controller design. Then, the observer is designed based on an auxiliary variable, which takes advantage of the attitude and modal information and the lumped disturbance. Based on this, a feedback control law employing the estimates of attitude and modal information and the lumped disturbance is developed. Furthermore, the sufficient condition for input MRCs can also be incorporated into controller design, to satisfy the actuator constraints. It is worth noting that the developed controller requires no prior information on inertia matrix, external disturbances and fault signals, and only information on upper bounds of the measurement errors, the first time derivative of lumped disturbance, intermediate variable, and its first time derivative. Theoretical proofs and simulation results reveal that the states of the closed-loop system can converge asymptotically to the vicinity of the expected states with very good performance. Comparisons with the mixed H_2/H_∞ controller also illustrate the superior performance of the proposed controller.

References

[1] He W, Meng T, He X, Ge SS. Unified iterative learning control for flexible structures with input constraints. Automatica 2018;96:326–36.
[2] Yuan Q, Liu Y, Qi N. Active vibration suppression for maneuvering spacecraft with high flexible appendages. Acta Astronautica 2017;139:512–20.
[3] Dai H, Jing X, Wang Y, et al. Post-capture vibration suppression of spacecraft via a bio-inspired isolation system. Mechanical Systems and Signal Processing 2018;105:214–40.
[4] Cao X, Yue C, Liu M, Wu B. Time efficient spacecraft maneuver using constrained torque distribution. Acta Astronautica 2016;123:320–9.
[5] Liu H, Li D, Xi J, et al. Robust attitude controller design for miniature quadrotors. International Journal of Robust and Nonlinear Control 2016;26(4):681–96.

[6] Yan R, Wu Z. Attitude stabilization of flexible spacecraft via extended disturbance observer based controller. Acta Astronautica 2017;133:73–80.
[7] Ma Y, Jiang B, Tao G, Cheng Y. Uncertainty decomposition-based fault-tolerant adaptive control of flexible spacecraft. IEEE Transactions on Aerospace and Electronic Systems 2015;51(2):1053–68.
[8] Liu C, Ye D, Shi K, Sun Z. Robust high-precision attitude control for flexible spacecraft with improved mixed H_2/H_∞ control strategy under poles assignment constraint. Acta Astronautica 2017;136:166–75.
[9] Liu H, Peng F, Lewis FL, et al. Robust tracking control for tail-sitters in flight mode transitions. IEEE Transactions on Aerospace and Electronic Systems 2019;55(4):2023–35.
[10] Liu H, Ma T, Lewis FL, et al. Robust formation control for multiple quadrotors with nonlinearities and disturbances. IEEE Transactions on Cybernetics 2018.
[11] Weiss A, Kolmanovsky I, Bernstein DS, et al. Inertia-free spacecraft attitude control using reaction wheels. Journal of Guidance, Control, and Dynamics 2013;36(5):1425–39.
[12] Sanyal A, Fosbury A, Chaturvedi N, et al. Inertia-free spacecraft attitude tracking with disturbance rejection and almost global stabilization. Journal of Guidance, Control, and Dynamics 2009;32(4):1167–78.
[13] Shen Q, Yue C, Goh CH, et al. Active fault-tolerant control system design for spacecraft attitude maneuvers with actuator saturation and fault. IEEE Transactions on Industrial Electronics 2019;66(5):3763–72.
[14] Hu Q, Shao X, Zhang Y, Guo L. Nussbaum-type function-based attitude control of spacecraft with actuator saturation. International Journal of Robust and Nonlinear Control 2018;28(8):2927–49.
[15] Su Y, Zheng C. Globally asymptotic stabilization of spacecraft with simple saturated proportional-derivative control. Journal of Guidance, Control, and Dynamics 2011;34(6):1932–6.
[16] You L, Dong Y. Near time-optimal controller based on analytical trajectory planning algorithm for satellite attitude maneuver. Aerospace Science and Technology 2019;84:497–509.
[17] Akella MR, Valdivia A, Kotamraju GR. Velocity-free attitude controllers subject to actuator magnitude and rate saturations. Journal of Guidance, Control, and Dynamics 2005;28(4):659–66.
[18] Liu C, Vukovich G, Sun Z, Shi K. Observer-based fault-tolerant attitude control for spacecraft with input delay. Journal of Guidance, Control, and Dynamics 2018;41(9):2041–53.
[19] Xiao B, Yin S, Kaynak O. Attitude stabilization control of flexible satellites with high accuracy: An estimator-based approach. IEEE/ASME Transactions on Mechatronics 2017;22(1):349–58.
[20] Di Gennaro S. Output stabilization of flexible spacecraft with active vibration suppression. IEEE Transactions on Aerospace and Electronic systems 2003;39(3):747–59.
[21] Cao X, Yue C, Liu M. Flexible satellite attitude maneuver via constrained torque distribution and active vibration suppression. Aerospace Science and Technology 2017;67:387–97.
[22] Sidi MJ. Spacecraft dynamics and control: a practical engineering approach. Cambridge, UK: Cambridge university press; 1997. p. 103–10.
[23] Wang H, Yang GH. Robust Mixed H_1/H_∞ filtering for affine fuzzy systems with measurement errors. IEEE Transactions on Cybernetics 2014;44(7):1100–10.
[24] Xia Y, Zhu Z, Fu M, Wang S. Attitude tracking of rigid spacecraft with bounded disturbances. IEEE Transactions on Industrial Electronics 2011;58(2):647–59.

[25] Zhang A, Hu Q, Zhang Y. Observer-based attitude control for satellite under actuator fault. Journal of Guidance, Control, and Dynamics 2015;38(4):806–11.

[26] Boškovic JD, Li SM, Mehra RK. Robust adaptive variable structure control of spacecraft under control input saturation. Journal of Guidance, Control, and Dynamics 2001;24(1):14–22.

[27] Boskovic JD, Li SM, Mehra RK. Robust tracking control design for spacecraft under control input saturation. Journal of Guidance, Control, and Dynamics 2004;27 (4):627–33.

[28] Huang S, Yang G. Fault tolerant controller design for T-S fuzzy systems with time-varying delay and actuator fault: a k-step fault-estimation approach. IEEE Transactions on Fuzzy Systems 2014;22(6):1526–40.

[29] You F, Li H, Wang F, Guan S. Robust fast adaptive fault estimation for systems with time-varying interval delay. Journal of the Franklin Institute 2015;352 (12):5486–513.

[30] Shi P, Liu M, Zhang L. Fault-tolerant sliding-mode-observer synthesis of Markovian jump systems using quantized measurements. IEEE Transactions on Industrial Electronics 2015;62(9):5910–18.

[31] Zhu JW, Yang GH. Fault accommodation for uncertain linear systems with measurement errors. International Journal of Robust and Nonlinear Control 2017;27 (10):1841–54.

[32] Tabatabaeipour SM, Bak T. Robust observer-based fault estimation and accommodation of discrete-time piecewise linear systems. Journal of the Franklin Institute 2014;51(1):277–95.

[33] Ge SS, Wang C. Adaptive neural control of uncertain MIMO nonlinear systems. IEEE Transactions on Neural Networks 2004;15(3):674–92.

[34] Zhang K, Jiang B, Shi P. Fast fault estimation and accommodation for dynamical systems. IET Control Theory & Applications 2009;3(2):189–99.

[35] Jiang B, Staroswiecki M, Cocquempot V. Fault accommodation for nonlinear dynamic systems. IEEE Transactions on Automatic Control 2006;51(9):1578–83.

[36] Liu C, Vukovich G, Shi K, Sun Z. Robust fault tolerant nonfragile H∞ attitude control for spacecraft via stochastically intermediate observer. Advances in Space Research 2018;62(9):2631–48.

[37] Jia S, Shan J. Optimal actuator placement for constrained gyroelastic beam considering control spillover. Journal of Guidance, Control, and Dynamics 2018;41 (9):2073–81.

[38] Chen T, Shan J. Rotation-matrix-based attitude tracking for multiple flexible spacecraft with actuator faults. Journal of Guidance, Control, and Dynamics 2018;42:1–8.

[39] Xu S, Sun G, Sun W. Takagi-Sugeno fuzzy model based robust dissipative control for uncertain flexible spacecraft with saturated time-delay input. ISA Transactions 2017;66:105–21.

[40] Yang CD, Sun YP. Mixed H_2/H_∞ state-feedback design for microsatellite attitude control. Control Engineering Practice 2002;10(9):951–70.

CHAPTER 10
Chaotic attitude tracking control

10.1 Introduction

Spacecraft have been widely used in widespread communications, remote sensing, and related scientific research in space. The attitude control system can have great influence on spacecraft pointing accuracy and stabilization precision. Generally, any spacecraft in orbit is influenced by several kinds of external disturbance torques, such as the aerodynamic drag torque, gravity gradient torque, solar radiation pressure, and magnetic torque caused by the Earth's magnetic field. Although the external disturbances are small compared to the weight of the spacecraft, it often consists of periodic and secular term, and the long-time disturbances on spacecraft may have significant influence on its real attitude motion [1]. When the inertia moment of spacecraft and external disturbances acting on spacecraft satisfy some conditions, it can lead to chaotic motion of the spacecraft [2]. Once one spacecraft with chaotic attitude motion is required to track the other spacecraft chaotic attitude plant to achieve angular velocity synchronization, a robust tracking controller becomes necessary.

The chaos phenomenon has been extensively studied by many researchers due to its unstable and complex behavior and wide range of applications in many industrial systems and different sciences [3,4]. The typical chaotic systems that the external disturbances acting on the spacecraft may induce are the Newton-Leipnik system [5], Lorenz system [6], Chen system [7], Lu system [8], Genesio-Tesi system [9], Rucklidge system [10], Liu system [11], and Rossler system [12]. Research on synchronization control in chaotic systems has recently become a subject of great interest, and considerable efforts have been made to study the control and synchronization problems of different chaotic systems, such as adaptive control [13], sliding mode control [14], linear matrix inequality techniques [15], fuzzy logic control [16], state observer control [17], active control [18], and passive control [19].

Many flight experiences during aerospace history have witnessed the unexpected behaviors in spacecraft attitude motion, which result from the external disturbances that had not been taken into consideration in

Spacecraft Attitude Control. DOI: https://doi.org/10.1016/B978-0-323-99005-9.00010-9

spacecraft design. Numerous theoretical studies have pointed out that spacecraft chaotic motion exists under the action of different kinds of external disturbances. Tong and Rimrott [20] studied the planar vibration of an asymmetric satellite in elliptic orbit under the action of gravity gradient torque. Meehan and Asokanthan [21,22] researched the chaotic motion of a spinning spacecraft which may result from circumferential nutation damper or unbalanced rotor or vibrations in the appendages. Salarieh and Alasty [23] have investigated the problem of synchronization between two chaotic gyros using a modified sliding mode control. Aghababa [24] adopted an adaptive finite-time controller to achieve the synchronization of two chaotic flywheel governor systems and verified its robustness. Beletsky et al. [25] investigated the numerical realization method to analyze the chaos in spacecraft attitude motion in circular polar orbit only influenced by the geomagnetic field. However, to the best of the authors' knowledge, the approach in the case that one spacecraft with chaotic attitude motion should track another spacecraft chaotic attitude plant to achieve angular velocity synchronization has not been investigated, and the main contribution of this chapter is to conduct theoretical research on the design of adaptive variable structure tracking controller for spacecraft chaotic motion to meet the requirement of angular velocity synchronization.

The remainder of this chapter is organized as follows. Section 10.2 introduces the spacecraft attitude dynamics equation and expands it in its component form. Besides, this section describes different kinds of chaotic phenomena in spacecraft attitude motion influenced by different external disturbance torque and describes the purpose of this work. Section 10.3 presents the robust adaptive variable structure controller based on adaptive control theory and variable structure control theory. Numerical simulations are given to illustrate the performance of the proposed technique in Section 10.4. Finally, some conclusions of this work are addressed.

10.2 Problem formulation

10.2.1 Chaotic attitude dynamics

The rigid spacecraft attitude dynamics equation is

$$\boldsymbol{I}\dot{\boldsymbol{\omega}} + \boldsymbol{\omega} \times (\boldsymbol{I}\boldsymbol{\omega}) = \boldsymbol{T}_c + \boldsymbol{T}_d \tag{10.1}$$

where $\boldsymbol{\omega} = [\omega_1 \quad \omega_2 \quad \omega_3]^T$ means the angular velocity of the spacecraft body reference frame with respect to the earth-centered inertial reference frame in the spacecraft body reference frame, $\boldsymbol{I}$ denotes the spacecraft inertia matrix, $\boldsymbol{T}_c$ means the control input torque; $\boldsymbol{T}_d$ is the external disturbance torque imposed on the spacecraft, which can be generally expressed in the following nonlinear form:

$$\boldsymbol{T}_d = \boldsymbol{D}\boldsymbol{\omega} + \boldsymbol{M} \tag{10.2}$$

where $\boldsymbol{D} = [d_{ij}]_{3\times 3} \in \boldsymbol{R}^{3\times 3}(i,j = 1,2,3)$, which can be a constant matrix or matrix varying with angular velocity; $\boldsymbol{M} = [m_i]_{3\times 1} \in \boldsymbol{R}^{3\times 1}(i = 1,2,3)$, which can be a constant matrix or matrix varying with angular velocity, or even periodic matrix or long-term matrix.

Take the three axes of the spacecraft body coordinate system for the inertial principal axes, then $\boldsymbol{I} = diag(I_1, I_2, I_3)$. In addition, use the Levi Civita symbol in three dimensions to express vector products, denoted as ε_{kij}, and the corresponding definition is given as follows:

$$\varepsilon_{kij} = \begin{cases} +1, & if\ (k,i,j)\ is\ (1,2,3),\ (2,3,1)\ or\ (3,1,2) \\ -1, & if\ (k,i,j)\ is\ (3,2,1),\ (1,3,2)\ or\ (2,1,3) \\ 0, & if\ i = j\ or\ j = k\ or\ k = i \end{cases} \tag{10.3}$$

For any two vectors $\boldsymbol{p} = [p_i]_{3\times 1}(i = 1,2,3)$, $\boldsymbol{q} = [q_j]_{3\times 1}(j = 1,2,3)$, one has that $\sum_{i,j} \varepsilon_{kij} p_i q_j = (\boldsymbol{p} \times \boldsymbol{q})_k$, where $(\ )_k$ represents the k-th component of the vector product.

Substitute Eq. (10.2) into Eq. (10.1) and expand it in its component form, one has

$$\begin{cases} I_1\dot{\omega}_1 - (I_2 - I_3)\omega_2\omega_3 = T_{c1} + d_{11}\omega_1 + d_{12}\omega_2 + d_{13}\omega_3 + m_1 \\ I_2\dot{\omega}_2 - (I_3 - I_1)\omega_1\omega_3 = T_{c2} + d_{21}\omega_1 + d_{22}\omega_2 + d_{23}\omega_3 + m_2 \\ I_3\dot{\omega}_3 - (I_1 - I_2)\omega_1\omega_2 = T_{c3} + d_{31}\omega_1 + d_{32}\omega_2 + d_{33}\omega_3 + m_3 \end{cases} \tag{10.4}$$

where I_1, I_2 and I_3 denote the three components of inertia matrix; ω_1, ω_2 and ω_3 denote the three components of angular velocity; T_{c1}, T_{c2} and T_{c3} denote the three components of control input torque.

Then, one has

$$\begin{cases} \dot{\omega}_1 = I_1^{-1}(I_2 - I_3)\omega_2\omega_3 + I_1^{-1}d_{11}\omega_1 + I_1^{-1}d_{12}\omega_2 + I_1^{-1}d_{13}\omega_3 + I_1^{-1}T_{c1} + I_1^{-1}m_1 \\ \dot{\omega}_2 = I_2^{-1}(I_3 - I_1)\omega_1\omega_3 + I_2^{-1}d_{21}\omega_1 + I_2^{-1}d_{22}\omega_2 + I_2^{-1}d_{23}\omega_3 + I_2^{-1}T_{c2} + I_2^{-1}m_2 \\ \dot{\omega}_3 = I_3^{-1}(I_1 - I_2)\omega_1\omega_2 + I_3^{-1}d_{31}\omega_1 + I_3^{-1}d_{32}\omega_2 + I_3^{-1}d_{33}\omega_3 + I_3^{-1}T_{c3} + I_3^{-1}m_3 \end{cases} \tag{10.5}$$

Define the relative inertia ratios as $a_1 = I_1^{-1}(I_2 - I_3)$, $a_2 = I_2^{-1}(I_3 - I_1)$, and $a_3 = I_3^{-1}(I_1 - I_2)$, and $[u_i]_{3\times 1} = [I_i^{-1}T_{ci}]_{3\times 1}(i = 1,2,3)$ is the angular acceleration generated by the control torque, and $[b_{ij}]_{3\times 3} = [I_i^{-1} d_{ij}]_{3\times 3}$, $[c_i]_{3\times 1} = [I_i^{-1} m_i]_{3\times 1} (i, j = 1, 2, 3)$.

Then, Eq. (10.5) can be transformed into

$$\begin{cases} \dot{\omega}_1 = a_1\omega_2\omega_3 + b_{11}\omega_1 + b_{12}\omega_2 + b_{13}\omega_3 + u_1 + c_1 \\ \dot{\omega}_2 = a_2\omega_1\omega_3 + b_{21}\omega_1 + b_{22}\omega_2 + b_{23}\omega_3 + u_2 + c_2 \\ \dot{\omega}_3 = a_3\omega_1\omega_2 + b_{31}\omega_1 + b_{32}\omega_2 + b_{33}\omega_3 + u_3 + c_3 \end{cases} \tag{10.6}$$

The matrix form of Eq. (10.6) is

$$\dot{\boldsymbol{\omega}} = \boldsymbol{B}\boldsymbol{\omega} + \boldsymbol{f}(\boldsymbol{\omega}) + \boldsymbol{u} \tag{10.7}$$

where,

$$\boldsymbol{B} = \begin{bmatrix} b_{11} & b_{12} & b_{13} \\ b_{21} & b_{22} & b_{23} \\ b_{31} & b_{32} & b_{33} \end{bmatrix}, \boldsymbol{f}(\boldsymbol{\omega}) = \begin{bmatrix} f_1(\boldsymbol{\omega}) \\ f_2(\boldsymbol{\omega}) \\ f_3(\boldsymbol{\omega}) \end{bmatrix} + \begin{bmatrix} c_1 \\ c_2 \\ c_3 \end{bmatrix} = \boldsymbol{A} \begin{bmatrix} \omega_2\omega_3 \\ \omega_1\omega_3 \\ \omega_1\omega_2 \end{bmatrix} + \boldsymbol{C},$$

$$\boldsymbol{A} = diag(a_1, a_2, a_3), \boldsymbol{C} = \begin{bmatrix} c_1 & c_2 & c_3 \end{bmatrix}^T, \boldsymbol{u} = \begin{bmatrix} u_1 & u_2 & u_3 \end{bmatrix}^T$$

10.2.2 Chaotic system characteristics and chaotic attractor

For uncontrolled spacecraft attitude system, that is, the control input torque satisfies $\boldsymbol{u} = 0$, different external disturbance torque may lead to different kinds of chaos in spacecraft attitude motion, such as the Newton-Leipnik system, Lorenz system, Chen system, Lu system, Genesio-Tesi system, Rucklidge system, Liu system, Rossler system, and so on. The analysis of these chaotic systems, their characteristics and chaotic attractors, and corresponding necessary condition for spacecraft properties can be seen in Table 10.1.

10.2.3 Tracking error dynamics and control objective

The divergence of $\dot{\boldsymbol{\omega}}$ is

$$\nabla \cdot \dot{\boldsymbol{\omega}} = \frac{\partial \dot{\omega}_1}{\partial \omega_1} + \frac{\partial \dot{\omega}_2}{\partial \omega_2} + \frac{\partial \dot{\omega}_3}{\partial \omega_3} \tag{10.8}$$

For one of the above eight typical chaotic systems when the spacecraft inertia moment and corresponding external disturbance torques satisfy given conditions, one can calculate $\nabla \cdot \dot{\boldsymbol{\omega}} < 0$, which indicates that the spacecraft attitude dynamics system is a dissipative system and its solution is bounded with increasing time.

Table 10.1 Chaotic system characteristics and chaotic attractor

Chaotic system	Characteristics	Spacecraft	Chaotic attractor
Newton–Leipnik system	$\boldsymbol{A} = diag(0,0,0), \boldsymbol{C} = \boldsymbol{0}$ $\boldsymbol{B} = \begin{bmatrix} -a & 1+10\omega_3 & 0 \\ -1+5\omega_3 & -0.4 & 0 \\ -5\omega_2 & 0 & b \end{bmatrix}$	$I_1 = I_2 = I_3$ $\boldsymbol{T}_d = \begin{bmatrix} I_1(-a\omega_1 + \omega_2 + 10\omega_2\omega_3) \\ I_2(-\omega_1 - 0.4\omega_2 + 5\omega_1\omega_3) \\ I_3(b\omega_3 - 5\omega_1\omega_2) \end{bmatrix}$	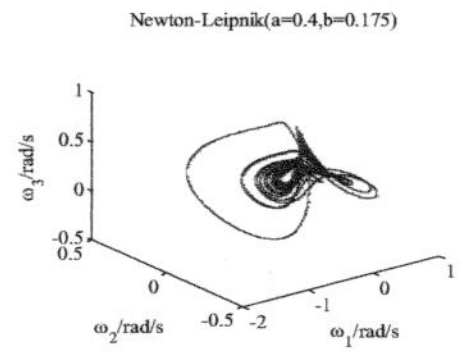
Lorenz system	$\boldsymbol{A} = diag(0,-1,1), \boldsymbol{C} = \boldsymbol{0}$ $\boldsymbol{B} = \begin{bmatrix} -a & a & 0 \\ c & -1 & 0 \\ 0 & 0 & -b \end{bmatrix}$	$I_1 = 2I_2 = 2I_3$ $\boldsymbol{T}_d = \begin{bmatrix} I_1 a(\omega_2 - \omega_1) \\ I_2(c\omega_1 - \omega_2) \\ -I_3 b\omega_3 \end{bmatrix}$	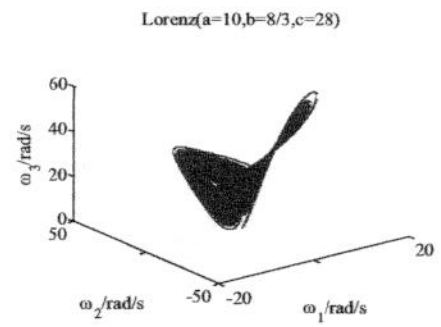
Chen system	$\boldsymbol{A} = diag(0,-1,1), \boldsymbol{C} = \boldsymbol{0}$ $\boldsymbol{B} = \begin{bmatrix} -a & a & 0 \\ c-a & c & 0 \\ 0 & 0 & -b \end{bmatrix}$	$I_1 = 2I_2 = 2I_3$ $\boldsymbol{T}_d = \begin{bmatrix} I_1 a(\omega_2 - \omega_1) \\ I_2(c\omega_1 - a\omega_1 + c\omega_2) \\ -I_3 b\omega_3 \end{bmatrix}$	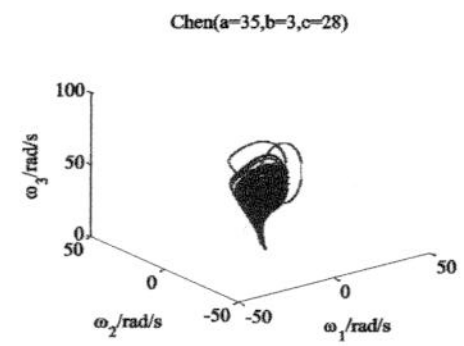

(Continued)

Table 10.1 (Continued)

Chaotic system	Characteristics	Spacecraft	Chaotic attractor
Lu system	$\boldsymbol{A}=diag(0,-1,1), \boldsymbol{C}=\boldsymbol{0}$ $\boldsymbol{B}=\begin{bmatrix}-a & a & 0\\ 0 & c & 0\\ 0 & 0 & -b\end{bmatrix}$	$I_1=2I_2=2I_3$ $\boldsymbol{T}_d=\begin{bmatrix}I_1a(\omega_2-\omega_1)\\ I_2c\omega_2\\ -I_3b\omega_3\end{bmatrix}$	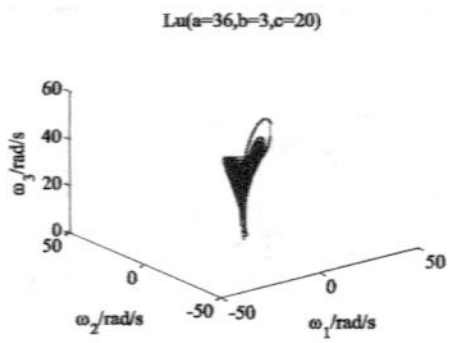
Genesio–Tesi system	$\boldsymbol{A}=diag(0,0,0), \boldsymbol{C}=\boldsymbol{0}$ $\boldsymbol{B}=\begin{bmatrix}0 & 1 & 0\\ 0 & 0 & 1\\ -a+\omega_1 & -b & -c\end{bmatrix}$	$I_1=I_2=I_3$ $\boldsymbol{T}_d=\begin{bmatrix}I_1\omega_2\\ I_2\omega_3\\ -I_3(a\omega_1-\omega_1^2+b\omega_2+c\omega_3)\end{bmatrix}$	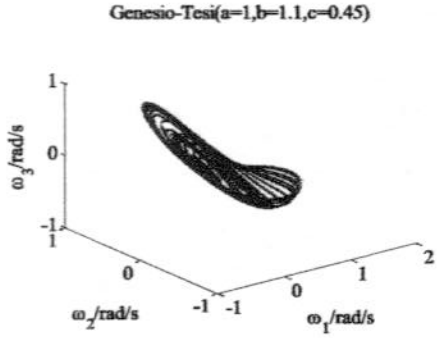
Rucklidge system	$\boldsymbol{A}=diag(0,0,0), \boldsymbol{C}=\boldsymbol{0}$ $\boldsymbol{B}=\begin{bmatrix}-a & b & -\omega_2\\ 1 & 0 & 0\\ 0 & \omega_2 & -1\end{bmatrix}$	$I_1=I_2=I_3$ $\boldsymbol{T}_d=\begin{bmatrix}I_1(-a\omega_1+b\omega_2-\omega_2\omega_3)\\ I_2\omega_1\\ I_3(\omega_2^2-\omega_3)\end{bmatrix}$	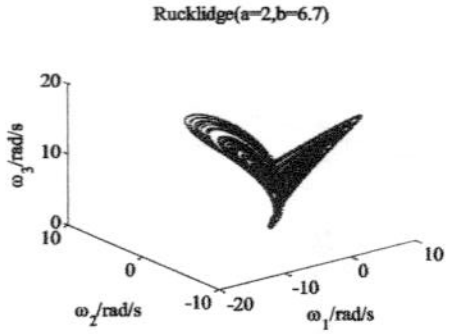

Liu system	$\boldsymbol{A} = diag(0,0,0), \boldsymbol{C} = \boldsymbol{0}$ $\boldsymbol{B} = \begin{bmatrix} -a & a & 0 \\ b & 0 & -k\omega_1 \\ h\omega_1 & 0 & -c \end{bmatrix}$	$I_1 = I_2 = I_3$ $\boldsymbol{T}_d = \begin{bmatrix} I_1 a(\omega_2 - \omega_1) \\ I_2(b\omega_1 - k\omega_1\omega_3) \\ I_3(h\omega_1^2 - c\omega_3) \end{bmatrix}$	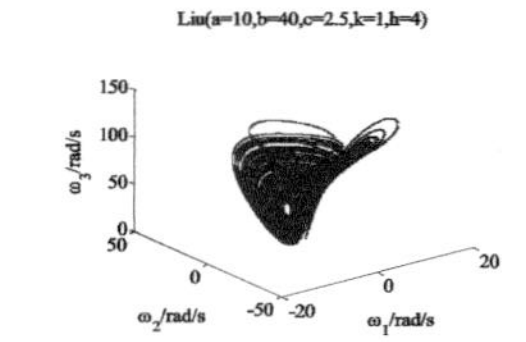
Rossler system	$\boldsymbol{A} = diag(0,0,0)$ $\boldsymbol{C} = \begin{bmatrix} 0 & 0 & b \end{bmatrix}^{\mathrm{T}}$ $\boldsymbol{B} = \begin{bmatrix} 0 & -1 & -1 \\ 1 & a & 0 \\ \omega_3 & 0 & -c \end{bmatrix}$	$I_1 = I_2 = I_3$ $\boldsymbol{T}_d = \begin{bmatrix} -I_1(\omega_1 + \omega_2) \\ I_2(\omega_1 + a\omega_2) \\ I_3(\omega_1\omega_3 - c\omega_3) + I_3 b \end{bmatrix}$	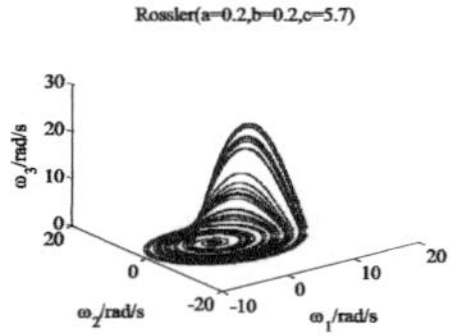

The desired reference angular velocity is defined in the following chaotic attitude equation

$$\dot{\hat{\omega}} = \hat{\boldsymbol{B}}\hat{\omega} + \boldsymbol{g}(\hat{\omega}) \tag{10.9}$$

where,

$$\hat{\boldsymbol{B}} = \begin{bmatrix} \hat{b}_{11} & \hat{b}_{12} & \hat{b}_{13} \\ \hat{b}_{21} & \hat{b}_{22} & \hat{b}_{23} \\ \hat{b}_{31} & \hat{b}_{32} & \hat{b}_{33} \end{bmatrix}, \boldsymbol{g}(\hat{\omega}) = \begin{bmatrix} g_1(\hat{\omega}) \\ g_2(\hat{\omega}) \\ g_3(\hat{\omega}) \end{bmatrix} + \begin{bmatrix} \hat{c}_1 \\ \hat{c}_2 \\ \hat{c}_3 \end{bmatrix} = \hat{\boldsymbol{A}} \begin{bmatrix} \hat{\omega}_2\hat{\omega}_3 \\ \hat{\omega}_1\hat{\omega}_3 \\ \hat{\omega}_1\hat{\omega}_2 \end{bmatrix} + \hat{\boldsymbol{C}},$$

$$\hat{\boldsymbol{A}} = diag(\hat{a}_1, \hat{a}_2, \hat{a}_3), \hat{\boldsymbol{C}} = \begin{bmatrix} \hat{c}_1 & \hat{c}_2 & \hat{c}_3 \end{bmatrix}^T$$

The tracking error is defined as $\boldsymbol{e} = \boldsymbol{\omega} - \hat{\boldsymbol{\omega}}$, with its components as $e_i\ (i = 1, 2, 3)$, and one can get the error system as

$$\begin{aligned} \dot{\boldsymbol{e}} &= \dot{\boldsymbol{\omega}} - \dot{\hat{\boldsymbol{\omega}}} = \begin{bmatrix} \dot{e}_1 & \dot{e}_2 & \dot{e}_3 \end{bmatrix}^T \\ &= \begin{bmatrix} \lambda_1(e_1, e_2, e_3) + u_1 & \lambda_2(e_1, e_2, e_3) + u_2 & \lambda_3(e_1, e_2, e_3) + u_3 \end{bmatrix}^T \end{aligned} \tag{10.10}$$

Assumption 10.1: There exists a constant $l > 0$ such that $\lambda_i(e_1, e_2, e_3) \leq l \min_i |e_i|\ \ (i = 1, 2, 3)$.

This assumption will be used in the stability analysis, and it is easy to check such an assumption in practice.

The tracking problem is solved if $\lim_{t \to \infty} \boldsymbol{e} = 0$, which is equivalent to the stabilization of $\boldsymbol{e}$, and the equation that governs the spacecraft's motion is given by

$$\dot{\boldsymbol{e}} = \boldsymbol{B}\boldsymbol{\omega} - \hat{\boldsymbol{B}}\hat{\boldsymbol{\omega}} + \boldsymbol{f}(\boldsymbol{\omega}) - \boldsymbol{g}(\hat{\boldsymbol{\omega}}) + \boldsymbol{u} \tag{10.11}$$

The control objective of this chapter is to design an adaptive variable structure controller for the plant (10.7), whose tracking error dynamics is given by Eq. (10.11), such that, for all physically realizable initial conditions, the following is achieved: $\lim_{t \to \infty} \boldsymbol{e} = 0$.

10.3 Adaptive variable structure control law

The controller design is based on the following sliding surface:

$$s_i = e_i + \int_0^t r e_i(\tau) d\tau \ (i = 1, 2, 3; r > 0) \tag{10.12}$$

When the system reaches the sliding surface and moves on it, the following conditions should be satisfied:

$$s_i = e_i + \int_0^t re_i(\tau)d\tau = 0 \tag{10.13}$$

$$\dot{s}_i = \dot{e}_i + re_i = 0 \tag{10.14}$$

From Eq. (10.14), one has

$$\dot{e}_i = -re_i \tag{10.15}$$

Therefore Eq. (10.15) is asymptotically stable, that is, $\lim_{t\to\infty} e_i(t) = 0\ (i = 1, 2, 3)$.

In this case, the controller is proposed in the form

$$u_i = -\sigma k_i |e_i| \mathrm{sgn}(s_i)\ (i = 1, 2, 3, \sigma > 0) \tag{10.16}$$

It is seen that the term on right-hand side contains a variable k_i that will be adjusted dynamically to ensure that $\boldsymbol{k}(t) > 0$ for all time given $\boldsymbol{k}(0) > 0$ and to guarantee asymptotic disturbance rejection, which will be discussed in the following part.

Choose the Lyapunov function candidate as

$$V = \frac{1}{2}\sum_{i=1}^{3} s_i^2 + \frac{1}{2\varsigma}\sum_{i=1}^{3}\sigma(k_i - k^*)^2 \tag{10.17}$$

where $k^* > (l + r)/\sigma$. Taking the first derivative of V along the motion of the error system yields

$$\begin{aligned}
\dot{V} &= \sum_{i=1}^{3} s_i\dot{s}_i + \frac{\sigma}{\varsigma}\sum_{i=1}^{3}(k_i - k^*)\dot{k}_i \\
&= \sum_{i=1}^{3} s_i(\dot{e}_i + re_i) + \frac{\sigma}{\varsigma}\sum_{i=1}^{3}(k_i - k^*)\dot{k}_i \\
&= \sum_{i=1}^{3} s_i(\lambda_i + u_i + re_i) + \frac{\sigma}{\varsigma}\sum_{i=1}^{3}(k_i - k^*)\dot{k}_i \\
&\le l\sum_{i=1}^{3}|s_i||e_i| + r\sum_{i=1}^{3} s_ie_i + \frac{\sigma}{\varsigma}\sum_{i=1}^{3}(k_i - k^*)\dot{k}_i - \sum_{i=1}^{3} s_i(\sigma k_i|e_i|\mathrm{sgn}(s_i)) \\
&= l\sum_{i=1}^{3}|s_i||e_i| + r\sum_{i=1}^{3} s_ie_i + \frac{\sigma}{\varsigma}\sum_{i=1}^{3}(k_i - k^*)\dot{k}_i - \sigma\sum_{i=1}^{3} k_i|e_i||s_i| \\
&= l\sum_{i=1}^{3}|s_i||e_i| + r\sum_{i=1}^{3} s_ie_i - \frac{\sigma}{\varsigma}\sum_{i=1}^{3} k^*\dot{k}_i + \frac{\sigma}{\varsigma}\sum_{i=1}^{3} k_i(\dot{k}_i - \varsigma|e_i||s_i|)
\end{aligned} \tag{10.18}$$

The adjustment law for $k(t)$ is now chosen as

$$\dot{k}_i = \varsigma |e_i||s_i| \tag{10.19}$$

This yields

$$\begin{aligned}\dot{V} \le l\sum_{i=1}^{3}|s_i||e_i| + r\sum_{i=1}^{3}s_ie_i - \sigma k^*\sum_{i=1}^{3}|e_i||s_i| \le l\sum_{i=1}^{3}|s_i||e_i| \\ + r\sum_{i=1}^{3}|s_i||e_i| - \sigma k^*\sum_{i=1}^{3}|e_i||s_i| = (l + r - \sigma k^*)\sum_{i=1}^{3}|s_i||e_i| < 0\end{aligned} \tag{10.20}$$

According to Lyapunov stability theory, the tracking error system is asymptotically stable.

10.4 Simulation test

To validate the effectiveness and feasibility of the proposed adaptive variable structure controller in this chapter, the Newton-Leipnik system is taken for example, which is shown below.

Choose the initial moment and external disturbance torque acting on the spacecraft as

$$I_1 = I_2 = I_3 = 1\text{kg m}^2,$$
$$\boldsymbol{T}_d = \begin{bmatrix} I_1(-a\omega_1 + \omega_2 + 10\omega_2\omega_3) \\ I_2(-\omega_1 - 0.4\omega_2 + 5\omega_1\omega_3) \\ I_3(b\omega_3 - 5\omega_1\omega_2) \end{bmatrix}$$

Assume the target system is

$$\dot{\hat{\boldsymbol{\omega}}} = \begin{bmatrix} -a & 1 & 0 \\ -1 & -0.4 & 0 \\ 0 & 0 & b \end{bmatrix}\begin{bmatrix} \hat{\omega}_1 \\ \hat{\omega}_2 \\ \hat{\omega}_3 \end{bmatrix} + \begin{bmatrix} 10\hat{\omega}_2\hat{\omega}_3 \\ 5\hat{\omega}_1\hat{\omega}_3 \\ -5\hat{\omega}_1\hat{\omega}_2 \end{bmatrix} \tag{10.21}$$

and the tracking system is

$$\dot{\boldsymbol{\omega}} = \begin{bmatrix} -a & 1 & 0 \\ -1 & -0.4 & 0 \\ 0 & 0 & b \end{bmatrix}\begin{bmatrix} \omega_1 \\ \omega_2 \\ \omega_3 \end{bmatrix} + \begin{bmatrix} 10\omega_2\omega_3 \\ 5\omega_1\omega_3 \\ -5\omega_1\omega_2 \end{bmatrix} + \boldsymbol{u} \tag{10.22}$$

Combining Eq. (10.21) and Eq. (10.22), the error system can be obtained as

$$\dot{\boldsymbol{e}} = \begin{bmatrix} -a & 1 & 0 \\ -1 & -0.4 & 0 \\ 0 & 0 & b \end{bmatrix} \begin{bmatrix} e_1 \\ e_2 \\ e_3 \end{bmatrix} + \begin{bmatrix} 10\omega_2\omega_3 - 10\hat{\omega}_2\hat{\omega}_3 \\ 5\omega_1\omega_3 - 5\hat{\omega}_1\hat{\omega}_3 \\ 5\hat{\omega}_1\hat{\omega}_2 - 5\omega_1\omega_2 \end{bmatrix} + \boldsymbol{u} \tag{10.23}$$

Choosing sliding mode $s_i = e_i + \int_0^t re_i(\tau)d\tau \ (i = 1, 2, 3; r > 0)$, the adjustment law for $\boldsymbol{k}(t)$ is

$$\begin{cases} \dot{k}_1 = \varsigma|e_1||s_1| \\ \dot{k}_2 = \varsigma|e_2||s_2| \\ \dot{k}_3 = \varsigma|e_3||s_3| \end{cases} \tag{10.24}$$

The adaptive variable structure tracking controller is

$$\begin{cases} u_1 = -\sigma k_1|e_1|\mathrm{sgn}(s_1) \\ u_2 = -\sigma k_2|e_2|\mathrm{sgn}(s_2) \\ u_3 = -\sigma k_3|e_3|\mathrm{sgn}(s_3) \end{cases} \tag{10.25}$$

The corresponding known parameters are chosen as $a = 0.4$, $b = 0.175$, $r = 4$, $\sigma = 0.052$, $\varsigma = 5$. The initial states are $\boldsymbol{k}_0 = \begin{bmatrix} 0.4 & 0.2 & 0.2 \end{bmatrix}^T$, $\boldsymbol{\omega}_0 = \begin{bmatrix} 0.3 & 0.1 & 0.1 \end{bmatrix}^T$, and $\hat{\boldsymbol{\omega}}_0 = \begin{bmatrix} 0.7 & 0.25 & 0.25 \end{bmatrix}^T$. The unit of angular velocity is rad s^{-1}. The simulation step is chosen as 0.001 second, and the corresponding simulation time is chosen as 300 seconds.

The response of the closed-loop system can be seen in Figs. 10.1–10.5 from which one can see that the control objective is achieved despite the presence of the external disturbances. The error $\boldsymbol{e}$ will converge to zero as time increases, and the control input torque will also converge to zero, which

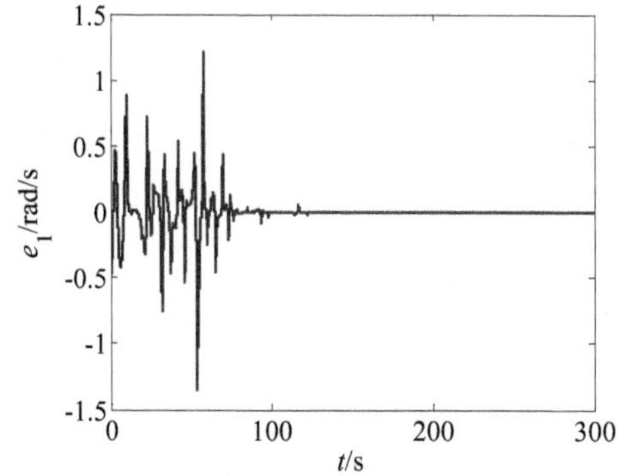

Figure 10.1 Response of the error e_1.

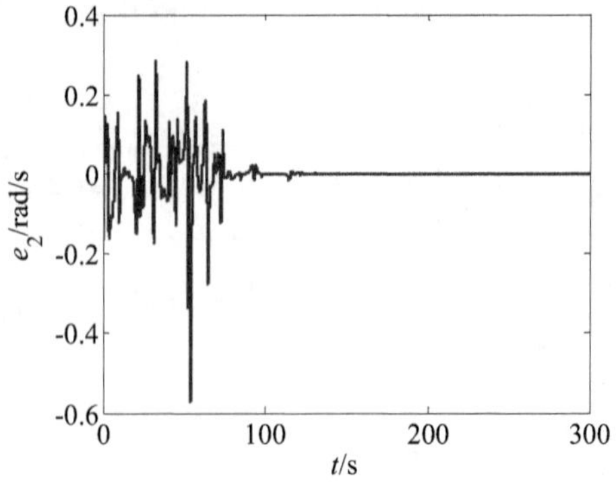

Figure 10.2 Response of the error e_2.

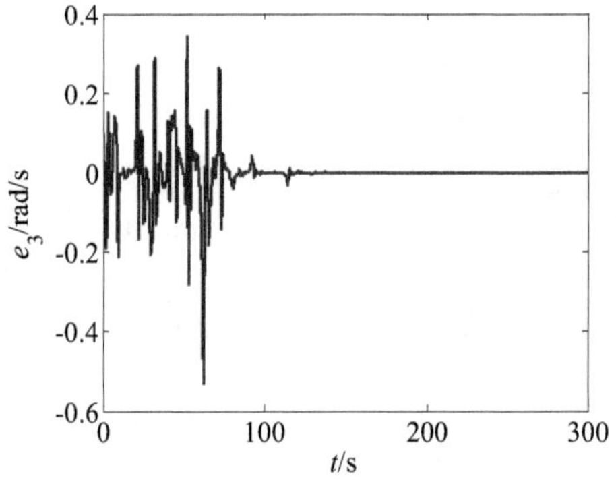

Figure 10.3 Response of the error e_3.

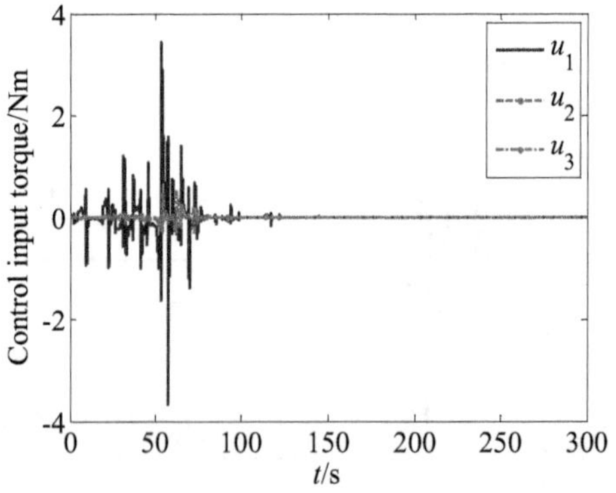

Figure 10.4 Response of the control input torque.

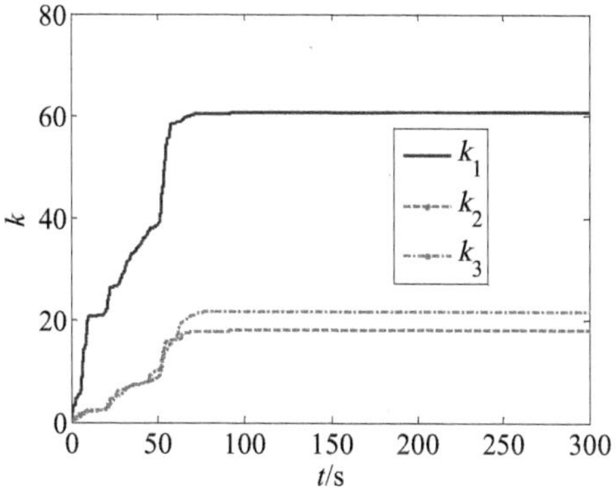

Figure 10.5 Response of $\boldsymbol{k}(t)$.

indicates that the tracking system will achieve synchronization with the target system, and the chaotic dynamics of spacecraft attitude motion do not change. It is also seen in Fig. 10.5 that $\boldsymbol{k}(t)$ increases during the transient, but then converges to a constant value.

10.5 Conclusions

For the tracking control problem of spacecraft chaotic attitude motion affected by external disturbances, this chapter presents a continuous globally stable tracking control algorithm which is based on adaptive control theory and variable structure control theory. In the proposed controller, the integral sliding surface is adopted, and a single vector is adjusted dynamically in such a fashion that the angular velocity error will tend to zero asymptotically. The stability proof is based on spacecraft error dynamics equation and a Lyapunov analysis. Numerical simulations also illustrate the tracking performance of spacecraft chaotic motion obtained using the proposed adaptive variable structure controller in this chapter.

References

[1] Jia FL, Xu W, Li HN, et al. Chaos control with unknown parameter of attitude motion of perturbed spacecraft. Acta Physica Sinica 2013;62(10):100503.

[2] Iñarrea M. Chaos and its control in the pitch motion of an asymmetric magnetic spacecraft in polar elliptic orbit. Chaos, Solitons & Fractals 2009;40(4):1637–52.

[3] Wen G, Yu W, Hu G, et al. Pinning synchronization of directed networks with switching topologies: a multiple Lyapunov functions approach. IEEE Transactions on Neural Networks and Learning Systems 2015;26(12):3239–50.

[4] Wen G, Chen MZQ, Yu X. Event-triggered master–slave synchronization with sampled-data communication. IEEE Transactions on Circuits and Systems II: Express Briefs 2016;63(3):304–8.

[5] Wang X, Tian L. Bifurcation analysis and linear control of the Newton–Leipnik system. Chaos, Solitons & Fractals 2006;27(1):31–8.

[6] Zhang R, Yang S. Chaos in fractional-order generalized Lorenz system and its synchronization circuit simulation. Chinese Physics B 2009;18(8):3295.

[7] Lu JG, Chen G. A note on the fractional-order Chen system. Chaos, Solitons & Fractals 2006;27(3):685–8.

[8] Lu JG. Chaotic dynamics of the fractional-order Lü system and its synchronization. Physics Letters A 2006;354(4):305–11.

[9] Park JH. Adaptive controller design for modified projective synchronization of Genesio–Tesi chaotic system with uncertain parameters. Chaos, Solitons & Fractals 2007;34(4):1154–9.

[10] Wang X. Si'lnikov chaos and Hopf bifurcation analysis of Rucklidge system. Chaos, Solitons & Fractals 2009;42(4):2208–17.

[11] Njah AN. Tracking control and synchronization of the new hyperchaotic Liu system via backstepping techniques. Nonlinear Dynamics 2010;61(1–2):1–9.

[12] Li C, Chen G. Chaos and hyperchaos in the fractional-order Rössler equations. Physica A: Statistical Mechanics and its Applications 2004;341:55–61.

[13] Liu H, Li SG, Sun YG, et al. Prescribed performance synchronization for fractional-order chaotic systems. Chinese Physics B 2015;24(9):090505.

[14] Razminia A, Baleanu D. Complete synchronization of commensurate fractional order chaotic systems using sliding mode control. Mechatronics 2013;23(7):873–9.

[15] Wang B, Xue JY, Zhu DL, et al. Control of Lorenz-Stenflo chaotic system via Takagi-Sugeno fuzzy model based on linear matrix inequality. International Journal of Control and Automation 2014;7(9):139–54.

[16] Yau HT, Shieh CS. Chaos synchronization using fuzzy logic controller. Nonlinear analysis: Real World Applications 2008;9(4):1800–10.

[17] Liu L, Liang D, Liu C. Nonlinear state-observer control for projective synchronization of a fractional-order hyperchaotic system. Nonlinear Dynamics 2012;69(4):1929–39.

[18] Agiza HN, Yassen MT. Synchronization of Rossler and Chen chaotic dynamical systems using active control. Physics Letters A 2001;278(4):191–7.

[19] Xiang-Jun W, Jing-Sen L, Guan-Rong C. Chaos synchronization of Rikitake chaotic attractor using the passive control technique. Nonlinear Dynamics 2008;53(1–2):45–53.

[20] Tong X, Rimrott FPJ. Numerical studies on chaotic planar motion of satellites in an elliptic orbit. Chaos, Solitons & Fractals 1991;1(2):179–86.

[21] Meehan PA, Asokanthan SF. Chaotic motion in a spinning spacecraft with circumferential nutational damper. Nonlinear Dynamics 1997;12(1):69–87.

[22] Meehan PA, Asokanthan SF. Control of chaotic instabilities in a spinning spacecraft with dissipation using Lyapunov's method. Chaos, Solitons & Fractals 2002;13(9):1857–69.

[23] Salarieh H, Alasty A. Chaos synchronization of nonlinear gyros in presence of stochastic excitation via sliding mode control. Journal of Sound and Vibration 2008;313(3):760–71.

[24] Aghababa MP. Design of an adaptive finite-time controller for synchronization of two identical/different non-autonomous chaotic flywheel governor systems. Chinese Physics B 2012;21(3):030502.

[25] Beletsky VV, Lopes RVF, Pivovarov ML. Chaos in spacecraft attitude motion in Earth's magnetic field. Chaos: an Interdisciplinary. Journal of Nonlinear Science 1999;9(2):493–8.

CHAPTER 11

Underactuated chaotic attitude stabilization control

11.1 Introduction

As space technologies have developed and progressed, advanced space missions have put forward higher requirements for spacecraft attitude control system to ensure rapid and accurate response in the presense of certain actuator failure, control input constraint, and external disturbances. A significant challege arises when these issues are considered simultaneously. Particularly, external disturbances are often constituted of periodic and secular term, once the spacecraft properties and external disturbances imposed on it satisfy certain condition, it can lead to chaotic attitude motion, which will cause great damage in the spacecraft. Consequently, the chaotic attitude control with certain actuator failure is extremely important in the space mission, which should be investigated in detail.

Many flight experiences throughout world aerospace history have witnessed the spacecraft attitude motion with unexpected behaviors, which results from the external disturbances that had not been taken into consideration in the process of spacecraft design. Numerous theoretical researches have pointed out that spacecraft chaotic motion exists under the influence of different external disturbances [1]. The Chen system [2] and Lu system [3] are two typical chaotic systems that the spacecraft attitude system may become. Classical chaotic control methods include adaptive control [4], sliding mode control [5], and fuzzy logic control [6] and so on. In recent years, considerable efforts have been made to study the attitude stabilization control and fault-tolerant control of spacecraft during rotational maneuvers and the theory of controller design has developed in a variety of directions [7–9]. The often used methods include H_∞ control [10,11], adaptive control [12], feedback linearization [13], sliding mode controller [14], and linear matrix inequality (LMI) techniques [15]. Among them, sliding mode control strategy is acknowledged as an effective way to withstand actuator failure and external disturbances, which has been widely used in aircraft or spacecraft attitude control system. In [16],

Spacecraft Attitude Control. DOI: https://doi.org/10.1016/B978-0-323-99005-9.00011-0

a fault-tolerant sliding mode control scheme was proposed for aircraft operating phase with actuator failure, where it allowed control allocation. According to [17], an adaptive fault-tolerant sliding mode control scheme was developed for attitude tracking of flexible spacecraft with partial loss of actuator effectiveness, where neural networks were introduced to cope with system uncertainties and on-line updating law was developed to estimate actuator failure bound. In [18], adaptive sliding mode control scheme with L_2 gain performance has been evaluated to perform attitude tracking control for flexible spacecraft with actuator failures, parameter uncertainties, and external disturbances. Ref. [19] pointed out that the system dynamics might be fragile to failures during the reaching phase. To cope with this problem, the idea of integral sliding mode control was presented [20,21]. This can ensure that the sliding surface begins from the initial time instant, which eliminates the reaching phase [22].

The main stabilization results for the case when certain actuator failures occur to the spacecraft have been presented in some literature. According to [23], sufficient and necessary conditions for the controllability of rigid spacecraft in the case of one, two, and three independent control torques were provided. It was pointed out that controllability was impossible with fewer than three devices in the case of momentum exchange devices. In addition, it has been pointed out that the rigid body system is never small-time locally controllable with only one control [24]. The angular velocity equations can be made asymptotically stable about the origin by means of two control torques along the principal axes [25]. Ref. [26] used one torque aligned with a principal axis to investigate the feedback stabilization problem of the zero solution of angular velocity equations. However, it is required that the moment of inertia of the rigid body along that principal axis should be the largest or the smallest. Ref. [27] also showed that a single control aligned with a principal axis cannot stabilize the rigid body system. We believe that, under given conditions, the spacecraft chaotic attitude system will be controllable when the failures of two actuators occur.

Motivated by the above research, an integral sliding mode-based control strategy was developed for spacecraft chaotic attitude motion with certain actuator failure and control input constraint, where the LMI technique is incorporated into the controller design. Also, the potential effects of external disturbances on system performance are explicitly investigated during the attitude dynamics and problem formulation process. Compared with most existing spacecraft attitude control approaches, the presented control scheme only requires a quantitative relationship of the moment of inertia without precise information, and the fault detection mechanism is not required to detect the actuator failure. Furthermore, the proposed control law could track the attitude motion trajectory designed in advance and use LMI techniques to obtain the unknown matrices introduced in integral sliding mode. Using the Lyapunov method, the stability analysis of the resulted closed-loop system is demonstrated in detail, and the effectiveness of the proposed control method is also analyzed via numerical simulations.

The remainder of this chapter is organized as follows. The next section introduces the spacecraft attitude dynamics equation and writes it in nonlinear equation form of three input. Besides, this section describes two kinds of chaos phenomena in spacecraft attitude motion influenced by external disturbances and transforms the nonlinear equation form of three inputs into that of a double input and single input under conditions of actuator failures. Then, the purpose of this work is addressed. Followed by the section that presents the integral sliding mode-based robust controller. Numerical simulations are performed to illustrate the performance of the controller. Finally, the conclusions of this work are drawn.

11.2 Problem formulation

11.2.1 Chaotic attitude system description

Assume that the spacecraft is a rigid body with actuators that provide control torques with respect to three mutually perpendicular axes which define a body reference frame. The attitude dynamics equation is given by

$$\boldsymbol{I}\dot{\boldsymbol{\omega}} + \boldsymbol{\omega} \times (\boldsymbol{I}\boldsymbol{\omega}) = \boldsymbol{T}_c + \boldsymbol{T}_d \tag{11.1}$$

where $\boldsymbol{\omega} = \begin{bmatrix} \omega_1 & \omega_2 & \omega_3 \end{bmatrix}^T$ means the angular velocity of the body reference frame with respect to the earth-centered inertial reference frame represented in the body reference frame, $\boldsymbol{I}$ denotes the spacecraft inertia matrix, $\boldsymbol{T}_c$ represents the control input torque, $\boldsymbol{T}_d$ means the external disturbance torque imposed on the spacecraft, which is generally expressed in the following nonlinear form:

$$\boldsymbol{T}_d = \boldsymbol{D}\boldsymbol{\omega} + \boldsymbol{M} \tag{11.2}$$

where, $\boldsymbol{D} = \left[d_{ij}\right]_{3\times 3} \in \boldsymbol{R}^{3\times 3} (i, j = 1, 2, 3)$, which can be a constant matrix or matrix varying with angular velocity; $\boldsymbol{M} = [m_i]_{3\times 1} \in \boldsymbol{R}^{3\times 1} (i = 1, 2, 3)$, which can be a constant matrix or matrix varying with angular velocity, or even periodic matrix or long-term matrix.

Combining Eqs. (11.1) and (11.2) yields

$$\begin{cases} I_1\dot{\omega}_1 - (I_2 - I_3)\omega_2\omega_3 = T_{c1} + d_{11}\omega_1 + d_{12}\omega_2 + d_{13}\omega_3 + m_1 \\ I_2\dot{\omega}_2 - (I_3 - I_1)\omega_1\omega_3 = T_{c2} + d_{21}\omega_1 + d_{22}\omega_2 + d_{23}\omega_3 + m_2 \\ I_3\dot{\omega}_3 - (I_1 - I_2)\omega_1\omega_2 = T_{c3} + d_{31}\omega_1 + d_{32}\omega_2 + d_{33}\omega_3 + m_3 \end{cases} \tag{11.3}$$

where I_1, I_2, and I_3 denote the three components of inertia matrix; T_{c1}, T_{c2}, and T_{c3} denote the three components of control input torque.

Then,

$$\begin{cases} \dot{\omega}_1 = I_1^{-1}(I_2 - I_3)\omega_2\omega_3 + I_1^{-1}d_{11}\omega_1 + I_1^{-1}d_{12}\omega_2 + I_1^{-1}d_{13}\omega_3 + I_1^{-1}T_{c1} + I_1^{-1}m_1 \\ \dot{\omega}_2 = I_2^{-1}(I_3 - I_1)\omega_1\omega_3 + I_2^{-1}d_{21}\omega_1 + I_2^{-1}d_{22}\omega_2 + I_2^{-1}d_{23}\omega_3 + I_2^{-1}T_{c2} + I_2^{-1}m_2 \\ \dot{\omega}_3 = I_3^{-1}(I_1 - I_2)\omega_1\omega_2 + I_3^{-1}d_{31}\omega_1 + I_3^{-1}d_{32}\omega_2 + I_3^{-1}d_{33}\omega_3 + I_3^{-1}T_{c3} + I_3^{-1}m_3 \end{cases} \tag{11.4}$$

Let

$$\boldsymbol{B}=\begin{bmatrix} I_1^{-1}d_{11} & I_1^{-1}d_{12} & I_1^{-1}d_{13} \\ I_2^{-1}d_{21} & I_2^{-1}d_{22} & I_2^{-1}d_{22} \\ I_3^{-1}d_{31} & I_3^{-1}d_{32} & I_3^{-1}d_{33} \end{bmatrix}, \boldsymbol{f}(\boldsymbol{\omega}) = \boldsymbol{A}\begin{bmatrix} \omega_2\omega_3 \\ \omega_1\omega_3 \\ \omega_1\omega_2 \end{bmatrix} + \hat{\mathbf{C}},$$

$$\boldsymbol{A}=diag(I_1^{-1}(I_2 - I_3), I_2^{-1}(I_3 - I_1), I_3^{-1}(I_1 - I_2)),$$

$$\hat{\mathbf{C}}=\begin{bmatrix} I_1^{-1}m_1 & I_2^{-1}m_2 & I_3^{-1}m_3 \end{bmatrix}^T, \boldsymbol{u}=\begin{bmatrix} I_1^{-1}T_{c1} & I_2^{-1}T_{c2} & I_3^{-1}T_{c3} \end{bmatrix}^T$$

Then, Eq. (11.4) can be transformed into the following form

$$\dot{\boldsymbol{\omega}} = \boldsymbol{B}\boldsymbol{\omega} + \boldsymbol{f}(\boldsymbol{\omega}) + \boldsymbol{u} \tag{11.5}$$

11.2.2 Two examples of Chen and Lu systems

First, two examples of spacecraft attitude motion are analyzed when the control torques that actuators provide are zero, that is, when $\boldsymbol{u}=0$.

Example 1: When the moment of inertia and the external disturbances imposed on the spacecraft satisfy

$$I_1 = 2I_2 = 2I_3$$

$$\boldsymbol{T}_d = \begin{bmatrix} I_1 a(\omega_2 - \omega_1) \\ I_2(c\omega_1 - a\omega_1 + c\omega_2) \\ -I_3 b\omega_3 \end{bmatrix}$$

Eq. (11.4) is transformed into the Chen system form,

$$\begin{cases} \dot{\omega}_1 = a(\omega_2 - \omega_1) \\ \dot{\omega}_2 = (c - a)\omega_1 + c\omega_2 - \omega_1\omega_3 \\ \dot{\omega}_3 = \omega_1\omega_2 - b\omega_3 \end{cases} \tag{11.6}$$

When $a = 35, b = 3, c = 28$, the system denoted by Eq. (11.6) is a chaotic system, and the chaotic attractor and its projection is shown in Fig. 11.1, from which one can obviously see that the spacecraft attitude motion is chaotic.

Example 2: When the moment of inertia and external disturbances imposed on the spacecraft satisfy

$$I_1 = 2I_2 = 2I_3$$
$$\boldsymbol{T}_d = \begin{bmatrix} I_1 a(\omega_2 - \omega_1) \\ I_2 c\omega_2 \\ -I_3 b\omega_3 \end{bmatrix}$$

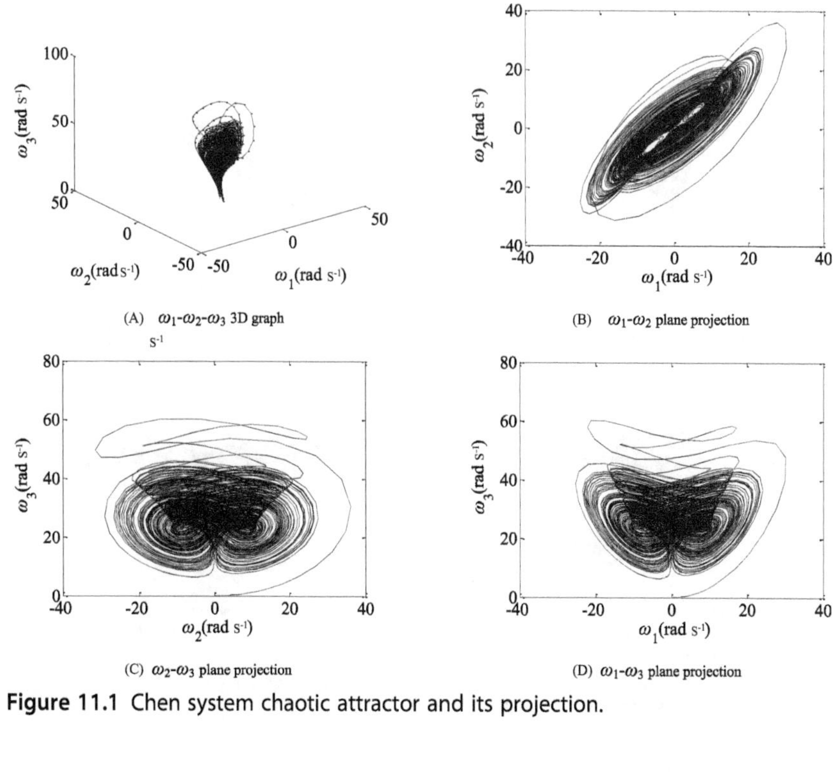

Figure 11.1 Chen system chaotic attractor and its projection.

Eq. (11.4) is transformed into the Lu system form,

$$\begin{cases} \dot{\omega}_1 = a(\omega_2 - \omega_1) \\ \dot{\omega}_2 = c\omega_2 - \omega_1\omega_3 \\ \dot{\omega}_3 = \omega_1\omega_2 - b\omega_3 \end{cases} \tag{11.7}$$

When $a = 36, b = 3, c = 20$, the system denoted by Eq. (11.7) is a chaotic system, and the chaotic attractor and its projection is shown in Fig. 11.2, from which one can also see that the spacecraft attitude motion is chaotic.

11.2.3 Control objective

Under conditions of certain actuator failure, sliding mode controllers can be designed. Two cases are considered, including failure of one actuator and failure of two actuators.

Case 1: (failure of one actuator): Two control inputs are added into the system, and take the failure of the actuator installed on the x-axis as an example.

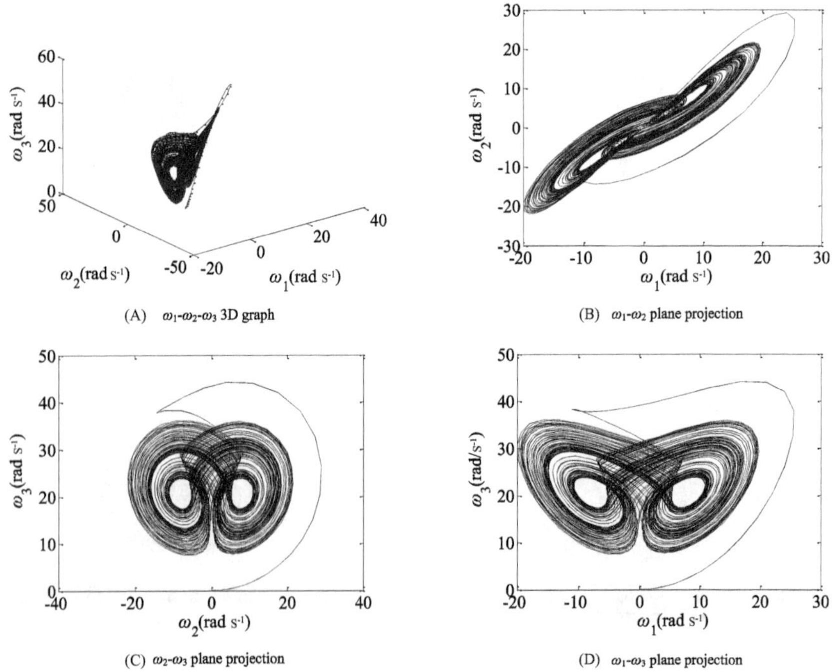

Figure 11.2 Lu system chaotic attractor and its projection.

In this case, one has $T_{c1} = 0$. The attitude dynamics system denoted by Eq. (11.5) can be transformed into the following nonlinear equation of double input:

$$\dot{\omega} = \boldsymbol{B}\omega + \boldsymbol{f}(\omega) + \begin{bmatrix} 0 \\ u_2 \\ u_3 \end{bmatrix} = \boldsymbol{B}\omega + \boldsymbol{f}(\omega) + \begin{bmatrix} 0 & 0 \\ 1 & 0 \\ 0 & 1 \end{bmatrix} \begin{bmatrix} u_2 \\ u_3 \end{bmatrix} = \boldsymbol{B}\omega + \boldsymbol{f}(\omega) + \boldsymbol{F}\boldsymbol{u}^* \tag{11.8}$$

where,

$$\boldsymbol{F} = \begin{bmatrix} 0 & 0 \\ 1 & 0 \\ 0 & 1 \end{bmatrix}, \boldsymbol{u}^* = \begin{bmatrix} u_2 \\ u_3 \end{bmatrix}$$

Case 2: (failure of two actuators): Only one control input is added into the system, take the failures of actuators installed on the x-axis and y-axis for example.

In this case, one has $T_{c1} = 0$, $T_{c2} = 0$. The attitude dynamics system denoted by Eq. (11.5) can be transformed into the following nonlinear equation of single input:

$$\dot{\omega} = \boldsymbol{B}\omega + \boldsymbol{f}(\omega) + \begin{bmatrix} 0 \\ 0 \\ u_3 \end{bmatrix} = \boldsymbol{B}\omega + \boldsymbol{f}(\omega) + \begin{bmatrix} 0 \\ 0 \\ 1 \end{bmatrix} u_3 = \boldsymbol{B}\omega + \boldsymbol{f}(\omega) + \boldsymbol{F}\boldsymbol{u}^* \tag{11.9}$$

where,

$$\boldsymbol{F} = \begin{bmatrix} 0 \\ 0 \\ 1 \end{bmatrix}, \boldsymbol{u}^* = u_3$$

Then, by analyzing Eq. (11.9), it is obviously seen that the spacecraft chaotic attitude system can be controllable if and only if the two initial angular velocities along the x-axis and y-axis are zero, that is, $\omega_{10} = \omega_{20} = 0$, which results in $\omega_1 = \omega_2 = 0$. We just need to design u_3 to stabilize the angular velocity along the z-axis.

The objective of this chapter is to design a control input $\boldsymbol{u}^*$ for spacecraft attitude dynamics system with chaotic attitude motion and certain actuator failure, such that, for all physically possible initial conditions, all $\boldsymbol{u} \leq \boldsymbol{u}_m$ and all $\boldsymbol{I} = \boldsymbol{I}^T > 0$, and the following is achieved within given finite time: $\lim \boldsymbol{\omega} = \boldsymbol{0}$.

11.3 Sliding mode control law

11.3.1 Reference trajectory design

To meet necessary observation missions or the requirement of control input constraint, the spacecraft angular acceleration should track a given reference trajectory. Assume that the maximum yaw angular acceleration is $\dot{\omega}_m > 0$, and the reference trajectory of yaw angular acceleration $\dot{\omega}_{dr}$ is shown in Fig. 11.3.

The corresponding mathematical expression is

$$\dot{\omega}_{dr}(t) = \begin{cases} -\dfrac{\dot{\omega}_m}{2}\left[1+\sin\left(\dfrac{\pi t}{2t_1}\right)\right] & 0 \le t \le t_1 \\ -\dot{\omega}_m & t_1 < t \le t_2 \\ -\dfrac{\dot{\omega}_m}{2}\left[1+\cos\left(\dfrac{t-t_2}{t_3-t_2}\pi\right)\right] & t_2 < t \le t_3 \\ 0 & t > t_3 \end{cases} \tag{11.10}$$

Then, the mathematical expression of reference angular velocity can be represented by Eq. (11.11), that is,

$$\omega_{dr}(t) = \begin{cases} \omega_{d0} - \dfrac{\dot{\omega}_m}{2}t + \dfrac{\dot{\omega}_m t_1}{\pi}\left(\cos\dfrac{\pi t}{2t_1} - 1\right) & 0 \le t \le t_1 \\ \omega_{d0} - \dfrac{\dot{\omega}_m}{2}\left(1+\dfrac{2}{\pi}\right)t_1 - \dot{\omega}_m(t-t_1) & t_1 \le t \le t_2 \\ \omega_{d0} - \dfrac{\dot{\omega}_m}{2}\left(1+\dfrac{2}{\pi}\right)t_1 - \dot{\omega}_m(t_2-t_1) - \dfrac{\dot{\omega}_m}{2}(t-t_2) - \dfrac{\dot{\omega}_m(t_3-t_2)}{2\pi}\sin\dfrac{\pi(t-t_2)}{t_3-t_2} & t_2 \le t \le t_3 \\ 0 & t \ge t_3 \end{cases} \tag{11.11}$$

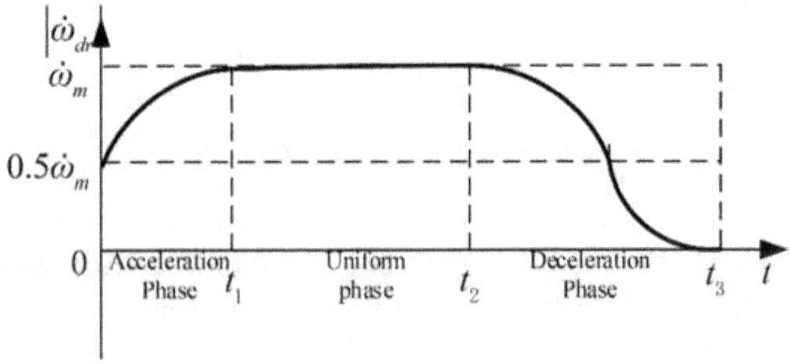

Figure 11.3 Reference trajectory of yaw angular acceleration.

where the maximum of angular acceleration depends on the characteristic time instant $t_i, i = 1, 2, 3$ and initial value of angular velocity ω_{d0}.

The maximum of yaw angular acceleration is calculated as

$$\dot{\omega}_m = \frac{2\pi\omega_{d0}}{2t_1 + \pi(t_2 + t_3 - t_1)} \tag{11.12}$$

11.3.2 Controller design

Assume the angular velocity error is $\boldsymbol{e}$, then the derivative of angular velocity error with respect to time is

$$\dot{\boldsymbol{e}} = \dot{\boldsymbol{\omega}} - \dot{\boldsymbol{\omega}}_d = \boldsymbol{B\omega} + \boldsymbol{f}(\boldsymbol{\omega}) + \boldsymbol{Fu}^* - \dot{\boldsymbol{\omega}}_d \tag{11.13}$$

The sliding mode control law is constituted of equivalent control $\boldsymbol{u}_{eq}$ and switching control $\boldsymbol{u}_{sw}$. With $\dot{\boldsymbol{s}} = 0$, the equivalent control term can be obtained, then let $\boldsymbol{u}^* = \boldsymbol{u}_{eq} + \boldsymbol{u}_{sw}$, by analyzing $\dot{\boldsymbol{s}}$ and substituting $\boldsymbol{u}^*$, such that $\boldsymbol{s}\dot{\boldsymbol{s}} < 0$ satisfies, the switching control term can be obtained. Equivalent control term ensures that the state of angular velocity error system is maintained on the sliding surface, and the switching control term ensures that the state does not leave the sliding surface.

Design sliding surface

$$\boldsymbol{s} = \boldsymbol{Ce} + \boldsymbol{K}\int_0^t \boldsymbol{e}dt \tag{11.14}$$

The definition of matrix $\boldsymbol{C}$ will be discussed later, then

$$\dot{\boldsymbol{s}} = \boldsymbol{C}\dot{\boldsymbol{e}} + \boldsymbol{Ke} = \boldsymbol{C}(\boldsymbol{B\omega} + \boldsymbol{f}(\boldsymbol{\omega}) + \boldsymbol{Fu}^*) - \boldsymbol{C}\dot{\boldsymbol{\omega}}_d + \boldsymbol{K\omega} - \boldsymbol{K\omega}_d \tag{11.15}$$

Let $\dot{\boldsymbol{s}} = 0$, the equivalent control term

$$\boldsymbol{u}_{eq} = -\boldsymbol{CB\omega} - \boldsymbol{Cf}(\boldsymbol{\omega}) + \boldsymbol{C}\dot{\boldsymbol{\omega}}_d + \boldsymbol{K\omega}_d - \boldsymbol{K\omega} \tag{11.16}$$

Substitute Eq. (11.16) into Eq. (11.13), one has

$$\dot{\boldsymbol{e}} = \boldsymbol{B\omega} + \boldsymbol{f}(\boldsymbol{\omega}) - \boldsymbol{FCB\omega} - \boldsymbol{FCf}(\boldsymbol{\omega}) + \boldsymbol{FC}\dot{\boldsymbol{\omega}}_d + \boldsymbol{FK\omega}_d - \boldsymbol{FK\omega} - \dot{\boldsymbol{\omega}}_d \tag{11.17}$$

In Case 1, when the failure of actuator installed on the x-axis is considered, $\boldsymbol{C} = \begin{bmatrix} c_1 & 1 & 0 \\ c_2 & 0 & 1 \end{bmatrix}$.

Then,

$$
\begin{aligned}
f(\omega)-FCf(\omega) &= \begin{bmatrix} 0 \\ -\omega_1\omega_3 \\ \omega_1\omega_2 \end{bmatrix} - \begin{bmatrix} 0 & 0 \\ 1 & 0 \\ 0 & 1 \end{bmatrix} \begin{bmatrix} c_1 & 1 & 0 \\ c_2 & 0 & 1 \end{bmatrix} \begin{bmatrix} 0 \\ -\omega_1\omega_3 \\ \omega_1\omega_2 \end{bmatrix} \\
&= \begin{bmatrix} 0 \\ -\omega_1\omega_3 \\ \omega_1\omega_2 \end{bmatrix} - \begin{bmatrix} 0 & 0 & 0 \\ c_1 & 1 & 0 \\ c_2 & 0 & 1 \end{bmatrix} \begin{bmatrix} 0 \\ -\omega_1\omega_3 \\ \omega_1\omega_2 \end{bmatrix} \\
&= \begin{bmatrix} 0 \\ -\omega_1\omega_3 \\ \omega_1\omega_2 \end{bmatrix} - \begin{bmatrix} 0 \\ -\omega_1\omega_3 \\ \omega_1\omega_2 \end{bmatrix} = \mathbf{0}
\end{aligned}
\tag{11.18}
$$

In Case 2, when the failures of actuators installed on the x-axis and y-axis are considered, $\omega_1=\omega_2=0$, $\boldsymbol{C}=\begin{bmatrix} c & 0 & 1 \end{bmatrix}$.

Then,

$$
\begin{aligned}
f(\omega)-FCf(\omega) &= \begin{bmatrix} 0 \\ -\omega_1\omega_3 \\ \omega_1\omega_2 \end{bmatrix} - \begin{bmatrix} 0 \\ 0 \\ 1 \end{bmatrix} \begin{bmatrix} c & 0 & 1 \end{bmatrix} \begin{bmatrix} 0 \\ -\omega_1\omega_3 \\ \omega_1\omega_2 \end{bmatrix} \\
&= \begin{bmatrix} 0 \\ -\omega_1\omega_3 \\ \omega_1\omega_2 \end{bmatrix} - \begin{bmatrix} 0 & 0 & 0 \\ 0 & 0 & 0 \\ c & 0 & 1 \end{bmatrix} \begin{bmatrix} 0 \\ -\omega_1\omega_3 \\ \omega_1\omega_2 \end{bmatrix} = \mathbf{0}
\end{aligned}
\tag{11.19}
$$

Consequently, Eq. (11.17) can be transformed into Eq. (11.20), that is,

$$
\begin{aligned}
\dot{e} &= B\omega - FCB\omega - FKe + FC\dot{\omega}_d - \dot{\omega}_d \\
&= (B-FCB)(e+\omega_d) - FKe + (FC - I_{3\times 3})\dot{\omega}_d \\
&= (B-FCB-FK)e + (B-FCB)\omega_d + (FC-I_{3\times 3})\dot{\omega}_d
\end{aligned}
\tag{11.20}
$$

Because $\lim \omega_d = \lim \dot{\omega}_d = \mathbf{0}$, the angular velocity error system is Hurwitz stabilizable if and only if the following LMI holds:

$$
P(B-FCB-FK) + (B-FCB-FK)^T P < 0 \tag{11.21}
$$

By choosing appropriate symmetric positive definite matrix $\boldsymbol{P}$, use YALMIP toolbox and one can obtain the unknown matrices $\boldsymbol{C}$ and $\boldsymbol{K}$.

Compare the above two cases, it can be seen that the forms of equivalent control term for Case 1 and Case 2 are the same, only the dimensions of matrices $\boldsymbol{F}$, $\boldsymbol{C}$, and $\boldsymbol{K}$ are different.

Lemma 11.1: Adopt a reaching law

$$\dot{\boldsymbol{s}} = -k\boldsymbol{s} - \kappa|\boldsymbol{s}|^{\alpha} sgn(\boldsymbol{s}), \ k > 0, \kappa > 0, 0 < \alpha < 1 \tag{11.22}$$

The error state $\boldsymbol{e}(t)$ will reach the sliding surface in finite time, denoted by t^*:

$$t^* = \frac{1}{k(\alpha - 1)} \ln \frac{\kappa}{k|\boldsymbol{s}(0)|^{1-\alpha} + \kappa} \tag{11.23}$$

Proof: With Eq. (11.22), one has

$$\begin{aligned} \boldsymbol{s}^T\dot{\boldsymbol{s}} &= -k\boldsymbol{s}^T\boldsymbol{s} - \kappa|\boldsymbol{s}|^{\alpha}\boldsymbol{s}^T sgn(\boldsymbol{s}) = -k|\boldsymbol{s}|^2 - \kappa|\boldsymbol{s}|^{\alpha-1}\boldsymbol{s}^T(|\boldsymbol{s}| sgn(\boldsymbol{s})) \\ &= -(k + \kappa|\boldsymbol{s}|^{\alpha-1})|\boldsymbol{s}|^2 = \frac{1}{2}\frac{d(\boldsymbol{s}^T\boldsymbol{s})}{dt} = \frac{1}{2}\frac{d|\boldsymbol{s}|^2}{dt} = |\boldsymbol{s}|\frac{d|\boldsymbol{s}|}{dt} \end{aligned}$$

that is,

$$-(k + \kappa|\boldsymbol{s}|^{\alpha-1})|\boldsymbol{s}| = \frac{d|\boldsymbol{s}|}{dt}$$

Then,

$$dt = -\frac{d|\boldsymbol{s}|}{(k + \kappa|\boldsymbol{s}|^{\alpha-1})|\boldsymbol{s}|} = -\frac{|\boldsymbol{s}|^{-\alpha}d|\boldsymbol{s}|}{k|\boldsymbol{s}|^{1-\alpha} + \kappa} = \frac{1}{\alpha - 1}\frac{d|\boldsymbol{s}|^{1-\alpha}}{k|\boldsymbol{s}|^{1-\alpha} + \kappa}$$

Integrating the above equation yields the reaching time as

$$\begin{aligned} t^* &= \frac{1}{\alpha - 1}\int_{|s(0)|}^{|s(t^*)|} \frac{d|\boldsymbol{s}|^{1-\alpha}}{k|\boldsymbol{s}|^{1-\alpha} + \kappa} = \frac{1}{k(\alpha - 1)}\ln(k|\boldsymbol{s}|^{1-\alpha} + \kappa)\Big|_{|s(0)|}^{|s(t^*)|} \\ &= \frac{1}{k(\alpha - 1)}\ln\frac{\kappa}{k|\boldsymbol{s}(0)|^{1-\alpha} + \kappa} \end{aligned}$$

When $t \geq t^*, \boldsymbol{s} = 0, \dot{\boldsymbol{s}} = 0$, the conclusion is proven.

Theorem 11.1: Design sliding mode control law

$$\boldsymbol{u}^* = -\boldsymbol{CB}\boldsymbol{\omega} - \boldsymbol{C}f(\boldsymbol{\omega}) + \boldsymbol{C}\dot{\boldsymbol{\omega}}_d + \boldsymbol{K}\boldsymbol{\omega}_d - \boldsymbol{K}\boldsymbol{\omega} - k\boldsymbol{s} - \kappa|\boldsymbol{s}|^{\alpha} sgn(\boldsymbol{s}) \tag{11.24}$$

Then, the error state starting from any initial point will reach the sliding surface in finite time. Consequently, the closed-loop system generated by Eqs. (11.13) and (11.24) is globally asymptotically stable. To avoid chattering, sign function sgn(s) can be replaced by continuous function $\theta(s)$.

$$\theta(s) = \frac{s}{|s| + \delta}$$

where δ is a very small positive constant.

Proof: Let a Lyapunov function candidate be of the form

$$V = \frac{1}{2}s^2$$

The first derivative along the solutions of Eq. (11.13) is

$$\begin{aligned} s^T\dot{s} &= s^T\left[C(B\omega + f(\omega) + Fu^*) - C\dot{\omega}_d + K\omega - K\omega_d\right] \\ &= s^T\left[CB\omega + Cf(\omega) - CB\omega - Cf(\omega) + C\dot{\omega}_d\right. \\ &\quad \left. + K\omega_d - K\omega - ks - \kappa|s|^\alpha sgn(s) - C\dot{\omega}_d + K\omega - K\omega_d\right] \\ &= s^T\left[-ks - \kappa|s|^\alpha sgn(s)\right] = -k|s|^2 - \kappa|s|^{\alpha+1} < 0 \end{aligned}$$

Therefore the control law can ensure that the state reach the switching surface in finite time. Once the state reach the switching surface, it will maintain on the switching surface. When the angular velocity error state is on the switching surface, the dynamics of the error system can be denoted by Eq. (11.20). According to Theorem 11.1, the error system denoted by Eq. (11.20) is asymptotically stable. Therefore the chaotic system denoted by Eq. (11.13) is asymptotically stable with the controller (11.24), such that lim$e$ = 0, and the proof is completed.

For limω_d = 0, one has limω = 0, so the chaotic system (11.8) is asymptotically stable with the controller (11.24). To guarantee the limit of the control input, one can choose the appropriate maximum boundary value u_m, and the criterion for judgement is whether the boundary value will influence the system stability.

11.4 Simulation test

This section will test the effectiveness of the proposed controller by numerical simulations. Based on the spacecraft chaotic attitude motion of

Chen system and Lu system, mathematical simulations are performed, respectively.

The moment of inertia of spacecraft in kilogram-square meter $\boldsymbol{I} = diag(2, 1, 1)(\text{kg m}^2)$, the reference angular velocity is given by $\boldsymbol{\omega}_d = [0.5\omega_{dr} \quad 0.6\omega_{dr} \quad \omega_{dr}]^T$, the reference angular acceleration is given by $\dot{\boldsymbol{\omega}}_d = [0.5\dot{\omega}_{dr} \quad 0.6\dot{\omega}_{dr} \quad \dot{\omega}_{dr}]^T$, the characteristic time instant is given by $t_1 = 15\text{s}, t_2 = 40\text{s}, t_3 = 60\text{s}$, and the initial value of yaw reference angular velocity is given by $\omega_{dr0} = 1.7189°/\text{s}$.

According to Eq. (11.12), one has $\dot{\omega}_m = 0.0364°/\text{s}^2$. Choose the maximum boundary value of control input $u_m = 0.88$, the symmetric positive definite matrix $\boldsymbol{P} = diag(0.01, 0.02, 0.03)$, and the related parameters of sliding mode controller $k = 0.04, \kappa = 0.01, \alpha = 0.4$.

11.4.1 Simulation results for the failure of one actuator

In this case, the initial value of angular velocity is chosen as $\boldsymbol{\omega}_0 = \begin{bmatrix} 1.7189 & 2.2918 & 1.4324 \end{bmatrix}^T °/\text{s}$, then, the solutions of Eq. (11.21) can be obtained using the YALMIP toolbox, that is,

$$\boldsymbol{C}_{\text{Chen}} = \begin{bmatrix} 0.2804 & 1 & 0 \\ 0 & 0 & 1 \end{bmatrix}, \boldsymbol{K}_{\text{Chen}} = \begin{bmatrix} 27.3126 & 32.5001 & 0 \\ 0 & 0 & 28.2085 \end{bmatrix}$$
$$\boldsymbol{C}_{\text{Lu}} = \begin{bmatrix} 0.2653 & 1 & 0 \\ 0 & 0 & 1 \end{bmatrix}, \boldsymbol{K}_{\text{Lu}} = \begin{bmatrix} 27.5526 & 33.2601 & 0 \\ 0 & 0 & 28.5418 \end{bmatrix}$$

The simulation results for Chen chaotic attitude system and Lu chaotic attitude system are shown in Figs. 11.4–11.13. Figs. 11.4 and 11.5 display the time histories of reference angular velocity and reference angular acceleration, which is in accordance with Fig. 11.3. Figs. 11.6 and 11.7 display the

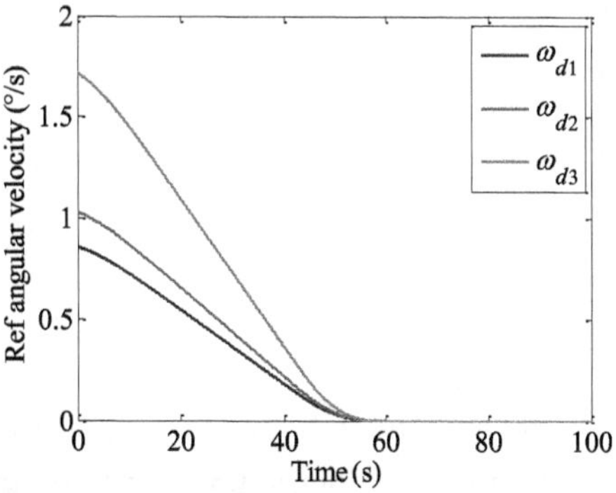

Figure 11.4 Reference angular velocity.

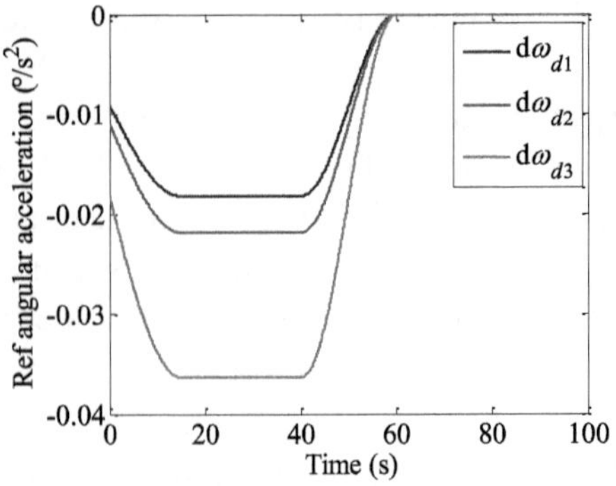

Figure 11.5 Reference angular acceleration.

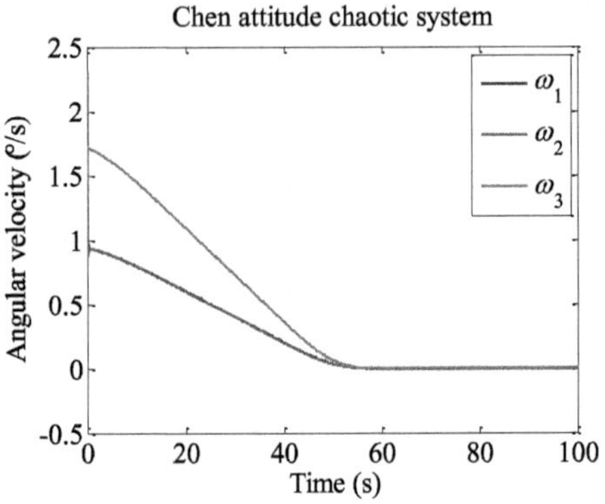

Figure 11.6 Angular velocity based on Chen system in Case 1.

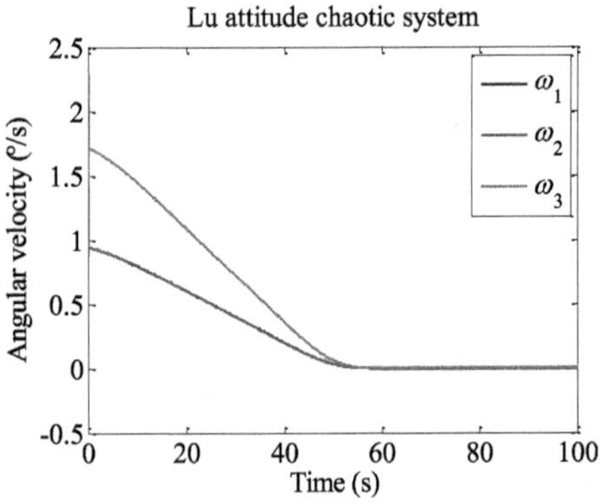

Figure 11.7 Angular velocity based on Lu system in Case 1.

variations of angular velocity based on the Chen system and Lu system, respectively. Figs. 11.8 and 11.9 display the variations of angular velocity error based on Chen system and Lu system, respectively. As can be seen, the

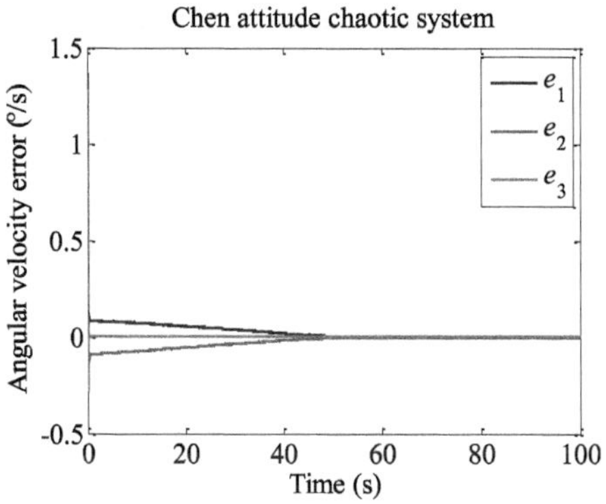

Figure 11.8 Angular velocity error based on Chen system in Case 1.

convergence time is 60 seconds, which is the same with the characteristic time instant t_3. These four figures indicate that the angular velocity and angular velocity error can rapidly converge in the presence of external disturbances which lead to chaotic motion.

Figs. 11.10 and 11.11 display the variations of control torque, and the value satisfies the given restraint condition. Figs. 11.12 and 11.13 display the external disturbance torque curves imposed on the spacecraft that leads to chaotic attitude motion. These four figures indicate that the control torque and external disturbance torque both rapidly converge to zero.

11.4.2 Simulation results for failure of two actuators

In this case, the initial value of angular velocity is chosen as $\boldsymbol{\omega}_0 = \begin{bmatrix} 0 & 0 & 1.4324 \end{bmatrix}^T$°/s, then, the solutions of Eq. (11.21) can be obtained using YALMIP toolbox, that is,

$$\begin{aligned} \boldsymbol{C}_{\text{Chen}} &= \begin{bmatrix} 0 & 0 & 1 \end{bmatrix}, \boldsymbol{K}_{\text{Chen}} = \begin{bmatrix} 0 & 0 & 3.9892 \end{bmatrix} \\ \boldsymbol{C}_{\text{Lu}} &= \begin{bmatrix} 0 & 0 & 1 \end{bmatrix}, \boldsymbol{K}_{\text{Lu}} = \begin{bmatrix} 0 & 0 & 6.3468 \end{bmatrix} \end{aligned}$$

The simulation results for the Chen chaotic attitude system and Lu chaotic attitude system are shown in Figs. 11.14–11.21. Figs. 11.14 and 11.15 display the time histories of angular velocity based on Chen system and Lu system, respectively. Figs. 11.16 and 11.17 display the variations of angular velocity error based on Chen system and Lu system, respectively. As can be seen, the convergence time is 60 seconds, which is the same with the characteristic time instant t_3. These four figures indicate that the angular velocity and angular velocity error could rapidly converge in the presence of external disturbances which lead to chaotic motion.

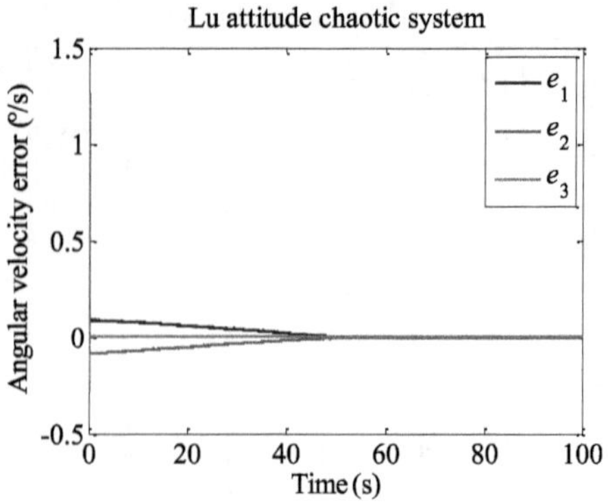

Figure 11.9 Angular velocity error based on Lu system in Case 1.

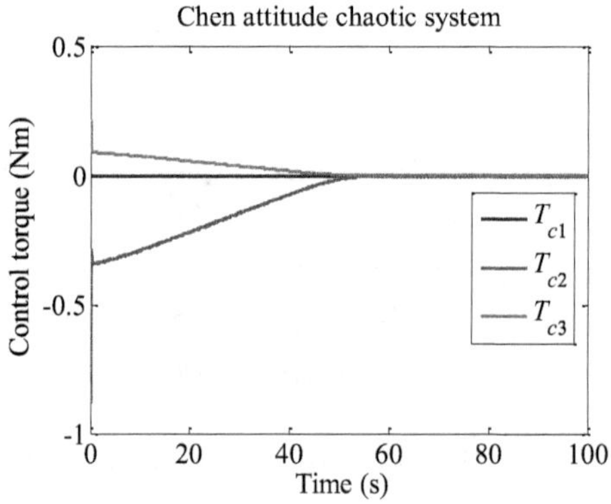

Figure 11.10 Control torque based on Chen system in Case 1.

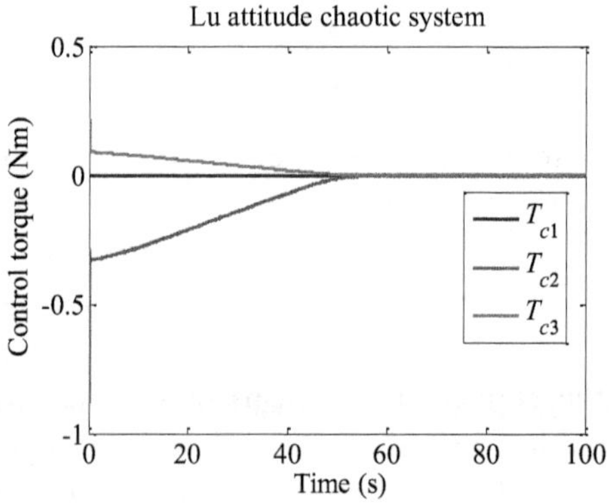

Figure 11.11 Control torque based on Lu system in Case 1.

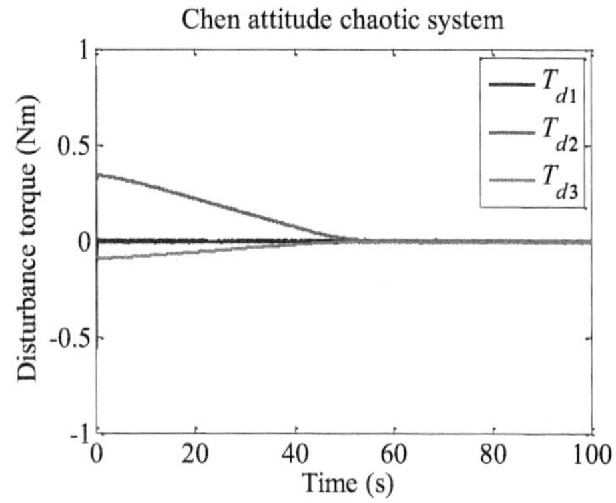

Figure 11.12 Disturbance torque based on Chen system in Case 1.

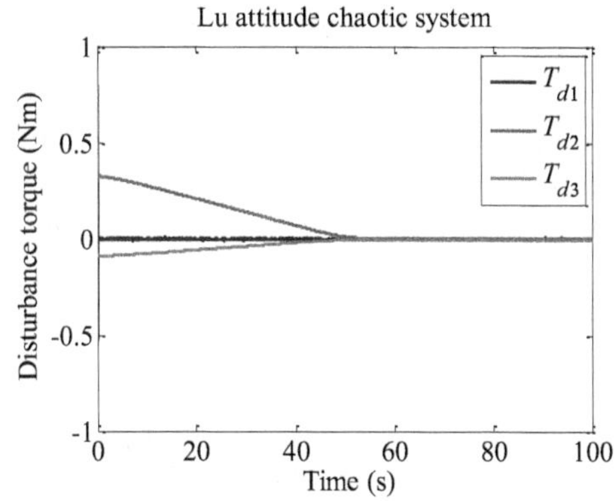

Figure 11.13 Disturbance torque based on Lu system in Case 1.

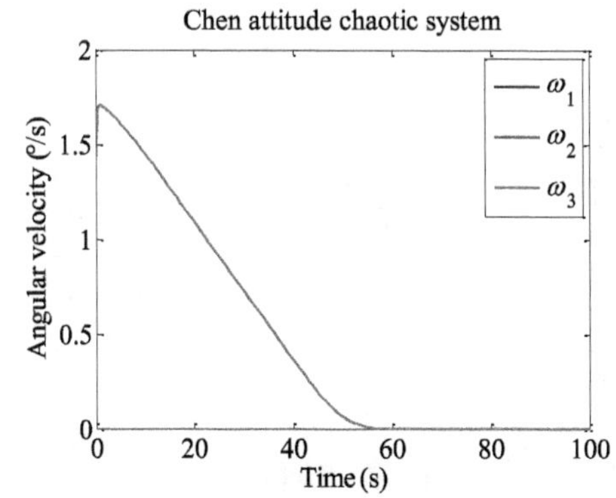

Figure 11.14 Angular velocity based on Chen system in Case 2.

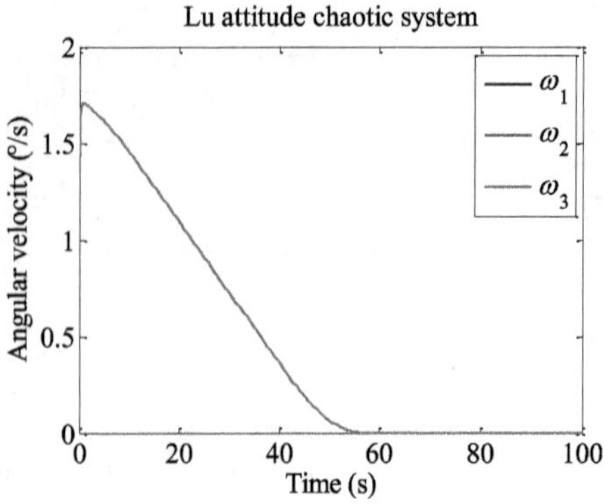

Figure 11.15 Angular velocity based on Lu system in Case 2.

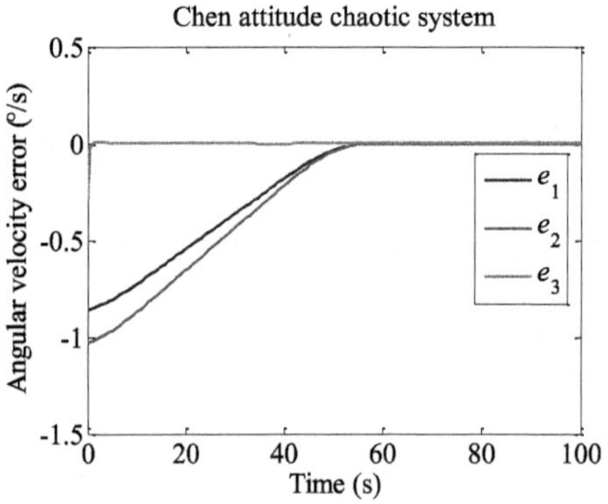

Figure 11.16 Angular velocity error based on Chen system in Case 2.

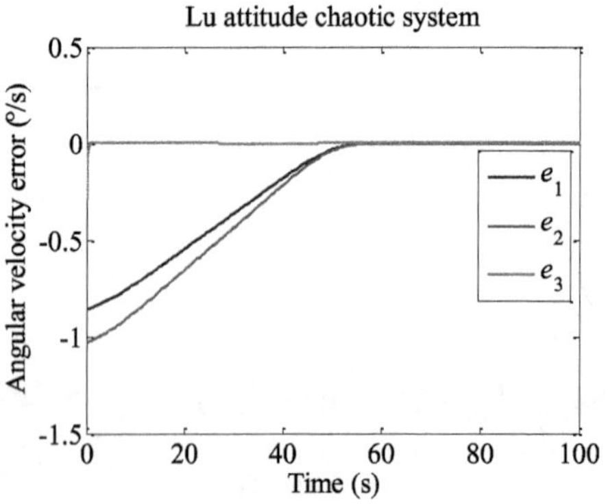

Figure 11.17 Angular velocity error based on Lu system in Case 2.

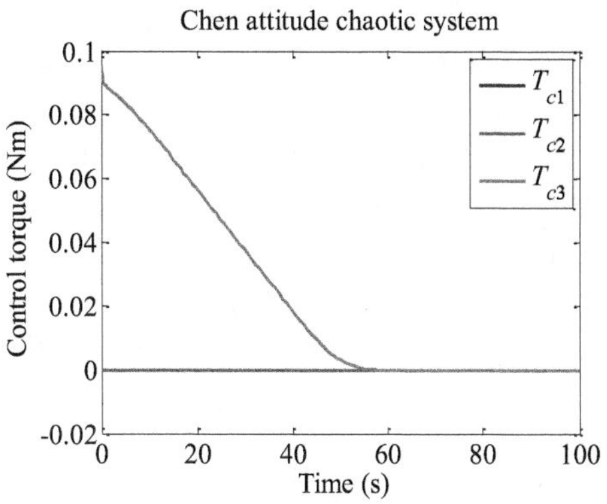

Figure 11.18 Control torque based on Chen system in Case 2.

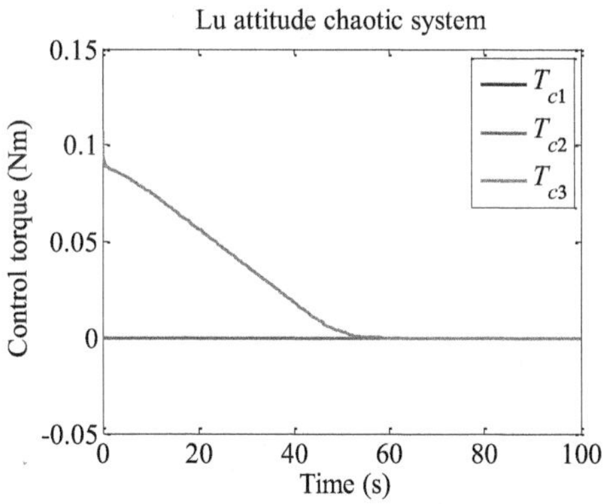

Figure 11.19 Control torque based on Lu system in Case 2.

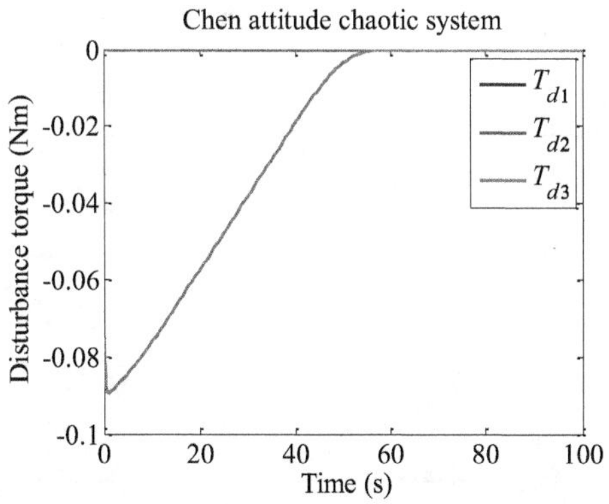

Figure 11.20 Disturbance torque based on Chen system in Case 2.

Figs 11.18 and 11.19 display the variations of control torque, and the value satisfies the given restraint condition. Figs. 11.20 and 11.21 display the external disturbance torque curves imposed on the spacecraft that leads to chaotic attitude motion. These four figures indicate that the control torque and external disturbance torque both rapidly converge to zero, and they are much smaller than those corresponding to Case 1.

If one of the initial angular velocities along the x-axis or y-axis is not strictly zero, that is, there is a very small perturbation. For example, $\omega_0 = \begin{bmatrix} 5.73 \times 10^{-7} & 0 & 1.4324 \end{bmatrix}^T$°/s, the variations of angular velocity based on Chen system and Lu system can be seen in Figs. 11.22 and 11.23, respectively. It indicates that the spacecraft attitude motion is still chaotic, and the designed controller becomes invalid. This is in accordance with the prerequisite for controller design.

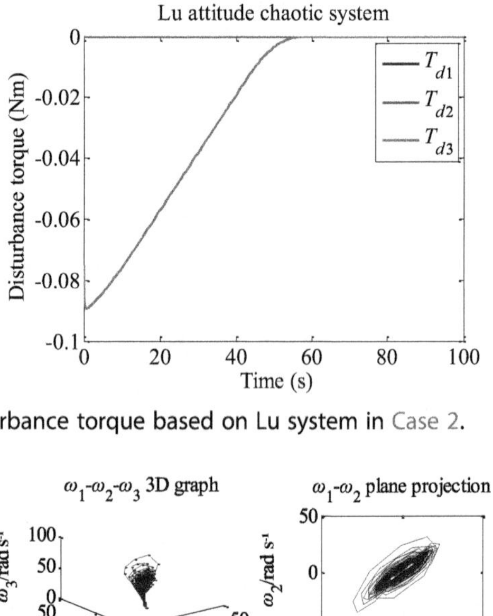

Figure 11.21 Disturbance torque based on Lu system in Case 2.

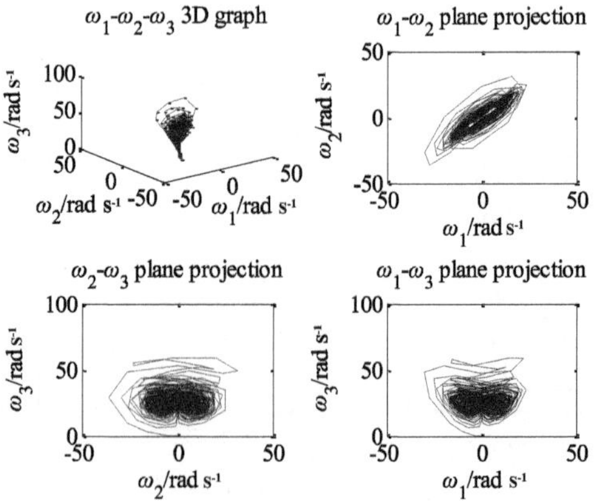

Figure 11.22 3D graph and its projection based on the Chen system.

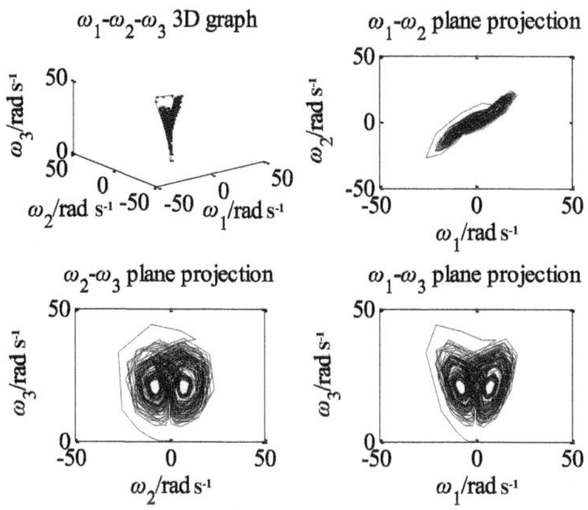

Figure 11.23 3D graph and its projection based on the Lu system.

11.5 Conclusions

A novel sliding mode control method with control input constraints has been proposed in this chapter. To deal with certain actuator failure and chaotic motion problem of spacecraft attitude system, integral sliding mode and angular velocity tracking ideas are incorporated into the controller, which consists of equivalent control term and switching control term. A detailed stability proof of the closed-loop system is also included based on Lyapunov analysis. When the failure of one actuator occurs, the angular velocities along three principal axes can be stabilized. Under given conditions, the spacecraft chaotic attitude system will be controllable when the failures of two actuators occur. Moreover, simulations are performed to assess the effectiveness and feasibility. The simulation results indicate that the controller presented in this chapter has the following characteristics: (1) elimination of chaotic attitude motion, (2) explicit consideration of control input constraint, (3) presupposition of attitude stabilization time, (4) track reference angular velocity trajectory designed in advance, (5) consideration of certain actuator failure, and (6) robustness to bigger external disturbance torque. Therefore the proposed sliding mode controller can be applied in practical stabilization control of spacecraft chaotic attitude motion under control input constraints.

References

[1] Aghababa MP, Aghababa HP. Robust synchronization of a chaotic mechanical system with nonlinearities in control inputs. Nonlinear Dynamics 2013;73(1):363–76.

[2] Lu JG, Chen G. A note on the fractional-order Chen system. Chaos, Solitons & Fractals 2006;27(3):685–8.
[3] Lu JG. Chaotic dynamics of the fractional-order Lü system and its synchronization. Physics Letters A 2006;354(4):305–11.
[4] Wei W. Synchronization of coupled chaotic Hindmarsh Rose neurons: an adaptive approach. Chinese Physics B 2015;24(10):100503.
[5] Ke Z, Zhi-Hui W, Li-Ke G, et al. Robust sliding mode control for fractional-order chaotic economical system with parameter uncertainty and external disturbance. Chinese Physics B 2015;24(3):030504.
[6] Yau HT, Shieh CS. Chaos synchronization using fuzzy logic controller. Nonlinear Analysis: Real World Applications 2008;9(4):1800–10.
[7] Zhang Y, Jiang J. Bibliographical review on reconfigurable fault-tolerant control systems. Annual Reviews in Control 2008;32(2):229–52.
[8] Zolghadri A. Advanced model-based FDIR techniques for aerospace systems: today challenges and opportunities. Progress in Aerospace Sciences 2012;53:18–29.
[9] Zhang R, Qiao J, Li T, et al. Robust fault-tolerant control for flexible spacecraft against partial actuator failures. Nonlinear Dynamics 2014;76(3):1753–60.
[10] Yang GH, Wang JL, Soh YC. Reliable H_∞, controller design for linear systems. Automatica 2001;37(5):717–25.
[11] Luo W, Chu YC, Ling KV. H-infinity inverse optimal attitude-tracking control of rigid spacecraft. Journal of Guidance, Control, and Dynamics 2005;28(3):481–94.
[12] Ma Y, Jiang B, Tao G, et al. Uncertainty decomposition-based fault-tolerant adaptive control of flexible spacecraft. IEEE Transactions on Aerospace and Electronic Systems 2015;51(2):1053–68.
[13] Wang Z, Wu Z. Nonlinear attitude control scheme with disturbance observer for flexible spacecrafts. Nonlinear Dynamics 2015;81(1):257–64.
[14] Alwi H, Edwards C, Stroosma O, et al. Fault tolerant sliding mode control design with piloted simulator evaluation. Journal of Guidance, Control, and Dynamics 2008;31(5):1186–201.
[15] Liao F, Wang JL, Yang GH. Reliable robust flight tracking control: an LMI approach. IEEE Transactions on Control Systems Technology 2002;10(1):76–89.
[16] Alwi H, Edwards C. Fault tolerant control using sliding modes with on-line control allocation. Automatica 2008;44(7):1859–66.
[17] Xiao B, Hu Q, Zhang Y. Adaptive sliding mode fault tolerant attitude tracking control for flexible spacecraft under actuator saturation. IEEE Transactions on Control Systems Technology 2011;20(6):1605–12.
[18] Hu Q. Robust adaptive sliding-mode fault-tolerant control with L2-gain performance for flexible spacecraft using redundant reaction wheels. IET Control Theory & Applications 2010;4(6):1055–70.
[19] Utkin V, Guldner J, Shi J. Sliding mode control in electro-mechanical systems. CRC press; 2017.
[20] Utkin V, Shi J. Integral sliding mode in systems operating under uncertainty conditions. Proceedings of the 35th IEEE Conference on Decision and Control 1996;4:4591–6.
[21] Rubagotti M, Estrada A, Castanos F, et al. Integral sliding mode control for nonlinear systems with matched and unmatched perturbations. IEEE Transactions on Automatic Control 2011;56(11):2699–704.
[22] Shen Q, Wang D, Zhu S, et al. Integral-type sliding mode fault-tolerant control for attitude stabilization of spacecraft. IEEE Transactions on Control Systems Technology 2015;23(3):1131–8.
[23] Crouch P. Spacecraft attitude control and stabilization: applications of geometric control theory to rigid body models. IEEE Transactions on Automatic Control 1984;29(4):321–31.

[24] Kerai E. Analysis of small time local controllability of the rigid body model. Proceedings of the IFAC symposium on System Structure and Control 1995;28:597–602.
[25] Brockett RW. Asymptotic stability and feedback stabilization. Differential Geometric Control Theory 1983;27(1):181–91.
[26] Aeyels D. Stabilization by smooth feedback of the angular velocity of a rigid body. Systems & Control Letters 1985;6(1):59–63.
[27] Aeyels D, Szafranski M. Comments on the stabilizability of the angular velocity of a rigid body. Systems & Control Letters 1988;10(1):35–9.

Index

Note: Page numbers followed by "*f*" and "*t*" refer to figures and tables, respectively.

D

E

F

G

H

I

K

L

O